Contents

Foreword v

Dedication xi

I. Geomorphic Significance of Subsurface Flow

1. *Hydrology, mechanics, and geomorphic implications of erosion by subsurface flow* 1
   Thomas Dunne

II. Above the Water Table; Water in the Vadose Zone

2. *Weathering, soil development, and landforms* 29
   C. R. Twidale

3. *The relation of subsurface water to downslope movement and failure* 51
   Donald R. Coates

4. *Piping and pseudokarst in drylands* 77
   Garald G. Parker, Sr., and Charles G. Higgins
   with case studies by
   Garald G. Parker, Sr., and Warren W. Wood

5. *Piping effects in humid lands* 111
   J.A.A. Jones

6. *Gully development* 139
   Charles G. Higgins
   with a case study by
   Barry R. Hill and André K. Lehre

7. *Surface and near-surface karst landforms* 157
   William B. White

III. At and Beneath the Water Table; Water in the Phreatic zone

8. *Groundwater processes in karst terranes* 177
   Arthur N. Palmer

9. *Permafrost and thermokarst; Geomorphic effects of
   subsurface water on landforms of cold regions* 211
   C. G. Higgins, and D. R. Coates
   with
   Troy L. Péwé, R.A.M. Schmidt, and Charles E. Sloan

10. *Land subsidence and earth-fissure formation caused by
    groundwater withdrawal in Arizona; A review* 219
    Troy L. Péwé

11. *Spring sapping and valley network development* 235
    Victor R. Baker
    with case studies by
    R. Craig Kochel, Victor R. Baker, Julie E. Laity,
    and Alan D. Howard

12. *Groundwater processes in the submarine environment* 267
    James M. Robb

13. *Erosion of seacliffs by groundwater* 283
    Robert M. Norris
    with a case study by
    William Back

14. *Seepage-induced cliff recession and regional denudation* 291
    Charles G. Higgins
    with case studies by
    W. R. Osterkamp and Charles G. Higgins

15. *Groundwater and fluvial processes; Selected observations* 319
    Edward A. Keller and G. Mathias Kondolf
    with case studies by
    D. J. Hagerty and G. Mathias Kondolf

IV. Geomorphic Controls of Groundwater Hydrology

16. *Geomorphic controls of groundwater hydrology* 341
    Donald R. Coates

Index 357

Groundwater Geomorphology; The Role of Subsurface Water in Earth-Surface Processes and Landforms

Edited by

Charles G. Higgins
Department of Geology
University of California
Davis, California 95616

Donald R. Coates
Department of Geological Sciences
State University of New York
Binghamton, New York 13901

SPECIAL PAPER
252
1990

© 1990 The Geological Society of America, Inc.
All rights reserved.

All materials subject to this copyright and included
in this volume may be photocopied for the noncommercial
purpose of scientific or educational advancement.

Copyright is not claimed on any material prepared
by government employees within the scope of their
employment.

Published by The Geological Society of America, Inc.
3300 Penrose Place, P.O. Box 9140, Boulder, Colorado 80301

Printed in U.S.A.

GSA Books Science Editor Richard A. Hoppin

Library of Congress Cataloging-in-Publication Data

Groundwater geomorphology : the role of subsurface water in earth
  -surface processes and landforms / edited by Charles G. Higgins,
  Donald R. Coates.
      p.   cm. — (Special paper ; 252)
    Some of the papers were originally presented at a symposium held
  at the 1984 Annual Meeting of the Geological Society of America.
    Includes bibliographical references and index.
    ISBN 0-8137-2252-7
    1. Hydrogeology—Congresses.   2. Geomorphology—Congresses.
  I. Higgins, Charles G., 1925-   . II. Coates, Donald Robert, 1922-
  . III. Geological Society of America. Meeting (1984 : Reno, Nev.)
  IV. Series: Special papers (Geological Society of America) ; 252.
  GB1001.2G765   1990
  551.3'5—dc20   90-13852
                                              CIP

Cover photo: Wave Rock, an exposed weathering front 10 to 12 m high, along the northern side of Hyden Rock, a granitic inselberg in Western Australia. It was initiated beneath the land surface by deep weathering around the base of the inselberg. Subsequent erosion of the weathered debris, perhaps a consequence of lowered base level, has exposed the indented bedrock slope. Photography by C. R. Twidale (see his "Origin of Wave Rock, Hyden, Western Australia," in Transactions of the Royal Society of South Australia, v. 92, 1968, p. 115–124. Copyright 1990 by C. R. Twidale.)

   This flared slope as well as the inselberg above it represent just one aspect of the effect of subsurface water on Earth-surface processes and landforms.

10   9   8   7   6   5   4   3   2

Foreword

Recognition of the many roles of subsurface water in Earth-surface processes and landform development has been a long time coming. Even now, appreciation of the full significance of underground water in geomorphology is just beginning. One function of the present volume is to stimulate awareness of its varied effects.

BACKGROUND

Geomorphology can be said to have come of age as an independent discipline in about 1877, this year of publication of Thomas Huxley's *Physiography: An introduction to the study of nature,* and G. K. Gilbert's *Land sculpture.* Even before then, however, what Gordon Davies (1969) has called "the fluvial doctrine" had begun to dominate geomorphic thought. This doctrine holds that subaerial erosion by rain and running water is, and has always been, the preeminent cause of regional denudation and the development of landforms and landscapes. This view reflects the ideas and the influence of some of the earliest observers, including John Playfair and George Greenwood in England, as well as James Dwight Dana, "the first great American fluvialist" (Chorley and Beckinsale, 1980, p. 132), and J. W. Powell in the American West. By the time of Huxley and Gilbert, fluvialism was well established, and it has continued to dominate geomorphological thought to the present day.

General recognition of the effects and importance of geomorphic agents and processes other than running water has come about gradually, and then only after special pleading by dedicated champions of these other agents. A notable example is C.F.S. Sharpe's monograph, *Landslides and related phenomena,* written at a time when "the importance of landslides and other types of mass-movement in the denudation of the lands has long been underestimated" (Sharpe, 1938, p. ix). This work broadened the scope of geomorphic thinking by alerting an entire generation of geologists and engineers to the geomorphic effects of creep, landslides, and other forms of downslope movement in land sculpture. Through the effects of Sharpe and his successors, the vital role of these processes in shaping the landscape is now generally recognized.

DEVELOPMENT OF GROUNDWATER GEOMORPHOLOGY

The effect of underground water on Earth-surface processes and its role in sculpting the land surface has similarly been long underestimated, minimized, discounted, or ignored in the geological, geographical, and engineering disciplines. An obvious exception is the solvent power of water in forming karst landscapes in terranes underlain by rock materials susceptible to dissolution. Pioneering studies such as the Rogers brothers' (1848) experiments on the solubility of various minerals and Jovan Cvijič' *Das Karstphänomen* (1893) helped to stimulate awareness of the dominance of solution in weathering and karst development, and its distinctive results. Consequently, most introductory texts in physical geography, geology, and geomorphology now contain brief discussions of groundwater corrosion and the formation of karst.

However, subsurface water plays a variety of other major roles in geomorphic processes, and the recognition of the importance of these, at least in some settings, is long overdue. This is not to say that there have not been visionary pioneers who have recognized these roles. In fact, some aspects of the work of subsurface water, especially in weathering and soil development, in downslope movement and slope failure, and in the influence of ground ice on landforms of cold regions, have long been recognized. Commonly, however, the vital role of water has tended to be mentioned only incidentally in discussions of such phenomena. Exceptions, such as G. N. Vysotski's statement that "water in the soil is as blood in the veins" (quoted by Glinka, 1927), have been rare.

In addition, some aspects of subsurface piping, seepage erosion, and spring sapping have been discussed in the engineering literature since the late 19th and early 20th centuries (e.g., Clibborn, 1909), and were reported in pioneering studies by W. W. Rubey, Kirk Bryan, Douglas Johnson, and others who are cited in the following chapters. Throughout, however, the work of subsurface water has been overshadowed in the literature by that of surface runoff, reflecting the continuing dominance of the fluvial doctrine. As a result, the geomorphic effects of underground water have rarely been mentioned in mainstream geomorphic studies. What has been particularly lacking is an overall or unified perspective of the various significant ways in which subsurface water can affect landform development. Such a synoptic view, which we call *groundwater geomorphology,* has finally begun to be appreciated only recently.

The last two decades have seen a revolution in concepts of the hydrology of hillslopes (e.g., Kirkby, 1978), and mathematical and geochemical modeling have effected dramatic advances in our understanding of subsurface hydrogeology. A great stride in the recognition of the geomorphic importance of subsurface water took place in September, 1982, when the 13th Annual "Binghamton" Geomorphology Symposium was convened to consider "Groundwater as a Geomorphic Agent." The symposium, held at Rensselaer Polytechnic Institute in Troy, New York, was organized by R. G. LaFleur, who also edited the volume of proceedings (LaFleur, 1984). Although the chief emphasis was on karst (eight of the 15 papers were devoted to karst processes and development), a number of authors discussed other aspects of the geomorphic roles of subsurface water—in weathering and soils (Foss and Segovia; Shilts; Twidale), in shallow subsurface erosion (Berger and Aghassy; Higgins), and in the erosional development of calcareous coasts (Back, Hanshaw, and Van Driel).

DEVELOPMENT OF THIS VOLUME

At the Troy meeting, William Back, one of the editors of the *Hydrogeology* volume in the Geological Society of America's Centennial Series (Back and others, 1988), invited Higgins to organize and edit for that volume a section on the role of groundwater in landform development. The result is two chapters: one on karst landforms (Ford and others, 1988); and one by 13 authors on more general relations between groundwater and landscape, divided among 13 aspects of the geomorphic role of subsurface water (Higgins and others, 1988).

In organizing those chapters, we were immediately confronted by the twin problems of scope and space. Groundwater geomorphology covers a broad spectrum of processes and landforms, as is clear in the contents of the present volume. How were we to encompass all this within the restricted space alloted to us in the *Hydrogeology* volume? Most of the authors found it very frustrating to try to distill the essence of each complex subject into just a few pages; all we could do was to write a kind of abstract of each topic. It then occurred to us that we could present a fuller treatment of the subject by preparing a separate volume in which each author could have adequate space to treat the subject in detail.

As part of the preparation of this volume, we organized a special symposium for the 1984 Annual Meeting of the Geological Society of America in Reno, Nevada. The outline for the symposium, called "Groundwater geomorphology," was somewhat expanded from that in *Hydrogeology,* with 14 papers by 18 authors. The coverage in the present volume is even broader, with much new material that was not included in the symposium. Some of the discussions are by authors who were not previously participants in the project.

ORGANIZATION OF THE VOLUME

For all of these ventures—the chapters for *Hydrogeology,* the symposium, and this volume—individual papers were not simply volunteered; instead, we made a deliberate attempt to provide a systematic survey of the whole field. In so doing, specific subjects were identified and particular knowledgable authors were asked to write about them. An additional feature of this volume is the inclusion in many chapters of significant case studies and illustrative case history reviews. Some of these are by the chapter authors; others are by guest authors.

There is some diversity of style and approach in the various chapters, in part because the varying types of subject matter do not lend themselves to a standardized format. Thus, the authors were given license to present the material in the most effective way to show what roles subsurface water can play in the specific geomorphic environments and phenomena under discussion. Despite this variability, most chapters provide an overview of the subject; a discussion, with examples and case studies of important research; and finally a summary or conclusions. Some chapters stress the broad survey, others emphasize current or original research on some specific aspects. Several suggest areas that may be ripe for future research. A few are mainly descriptive, relying heavily on Tom Dunne's introductory chapter to provide basic information on the hydraulics of subsurface flow and the mechanics of the resulting erosion. Other chapters are more quantitative or contain their own discussions of principles. All, however, should give the reader a sense of how each subject is related to the overall theme of groundwater geomorphology.

In Part I, Tom Dunne's Chapter 1 provides a re-examination of the hydrology and dynamics of subsurface flow processes as background and foundation for the chapters that follow, especially Chapters 4, 5, 6, 10, 11, and 14. The remainder of the volume is divided into three parts, in which the geomorphic roles and effects of subsurface water are discussed from the ground surface down. That is, Part II concerns water above the water table, beginning with the importance of near-surface water in weathering and soil development (Chapter 2) followed by a consideration of its role in downslope movement and slope failure (Chapter 3), piping and pseudokarst development, both in drylands (Chapter 4) and in humid regions (Chapter 5), and its influence on the development of some gullies (Chapter 6). Part II concludes with William White's examination of near-surface karst processes and landforms (Chapter 7).

Part III concerns groundwater *sensu stricto,* at and beneath the water table. The geomorphic effects of groundwater in the phreatic or "saturated" zone tend to be more pronounced than those of water above the water table, in part because the reservoir is larger and the circulation is generally more continuous, more readily detectable, and hence more intensively studied. The features that result vary according to such factors as the environment and the degree to which subsurface flow is concentrated or diffuse. Part III surveys some of this variety, beginning with a consideration of deeper karst processes (Chapter 8).

Chapter 9 reviews the influence of permafrost and ground ice on the thermokarst landforms of cold regions. Although this is essentially an arctic phenomenon, it affects as much as 20 percent of the Earth's land surface, and thus constitutes an important aspect of the geomorphic effects of subsurface water. The late Robert F. Black was to have treated this subject for the *Hydrogeology* volume (see the Dedication of the present volume). After his untimely death, Charles Sloan stepped in to write that part of the *Hydrogeology* chapter. The present brief overview is a collage assembled by the editors from Sloan's text and other sources.

Chapter 10 is chiefly an extended case study of land subsidence and earth fissures resulting from groundwater withdrawal in Arizona. Chapters 11 through 14 explore various effects of erosion by spring sapping and groundwater seepage, first in the development of valley networks (Chapter 11), then in submarine landforms of the continental shelves and slopes (Chapter 12), in coastal erosion and seacliff retreat (Chapter 13), and in regional cliff sapping, scarp retreat, and denudation (Chapter 14). Part III concludes with a consideration of some inluences of groundwater on surface-water processes and fluvial channel form (Chapter 15).

In Part IV (Chapter 16) Coates reverses the perspective to survey some of the influences that landscape and geomorphic history may have on groundwater hydrology. By means of this organization we hope to have stressed the interdependence of the many ways in which both near-surface and deeper subsurface waters can affect Earth-surface processes and vice versa.

NOMENCLATURE

Terminology is a recurring problem in the scientific literature, and misunderstandings have been generated by imprecise usage. Throughout this volume we and the other authors have tried to use the most accurate and appropriate expressions for features and processes. In most cases the nomenclature is unequivocal. For example, although the names of karst features may vary from country to country (see Coleman and Balchin's "selection" of 50 regional synonyms and approximations for "sinkholes" [1959, p. 292]), there seems to be general agreement on the meanings of the terms, particularly those of the classical Karst region of Yugoslavia.

There is far less agreement about nomenclature for some other domains and effects of subsurface waters. For example, the terms "vadose" and "phreatic," referring respectively to the zones above and below the permanent water table, are currently out of favor among many hydrogeologists; the preferred usage is "unsaturated" (or "aerated") and "saturated." However, although the reader may find all of these terms used in the following chapters, we have retained the older terminology in the headings for the major parts of the book. Part of our rationale for this is that some parts of the zone above the water table may become saturated with water at times, as Tom Dunne makes clear in Chapter 1. Clearly, the term "unsaturated" for this zone can be misleading, so we prefer "vadose." Derived from the Latin *vadosus,* simpy meaning shallow, it holds no connotation of the conditions of saturation. In Chapter 8, Arthur Palmer gives some other reasons to retain the older terms.

There is also a lack of general agreement on the meaning of some other words. This is especially true of the erosional effects of seepage and spring outflow, for which a uniform terminology has been particularly lacking. As Dunne states in Chapter 1, some of the existing terms, such as "piping" and "sapping," are derived from civil and military engineering, where their meanings may be at odds with their later applications in hydrology. Dunne would recommend replacing these words with others that do not carry such a weight of varying meanings. However, some may now be too well established to be easily replaced. A case is made in Chapter 4 for retaining "piping" or "soil piping" for both the process and the result of the formation of open conduits, or "soil pipes." We recognize that there are some legitimate objections to these terms, but we also note that they have long been used in the literature of hydrology and geomorphology, and there are no well-known alternatives. Consequently, we have not arbitrated these nomenclatural conflicts, and the reader will find both "piping" and "soil piping" as well as the equivalent process term "tunnel erosion" in the following pages. Similarly, we have not discouraged the use of "sapping" for the undermining of a stream headcut or cliff, but instead have encouraged the authors to clarify the meanings of all terms by careful application of them and explication of the processes involved. As a result, a new expression, "seepage weathering," is here defined and used for certain effects of seepage in consolidated rock. The related process in unconsolidated materials is redefined as "seepage erosion." We thereby hope to clarify the nomenclature of ground-water geomorphology rather than to perpetuate old confusions.

ACKNOWLEDGMENTS

In addition to the individuals who are specifically mentioned in the authors' acknowledgments, many others gave freely of their time and knowledge in serving as referees, first of abstracts for the symposium, and later of chapters for this book. Some, who may not be listed here, gave informal advice to individual authors, while others served as GSA reviewers. Among them, we thank the following persons who reviewed abstracts for the symposium in Reno:

Donald W. Ash*, Department of Geography and Geology, Indiana State University, Terre Haute, Indiana 47809

James C. Brice, U.S. Geological Survey, Mail Stop 66, 345 Middlefield Road, Menlo Park, California 94025

Alan J. Busacca, Department of Agronomy and Soils, Washington State University, Pullman, Washington 99164

Derek C. Ford, Department of Geography, McMaster University, Hamilton, Ontario L85 4K1 Canada

Jennifer W. Hardin, U.S. Geological Survey, 345 Middlefield Road, Menlo Park, California 94025

John Hubbard, Department of Earth Sciences, SUNY College at Brockport, Brockport, New York 14420

Robert G. LaFleur, Department of Geology, Rensselaer Polytechnic Institute, Troy, New York, 12181

Michael C. Malin, Department of Geology, Arizona State University, Tempe, Arizona 85287

Robert A. Matthews, Department of Geology, University of California, Davis, California 95616

Wilton, N. Melhorn, Department of Geosciences, Purdue University, West Lafayette, Indiana 47907

Marie Morisawa, Department of Geological Sciences, SUNY Binghamton, Binghamton, New York 13901

Richard P. Novitski, U.S. Geological Survey, Water Resources Division, 521 West Seneca Street, Ithaca, New York 14850

Allan D. Randall, U.S. Geological Survey, Water Resources Division, P.O. Box 1669, Albany, New York 12201-0744

John Thrailkill, Department of Geology, University of Kentucky, Lexington, Kentucky 40506

Bennie W. Troxel, 2961 Redwood Road, Napa, California 94558

We also thank John Bell and Burt Slemmons, the Joint Technical Program Committee, for accepting our symposium and making a space for it at the Reno meeting.

The present work has benefited from rigorous reviews in which each chapter was assessed by at least two external reviewers as a guide to the authors' revisions. We particularly wish to acknowledge the assistance of:

Robert H. Belderson, Institute of Oceanographic Sciences, Brook Road, Wormley, Godalming, Surrey GU8 5UB England

Arthur L. Bloom, Department of Geological Sciences, Cornell University, Ithaca, New York 14853

James C. Brice, U.S. Geological Survey, Mail Stop 66, 345 Middlefield Road, Menlo Park, California 94025

Alan J. Busacca, Department of Agronomy and Soils, Washington State University, Pullman, Washington 99164

Rorke B. Bryan, Department of Geography, Scarborough Campus, University of Toronto, 1265 Military Trail, Scarborough, Ontario M1C 1A4 Canada

Richard J. Chorley, Department of Geography, University of Cambridge, Cambridge CB2 3EN England

Ian Douglas, School of Geography, University of Manchester, Manchester M13 9PL England

K. O. Emery, Department of Geology and Geophysics, Woods Hole Oceanographic Institution, Woods Hole, Massachusetts 02543

Derek C. Ford, Department of Geography, McMaster University, Hamilton, Ontario L85 4K1 Canada

Hugh M. French, Department of Geology, University of Ottawa, Ottawa K1N 6N5 Canada

Kenneth J. Gregory, Department of Geography, University of Southampton, Southampton SO9 5NH England

Thomas C. Gustavson, Bureau of Economic Geology, University of Texas, Box X, Austin, Texas 78712

Bruce B. Hanshaw, U.S. Geological Survey, Mail Stop 104, 917 National Center, Reston, Virginia 22092

Burchard H. Heede, Forestry Science Laboratory, Arizona State University, Tempe, Arizona 85287

Thomas L. Holzer, U.S. Geological Survey, Mail Stop 977, 345 Middlefield Road, Menlo Park, California 94025

Anton C. Imeson, Fysisch Geografisch en Bodenkundig Laboratorium, Universiteit van Amsterdam, Dapperstraat 115, 1093 BS Amsterdam, Netherlands

Richard M. Iverson, U.S. Geological Survey, Cascade Volcano Observatory, 5400 MacArthur Blvd., Vancouver, Washington 98660

Michael J. Kirkby, School of Geography, University of Leeds, Leeds LS2 9JT England

Robert G. LaFleur, Department of Geology, Rensselaer Polytechnic Institute, Troy, New York 12181

Thomas E. Lisle, U.S.D.A. Forest Service, Redwood Sciences Laboratory, 1700 Bayview Drive, Arcata, California 95521

Luna B. Leopold, Department of Geology and Geophysics, University of California, Berkeley, California 94720

John Ross Mackay, Department of Geography, University of British Columbia, Vancouver, B.C. V6T 1W5 Canada

Michael C. Malin, Department of Geography, Arizona State Univesrity, Tempe, Arizona 85287

Ted A. Maxwell, Center for Earth and Planetary Studies, National Air and Space Museum, Smithsonian Institution, Washington, D.C. 20560

Brainerd Mears, Jr., Department of Geology and Geophysics, University of Wyoming, Laramie, Wyoming 82077

Wilton N. Melhorn, Department of Geosciences, Purdue University, West Lafayette, Indiana 47907

Malcolm D Newson, Department of Geography, University of Newcastle-upon-Tyne, Newcastle NE1 7RO England

David C. Pieri, Jet Propulsion Laboratory, California Institute of Technology, 4800 Oak Grove Drive, Pasadena, California 91109

Allan D. Randall, U.S. Geological Survey, WRD, P.O. Box 1669, Albany, New York 12201

Stanley A. Schumm, Department of Geology, Colorado State University, Fort Collins, Colorado 80523

Mitchell L. Swanson, Philip Williams and Associates, Pier 35,

*Deceased.

The Embarkadero, San Francisco, California 94133

John Thrailkill, Department of Geology, University of Kentucky, Lexington, Kentucky 40506

Brian E. Tucholke, Department of Geology and Geophysics, Woods Hole Oceanographic Institution, Woods Hole, Massachusetts 02543

David J. Varnes, U.S. Geological Survey, Mail Stop 966, Denver Federal Center, Box 25046, Denver, Colorado 80225

H. Wopfner, Geologisches Instutut, Univeristät zu Köln, Zulpicher Strasse 49, D-5000 Köln 1, West Germany

We also wish to express our thanks and appreciation to the other reviewers and consultants not listed above, to the authors of the chapters and case studies, to Campbell Craddock and Richard Hoppin, GSA Book Editors, and to Lee Gladish and the staff of the GSA Books Department. Their contributions have made the present work possible.

This volume is much more than the record of our symposium. We have aimed to make it a state-of-the-art report on the many roles of groundwater in Earth-surface processes and the development of landforms. Our plan has been to map the territory of a heretofore unrecognized field of geomorphology. This field has not been completely unknown; parts of it, such as karst studies, have long been somewhat familiar to most geologists and geographers, but the whole subject area has not previously been presented in its entirety. This was the purpose of our symposium, and we hope that this volume will further this objective, and do for groundwater geomorphology what Leopold, Wolman, and Miller's book did for fluvial geomorphology 25 years ago. We particularly hope that the reader will have been given some new insights and will have gained a greater understanding of the importance of subsurface water in Earth-surface processes and landforms.

Charles G. Higgins
Donald R. Coates

REFERENCES CITED

Back, W. Rosenshein, J. S., and Seaber, P. R., 1988, Hydrogeology: Boulder, Colorado, Geological Society of America, The Geology of North America, v. O-2, 524 p.

Chorley, R. J., and Beckinsale, R. P., 1980, G. K. Gilbert's geomorphology, in Yochelson, E. L., The scientific ideas of G. K. Gilbert: Geological Society of America Special Paper 183, p. 129–142.

Clibborn, J., 1909, Experiments made on the passage of water through the sand of the Chenab River from the Khanki Weir site, in Clibborn, J., Roorkee treatise on civil engineering; Irrigation work in India: Allahabad, United Provinces, Government Press, Appendix D, p. xxv–xxxviii.

Coleman, A. M., and Balchin, W.G.V., 1959–60, The origin and development of surface depressions in the Mendip Hills: Proceedings of The Geologists' Association, v. 70, p. 291–309.

Cvijič, J., 1893, Das Karstphänomen: Albrecht Penck's Geographische Abhandlungen, v. 5, no. 3, p. 217–329.

Davies, G. L., 1969, The Earth in decay; A history of British geomorphology, 1578–1878: New York, American Elsevier, 390 p.

Ford, D. C., Palmer, A. N., and White, W. B., 1988, Karst landform development, in Back, W., Rosenshein, J. S., and Seaber, P. R., eds., Hydrogeology: Boulder, Colorado, Geological Society of America, The Geology of North America, v. O-2, p. 401–412.

Gilbert, G. K., 1877, Land sculpture, in Gilbert, G. K., Report on the geology of the Henry Mountains; U.S. Geographical and Geological Survey of the Rocky Mountain region: Washington, D.C., U.S. Government Printing Office, p. 99–150.

Glinka, G. D., 1927, Dokuchaeiv's ideas in the development of pedology and cognate sciences; Russian pedological investigations: Leningrad, Academy of Sciences of the USSR, 32 p.

Higgins, C. G., and 12 others, 1988, Landform development, in Back, W., Rosenshein, J. S., and Seaber, P. R., eds., Hydrogeology: Boulder, Colorado, Geological Society of America, The Geology of North America, v. O-2, p. 383–400.

Huxley, T. H., 1877, Physiography; An introduction to the study of nature: London, Macmillan, 382 p.

Kirkby, M. J., ed., 1978, Hillslope hydrology: Chichester, Wiley, 389 p.

LaFleur, R. G., 1984, Groundwater as a geomorphic agent: Boston, Massachusetts, Allen and Unwin, 390 p.

Rogers, W. B., and Rogers, R. E., 1848, On the decomposition and partial solution of minerals, rocks, etc., by pure water, and water charged with carbonic acid: American Journal of Science, 2nd series, v. 5, p. 401–405.

Sharpe, C.F.S., 1938, Landslides and related phenomena: New York, Columbia University Press, 137 p.

Manuscript Accepted by the Society November 14, 1989

Dedication

Robert F. Black, 1918–1983

We dedicate this volume in fond memory of Robert F. Black, whose untimely death prevented his participation in this volume as an author on his favorite subject.

Chapter 1

Hydrology, mechanics, and geomorphic implications of erosion by subsurface flow

Thomas Dunne
Department of Geological Sciences, University of Washington, Seattle, Washington 98195

GEOMORPHIC SIGNIFICANCE OF SUBSURFACE FLOW

Hydrologic processes drive erosion by water; therefore, the study of groundwater geomorphology requires a review of the hydrology of subsurface flow and the physical constraints on the erosion that it accomplishes. The principles of groundwater hydrology are described in many textbooks (e.g., Freeze and Cherry, 1979), and only those aspects most specifically related to erosion by flowing groundwater will be reviewed here. The role of subsurface water in mass wasting will be reviewed in Chapter 2. However, in the past 20 years there has been an explosion of research on shallow subsurface flow, with implications for the understanding of erosion processes and landforms. Kirkby and Chorley (1967) and Kirkby (1978) pointed out that the evolution of concepts describing shallow subsurface flow and its relation to surface runoff has implications for understanding erosion processes. This recent research will also be reviewed in the present chapter. The review of erosion mechanisms will be restricted to processes operating in rocks which, although subject to chemical weathering, are less soluble than those producing karst landscapes. The mechanisms and geomorphic role of solution are reviewed in Chapters 6 and 7.

GROUND-WATER STORAGE AND MOVEMENT

Recharge

Although some groundwater originates as connate or juvenile water, the vast majority is meteoric, and the flux of water across the water table is defined in hydrology as the groundwater recharge. Thus, most geomorphological effects of groundwater require recharge by infiltration and percolation to the water table, even if the recharge was remote from the current time and place of interest (Lloyd and Farag, 1978). The magnitude of recharge is limited first of all by the amount of rain that falls or snow that melts at rates lower than the infiltration capacity of the soil. Water delivered to the soil at higher rates runs over the surface as Horton overland flow and is responsible for surface erosion, as discussed in other geomorphological texts (Carson and Kirkby, 1972; Selby, 1982), which review the climatic, edaphic, and biotic controls on infiltration, runoff, and the resulting sheetwash and gully erosion. Some of this runoff may accumulate later in valley floors and recharge the groundwater, particularly in arid landscapes. For many years the geomorphological literature concentrated almost entirely on the action of surface runoff, except in soluble rocks. However, it has long been reported in early literature from Indonesia reviewed by Roessel (1950) and the writings of Hursh (1944) and his colleagues from the forested southern Appalachians, for example, that, over large areas of Earth's surface, infiltration capacities almost always exceed rainfall intensity, and virtually all runoff travels underground.

This subsurface flow percolates through a shallow (vadose) zone, which may be saturated periodically but is unsaturated (i.e., the voids are only partly filled with water) most of the time. In this zone the co-existence of air and water in void spaces is responsible for menisci and the resulting negative pressures (capillary effects) in the pore water. At greater depths lies the saturated zone where all voids are filled with water. In coarse-grained porous media and jointed rocks, the pore-water pressures in this saturated zone are positive with respect to atmospheric pressure, and the boundary between the saturated and unsaturated zones is defined as the water table, where the water pressure is equal to atmospheric pressure. In fine-grained porous media, however, capillary effects fill pores with water to some height above the water table, forming a capillary fringe or tension-saturated zone. In this case, the water table is still defined as the level at which atmospheric pressure exists in the pore water; alternatively, it is defined as the height to which water stands in a well perforated throughout its length. Because water can only enter a well below the water table where pressures exceed atmospheric pressure, this zone is sometimes called the phreatic zone. This term, as well as "vadose zone," are falling into disuse, however, because seasonal fluctuations of the water table and the development of perched saturated zones over aquitards at shallow depth (see later) make the definitions confusing and useless.

Dunne, T., 1990, Hydrology, mechanics, and geomorphic implications of erosion by subsurface flow, *in* Higgins, C. G., and Coates, D. R., eds., Groundwater geomorphology; The role of subsurface water in earth-surface processes and landforms: Boulder, Colorado, Geological Society of America Special Paper 252.

As water infiltrates the soil, a portion of it is stored in raising the soil-water content, and is later evaporated back to the atmosphere. The difference between the precipitation input and evaporative loss sets the magnitude of the groundwater recharge, which therefore reflects climatic, edaphic, and biotic factors that are themselves linked. However, many aspects of erosion resulting from subsurface flow are controlled not by the total annual recharge, but by short-term rates of recharge during rainstorms, melts, or wet seasons. Freeze (1969) used a mathematical model to examine the conditions that favor recharge of the groundwater and, therefore, water-table rise. Recharge is enhanced by: (1) rainfalls or melts of low intensity (relative to the saturated hydraulic conductivity of the soil) and long duration rather than short, intense storms; (2) shallowness of the unsaturated zone above the water table; (3) wet antecedent conditions; and (4) soils with high conductivity, low rate of increase of moisture content with pressure, or high moisture content over a considerable range of pressure-head values.

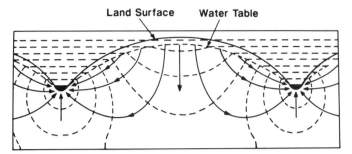

Figure 1. Groundwater flow net beneath a landscape of evenly spaced river valleys with channels at a single elevation in a homogeneous isotropic aquifer (modified after Hubbert, 1940).

Percolation

The vertical motion of water in a homogeneous and isotropic soil is described by Darcy's Law (Rubin, 1966):

$$I(z) = -K(w) \frac{\partial \Phi}{\partial z} = -K(w) \left[\frac{\partial \Psi}{\partial z} + 1 \right] \qquad (1)$$

where $I(z)$ is the vertical flux rate at any depth, z; $K(w)$ is the hydraulic conductivity, a function of the moisture content, w; Φ is the total hydraulic head and Ψ is the pressure head, both functions of w. A complete list of the symbols used in this chapter and their definitions is found in Appendix 1. As the soil is wetted to progressively greater depths, the vertical moisture gradient, and therefore the pressure gradient, $\partial \Psi / \partial z$, declines, and $I(z)$ converges on $K(w)$. Thus, the maximum value that $I(z)$ can attain after a long period of rainfall (typically 10 to 60 minutes) is equal to the maximum value that K can attain, which is the saturated hydraulic conductivity. If the vertical flux rate is imposed at the surface by a fixed rainfall intensity less than this saturated conductivity, the moisture content, w, will increase until $K(w)$ is sufficient to pass the water at the applied rate. The exact flow path depends on the degree to which the hydraulic conductivity is homogeneous and isotropic (Zaslavsky and Rogowski, 1972), but in many simple stratigraphic conditions is vertical. The rainfall intensity controls the flux rate at the surface from the beginning of the storm and at increasing depths later in the storm as the wave of moisture of constant w penetrates deeper into the soil. At the end of the rainstorm, the surface influx stops, but water continues to drain out of the upper layers of the soil, so that the wave of vertical moisture flux is damped as water moves to lower levels (Youngs, 1958; Dunne, 1978; Sakura, 1983).

A succession of rainstorms generates waves of percolation, each becoming damped as it travels deeper into the unsaturated zone. Inter-storm differences of intensity, duration, and the intervening evaporation usually mask this simple picture, but it is useful for illustration. At sufficient depth, such as in the saturated zone in a deep, permeable aquifer, one might observe only slow, seasonal changes in recharge, representing the highly damped influence of many rainstorms. In other cases, a saturated zone at a depth in the range 1 to 10 m may be recharged by a single large rainstorm, but many hours after the end of the storm and at a rate less than the maximum infiltration rate (for examples see Dunne, 1970, p. 27; 1978, p. 287). Open fractures in bedrock or long, large openings in soil profiles may reduce both the delay and damping of these single waves of recharge.

Regional groundwater flow

When the percolating water reaches the saturated zone, its movement continues to reflect the spatial variation of the hydraulic potential, as expressed in the general form of Darcy's Law:

$$Q = -K \nabla \Phi \qquad (2)$$

where Q is the specific flux (discharge per unit cross-sectional area), which in an isotropic homogeneous medium is parallel and opposite to the gradient of the hydraulic head Φ. The saturated hydraulic conductivity, K, is constant at saturation (in most rocks and soils), but may vary spatially and with direction as a result of lithologic and pedologic characteristics. The hydraulic head is the sum of the elevation head, z, measured with respect to some arbitrary datum, and the pressure head, Ψ, measured as the height to which water rises inside a piezometer. Surfaces or contours of equal Φ can thus be defined, and are called equipotentials. Equation 2 implies that flow lines cross these equipotentials at angles that depend on the degree to which K is homogeneous and isotropic but are 90° for the simplest case.

Boundary conditions such as the presence of underlying aquitards, surface topography, and the locations of drainage outlets in stream depressions, as well as heterogeneous and anisotropic hydraulic properties of the aquifer resulting from lithologic, structural, and stratigraphic characteristics, cause the flow lines to diverge from the vertical and to follow arcuate paths to a drainage outlet. Figure 1 indicates the arcuate flow paths perpendicular

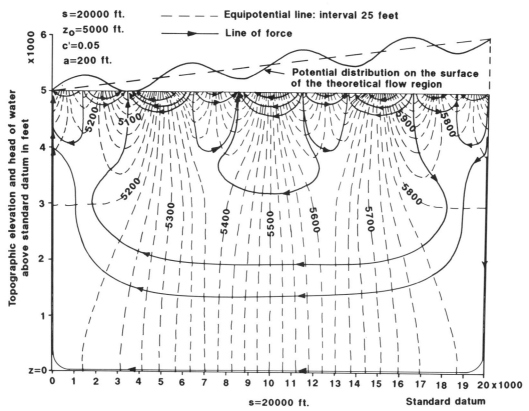

Figure 2. Regional groundwater flow in an isotropic homogeneous aquifer beneath a sinusoidal topography with drainage outlets at different elevations (Toth, 1963).

to equipotential lines in a homogeneous isotropic aquifer beneath a simple periodic topography, underlain by an impermeable boundary. The underlying aquiclude and the vertical flow divides beneath the streams and ridge crests form the boundary conditions, which together with the uniform distribution of recharge along hillslopes, force the flow lines to curve. For simple boundary conditions and uniform hydraulic characteristics, there are analytical methods for predicting flow velocities and equipotentials. In more complicated situations, the flow net can usually be defined graphically (Cedergren, 1967), although complex problems are solved by computer methods (Wang and Anderson, 1982).

If the topography has a regional slope, with drainage lines at different elevations, the regional groundwater body develops a nested set of flow subsystems (Fig. 2), as illustrated by Toth's (1962, 1963, 1966) mathematical analysis and interpretation of field measurements. As the amplitude of the topography increases, more of the groundwater flow is included in local systems between the ridge and the nearest depression, and less travels long distances to regional drainage lines. Freeze and Witherspoon (1967) used finite-difference computations to illustrate how geology can complicate these simple patterns.

In most discussions and applications of Darcy's Law it is assumed that the water migrates through intergranular pores at velocities that keep the flow behavior well within the viscous range. Fractured rocks, on the other hand, allow flow through cracks, joints, fissures, and large solutional cavities. If the fractures are dense, narrow, and of roughly uniform size, Darcy's Law is usually adequate for describing the flow if it is slow enough and if the hydraulic properties are specified for a sufficiently large, representative volume, although strong heterogeneities and anisotropies may be imposed by fracture densities and orientations. Where groundwater flows through only a few large, widely spaced fissures, the continuum assumption implicit in Darcy's Law is not valid, and it may be necessary to consider the role of individual conduits. If the fractures are sufficiently wide, flow behavior deviates from the linear behavior specified by Equation 2, and new empirical flow laws are presently being investigated for such complex, rough conduits. Unfortunately, for many aspects of groundwater geomorphology in massively jointed rocks or conduits in soils, such non-Darcian aspects of the flow are paramount, and modeling based on Darcy's Law is not applicable.

Shallow subsurface flow

In many landscapes, the hydraulic conductivity of the near-surface zone decreases from the topsoil to the unweathered parent material. This decrease may be gradual or abrupt, and is not necessarily monotonic, although for the sake of simplicity it will

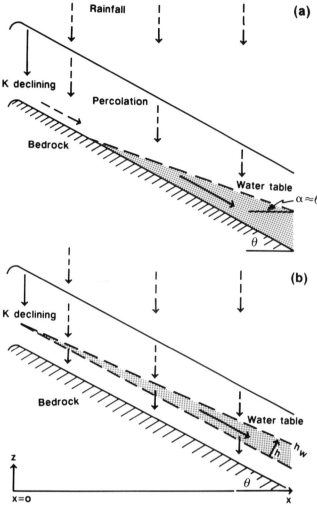

Figure 3. In a soil with a hydraulic conductivity that decreases with increasing depth, the rainfall intensity may exceed the saturated conductivity at a level that depends on rainfall intensity. In (a), a low rainfall intensity can be conveyed to the bedrock surface and then downslope as unsaturated flow (dashed arrows). At some distance along the hillside, the increase of discharge caused by downslope flow and vertical percolation may saturate a thin layer (shaded and solid arrows). In (b), a higher rainstorm intensity generates a saturated layer at a higher level in the soil, and saturated flow occurs into the subsoil, as well as downslope.

vertical percolation rate, initial moisture content, hydraulic properties of the medium, the gradient of the impeding layer, and hillslope length.

Kirkby and Chorley (1967) called this subsurface flow "throughflow," which Chorley (1978, p. 374) defined as "downslope flow of water occurring physically within the soil profile, usually under unsaturated conditions except close to flowing streams, occurring where permeability decreases with depth." Whipkey and Kirkby (1978) give a thorough discussion of throughflow under saturated conditions, which is of greater significance for geomorphology. Because the definition of throughflow is no more specific than the older term "interflow," which had been rejected by hydrologists as being poorly defined (e.g., Hewlett and Hibbert, 1967, p. 275), I will use only the term "subsurface flow" and will designate whether the flow is saturated or unsaturated, and whether the flow path lies through soil or rock at shallow depth or deeper, when such issues are important.

In the simplest case of steady-state percolation down a long, straight hillslope, the slope of the water table in Figure 3 is approximately equal to the slopes of the ground surface and the base of the saturated zone. The flow lines, thus, are approximately parallel to the water table, but the saturated hydraulic conductivity (K) may vary with height (h) normal to the base of the saturated zone. Darcy's Law reduces to

$$q_1 = \int_o^x I\,dx = \sin\alpha \int_o^{h_w} K(h)\,dh$$

where q_1 is the downslope discharge of subsurface flow per unit width of hillslope, I is the recharge rate, x is the horizontal distance, α ($\approx \theta$, the hillslope angle) is the angle of the water table, and h_w is the thickness of the saturated zone. Thus, the saturated layer thickens downslope, at a linear rate if K is constant with depth:

$$h_w = \frac{I\,x}{K \sin\theta} \tag{3a}$$

and more slowly if K varies, for example, with h^2:

$$h_w = \left(\frac{3I\,x}{\sin\theta}\right)^{1/3} \tag{3b}$$

Since flow lines parallel to the slope imply equipotentials normal to the slope (for an isotropic and homogeneous soil), the distributions of hydraulic head and pore pressure within the saturated zone can be obtained as shown in Figure 4. Beven (1982) and Iida (1984) demonstrated the prediction of water-table elevations on planar hillslopes at any time before steady state is reached. Iida also illustrated the effects of nonplanar hillslopes on the water-table and pore pressure. Dietrich and others (1986) used the Iida method to calculate pore pressures in colluvium for their study of the effect of subsurface flow on mass wasting.

These methods predict that the saturated thickness increases monotonically along the slope, and take no account of the drawdown of the water table by drainage out of a seepage face at the

be treated as such here. As the downward-percolating rainwater encounters zones of diminishing vertical hydraulic conductivity, the moisture content of successive layers must be raised to ever-increasing levels to pass the water at the applied percolation rate. At some depth, moisture content will be raised to saturation. At this depth a "perched" zone of saturation develops, either above an obvious impeding horizon, or simply at some depth (varying between rainstorms) at which the saturated conductivity is less than the rainfall intensity (Fig. 3). At these depths, water is diverted laterally and percolates downslope through the colluvium or bedrock to stream channels. The thickness of the resulting saturated zone and the pressure of water in it depend on the

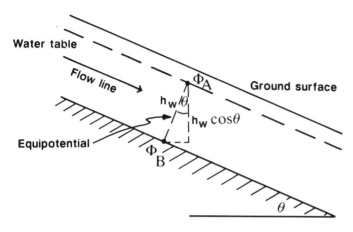

Figure 4. Relation of hydraulic head and pressure head to flow lines and water-table elevations in a homogeneous isotropic soil on an infinite slope. Φ is the hydraulic head.

downstream end of the flow system. A very approximate solution for the shape of the steady-state water table above a sloping impermeable boundary under uniform recharge, I, can be obtained by using the Dupuit-Forchheimer assumptions (Freeze and Cherry, 1979) that the aquifer is homogeneous and isotropic, flow lines are horizontal, and the hydraulic head gradient is equal to the slope of the water table. Then, in Figure 5, the discharge per unit width at some horizontal distance, x, is

$$q_x = Ix = -K\zeta \frac{dz}{dx}$$

where ζ is the saturated thickness of the aquifer and z is the elevation of the water table. It is not necessary in this analysis that $x = 0$ be set directly beneath the surface drainage divide if there is a stratigraphic control on the recharge area.

Since $z - \zeta = (L-x) \tan \beta$

$$q_x = Ix = -K[z + (x-L) \tan \beta] \frac{dz}{dx}$$

and the slope of the water table is given by

$$\frac{dz}{dx} = -\frac{Ix}{Kz - KL \tan \beta + Kx \tan \beta}. \qquad (4)$$

Except for the case of $\beta = 0$, which implies that the water table is parabolic, the solution of the differential equation is complicated and does not yield much physical insight to this writer. However, for any β, Equation 4 indicates that the water-table slope, and therefore the magnitude of the hydraulic head gradient (i) at the base of the hillslope ($x = L$) where $z = z_s$ (the height of the seepage face), is

$$i = \left| \frac{dz}{dx} \right|_{x=L} = \left| -\frac{IL}{Kz_s} \right|. \qquad (5)$$

The Dupuit-Forchheimer assumption of horizontal flow lines is a useful approximation only where the water-table gradient is small, and thus Equation 4 is particularly inaccurate for predicting the water table near the seepage face where it predicts unrealistically large water-table gradients. However, it incorporates the essential physics of the steady-state water-table response to recharge, and thus is used here for illustration of the effects of the major controls on the hydraulic-head gradient near the seepage face. It indicates that the hydraulic-head gradient at the seepage face is directly proportional to the recharge rate and drainage area, and inversely proportional to hydraulic conductivity and height of the seepage face, which in Figure 5 is equated to the depth of streamflow. The gradient of the underlying flow base, β, is not included in the steady-state form of Equation 4 at $x = L$, but this is partly a limitation imposed by the assumption of horizontal flow lines. In the field, the magnitude of β sets a lower limit on the angle of the water table at the seepage face. Also, the time required for attaining steady state, and therefore the peak discharge and water-table gradient, in shorter events would increase with β. Iida (1984) showed that in any recharge event shorter than the equilibration time for subsurface flow in a shallow colluvial aquifer of uniform depth

$$q_{pk} = \frac{IK}{p} \sin \beta \cos \beta \, t \qquad (6)$$

where q_{pk} is the peak discharge per unit width, p is the porosity, and t is duration of constant recharge entering the saturated layer. Beven (1982) added the duration of unsaturated percolation to the analysis.

If the soil and rock are sufficiently deep and permeable to conduct all the infiltrated water to the stream channel, the flow path lies entirely below the ground surface. Field experiments describing such subsurface flow were summarized by Dunne

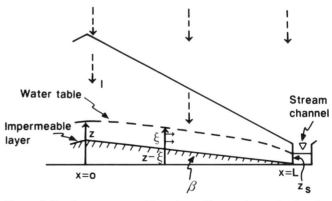

Figure 5. Steady-state water table under uniform recharge, I, using the Dupuit-Forchheimer assumption of horizontal flow lines over a planar aquiclude.

(1978) and by Whipkey and Kirkby (1978), who summarized the hydrological principles governing the results. More recently there has been some debate about the relative contributions of rapid flow through large voids, called macropores (see later), and more diffuse, slower flow through intergranular and interpedal voids (Pearce and others, 1986; Sklash and others, 1986). The answer, suggested by soil physics (Beven and Germann, 1981, 1982) and by experiment (Horton and Hawkins, 1965), is likely to be that the relative importance of macropore flow increases with rainfall intensity, antecedent soil moisture, and the linearity of the moisture content–tension–conductivity relation, and that it decreases with increasing depth of soil. In turn, these parameters vary with physical geography, and the field studies of Pearce and colleagues, Jones (1981), Gilman and Newson (1980), and others are currently mapping the geography of these relative contributions. The issue can become very important for understanding erosion by subsurface flow in some regions, as described in later sections of this chapter.

On footslopes and in valley bottoms where the water table is close to the ground surface at the beginning of the storm (Fig. 6), the tension-saturated zone may intersect the ground surface, and the addition of only a small amount of water to the surface converts the soil-water pressures in this zone to positive values, so that the water table rises rapidly. As it rises, the water table slopes steeply toward the channel, driving an increase in subsurface flow to the channel, often through a highly permeable soil or sediment with abundant macropores. Ragan (1968) and Dunne (1978) demonstrated by piezometric, water-table, and soil-moisture measurements that such rapid rises of the regional and local water table outcropping along a valley floor or footslope could cause the water table to rise in a ridge-like pattern, sloping both toward and away from the stream. Sklash and Farvolden (1979) and Sklash and others (1986) have provided isotopic evidence of the relative contributions of runoff from this source to storm runoff, while Sakura and Taniguchi (1983) and Gillham (1984) have studied the process experimentally.

There is a limit to the amount of subsurface flow that can pass downhill within a soil profile. The limit is governed by the hillslope gradient, the hydraulic conductivity, and the soil depth, as expressed in Equation 3. When h_w coincides with the soil surface, this limit is reached, and any downslope decrease in this capacity or further additions of moisture by rainfall must be accommodated by overland flow (Fig. 6). Downslope decrease in the capacity of the soil to transmit subsurface flow causes water to emerge from the ground surface as return flow (Musgrave and Holtan, 1964; Dunne and Black, 1970b), also referred to as exfiltration (Freeze, 1974). The water emerging from this fully saturated soil is augmented by direct rainfall or melt onto the saturated zone, and the sum is referred to as saturation overland flow (Kirkby and Chorley, 1967). This form of runoff is most likely to occur where topographic or stratigraphic conditions force flow to converge, or where gradient, soil thickness, or hydraulic conductivity decrease (Kirkby and Chorley, 1967; Dunne and Black, 1970a, b; Tanaka, 1982; Wilson and Dietrich,

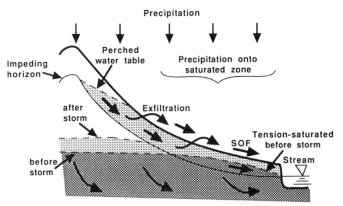

Figure 6. Shallow subsurface flow developing in a perched saturated zone above the regional water table, which outcropped only at the stream before the rainstorm. Downslope percolation in the perched saturated zone raises the perched water table to the soil surface on the footslope so that water exfiltrates and, augmented by rain falling on the expanding saturated area, travels to the stream as saturation overland flow (SOF). The expanding saturated zone is formed partly by conversion of the regional near-stream tension-saturated zone to a zone of positive pressure by local rainfall, and partly by lateral inflow from upslope.

1987). Hence, the emergence of shallow subsurface flow is most common on footslopes, in swales, and on thin soils with low conductivity. The emergence is associated with a lateral or downward component of the hydraulic head gradient in the water, exerting an oppositely directed stress on soil particles, which may thus be eroded and carried downslope by the saturation overland flow. The erosion mechanisms will be considered later in this chapter.

Subsurface flow in soils and near-surface sediments above the regional water table also includes some non-Darcian components as a result of large passages, which modify the flow and pressure fields and in turn are altered by them. Such passages include root holes, the burrows of soil fauna, shrinkage cracks, cracks around landslide blocks, and tectonic joints. The origin and hydraulic characteristics of "macropores" in soils have been described by Jones (1981) and Beven and Germann (1982). Figure 7, developed from an original example in the latter paper, summarizes the influence of macropores on overland and subsurface flow and their interaction. Some larger passages develop as a result of the subterranean erosion processes described later, and may attain diameters of up to tens of meters (Parker and others, 1964; Carey, 1976). In these passages, the flow process is turbulent streamflow, and is similar to conduit flow in large solutional cavities described in Chapter 7. The complex geometry of these tortuous conduits has so far defeated any study of their hydraulics, but an indication of their roughness is given by a measurement of flow speed in an elliptical soil pipe (Tanaka and others 1982, p. 34) from which a value of approximately 4 $m^{1/2}s^{-1}$ can be calculated for the Chézy C-value for the smoothness of pipe flow. Jones (1982) presents a thorough review of pipeflow measurements in Britain.

The pressure at which water can flow from a porous medium and enter a macropore or a conduit such as an open joint depends on the size of the passage, but for the large cracks and root holes of interest here, water in the porous medium must be under positive pressure relative to atmospheric pressure. Thus, overland flow, which occurs over a saturated soil surface, may enter such a passage, falling freely or trickling slowly down the sides, depending on its surface velocity and the morphology of the passage. If it falls freely, it may travel far down the macropore before entering the porous medium. If the flow trickles slowly down the margins of the passage through an unsaturated soil or rock, a portion of it will be drawn into the unsaturated matrix by the resulting pressure gradient (Horton and Hawkins, 1965; Bouma and others, 1978). For flow to continue along the macropore, the influx must exceed the lateral loss, which decreases through time as the matrix becomes wet and the pressure gradient decreases. Thus, Heede (1971) noted that intense rainstorms in Colorado did not generate flow in tunnels; only longer snowmelt periods could wet the tunnel margins sufficiently to allow discharge.

Water that infiltrates the porous medium under unsaturated conditions migrates within the matrix, avoiding macropores; but if the soil becomes saturated, even locally, either from the surface, or above some horizon that impedes vertical drainage and forms a perched zone of saturation, or at the roof of the macropore if vertical percolation supplies water to that point faster than unsaturated percolation can convey it around the cavity, the pressure gradient is reversed, and water flows from the matrix to the macropore (Fig. 7), along which it can travel much faster than the matrix flow. By this means, large passages provide paths through which water can by-pass slow flow through the soil matrix and, especially important, through horizons of low conductivity. Even thin, discontinuous zones of low conductivity in the soil or rock may force water into passages of various size and origin. Pierson

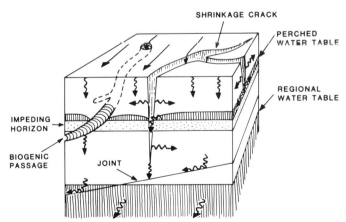

Figure 7. Paths of water into and out of macropores (modified after Beven and Germann, 1982). Saturated zones are shaded with vertical lines. Wiggly arrows indicate percolation. Straight arrows indicate overland flow.

(1983) has demonstrated experimentally that the local increase in conductivity associated with a macropore may cause an abrupt rise in pore pressure at the end of a blocked passage. Such a pressure increase could lead to slope failure or seepage erosion under favorable conditions.

Groundwater discharge

Subsurface flow discharges from the ground surface where the topography is intersected by the water table in an unconfined aquifer, or lies below the piezometric surface of a confined aquifer that has become connected to the surface by some breach (such as a fault) in the overlying aquiclude. Thus, topography, geologic structure and stratigraphy, and the recharge rate and size of the flow system determine the spatial distribution of seepage. The terms "seepage zone" or "seepage face" are applied to the area of ground surface over which the discharge occurs. These zones may be active perennially or ephemerally, depending on the timing of recharge and the geometry of the flow system as described above.

If one walks along a seepage zone, it is usually possible to discern variations in local discharge rates along a contour. Where the discharge is sufficiently concentrated, the feature is informally called a spring. Fetter (1980), p. 165–167) illustrates some of the topographic and geologic conditions that localize springs. Large springs, usually occurring at a stratigraphic and/or structural inhomogeneity, are easy to recognize, and the fact that they distort the near-surface flow field and drain large volumes of aquifer cause them to dry out the neighboring hillside, but there is a spectrum of spring magnitudes that grades imperceptibly into seepage zones.

Seepage is rarely uniform along a contour because inhomogeneities and anisotropy of aquifer properties, particularly porosity and hydraulic conductivity, cause concentrations of flow within the aquifer and near the seepage face. The most obvious of these heterogeneities result from fractures that can have a range of widths from micro-cracks to master joints and solution conduits. The effect of these properties on groundwater flow fields is beginning to receive study, but their consequences for flow near seepage faces has not yet been examined. It is now known, however, that a formation previously characterized as spatially uniform is better thought of as having random spatial variations of hydrogeologic parameters (Freeze, 1975). Thus, a "homogeneous" formation has highly correlated frequency distributions of conductivity and porosity, the former usually being lognormally distributed and the latter being normally distributed. The formation is considered to be homogeneous if these frequency distributions do not change through space. If they do vary spatially, and especially if they are multimodal, the formation is considered to be heterogeneous. If the properties are strongly autocorrelated spatially (e.g., if master joints are spaced uniformly), the concentrations of flow will be strong. Ineson (1963), for example, showed that large differences of transmissivity (the product of saturated thickness and hydraulic conductivity) occurred along the strike of the Chalk aquifer in southeastern England. The formation is a soft limestone with systems of fine fractures, but small faults and zones of concentrated fracturing along the axes of gentle anticlines transverse to the regional dip locally increase the transmissivity more than tenfold. These maxima have a roughly uniform spacing of 10 to 15 km, and valleys coincide with them.

Howard (1986b) incorporated random variability of the hydraulic conductivity into his two-dimensional mathematical model of groundwater flow and resulting erosion, but he did not utilize any spatial autocorrelation parameter to reflect repetitive aquifer properties such as joint systems. Nevertheless, the alteration of the shape of the seepage boundary by erosion in his mathematical and laboratory models caused flow concentrations with a characteristic spacing as the headward-eroding channels competed in draining water out of the aquifer (see later).

Local increases in groundwater flux and exfiltration can also result from the presence of buried features such as an anticline or other protuberance of low-conductivity rock buried by more permeable sediments. Such disturbances of the flow field often cause sufficient groundwater concentration to affect features such as springs, seepage zones, or channel networks, and the presence of buried structures is often recognized from such drainage features during map or photo interpretation. Concentrated sources of groundwater inflow can also produce concentrations of outflow nearby, although the diffusion of groundwater tends to reduce the concentration over greater flow distances. For example, near Bananal in southeastern Brazil, I have seen large gullies (voçorocas), which are eroding headward through a valley fill (Coelho Netto and others, 1987) branch and head toward concentrations of subsurface flow emanating from bedrock hollows partially filled with deep colluvium on the valley sides.

In a later section of this chapter it will be emphasized that the flow direction and speed, and therefore the hydraulic potential gradient in the outflow zone, are important characteristics governing the mechanics of erosion by water emerging from the ground. Construction of flow nets for regional or local flow fields (using piezometric measurements or the graphical or mathematical modeling techniques illustrated by Cedergren, 1967; Freeze and Cherry, 1979; and Wang and Anderson, 1982) indicates that these flow speeds and directions, as well as the location and extent of the discharge area, depend on: recharge rate; topography, and the stratigraphy, thickness, lateral extent, and hydraulic properties of rock formations. Of particular importance is the transient response to high recharge rates, which is exceedingly complex to compute (Freeze, 1971) for realistic landscape geometries, and thus is usually defined by piezometric measurements. The timing and magnitude of the increased hydraulic gradients in the discharge zone depend on the timing and magnitude of widespread water-table rise in the recharge zone. Local flow systems with shallow water tables and high hydraulic conductivity respond quickly, and the highest hydraulic gradients in the discharge zone occur during relatively intense, individual rainstorms. By contrast, discharges from large flow systems and those in less permeable rocks respond only slowly to recharge, so that a long rain or snowmelt season may be required to cause widespread water-table rise and the maximum hydraulic gradients at the outlet of such a system.

The geometry of the outflow boundary, as well as stratigraphic or structural inhomogeneities in the aquifer near the boundary, strongly affect the flow and pressure fields in the critical region where emerging groundwater has a potential for erosion. Figures 1 and 11 indicate that wherever the water table outcrops at a concave boundary, flow will converge on the concavity. It will be established later in this chapter that such a flow concentration produces an increase in hydraulic head gradient and a concentration of seepage erosion. At a three-dimensional concave boundary such as a stream head, the concentration is particularly strong. Rulon and others (1985) have used numerical modeling to examine the development of several seepage faces on slopes in which layers with low hydraulic conductivity force a horizontal component on the flow field. The hydraulic-head distributions and water-table configurations depend strongly on the position of the impeding layers and their hydraulic properties relative to those of the intervening aquifers. The horizontal component of flow increases with the recharge rate and with the ratio of the conductivities of aquifer to aquitard, and it increases as the highest aquitard lies at greater depth in the section.

Stratigraphic inhomogeneities may cause the development of "perched" saturated zones above the regional groundwater body. These perched zones may be saturated perennially or ephemerally, and in the latter case they may be difficult to recognize, despite their hydrological and geomorphological significance. However, the distinction between perched, unconfined, or confined groundwater bodies is sometimes only an artifact of the choice of measurement point (Davis and DeWiest, 1966, p. 44–45), and the mechanics of flow and erosion in perched aquifers is exactly the same as in deeper, regional groundwater bodies. Therefore, the following discussion of erosion mechanics makes no distinction between the vadose and phreatic zones, which themselves are poorly defined and confusing terms, or between perched, confined, or unconfined aquifers. For example, in a zone that is usually defined to be vadose, erosion may occur only during wet years when a perched water table develops there, or when the regional water rises to that elevation, establishing phreatic conditions.

EROSION BY SUBSURFACE FLOW

Introduction

Subsurface water emerging either at a free face (such as a stream bank, gully head, or cliff) or along a valley floor or hillside has a potential for erosion if the outflow is sufficient to mobilize particles and then overcome local flow resistance and maintain the velocity necessary to transport mobilized particles away from the site. Such erosion must begin from the downstream end of a subterranean flow path and extend headward, either underground or at the surface. This is required because sediment eroded along a flow path must be evacuated downstream along a previously formed conduit or channel.

Many studies of erosion by subsurface flow have involved detailed examination of channel or valley networks, but little consideration of process. The following text reviews the status of knowledge about the processes responsible for erosion by groundwater and the hydrological constraints on these processes. Attention is paid to the important questions of how the erosion is accomplished and what governs the relation between process and form in a spectrum of landscapes with varying degree of effectiveness of erosion by groundwater relative to erosion by surface agents.

Terminology

Subsurface flow can erode in two ways: (1) through the development of a critical body force or drag force that entrains particles in water seeping through and out of a porous medium, causing either liquefaction or Coulomb failure; and (2) through the application of a shear stress to the margins of a macropore, which may have originated independently of the water flow. Parker (1963, p. 104–106) made a distinction between these two processes contributing to the formation of conduits. He also concluded that the first is characteristic of unconsolidated materials, whereas the latter is found in consolidated materials, especially where cracks provide the main avenues for water flow and where the critical shear stress for erosion is lowered by the physical chemistry of the interstitial cement or pore water. The critical stress required for entrainment in either case may be reduced by weathering caused by the exfiltrating water, but landform development through the process referred to as seepage weathering (Higgins, this volume) requires that the weathered residuum be removed by one of the erosion processes referred to above, even if in the limit the mobilizing process is simply mass failure as the seepage intensity declines to a low value.

There is confusion in the geomorphological terminology referring to these processes. For example, Dunne (1970, 1980) followed the engineering literature (e.g., Terzaghi, 1943; Zaslavsky and Kassiff, 1965) in referring to the first of the abovementioned processes as "piping," which caused "spring sapping" (i.e., undermining of the spring head when a critical seepage force is generated). Higgins (1982, 1984) made a distinction between "piping" and "sapping," but his use of "piping" followed that of soil scientists, agronomists, and hydrologists (e.g., Fletcher and others, 1954; Jones, 1981), who used the term either without reference to a particular erosion process or with reference to processes unrelated to seepage forces. Chorley (1978, p. 370) simply defined piping as "the formation of natural pipes in soil or other unconsolidated deposits by eluviation or other processes of differential subsurface erosion." In the papers edited by Bryan and Yair (1982), the term "piping" is also commonly used to imply only pipe formation, and often by the second of the erosion processes mentioned above, which had formerly been called "tunnel erosion" (Bennett, 1939; Buckham and Cockfield, 1950).

The uncertainty is well illustrated in the literature reviewed by Jones (1981).

Terminology that causes so much confusion of the important issue of process should be abandoned. I suggest that the two erosion processes referred to above be called, respectively, seepage erosion (following Hutchinson, 1968, p. 691) and tunnel scour. If "pipe formation" and "tunnel erosion" must be used, they should be employed only in the morphogenetic sense used by Chorley (1978), rather than specifying a process. I suggest that the term "piping" no longer be used. Sapping should be restricted to its dictionary definition as "the undermining of a foundation by digging or eroding" (Concise Oxford Dictionary, 1951), or "the extension of a trench from within the trench itself" (Webster's New Collegiate Dictionary, 1981). Thus, sapping can be induced by seepage erosion or tunnel scour, and by mass failure or undercutting by a stream. If seepage is concentrated at a spring, trench-like extension or spring sapping is likely; if the seepage is more diffuse or if springs are close together, an entire hillside may gradually retreat by general sapping of its footslope. The terms "sapping" and "pipe formation" give no sense of the process responsible for the undermining or the conduit erosion.

The processes of seepage erosion and tunnel scour are not mutually exclusive. For example, seepage erosion is likely where percolation converges on the head of a biogenic macropore (Jones, 1987, p. 232). Parker (1963) described how seepage erosion of gully walls could produce a conduit that could be rapidly enlarged by corrasion and caving of the walls and roof. In the limit, however, the two processes of seepage erosion and tunnel scour are governed by different mechanics and therefore different controls on the resulting morphology and density of channels. Both seepage erosion and tunnel scour can cause retreat of hillslopes by basal erosion or the formation of tunnels or pipes. These conduits are ephemeral, however, on a geological time scale, and they eventually collapse, causing hillslope retreat or channel formation with new bounding hillslopes.

Mechanics of seepage erosion

Seepage erosion is the entrainment of soil or rock resulting from water flowing through and emerging from a porous medium. Individual grains or large masses of soil or fractured rock may be involved. The mobilization may occur by flowage if the fluid stresses cause the particles to lose their frictional strength, or by Coulomb failure allowing sliding or avalanching. The latter situation is more likely on steep slopes where seepage erosion may trigger or grade into landsliding or debris flow (Iverson and Major, 1986).

The entrainment process may be preceded by a long period of weathering concentrated at the seepage face, reducing the cohesion of the geologic material. Although there have been few detailed instrumental studies of the phenomenon, even casual field observation of seepage faces often reveals more thorough mineral decomposition there than in the surrounding rock (Howard, 1986a). These weathering processes are usually necessary for reducing the cohesion of intact rock before sapping can occur by seepage erosion or small mass failures. The exfiltrating water is responsible both for concentrating the weathering and for provoking mass failure and seepage erosion, and therefore factors controlling the magnitude and direction of flow and pore-pressure gradients are important to all of these processes. Laity (1983) made a microscopic study of the deposition by exfiltrating water of calcite within the pores of the weakly cemented Navajo Sandstone in southern Utah. The calcite eventually wedges apart the sand grains, so that the strength of the weathered zone depends on cohesion within the vesicular calcite. This layer of calcite-separated grains thickens until it can no longer adhere to the stronger sandstone, and it falls off as spalls several centimeters thick. The spalling could be triggered by frost-action, or by the weight of icicles formed by freezing groundwater discharge, or by the seepage pressure itself. This type of detailed study, which is relevant to dry climates where evaporation at the seepage face can bring percolating solutions to saturation, should be complemented by similarly detailed studies of weathering in other climatic and lithologic environments. Sharp (1976) made qualitative observations of frost action on rocks in a zone of groundwater discharge. The growth of ice crystals lifted some rock particles away from the seepage face, while others forming in the interstices of the rock reduced its hydraulic conductivity, and would therefore be expected to increase the seepage pressure and favor spalling. Instrumental studies of concentrated chemical decomposition in seepage zones remain to be done.

Whether acting on decomposed or unaltered geologic materials, seepage erosion results from exfiltrating water imposing a force associated with a hydraulic head gradient across a grain or a larger volume of rock or soil. If seepage is responsible for concentrated weathering, it is also imposing a force on the weathering particles. Therefore, the slope angle required for the particles to collapse or the degree of strength reduction required to allow the particles to fall away are reduced in the presence of a seepage force. As the seepage intensity diminishes (for example, where the exfiltrating water evaporates at the free face), seepage erosion grades into pure seepage weathering with removal by gravitational mass wasting. However, most mass wasting is affected to some degree by seepage.

Writers concentrating on the role of seepage pressure in eroding and undermining hillslopes have expressed this effect as a body force acting on some representative volume of the porous medium to reduce its internal resistance to rupture (e.g., Terzaghi, 1943, p. 258; Zaslavsky and Kassiff, 1965; Iverson and Major, 1986). Kochel and others (1985), emphasizing the continuity between seepage erosion and the fluvial transport that must evacuate the mobilized sediment to maintain a channel, considered seepage erosion as the result of the fluid drag of emergent groundwater applying a torque to particles at the ground surface. Both the body force and drag force are proportional to the near-surface gradient of the hydraulic potential so that the numerical results are equivalent. However, some of the characteristics of the resulting failure and detailed form of the channel head and seepage face may differ, depending on whether the erosion comprises fluid particulate transport or mass failure, including debris-flow initiation. Both forms of seepage erosion can be observed at seepage faces, particularly at channel heads, but little fieldwork has been done to define the conditions under which each occurs.

Terzaghi (1943) defined the critical conditions required for the simplest case of instantaneous seepage erosion (he called it piping) of a relatively large mass of cohesionless material by the vertical component of seepage beneath a horizontal surface. In Figure 8, the specific water flux, q, is represented by a vector that lies in the x–z plane and makes an angle λ with the vertical. This seepage is driven by a gradient in the potential energy per unit weight of water, which is called the hydraulic head Φ. The potential gradient creates a force per unit area of the porous medium normal to the flow line indicated by q. However, in this simplest case the surface is horizontal, and we are concerned with vertical forces that act to move particles outward from that surface. The force per unit area of the horizontal cross section $\Delta x\,\Delta y$ is provided by the vertical component of the water flux, and thus the vertical component of the hydraulic head gradient. With reference to Figure 8, the force associated with the vertical flux is

$$\frac{\text{Force}}{\text{Area}} \times \text{Area} = [\rho_f\,g\,(\Phi_2 + \Delta\Phi) - \rho_f\,g\,\Phi_2]\,\Delta x\,\Delta y$$

where $\Delta\Phi$ is given by $\cos\lambda\,(\Phi_1 - \Phi_2)$. This seepage force is distributed through the volume of the body, to yield

$$\frac{\text{Force}}{\text{Volume}} = \frac{\rho_f\,g\,\Delta\Phi\,\Delta x\,\Delta y}{\Delta x\,\Delta y\,\Delta z} = -\rho_f\,g\,\frac{\partial\Phi}{\partial z} \quad \text{as } \Delta z \to 0$$

where $\partial\Phi/\partial z$ is the vertical component of the hydraulic head gradient.

The vertical seepage force is resisted by the immersed weight of the grains, aggregates, or larger quantity of rock or soil in the control volume:

$$\frac{\text{Force}}{\text{Volume}} = (\rho_s - \rho_f)\,g\,(1 - p)$$

where p is the porosity. If the force due to the upward vertical component of the seepage exceeds this immersed weight, the particles will be buoyed until the effective stress, and therefore

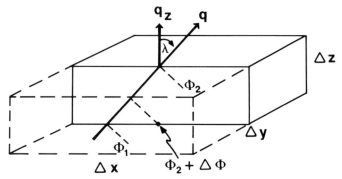

Figure 8. Imposition of a seepage force on a volume of soil by the vertical component of seepage, q_z.

effective strength, decline to zero; the material will undergo static liquefaction. Thus, the critical condition required for liquefaction is:

$$-\rho_f g \frac{\partial \Phi}{\partial z} \geq (\rho_s - \rho_f) g (1 - p)$$

or, if the magnitude of the hydraulic head gradient is represented by i, which has a vertical component, i_z, then

$$i_z > \frac{(\rho_s - \rho_f)}{\rho_f} (1 - p) \qquad (7)$$

In turn, the magnitude of i is related to the seepage discharge by

$$i = \frac{Q}{K}. \qquad (8)$$

If the flow field were three-dimensional, then i would be equal to $|\nabla \Phi|$ in Equation 2.

When the condition described in Equation 7 occurs, the material may dilate and increase its permeability rapidly over a relatively large area, or a few smaller grains may be lifted first, and outflow may converge on the resulting cavities, increasing the water flux and hydraulic gradient and extending the cavities to form ephemeral pipes (Kälin, 1977). The material loses all of its frictional strength and deforms by flowing. Figures 1 and 6 show that vertical components of seepage and potential gradients may exist in valley floors and on footslopes. Sakura and others (1987) have measured equipotential fields and hydraulic gradients in laboratory model and showed that liquefaction and valley-head failure occurred when the normal component of the hydraulic gradient exceeds a critical value. During the snowmelt season in Vermont, I have observed highly turbid water seeping from valley floors, and have measured vertical components of pore-pressure gradients of the same magnitude as those predicted by Terzaghi's theory to be required for the observed occurrence of erosion by emergent groundwater. In these cases the dislodged grains were carried away by saturation overland flow and by channel flow, but if large masses of material suddenly lose their effective strength in this way a debris flow may result. In either case, liquefaction proceeds downward from the surface because it is there that the effective strength of the material declines to zero (Bear, 1972, p. 188).

If the geologic material in Figure 8 were cohesive, an extra force would be acting across its base to resist the vertical separation. Zaslavsky and Kassif (1965) pointed out that seepage erosion would then occur as a tensile failure as cracks parallel to the surface become pressurized with the fluid. The balance of forces in Equation 7 would then be altered to

$$i_z \rho_f g \geq (\rho_s - \rho_f) g (1 - p) + \frac{c \Delta x \Delta y}{\Delta x \Delta y \Delta z} \qquad (9)$$

where c is the cohesion per unit area. The critical hydraulic potential gradient thus depends on the thickness of the volume that eventually separates. Near the surface, Δz may be so small relative to c that the difference in hydraulic head (and therefore the seepage force) is insufficient for erosion. With increasing depth (Δz in Fig. 8), the volume being mobilized by the seepage force (per unit volume), $i_z \rho_f g$, becomes so large that its movement cannot be resisted by the tensile strength of the material, which remains constant as Δz increases. Failure occurs at a critical depth

$$\Delta z_c = \frac{c}{i_z \rho_f g - (\rho_s - \rho_f) g (1 - p)}. \qquad (9a)$$

However, if c is large, realistic values of i lead to thick failed layers, which would be unlikely to flow. On the other hand, as Δz_c declines to the size of fragments that commonly flow, the seepage gradient would have to be unrealistically high to cause liquefaction of material with a significant cohesion. Thus, although Equation 9a describes failure, it seems likely that the liquefaction of cohesive materials requires that the cohesive bonds first be weakened considerably by weathering, gravity, or other forces near the seepage face, and the erosion process depends mainly on the condition summarized in Equation 7. The resistance of cohesive materials to seepage erosion is probably the reason for Parker's (1963, p. 107) observation that tunnel scour is the most common mode of conduit formation in consolidated materials.

Static liquefaction may also be impeded by plant roots, which can be viewed as resisting the vertical separation in the manner of a cohesion, or as providing an extra vertical load when roots tied laterally or vertically to stable ground are extended by dilation of the ground surface. This effect may be the reason why artesian pore pressures and swelling of the ground surface have been noticed before colluvium liquefied to produce debris flows in some forested bedrock hollows.

Iverson and Major (1986) generalized the analysis of seepage forces to account for static liquefaction and Coulomb failure due to water flowing in any direction within colluvium on hillslopes. Thus, they took into account the conditions that produce mass failure, debris-flow initiation, and seepage erosion, which grade into one another and extend channels. Their analysis was strictly limited to homogeneous, isotropic, and cohesionless soils subject to steady, uniform seepage on a planar hillslope, but it illustrates important conditions that are relevant to interpreting most field situations. Iverson and Major analyzed the influence of seepage on liquefaction of a hillside defined in Figure 9a, where the seepage force is proportional to the hydraulic gradient, defined in terms of its magnitude i and the angle λ, which it makes with the surface normal. Liquefaction occurs under the condition summarized in Equation 7, which requires that

$$i = \frac{i_z}{\cos(\lambda + \theta)} > \frac{(\rho_s - \rho_f)}{\rho_f} \frac{(1 - p)}{\cos(\lambda + \theta)}. \qquad (10)$$

The magnitude i required for seepage erosion by liquefaction is therefore a minimum where $\lambda = -\theta$, which implies vertically upward seepage. However, if i is sufficiently large, liquefaction can occur at other combinations of λ and θ, as indicated by Equation 10 and summarized graphically in the original paper.

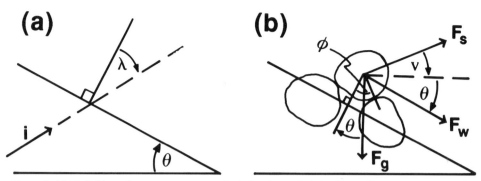

Figure 9. (a) Definition of seepage vector magnitude for the analysis by Iverson and Major (1986). (b) Force balance on a particle lying on a sloping surface subject to a seepage force (F_s), a shearing force due to surface runoff (F_w), and gravity (F_g). Note that $\nu = (90 - \lambda - \theta)$ in Figure 9a. (After Kochel and others, 1985.)

Iverson and Major showed that the direction and magnitude of the seepage gradient also affect the susceptibility of a hillslope to Coulomb failure, and that, even when water is emerging from the surface, this form of mass failure often preempts liquefaction or occurs when the hydraulic head gradient is insufficient for liquefaction. They demonstrate that the flow of water through colluvium and the attendant head gradient alter both the effective stress, which controls the frictional resistance, and the alongslope driving force. If the flow vector is pointed outward and downslope, for example, it reduces the frictional resistance and increases the downslope stress, whereas downslope flow parallel to the surface increases the driving stress but not the normal stress, yielding the usual "infinite-slope" stability equation (Carson and Kirkby, 1972, p. 156). Other combinations are obvious from Figure 9, or from Equation 3 of the original paper. Modification of the force balance of saturated colluvium on a hillslope to take into account the effects of seepage magnitude and direction led Iverson and Major to summarize the condition for Coulomb failure of a cohesionless material as:

$$i > \frac{(\rho_s - \rho_f)}{\rho_f}(1 - p)\frac{\sin(\phi - \theta)}{\sin(\lambda - \phi)} \quad (11)$$

where ϕ is the angle of internal friction of the porous medium. One particularly interesting result of this equation is that a specific (not vertical) seepage direction ($\lambda = 90 - \phi$) requires the minimum seepage magnitude to provoke failure, or yields the minimum stable hillslope angle for a given magnitude of i and Q. The original paper investigates all conceivable combinations of hydrologic and geotechnical parameters of the model.

Kochel and others (1985) and Howard and McLane (1988), on the other hand, emphasized the continuity between seepage erosion and fluvial transport. They constructed a two-dimensional sapping chamber (Fig. 10) with walls 5 cm apart. The intensity of groundwater flow through the sand hillslopes constructed in the tank could be altered by adjusting the water levels in the reservoir at the rear of the tank and at the outflow, as well as the position of the screen separating the sand from the reservoir, which sets the length of the flow system. The authors reported that sapping over the seepage zone involved an intimate mixture of seepage erosion and overland transport. Seepage erosion dominated at the upper end of the seepage zone where the surface slope was steep, hydraulic gradients in the emergent groundwater were strong, and the surface flow was weak. Removal of sediment from this zone triggered mass failure from the zone upslope. Further downstream, the hydraulic gradient decreased in magnitude, but the overland flow and resultant fluvial transport increased. Sapping was most intense at the upper end of the seepage face, causing headward erosion by slumping of sand, which then had to be transported downstream before the backwearing could continue. The fluvial transport required the development of a surface gradient proportional to the amount of sediment eroding from the model landscape, and inversely proportional to the water discharge and thus to the average hydraulic gradient through the hillslope.

Kochel and others (1985) and Howard and McLane (1988) analyzed the interaction of surface runoff and emergent groundwater by considering the balance of torques (Fig. 9b) acting on a particle that must roll out of an intergranular "pocket" over a rotation point. They considered four forces to be acting on the particle: gravity (F_g), cohesion (F_c), the drag imposed by surface runoff parallel to the slope (F_w), and the drag imposed by emer-

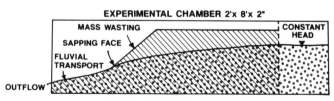

Figure 10. Longitudinal section of a two-dimensional experimental seepage chamber. Ruled lines indicate sand, circles show saturated zone. (Kochel and others, 1985.)

gent seepage (F_s). The forces are defined with respect to particle characteristics as:

$$F_g = C_1 (\rho_s - \rho_f) g D^3 \quad (12)$$

$$F_s = C_2 \rho_f g i D^3 \quad (13)$$

$$F_w = C_3 \tau D^2 \quad (14)$$

$$F_c = C_4 D^2 \quad (15)$$

where C_1, C_2, and C_3 represents the effects of particle shape and packing, and C_4 requires a specific measurement for cohesive soils. The boundary shear stress τ exerted on the grains by surface flow and the hydraulic gradient i must also be measured or estimated independently from hydrologic considerations.

At incipient motion, the rotational point makes an angle, ϕ, with the normal to the hillslope surface about the center of the particle to be moved (Fig. 9b). The average value of this angle for the loosely packed, granular material is approximately equal to the angle of internal friction used in the formulations based on mass wasting (see above). The balance of torques as erosion begins is defined by:

$$F_c D = F_w D\cos \phi + F_s D\cos (\theta + \nu - \phi) - F_g D\sin (\phi - \theta). \quad (16)$$

The authors then showed that for cohesionless sediment with an angle ϕ and negligible surface flow, the maximum stable slope θ depends once again on the direction ν and magnitude i of the hydraulic gradient. Thus, for cohesionless sediment with an angle ϕ and negligible surface flow, failure occurs when:

$$i > \frac{C_1}{C_2} \frac{(\rho_s - \rho_f)}{\rho_f} \frac{\sin (\phi - \theta)}{\cos (\theta + \nu - \phi)} \quad (17)$$

The magnitude of i required for such failure is minimized when $\nu = (\phi - \theta)$, which is equivalent to the condition in Equation (11) because, between Figures 9a and b, such a condition implies that $\lambda + \phi = 90°$, which also minimizes i. The results of Kochel and others (1985) agree with those of Iverson and Major (1986) to the effect that a specific, nonvertical seepage vector will provide the critical condition governing erosion and retreat of the hillslope or spring head. This seepage vector is governed, in turn, by the geometry of the groundwater flow net, surface topography, hydraulic conductivity, and recharge rate.

Morphological expression of seepage erosion

Introduction. Other chapters in this volume describe the characteristics of landforms generated by seepage erosion. However, there are some generalizations about morphology that follow from the principles of groundwater flow and seepage erosion mechanics reviewed in this chapter. These generalizations will be outlined here. Many recent studies have been motivated by a search for terrestrial analogs of Martian landscapes where, it is hypothesized, groundwater has been an important landforming agent. Thus, many investigators have focused their attention on terrestrial landscapes where the geomorphic role of groundwater is relatively free from complications by other influences. The field studies have concentrated on gently dipping aquifers in highly permeable or arid zones where surface erosion and shallow mass wasting are relatively slow, and where there is relatively strong structural control of landform evolution by major vertical joints, bedding planes, and strong contrasts in hydraulic conductivity. The landscapes studied include beaches (Higgins, 1982), the Colorado Plateaus (Laity and Malin, 1985; Howard, 1986a), Hawaiian volcanoes (Baker, 1986; Kochel, 1986), and the Libyan desert (Maxwell, 1982).

It is likely that the geomorphic effectiveness of groundwater is also widespread outside of these "model" landscapes, but that in other regions groundwater interacts with the effects of hillslope erosion, or is not quite so strongly controlled by structural features. Thus, some landscape characteristics interpreted by Laity and Malin (1985), Howard (1986a), and others to be diagnostic of the landforming role of groundwater may simply be the effects of groundwater in particular geologic settings, such as gently dipping, massive, permeable rocks with little weathered residuum. These characteristics include: elongate channel networks; theater-headed valleys; short, stubby tributaries with large junction angles; relatively constant valley width from source to mouth; and strong structural control of channel orientation, wall morphology, and longitudinal profiles. They may be found only in landscapes formed by emerging groundwater in certain geologic settings, but seepage erosion may also develop channel networks without these characteristics in other geologic regions. For example, Dunne (1970, 1980) proposed that less distinctive Appalachian channel networks, with only slight structural control (Newell, 1970), were also formed as a result of groundwater emergence that promoted seepage erosion of weathered rock and colluvium. DeVries (1976) proposed that the stream network of the Netherlands was formed as a result of the outcropping of regional water tables as groundwater invaded Pleistocene deposits after deglaciation. Carlston (1963, p. C5) proposed a similar reason for the occurrence of streams. In these cases the geologic structure, climate, and hillslope erosion processes favored the development of a landscape without some of the diagnostic characteristics listed above.

Channel initiation and network growth. If a groundwater flow field, of any extent or depth, is subject to a perturbation as the water emerges from the ground surface, there will be a tendency to focus weathering and erosion, and to notch the surface, creating the potential for elongation of the notch into a channel and eventually a valley. The perturbation may result from erosion of the boundary by some other process, such as river erosion (Fig. 11a), or it may result from the establishment of a new or larger groundwater flow field within a geologic deposit with irregular boundaries formed by some other agent, such as a glacier, a river, or wind. DeVries (1976) described, for example, the expansion of groundwater seepage zones in Pleistocene fluvial

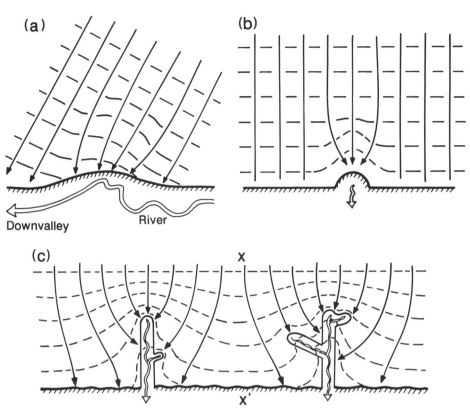

Figure 11. Plan view of the perturbations of a groundwater flow net that lead to the extension of spring heads to form a drainage network. Solid arrows are flow lines, dashes indicate equipotentials. (a) Concentration of flow at a boundary perturbed by an outside agent. (b) Concentration of flow caused by a small seepage failure localized by a lithological heterogeneity. (c) Increased convergence of flow lines around neighboring spring heads that have retreated into a land mass. Tributary valleys form as a result of secondary perturbations of the flow field due to the same geometric or hydrogeologic factors. Convergence on the spring heads leads to divergence of flow between the valleys (After Dunne, 1980.)

and eolian sediments as precipitation increased during the Holocene. The flow lines converge on the re-entrants in the boundary as shown in Figures 1 and 11a. Convergence of flow on the initial or eroded, concave boundary increases the water flux, Q, in Equation 8, and therefore the hydraulic gradient and seepage forces in Equations 7, 10, 11, and 17. Thus, the probability of attaining the critical conditions for seepage erosion during a particular storm or season is increased by disturbance of the boundary.

Dunne (1970, 1980) also proposed that spatial variation in the hydrogeological properties of the aquifer could cause either local increases of specific discharge, Q, or local reductions in resistance to weathering and seepage erosion. For example, there are autocorrelated spatial variations of both hydraulic conductivity and porosity in even homogeneous isotropic aquifers, as previously described. These variations result from random and periodic differences in the factors controlling petrogenesis, diagenesis, and deformation, which cause variations in grain size, pore fillings, and in fracture widths and spacing. Other spatial variations in mineralogic composition result in variable resistance to chemical decomposition and therefore in changes of porosity, hydraulic conductivity, and tensile strength.

Each of the spatial variations in hydrogeologic properties affects the flow field, causing local concentrations of discharge at a seepage face. The most obvious examples of this effect occur where master joints provide narrow zones of high and anisotropic conductivity, which localize outflow and generate major springs (Schick, 1965; Campbell, 1973; and Fig. 12), but in many other field situations it is difficult to identify visually the hydrological parameter responsible for the concentration, and as far as I know, detailed instrumental measurements that might identify the effective spatial patterns have not been made. On a larger scale, however, Ineson (1963) identified small faults and fracture zones along the axes of gentle anticlines as the primary controls on the distribution of transmissivity in the English Chalk. The local concentration of outflow either accelerates weathering, lowering the resistance to seepage erosion, or increases the hydraulic gradient, increasing the seepage force (Equations 7, 10, 11, and 17). If seepage erosion occurs, it perturbs the boundary, and again increases the flow convergence (Fig. 11b), increasing the probability of further seepage erosion. As the head of the embayment retreats into the land surface or hillslope, this flow convergence will increase (Fig. 11c), setting up a positive feedback loop that affects the growth of a new valley.

The initial seepage erosion may produce only a subterranean conduit rather than a valley. Howard (1986b, p. 134) observed the formation of temporary conduits in his experiments on seepage erosion, presumably as a result of cohesion imposed by intergranular water in the tension-saturated zone. It is difficult to point unequivocally to published reports of conduits formed by seepage erosion alone, because most reports do not consider the pipe-forming process in detail, and perhaps because cohesion resists seepage erosion, as indicated by Equation 9. However, it is likely that in some cases the heads of conduits are extended by seepage erosion, even if their roofs are maintained by cohesion and they are eroded by tunnel scour along most of their courses. Zaslavsky and Kassif (1965, p. 315–316) emphasized that survival of an arch over a pipe or tunnel requires cohesive strength, which presumably could be provided by both interparticle forces and plant roots. However, over long periods of geologic time, as the land surface is lowered above them, the roofs of conduits are ephemeral and they collapse to form gullies and valleys. In many situations, pipes do not form; the channel or valley forms as a direct result of seepage erosion and mass wasting into the sapped zone.

It is a common observation that spring heads or valleys formed by seepage erosion at a range of scales down to the size of laboratory models (e.g., Kochel and others, 1985; Howard, 1986b) exhibit a roughly uniform spacing. In some cases (e.g., the valleys studied by Campbell [1973], Howard [1986a], and Laity and Malin [1985] on the Colorado Plateaus or some of the large springs draining basalts on the Columbia and Snake River Plateaus), it is obvious that the uniformity is imposed by the roughly equal spacing of master joints. However, there have been no formal quantitative studies of the relation between the hydrogeologic properties, the length scales of patterns in these properties, and the average spacing of channels formed by seepage.

The most general influence on the relatively uniform spacing of channels, occurring even in the absence of structural controls, is the competition of growing channel networks for subsurface flow. Figure 11c indicates that convergence of flow lines toward indentations or alcoves in a boundary or at the heads of channels is compensated by divergence between these features. Divergence of the flow lines causes a decrease in the local water flux, and therefore a lower hydraulic gradient and probability of erosion, stabilizing the surface between the indentations. The growing indentations along the boundary compete with one another in draining groundwater from the aquifer, and this competition sets the drainage density of the consequent streams in a manner that has not yet been derived mathematically. However, on the basis of the foregoing discussion of mechanics it seems reasonable to speculate that high recharge rates, low hydraulic conductivity, high porosity, and susceptibility to weathering would all tend to increase drainage density, as would the presence of a steeply sloping aquiclude beneath the aquifer (see below).

The extending valley eventually intersects another zone where the combination of hydrogeologic characteristics causes a sufficiently strong concentration of discharge to cause seepage erosion on the valley wall (Fig. 11c). Alternatively, the consequent stream draining the valley may undercut the valley wall, creating an embayment (Figure 11a). Either perturbation causes a headward-sapping tributary to form, along which the process can be repeated to generate a branched heirarchical drainage network (Fig. 13).

If neither the flow field nor distribution of rock properties is affected by strong heterogeneities or anisotropy, one might expect the junction angle to depend on the ratio of the general gradient of the water table (along x-x' in Fig. 11c) to the lateral perturbation of the water-table slope imposed by the formation of the consequent stream, assuming that the resultant of these two components is sufficiently steep to trigger seepage erosion. Both of these components of the local hydraulic gradient depend on the recharge rate and hydraulic conductivity, and the lateral component depends also on the spacing and time-varying depth of inci-

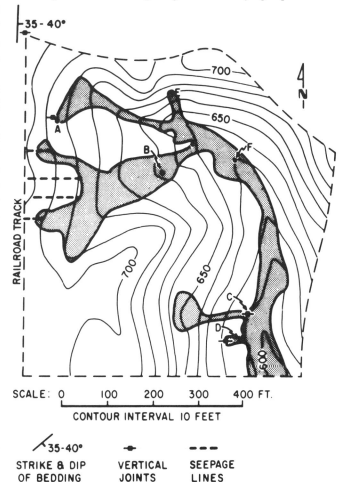

Figure 12. Hydrologic and structural features of a small drainage basin in Vermont in which major joint systems and bedding surfaces localize the outflow of groundwater at springs (A through D) and seepage lines throughout the valley floor, and, during large storms or snowmelt, in topographic hollows. The valley is developed in a fluvioglacial terrace lying against a bedrock valley wall overlain by a thin layer of till. The overall orientation of the valley network reflects the major joint system in the underlying bedrock. The shaded area indicates the extent of saturation overland flow during the snowmelt season. (Source: Dunne, 1980.)

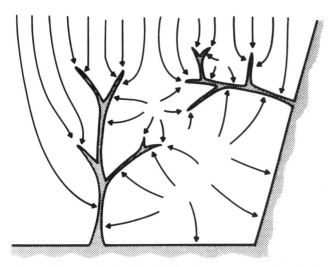

Figure 13. A network of valleys developed by seepage erosion at spring heads. Flow lines (arrows) converge on the heads of valleys, and the pattern stabilizes when the declining drainage area of each spring head is no longer large enough (under the prevailing climate and with the hydrogeologic properties of the aquifer) to yield enough groundwater to cause seepage erosion in the weathered bedrock.

sion of the consequent valleys. If all of these variables were held constant except for the general gradient of the water table along x–x′, and if the resultant local gradient were sufficiently steep, the junction angles would become more acute as the gradient increased. Such a high gradient could be imposed, for example, by a sloping aquiclude beneath the formation being sapped. On the other hand, various degrees of preferred orientation can be imposed on junction angles, channel reaches, and valley bends if either the flow field or the distribution of weaker zones is controlled by rock structure. Schick (1965), Campbell (1973), and Laity and Malin (1985, p. 211–216) have documented examples of the control of stream and valley orientations by rock fractures. Figure 14 shows some examples of interpretations by Laity and Malin (1985). Figure 12 shows another example.

The possibilities for structurally controlled valley networks range from rectangular, trellised patterns to dendritic patterns, in which the influence of structure can only be recognized through systematic map and field measurements. For example, Newell (1970) conducted a series of joint surveys in calcareous schist in the region that includes the drainage basin in Figure 12. He plotted poles to joint surfaces on an equal-area graticule and contoured the results to define sets of high-angle joints. He then measured hillslope gradients and aspects and found a rough parallellism between the preferred orientation of the stream valleys orthogonal to the slopes and that of the fracture pattern. Newell (1970, p. 111) interpreted the parallelism as being due to the concentration of groundwater flow and chemical weathering:

Ideally, such flow in bedrock would proceed with greatest ease along the set of joint planes most nearly normal to the force equipotential surface of the water table. Under some preexisting land surface, leaching along such preferred zones would gradually lower the bottoms of valleys and reorient them with respect to the trend of the joints being followed. The resulting cover of weathered residuum provided the course of least resistance to headward eroding streams. Proceeding from the initial topography, low areas were lowered more rapidly than the contiguous slopes and divides. Under these differential rates a fabric of valleys and relict interfluves developed which reflect the structural permeability of the underlying bedrock.

Although it is not clear that leaching alone would lower the surface, rather than producing a cohesionless residuum with low porosity, chemical weathering would reduce the resistance of the rock, and thus Newell's regional interpretation agrees well with the interpreted role of groundwater based on the instrumental observations of pore pressures, seepage erosion, and local topography in the small basin shown in Figure 12.

Whether or not their orientations are controlled by rock structure, the headward growth of channels progressively disrupts the two-dimensional groundwater field shown in Figure 11, producing a network of valleys (Fig. 13). As the number of spring heads increases, each spring drains groundwater from a smaller area. The spring or valley heads compete for groundwater, and if one is at a lower elevation than its neighbors, it can grow faster and leave them dry as a result of subsurface piracy. Reduction of the drainage area diminishes the groundwater flux and therefore the hydraulic gradient (Equation 8). The frequency of groundwater recharge events large enough to cause seepage erosion of weathered rock will decrease as the drainage area diminishes, and eventually only weak colluvium traveling to the valley head will be removed.

This limiting condition for channel extension by seepage erosion of a weathered residuum can be illustrated approximately by generalizing the Dupuit-Forchheimer analysis in Equation 4 to the case of steady-state seepage from a conical drainage area in colluvium-mantled bedrock converging on the head of a channel (Fig. 15). The limitations of this analysis were discussed previously, but again it includes the essential physics of the way in which the drainage area, recharge rate, and hydraulic conductivity control the hydraulic head gradient at the seepage face.

In some increment of radial distance, dr, the drainage area converging on the 1-m strip of the perimeter of a channel head generates an increase in discharge

$$dq = I \, \epsilon \, r \, dr$$

where r is the radial distance and ϵ is the convergence angle of the drainage area (in radians). Integrating this differential equation, and taking into account the boundary condition that $q = 0$ at $r = r_D$, yields the specific discharge at any distance, r:

$$q_r = \frac{-I \epsilon}{2} (r_D^2 - r^2)$$

where the first negative sign on the right-hand side implies that

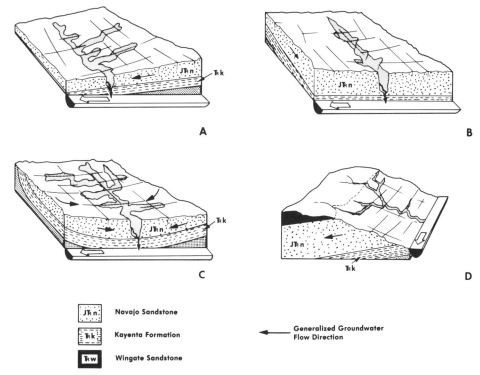

Figure 14. Relations between geologic structure, valley morphology, and network pattern in the Colorado Plateau. Pattern and form depend primarily on the direction of groundwater flow in the Navajo Sandstone relative to the orientation of the valley (assuming a more or less orthogonal joint set). A and B are developed on a monocline: A has a strongly asymmetric network and B has a stubby, branched, symmetric network. C has a more elongate, symmetrical network because the gently plunging syncline focuses more groundwater on the heads of the tributaries. In D, groundwater flows away from valley walls and streamflow is generated by overland flow, which during its convergence on the channel erodes tapered valley heads. (Source: Laity and Malin, 1985.)

discharge increases in the direction of decreasing r. The Dupuit-Forchheimer approximation yields

$$q_r = -K \epsilon r \zeta \frac{dz}{dr}$$

where, in Figure 15

$$\zeta = z - r \tan \beta.$$

Combination of the three previous equations yields the water-table slope:

$$\frac{dz}{dr} = \frac{I r_D^2}{2K} \left[\frac{1}{r(z - r \tan \beta)} \right] - \frac{I}{2K} \left[\frac{r}{(z - r \tan \beta)} \right], \quad (18)$$

which does not have a simple solution, except for the case where $\beta = 0$:

$$z^2 - z_s^2 = \frac{I}{K} \left[r_D^2 \ln \left(\frac{r}{r_s} \right) - \frac{(r^2 - r_s^2)}{2} \right]. \quad (19)$$

Since

$$r_s = \frac{1}{\epsilon},$$

the steady-state specific discharge at the channel head is

$$q_s = \frac{-I \epsilon}{2} (r_D^2 - r_s^2). \quad (20)$$

This discharge entering the stream channel would determine the depth of streamflow, and therefore the height of the seepage face, z_s, in Equation 19. For example, if the 1-m perimeter of the channel head has a drainage area with a convergence angle of 30°, the channel width (b) would be 0.27 m (Fig. 16). If the gradient of this channel (s) is 0.02 and its Manning roughness (n) is 0.05, the depth of flow, and therefore z_s at steady-state outflow, would be

$$z_s = \left[\frac{|q_s| n}{b \, s^{1/2}} \right]^{0.6} = 1.18 \, |q_s|^{0.6} \quad (21)$$

where q_s is given by Equation 20 with $\epsilon = \pi/6$. Substitution of Equation 21 into Equation 18 for $r = r_s$ yields an approximate expression for the hydraulic head gradient at the seepage face:

$$i = \left| \frac{dz}{dr} \right|_{r = r_s} = \frac{I r_D^2}{2K} \left[\frac{1}{r_s (z_s - r_s \tan \beta)} \right] - \frac{I}{2K} \left[\frac{r_s}{z_s - r_s \tan \beta} \right]. \quad (22)$$

This gradient is equivalent to i in Equations 11, and 17, and

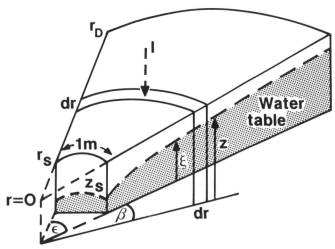

Figure 15. Steady-state water table under uniform recharge on a conical drainage area with a slope β and convergence angle ϵ using the Dupuit-Forchheimer approximation of horizontal flow lines.

therefore determines whether seepage erosion by Coulomb failure will occur. Equation 22 cannot predict liquefaction directly because the Dupuit-Forchheimer assumption precludes upward flow components, but it still indicates which factors govern the magnitude one would expect for those upward components if they were to be defined by graphical or numerical methods. Conversely, Equation 22 and its subsidiary equations summarize how the drainage area (represented by ϵ and r_D) must diminish to prevent seepage erosion for a given set of the other parameters in the hydrologic and erosion equations. The equation therefore summarizes the controls on the drainage area required for channel initiation, which in turn controls the drainage density, the governing factors of which will be discussed in the next section. Equation 22 is used only as a summary of the interactions that govern drainage density. It cannot be used quantitatively because of the limitations of the Dupuit-Forchheimer model, and because the equation does not include such potential influences as changes in hydraulic conductivity due to weathering as the seepage face is approached.

If recharge events competent to cause seepage erosion of even colluvium occur only infrequently, the head of the valley may be partially filled and evolve into an unchanneled swale, floored by a zone subject to saturation overland flow or, if the colluvium is deep enough, by a well-drained slope generating only subsurface stormflow. This accumulation can then be scoured from the bedrock and a channel reestablished during rare seepage erosion (Dietrich and others, 1987, Fig. 4). Thus, the drainage density may vary over time, but the long-term valley head is localized by seepage erosion in the weathered bedrock. If the upper end of the valley is sufficiently steep, the seepage erosion will occur in the form of landsliding, and the headward sapping of the channel grades into the repeated failure and refilling of funnel-shaped bedrock hollows, as described by Dietrich

and Dunne (1978). Dietrich and others (1986) have shown how the stable drainage density of channels forming at the downslope end of long, trough-shaped hollows in California can be rationalized in terms of the role of groundwater seepage in landslide mechanics.

Little work has yet been devoted to mathematical modeling of the evolution of valley networks by seepage erosion. Howard (1986b) has made the most progress in this direction. He developed a finite-difference model of steady two-dimensional flow through a porous medium with a spatially random hydraulic conductivity. As flow emerged from a scarp face, the local rate of erosion was assumed to be proportional to the specific discharge rate, q, minus a critical discharge rate. The erosion altered the boundary and therefore the flow field, and the computation was repeated for the next time step. The model simulates the evolution of valley networks and groundwater flow fields, although the mechanics of seepage erosion are not represented explicitly. It illustrates that even where aquifer properties are randomly distributed, a roughly uniform spacing is imposed on the valley networks by competition for groundwater flow between the lengthening valleys.

Drainage density. Until a rigorous mathematical model is developed for seepage erosion resulting from two- (or three-) dimensional, transient groundwater flow, it will not be possible to specify in an exact, quantitative manner the controls on the drainage density of channels formed by this process. However, by reviewing the components of Equation 22 in conjunction with those of Equations 7, 10, 11, and 17 describing seepage erosion, it

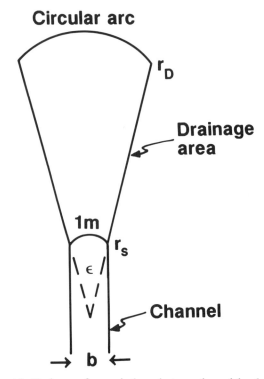

Figure 16. Drainage of a conical swale to a channel head with an upstream perimeter of 1 m and a width of $2\epsilon \sin(\epsilon/2)$ m.

is possible to suggest the most probable controlling factors, even if the exact functional relations cannot yet be defined. The drainage density, which is related in a rather complex way to the drainage area required for channel initiation, is represented in Equation 22 by r_D, r_s, and by ϵ, implicit in z_s.

The most obvious control involves the amount of water (I) available for seepage erosion. If the rock is significantly weakened by weathering before seepage erosion occurs, the total annual runoff may be an important control through its influence on rock decomposition (seepage weathering). However, the removal of the weathered residuum by seepage erosion is likely to occur during some shorter period of high discharge and hydraulic gradient. Thus, total wet-season rainfall or spring snowmelt might be the critical variable in large groundwater flow systems. The deeper and more extensive the flow system controlling the stream network density, the more sustained would the water input have to be to generate seepage erosion at the outlet. Shallower and smaller flow systems develop critical flow rates and hydraulic gradients in shorter, more intense rainstorms. In Figure 12, the bedrock flow system feeding the springs at A, C, and D and upwelling beneath the channel at F generated maximum pore pressures during the spring snowmelt and sustained them for several weeks; in the shallower flow systems in soils at sites B and E, peak hydraulic gradients directed into the slope occurred in response to daily snowmelt cycles and large rainstorms (Dunne, 1970).

It is difficult to isolate which hydrologic input is the dominant control of channel formation and drainage density by statistical analysis of morphometric data because of the high degree of correlation among various measures of rainfall intensity over different time scales at one location. The relevant measure probably varies with the geometry and hydrogeological parameters (mainly hydraulic conductivity and porosity) of the flow system, since these are the factors affecting the transient response of groundwater flow (Freeze and Cherry, 1979, p. 65). The fact that drainage density increases with some measure of precipitation in regions where runoff is dominated by subsurface flow has been established by Chorley (1957), Chorley and Morgan (1962), Cotton (1964), and DeVries (1976).

The intensity and form of weathering and its effect on resistance to seepage erosion (c, ϕ, p, and ρ_s in Equations 7, 9, and 10) must also affect channel formation and drainage density. These processes are controlled in turn by climate, hydrology, and lithology, as discussed in Chapter 1. The hydrogeologic characteristics of the aquifer are important in setting the drainage density. For a fixed rate of groundwater recharge (I) per unit area of catchment, the hydraulic gradient (i) at a seepage face intensifies as the hydraulic conductivity (K) declines (Equation 22), decreasing the critical discharge and therefore drainage area necessary for seepage erosion. The effect of spatial variations in hydraulic conductivity was also emphasized earlier as a factor controlling gradients of hydraulic head (Rulon and others, 1985).

Finally, relief and gradient should be important controls on drainage density. In a homogeneous isotropic aquifer the water table is a subdued replica of the ground surface (Fig. 1), and strong local relief causes high hydraulic gradients between channels and the intervening drainage divides (Freeze and Cherry, 1979, p. 193–199). Also, a sloping aquiclude beneath an aquifer steepens the hydraulic gradient for a fixed recharge rate and conductivity (Equations 18 and 22). Equations 10, 11, and 17 indicate that the critical hydraulic gradient required for seepage erosion also decreases as the hillslope angle increases, but this angle is not, in general, independent of the other controls discussed here. However, it is possible to conclude that both relief and the dip of the aquiclude should be positively correlated with drainage density. DeVries (1976) pointed out that on Pleistocene sandy deposits in the Netherlands, where he interprets the stream networks to result from the outcropping of groundwater, drainage density is inversely proportional to relief. This result does not conflict with the assertions made above because DeVries does not call on seepage erosion to form the channels and valleys. Although there is no explicit description of process mechanics in his paper, he seems to interpret the stream net as resulting from surface sediment transport (by F_w in Equation 16) wherever the recharge rate, hydraulic conductivity, and initial surface geometry have caused the water table to outcrop. For a fixed recharge rate and hydraulic conductivity, low relief and gentle gradients would increase the area of the seepage zone and the number of depressions intersected by the water table.

Channel networks or scarp retreat. Another problem relates to whether concentrated seepage erosion at a spring head will produce a valley or simply sap the hillside along a broad, irregular front to produce an escarpment. As channels or valleys are eroded headward, as shown in Figure 11c, the intervening, unchanneled surface may also retreat as a result of mass failure or undercutting by waves or streams. Up to this point it has been assumed that the former retreat rate far exceeds the latter, so that a channel network forms. This result is more likely where there are strong lateral gradients in hydraulic conductivity, porosity, or susceptibility to weathering, and where there is some dominant wavelength to the spatial pattern (such as regularly spaced master joints) so that flow concentrations are strong. Concentration of erosion at the valley heads is also favored if the structural and lithologic properties of the rock make it susceptible to a radical increase in the frequency and magnitude of sapping in response to the increase in discharge resulting from flow concentration.

However, if the spurs between valleys are aggressively undermined by lateral stream erosion, wave action, or mass wasting, the spurs and valley heads could retreat at approximately the same rate. The land surface then does not become as crenulated as shown in Figure 11c or Figure 13; retreating escarpments develop instead. For example, if the rock is prone to rapid mass failure in the presence of only a small amount of exfiltration over some mechanically weak layer, then cliff retreat may be rapid even where the outflow is diffuse. In such a case one would expect valleys to be wide. Laity and Malin (1985) mapped several examples of this condition (e.g., Fig. 17) and pointed out that the rate of lateral weathering approaches that of headward retreat as

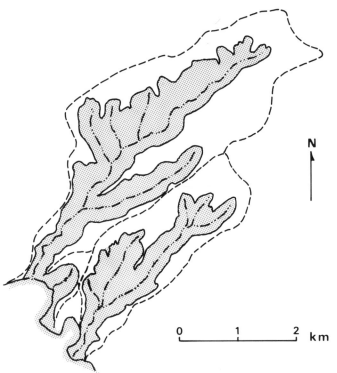

Figure 17. Valley networks tributary to the Escalante River, Utah, where the Navajo Sandstone aquifer (unshaded) has retreated to expose the underlying silt and clay aquiclude of the Kayenta Formation (shaded) over most of the drainage area. (Source: Laity and Malin, 1985.)

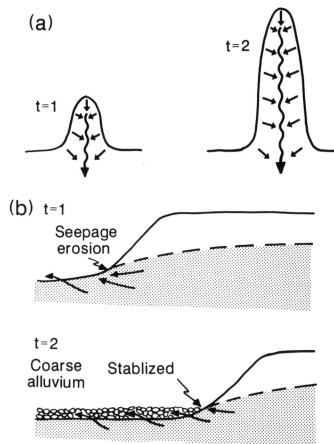

Figure 18. Two ways in which the extension of valley networks may be resisted by sediment influx. (a) As the valley is extended between times 1 and 2, there is an increase in the area of hillslopes shedding sediment to the channel. The increased sediment supply may eventually tax the sediment transport capacity of the river and fill the valley floor. (b) Cliff retreat through seepage erosion may release very coarse colluvium, which exceeds the competence of the stream and accumulates as an apron at the foot of the valley head and along the valley floor. This colluvium loads the toe of the slope, reducing seepage erosion, and conveys the groundwater downvalley as subsurface flow.

the drainage area of the spring head declines (p. 210). Some sequences of geologic materials are so prone to mass failure where groundwater emerges that a single retreating escarpment cannot be maintained; the seepage zone becomes a chaotic terrain of blocks and connected depressions.

It should also be remembered that there is a set of diffusive erosion processes that tend to fill and eradicate the pipes, channels, and valleys produced by seepage erosion at a range of scales. For example, soil pipes formed by seepage erosion are filled or disrupted by soil falls from the perimeter of the conduit and by bioturbation due to root growth, tree-throw, and animal burrowing. The conduit survives only if the fallen soil is evacuated by the outflowing water and if diffusion of soil toward the pipe is less than or equal to the volume evacuated. The more probable condition over time scales exceeding decades is that collapse and evacuation prevail and the roof eventually collapses, or that disruptive processes such as tree-throw fill the pipes and new ones develop in the vicinity.

On the scales of channel or valley networks, the growth of the network causes the expansion of eroding hillslopes and therefore an increased flux of sediment (Fig. 18a), which the stream must evacuate if it is to sustain the new valley head. The resulting balance between sediment supply from the eroding hillslopes and the sediment transport capacity of the runoff produced by seepage would stabilize the valley head (Kirkby, 1986). Although I do not know of any documented examples, it is possible in principle that an increased sediment supply from the crenulated land margin could cause the accumulation of coarse sediment in the lengthening valley floor (Fig. 18b). If sufficiently thick, this rampart of sediment could eventually load the seepage face, preventing seepage erosion (it would increase F_g in Equation 16), and convey the groundwater emerging from the bedrock downstream without causing erosion. Such a situation is more probable where the developing hillslopes are shedding coarse colluvium.

Forms of valley heads. Many investigators (e.g., Laity and Malin, 1985; Baker, 1986; Howard, 1986a) point to the steep, blunt, amphitheater-head valleys commonly produced by sapping in areas where the effects of seepage erosion have been studied.

Kochel and Piper (1986) propose that detailed morphometric analyses could successfully resolve valleys dominated by sapping from those dominated by "runoff processes" (presumably connected with Horton overland flow). They go on to emphasize both in their laboratory simulations and in terrestrial and Martian valleys the presence of amphitheater heads and alcoves. Sakura and others (1987) also found that the valley head was an amphitheater if only exfiltration occurred and more gradual when erosion by surface flow exceeded the effects of seepage erosion. Thus, sapping by seepage erosion and related mass wasting can produce amphitheater-headed valleys, although Laity and Pieri (1986) have challenged the value of these forms as diagnostic criteria of groundwater sapping on Mars. However, there are many other valley systems in the world that seem to have been generated by seepage erosion in the absence of Horton overland flow, but which have tapered valley heads with gentler gradients than those emphasized in the literature.

The form of the hillslopes bounding and defining a valley head depends on either: (a) stability with respect to mass failure; or (b) the balance between erosion and transport processes tending to undermine and steepen the slope and those tending to fill any incision such as that caused by sapping. For example, if seepage erosion removes the basal support from beneath jointed, gently dipping rock (as Laity, 1983, demonstrated on the Colorado Plateau), the hillslope erodes by slab failure and toppling, and stands at metastable angles close to vertical. Even unconsolidated sediments with low cohesion can stand temporarily at steep angles (Lohnes and Handy, 1968; Kochel and others, 1985). However, if the jointing of the rock, or other factors affecting its cohesion and internal friction cause collapse to lower slope angles, or if soil creep, talus accumulation, and debris-flow deposition spread colluvium toward the center of the valley, a more conical, gently sloping valley head results. The valley head can take on an even gentler slope if the water table and the zone of most intense seepage erosion outcrop at different points along the valley in different years. A range of conditions is summarized in Figure 19.

If seepage erosion is concentrated along the base of the valley head, but has achieved a stable drainage density by transporting colluvium at a rate that can be equaled by weathering and soil creep, raveling, debris flow, or other processes of hillslope retreat, the form of the valley head depends on this balance of sediment transport. The result may be a conical or trough-shaped bedrock hollow from which colluvium is periodically evacuated by debris flows (Dietrich and others, 1986) or a gently sloping swale into which soil is transported by creep between rare catastrophic scouring resulting from seepage erosion. Where the role of seepage erosion is to cause a slight increase in the competence and capacity of saturation overland flow or channel flow (Equation 16, Fig. 10), the gradient at the valley head may be very slight. The shapes of these valley heads grade into the runoff-dominated forms referred to in the literature.

It seems appropriate to conclude that valley heads such as those shown in Figures 12 and 19c are just as typical of the results

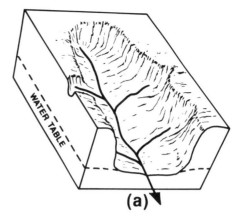

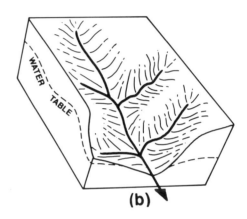

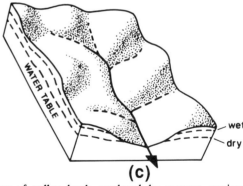

Figure 19. Forms of valley heads produced by seepage erosion: (a) amphitheater-head valley produced by intense seepage erosion in rocks with a high angle of internal friction with the fallen blocks removed between failures; (b) gradually converging valley head eroded by less intense seepage in rocks with a low angle of internal friction or in rocks susceptible to rapid weathering and mass wasting; (c) gradually converging valley head in which the locus of seepage erosion moves up and down valley during fluctuations of climate. Dashed lines represent ephemeral channels and different water-table positions during wet and dry periods. Soil creep and debris flows cause the accumulation of long footslopes of colluvium at the valley head when the water table is low, and extension of seepage erosion causes later evacuation of sediment from swales.

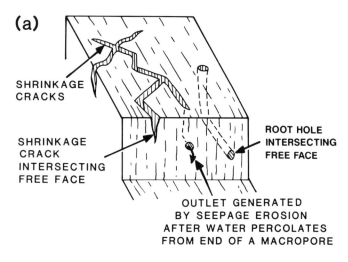

Figure 20. Paths along which water may transport sediment to enlarge macropores by tunnel erosion.

of spring sapping by seepage erosion as are the amphitheater heads and alcoves occurring in the Colorado Plateaus and Hawaii (Fig. 19a). These latter characteristics, along with some others such as low drainage density and stubby tributaries, are diagnostic only of a more restricted set of conditions, including some combination of gently dipping, vertically jointed, highly permeable rocks overlying single, well-defined aquitards, or arid regions, or valleys in which the rate of incision and headward retreat are so high that the valley walls have not yet adjusted to the sudden relief generation, or some other condition that allows little colluvium to accumulate as gently sloping footslopes. The geomorphic effectiveness of seepage erosion is more widespread than the presence of the supposedly diagnostic forms popularized by recent literature.

Mechanics of tunnel scour

Water may enter a hole or crack from some saturated zone within or at the surface of a soil or sediment (Fig. 7). The initial conduit may have a biogenic origin (for examples see Fletcher and others, 1954; Jones, 1981; Beven and Germann, 1982), or be due to tensional cracking along the margins of gullies (Heede, 1971) and landslide blocks, or to shrinkage in certain clay-rich soils under a seasonal rainfall regime (e.g., Fletcher and others, 1954; Jones, 1971; Crouch, 1976), or to seepage erosion, as described above. Many authors have documented evidence of subsurface flow along cracks and tunnels in sodium-rich soils with a high shrink-swell potential occurring in regions with a strongly seasonal hydrologic regime (Parker, 1963; Heede, 1971; Crouch, 1976; Bryan and others, 1978; Imeson, 1986). These soil conditions provide not only the initial subsurface passageways for runoff and sediment transport, but also the low hydraulic conductivity that promotes saturation and enhances flow in the passages.

Thus, factors that promote surface runoff or the development of saturation at shallow depth in the presence of even initially narrow cracks, rootholes, or burrows will increase flow in these conduits. For example, Brown (1962) emphasized the frequency of reports associating the erosion of tunnels with intense rainfall and with soil packing and devegetation resulting from heavy grazing. On the other hand, if the peds between cracks have a dense vegetation cover and root mat, so that their infiltration and water-holding capacities are higher than the intensities and volumes of most rainstorms in the region, the peds will remain unsaturated, and little or no runoff will enter the interpedal cracks or animal burrows.

If surface runoff falls into the conduit, it shears soil at the base in a plunge pool. As the water flows along the conduit, a shear stress is applied to the margins, which may thus be eroded. The intricate geometry of natural conduits precludes a useful computation of the magnitude of this fluid shear. However, it is possible to say in a general way, on the basis of the derivation of the Chézy pipe-flow formula, that the eroding stress is proportional to the square of the average flow velocity, and therefore to the head gradient, the diameter or width of the conduit, and its hydraulic roughness, which in turn depends on the various scales of roughness elements along its path, ranging from grains and aggregates to the tortuosity of the entire conduit. The shear stress can be particularly high in tunnels on a steep hillslope or close to the bank of a stream or gully, where the head gradient is high (Parker, 1963; Heede, 1971).

If the shear stress is sufficiently high, grains are eroded from the margins of the conduit. Erosion of the dry boundary may initially be relatively rapid, but may diminish quickly in cohesive soils (Parthenaides, 1971), and material may soften and fall from the roof and walls as it becomes wet. The subsequent fate of the eroded grains depends on the velocity of flow, and therefore whether the conduit has an outlet from which water can drain freely. Outlets of various origin are shown in Figures 20 and 21. If the conduit is blocked, at the end of a root-hole for example, the eroded sediment will be deposited and the passage will be filled. If the passage is near a free face, such as a riverbank, or on a steep slope, the local increase of pore pressure around the conduit may cause seepage erosion (Fig. 20a).

An open connection downslope to the soil surface and sufficient head gradient can maintain a velocity high enough to transport the eroded sediment. Water can enter the passage from the surrounding matrix (if the latter is locally saturated), or from smaller tributary cracks, or from surface runoff. The outlet may be at a free face or on a gentler hillslope (Fig. 20b). It may intersect such a surface because of headward seepage erosion, or because the original conduit intersected the surface as a biogenic

Figure 21. Conduits of various size intersecting the walls of a gully in thoroughly weathered volcaniclastic sediments near Karatu, northern Tanzania.

passage or shrinkage crack. If the passageway began as a shrinkage crack, wetting of the soil surface usually closes the cracks late in the wet season, leaving a tunnel enlarged by erosion (Fig. 22). Seasonal repetition of this process can form networks of tunnels, bridged by an arch of cohesive soil. This cover eventually collapses, producing pits at various points along the tunnel, or continuous gullies with sharp, irregular boundaries (Rubey, 1928; Löffler, 1977; Swanson, 1983; Fig. 21). However, if the subsurface flow exploits deeper cracks, joints, and bedding planes, the larger subterranean caverns referred to as pseudokarst may develop by the same process of shearing the margins of a conduit through otherwise unsaturated geologic materials (Parker and others, 1964; Carey, 1976).

The critical shear stress required to erode the margins of a tunnel is radically reduced if the geologic material contains a dispersive clay, although this is not a necessity for tunnel scour if other conditions are favorable. When wetted, some earth materials swell and eventually disperse due to a complex set of processes, including hydration and deaeration, fracturing, and osmosis (Holmgren and Flanagan, 1977). Many authors have noted the association between tunnel scour and high concentrations of salts (especially of sodium) in the soil (e.g., Heede, 1971; Crouch, 1976; Bryan and others, 1978; Stocking, 1981; Imeson, 1986). Heede (1971) found that in deeply cracked, montmorillonitic clays along gully banks, tunnel scour did not occur unless the exchangeable sodium percentage exceeded a value of about 1.0. In the soils studied by Stocking (1981), this critical value was closer to 10, but with some values as low as 4.

Several laboratory studies of the erosive behavior of consolidated earth materials, reviewed by Arulanandan and Heinzen (1977) and Heinzen and Arulanandan (1977), have shown that the shear stress required to initiate erosion is affected by the amount and type of clay, and the concentration of solutes in the pore and eroding fluids. The osmotic potential gradient set up at the surface of the clay due to the concentration gradient between pore fluid and eroding fluid, together with the capillary gradient,

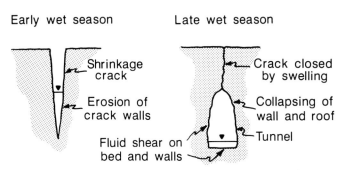

Figure 22. Enlargement of a shrinkage crack by tunnel erosion.

cause flow into the porous medium. Dilute water traveling along conduits is drawn into pores and cracks, causing swelling, a reduction of interparticle bonding forces, fracturing into aggregates, and dispersion of particles. The degree of swelling and dispersion depends on the sodium-absorption ratio and the solute concentration in the pore fluid (Crouch, 1976). As the concentration increases, the critical sodium-absorption ratio required for dispersion increases. The physical chemistry of these effects is described by Shainberg and Letey (1984). Clay mineralogy and soil pH also affect the degree of dispersion, while less soluble cements such as calcium carbonate suppress it. Crouch (1976) summarized studies on the dispersive effects of organic matter and concluded that it usually stabilizes soil structure and reduces the degree of dispersion, but that organic compounds of low molecular weight may exacerbate dispersion in some soils.

Heinzen and Arulanandan (1977) demonstrated that the critical tractive stress for a clay decreases as the solute concentration of the eroding fluid declines relative to that in the pore water. A true dispersive clay is one for which the critical tractive stress equals zero. All that is then required for erosion is a flow velocity sufficient to transport the resulting colloidal suspension. Dispersion can also reduce the critical pore-pressure gradient required for seepage erosion by reducing the cohesion in Equation 9. For example, if dispersible silty sediments, with a sufficiently high conductivity to allow significant saturated outflow, become connected with a source of dilute groundwater, dispersive seepage erosion can be particularly rapid. However, more typical is the experience of Sherard and Decker (1977, p. 5) that most dispersive clays are so impervious that little matrix flow and seepage erosion occur, and in order to initiate erosion it is necessary to provide a path for concentrated leakage. The low conductivity then forces flow from the surface or a shallow permeable horizon into cracks that are rapidly enlarged by dispersion-accelerated shear. Many of the field situations described in the papers edited by Bryan and Yair (1982) appear to be of this type. Imeson and others (1982, p. 68) provide a detailed description of one example.

Stocking (1981) presents a particularly interesting interpretation of results from Zimbabwe, where he suggests that high exchangeable sodium concentrations develop in soils where sodium ions released by feldspar weathering accumulate on clay minerals in a shallow, seasonally saturated soil horizon. The deflocculated clays block soil pores and vertical drainage. However, there is groundwater flow in the now-confined aquifer beneath the impeding layer. If gullying incises the confined aquifer, headward erosion begins at the gully wall (presumably by either seepage erosion or tunnel scour) and proceeds beneath the impeding layer to form a tunnel with a low gradient. Once the tunnel is formed, the impeding layer forming the roof is breached, possibly by cracking due to gravitational settling, or because the formation of the tunnel increases the vertical head gradient and therefore the vertical flow through pores or cracks. This vertical flow then erodes vertical tunnels tributary to the horizontal tunnel, weakening the roof and eventually causing its demise.

The presence of tunnels in a porous earth material increases the efficiency with which water and sediment can be evacuated from the landscape. Flow within a crack or a tunnel is faster than percolation through the porous matrix, and the presence of steep or overhanging faces of bare, often dispersive material without the binding action of roots provides conditions that favor rapid erosion and sediment transport. The complexity of interaction between the fluid shear (with or without dispersion) and of gravitational collapse of the tunnel margins often causes rapid fluctuations in sediment concentration in the outflow from tunnels. Temporary blockage of pipes, or their joining, can also cause pulsations in flow. Yair and others (1980) attributed pulsations of flow and sediment concentration in runoff to tunnel collapse and formation of mudflows. Bryan and Harvey (1985) measured rapidly fluctuating sediment concentrations of up to 97,000 mg/l emanating from tunnels in Na-montmorillonite-rich clay badlands. Swanson (1983) compared sediment concentrations in tunnel discharge from a sodium-rich colluvium with synchronous concentrations in overland flow from the grassy surface. Concentrations were roughly correlated with discharge, but those in the overland flow ranged between 100 and 1,100 mg/l, while the tunnel discharge had concentrations of 3000 to 30,000 mg/l and flow at the site of a recent roof collapse a few meters upstream of the headcut had concentrations of 800 to 8,000 mg/l. Swanson and others (1989) found that 95 percent of all the sediment eroded from this small drainage basin during a 25-yr, 6-hr rainstorm originated by tunnel scour.

Morphological significance of tunnel scour

The following conditions are necessary for the occurrence of tunnel scour:

(1) Saturation at the land surface, above a subsurface horizon, or at the roof of a macropore being heavily recharged by percolation from above.

(2) An initial passage with an outlet, into which water from the saturated zone can flow.

(3) A sufficiently high shear stress on one boundary of the passage to exceed the critical tractive force of the boundary material. Dispersion of clay minerals may reduce this critical tractive force virtually to zero, and such minerals are more likely to form by weathering in subhumid climates, or in certain lithologies such as marine clays or glass-rich volcanic rocks in a range of climates. Mass failure of walls and roofs may also mobilize sediment.

(4) Sufficient transport capacity and therefore sufficient flow and hydraulic gradient throughout the passage to discharge all eroded sediment.

(5) Sufficient cohesion to maintain walls and a roof for some time.

The origin of such controlling factors is reviewed above.

The formation of a passage also stimulates the tendency to fill it. Thus, tunnels in forest soils, such as those described by Jones (1971) and Pierson (1983) accelerate the inward diffusion of soil particles by creep, or are obliterated by bioturbation.

Small, seasonally enlarged cracks in cracking-clay soils are filled by swelling during the subsequent wet season. Larger tunnels in dispersive soils tend to fill by collapse of sidewalls and roofs until the ground surface is breached (Heede, 1971; Löffler, 1977; Swanson, 1983).

The survival of tunnels depends on the relative rates of the scouring and filling processes, within the constraint set by the survival of roofs the breaching of which converts the tunnel into a gully. Thus, in a steep, windy, forested environment where treethrow is vigorous, tunnels, especially in well-aggregated soils, would not be expected to survive for long or to grow large, except below the root zone in colluvium-filled bedrock hollows (Pierson, 1983), or in peat layers with low biogenic activity (Gilman and Newson, 1980), or beneath shallow-rooted vegetation in percolines (Jones, 1971). Despite the claim of Jones (1971), I am not aware of any proof that such tunnels usually evolve into stream channels, rather than the ensemble of them being a steady-state feature of the landscape with an average rate of production and destruction. At the other extreme, dispersive sedimentary rocks described by Parker (1963) and Parker and others (1964) may erode along tunnels that are too deep to be disrupted by bioturbation or any other filling process except collapse, which is resisted by the high cohesive strength of the unwetted parts of the formations. Therefore, they can survive for long enough to develop the large intricate features known collectively as pseudokarst, described in Chapter 3.

Whatever their depth, dimensions, and age, tunnels are ephemeral on a geological time scale, and the ground is eventually lowered to them by surface erosion or internal failure. Thus, the dominant geomorphological expression of tunnel scour at the surface results from collapse (Rubey, 1928; Buckham and Cockfield, 1950; Löffler, 1977, p. 158). The resulting channel and valley form tends to be sharp-edged (Fig. 21). Collapsed blocks disperse and are eroded by even shallow channelized runoff. Sinkholes, vertical shafts, and subhorizontal tunnels occur upstream of the heads of subaerial channels, and the local hydrology and sediment transport may be exceedingly complex (Bryan and Harvey, 1985).

The initial conduits may be localized by very subtle sedimentary structures or stress cracks, so that it is difficult to predict the next phase of tunnel extension. Therefore, there is little prospect of predicting the magnitudes of flows and shear stresses to which the tunnel margins will be subject. This uncertainty, together with the difficulty of specifying a critical tractive force in heterogeneous, dispersible material makes the prediction of a stable drainage density intractable at present. Drainage density should be positively correlated with rainstorm intensity, land surface gradient, local relief, and the dispersibility of the geologic material, and in truly dispersive materials (in which the critical tractive force declines to zero), an intricate network of tunnels should eventually consume the material. However, if the critical tractive force is not zero, and if the processes tending to fill the tunnel are significant, the drainage density is stabilized by the balance between tunnel scour and infilling. Examples of balanced and unstable tunnel systems are described in Chapter 3 and the papers referred to in this section.

SUMMARY

Subterranean water can erode geologic materials: (1) as it flows through and emerges from the matrix; and (2) as it flows along conduits of various origin. The former process is here called seepage erosion; the latter is called tunnel scour. The processes are not mutually exclusive, but the conditions favoring each are different. The morphogenetic significance of seepage erosion in establishing valley systems in weathered bedrock over geological time scales is probably greater than that of tunnel scour, which seems to have a large-scale morphological expression only in a rather restricted, though widespread, range of geologic and climatic conditions. The role of seepage erosion in forming valley systems is probably much more widespread than the restricted set of conditions in which "groundwater sapping" has been studied. Conversely, in some landscapes, such as Hawaii, in which groundwater sapping has been called on as the dominant erosive agent, mechanisms such as undermining by plunge-pool activity have not yet been ruled out. Detailed field studies of the dominant erosion processes have been conducted only at a few localities.

Understanding the mechanics of each erosion process requires study of the governing subsurface hydrology. Important areas of uncertainty remain in:

(1) quantitative prediction of flow rates and stresses in subterranean runoff as they are affected by the complex geometries and evolving hydrogeological properties at eroding sites;

(2) analysis of the roles of structural features of rocks and soils, which affect both the subsurface flow and resistance to erosion;

(3) quantitative prediction of the stabilization of valley or channel networks and their drainage density as a function of the controlling variables for each erosion process:

(4) analysis of the response of hillslopes, tunnel margins, and other surfaces formed by expansion of the valley or channel networks in order to understand how the formation of channelized forms stimulates the tendency to fill them.

ACKNOWLEDGMENTS

This chapter was written with the support of a grant from the National Science Foundation, EAR-8313172. It was improved significantly as a result of reviews by R. J. Chorley, W. E. Dietrich, C. G. Higgins, M. J. Kirkby, and J. E. Laity. R. M. Iverson was particularly helpful in clarifying my understanding of seepage erosion in cohesive materials.

APPENDIX 1.

Symbols.

b	channel width
c	cohesion
g	gravitational acceleration
h_w	thickness of saturated layer normal to slope
I	infiltration or recharge rate
i	magnitude of the hydraulic head gradient
i_z	magnitude of the vertical component of the hydraulic head gradient
K	saturated hydraulic conductivity
$K(w)$	unsaturated hydraulic conductivity, varying with moisture content
L	horizontal length of a hillslope
n	Manning's roughness coefficient
p	porosity
Q	subsurface flux per unit cross-sectional area
q	subsurface flux per unit width (direction, distance, etc., indicated by subscript)
r	radial distance
s	channel gradient
t	time
w	moisture content
x, y, z	orthogonal coordinates
z_s	height of a seepage face
α	angle of water table
β	angle of underlying aquiclude
ϵ	concentration angle of a conical hollow
ζ	water-table height above aquiclude
θ	hillslope angle
λ	angle between seepage vector and surface normal vector
ν	angle between seepage vector and horizontal
ρ_f, ρ_s	densities of fluid, solid
Φ	total hydraulic head
ϕ	angle of internal friction
Ψ	pressure head

REFERENCES CITED

Arulanandan, K., and Heinzen, R. T., 1977, Factors influencing erosion in dispersive clays and methods of identification: International Association of Scientific Hydrology Publication 122, p. 75–81.

Baker, V. R., 1986, Evolution of valleys dissecting volcanoes on Mars and Earth, *in* Howard, A. D., Kochel, R. C., and Holt, H. E., eds., Proceedings and Field Guide for the NASA Groundwater Sapping Conference, November 1985, Flagstaff, Arizona: National Aeronautics and Space Administration, p. 187–189.

Bear, J., 1972, Dynamics of fluids in porous media: New York, American Elsevier, 784 p.

Bennett, H. H., 1939, Soil conservation: New York, McGraw-Hill Book Co., 993 p.

Beven, K., 1982, On subsurface stormflow; Predictions with simple kinematic theory for saturated and unsaturated flows: Water Resources Research, v. 18, p. 1627–1633.

Beven, K., and Germann, P., 1981, Water flow in soil macropores; 2, A combined flow model: Journal of Soil Science, v. 32, p. 15–29.

—— , 1982, Macropores and water flow in soils: Water Resources Research, v. 18, p. 1311–1325.

Bouma, J. L., Dekker, L. W., and Wösten, J.H.M., 1978, A case study on infiltration into dry clay soil; 2, Physical measurements: Geoderma, v. 20, p. 41–51.

Brown, G. W., 1962, Piping erosion in Colorado: Journal of Soil and Water Conservation, v. 17, p. 220–222.

Bryan, R. B., and Harvey, L. E., 1985, Observations on the geomorphic significance of tunnel erosion in a semi-arid ephemeral drainage system: Geografiska Annaler, v. 67A, p. 257–272.

Bryan, R. B., and Yair, A., eds., 1982, Badland geomorphology and piping: Norwich, U.K., GeoBooks, 408 p.

Bryan, R. B., Yair, A., and Hodges, W. K., 1978, Factors controlling the initiation of runoff and piping in Dinosaur Provincial Park badlands, Alberta, Canada: Zeitschrift für Geomorphologie, Supplement Band 29, p. 151–168.

Buckham, A. F., and Cockfield, W. E., 1950, Gullies formed by sinking of the ground: American Journal of Science, v. 248, p. 137–141.

Campbell, I. A., 1973, Controls of canyon and meander forms by jointing: Area, v. 5, p. 291–296.

Carey, D. L., 1976, Forms and processes in the pseudokarst topography of Arroyo Tapiado, Anza-Borrego Desert State Park, San Diego County, California [M.S. thesis]: Los Angeles, University of California, 132 p.

Carlston, C. W., 1963, Drainage density and streamflow: U.S. Geological Survey Professional Paper 422–C, 8 p.

Carson, M. A., and Kirkby, M. J., 1972, Hillslope form and process: London, Cambridge University Press, 473 p.

Cedergren, H. R., 1967, Seepage, drainage, and flow nets: New York, John Wiley & Sons, 489 p.

Chorley, R. J., 1957, Climate and morphometry: Journal of Geology, v. 65, p. 628–638.

—— , 1978, The hillslope hydrological cycle, *in* Kirkby, M. J., ed., Hillslope hydrology: Chichester, U.K., John Wiley and Sons, p. 1–42.

Chorley, R. J., and Morgan, M. A., 1962, Comparison of morphometric features, Unaka Mountains, Tennessee and North Carolina, and Dartmoor, England: Geological Society of America Bulletin, v. 73, p. 17–34.

Coelho Netto, A. L., Fernandes, N. F., and Moura, J. R., 1987, Drainage network expansion by gully advance through rampa deposits in S.E. Brazilian Plateau, Bananal, S.P.: EOS Transactions of the American Geophysical Union, v. 68, p. 1273.

Cotton, C. A., 1964, The control of drainage density: New Zealand Journal of Geology and Geophysics, v. 7, p. 348–352.

Crouch, R. J., 1976, Field tunnel erosion; A review: The Journal of the Soil Conservation Service of New South Wales, v. 32, p. 98–111.

Davis, S. N., and DeWiest, R.J.M., 1966, Hydrogeology: New York, John Wiley and Sons, 463 p.

DeVries, J. J., 1976, The groundwater outcrop-erosion model; Evolution of the stream network in the Netherlands: Journal of Hydrology, v. 29, p. 43–50.

Dietrich, W. E., and Dunne, T., 1978, Sediment budget for a small catchment in mountainous terrain: Zeitschrift für Geomorphologie, Supplement Band 29, p. 191–206.

Dietrich, W. E., Wilson, C. J., and Reneau, S. L., 1986, Hollows, colluvium, and landslides in soil-mantled landscapes, *in* Abrahams, A. D., ed., Hillslope processes: Winchester, Massachusetts, Allen and Unwin Inc., p. 361–388.

Dietrich, W. E., Reneau, S. L., and Wilson, C. J., 1987, Overview; "Zero-order basins" and problems of drainage density, sediment transport, and hillslope morphology: International Association of Hydrological Sciences Publication 165, p. 27–37.

Dunne, T., 1970, Runoff production in a humid area: U.S. Department of Agriculture Report ARS 41-160, 108 p.

—— , 1978, Field studies of hillslope flow processes, *in* Kirkby, M. J., ed., Hillslope hydrology: Chichester, U.K., John Wiley and Sons, p. 227–293.

——, 1980, Formation and controls of channel networks: Progress in Physical Geography, v. 4, p. 211–239.
Dunne, T., and Black, R. D., 1970a, An experimental investigation of runoff production in permeable soils: Water Resources Research, v. 6, p. 478–490.
——, 1970b, Partial area contributions to storm runoff in a small New England watershed: Water Resources Research, v. 6, p. 1296–1311.
Fetter, C. W., 1980, Applied hydrogeology: Columbus, Ohio, Merrill Publishing Company, 488 p.
Fletcher, J. E., Harris, K., Peterson, H. B., and Chandler, V. N., 1954, Piping: EOS Transactions of the American Geophysical Union, v. 35, p. 258–263.
Freeze, R. A., 1969, The mechanisms of natural groundwater recharge and discharge; 1, One-dimensional, vertical, unsteady, unsaturated flow above a recharging or discharging groundwater flow system: Water Resources Research, v. 5, p. 153–171.
——, 1971, Three-dimensional, transient, saturated-unsaturated flow in a groundwater basin: Water Resources Research, v. 7, p. 347–366.
——, 1974, Streamflow generation: Reviews of Geophysics and Space Physics, v. 12, p. 627–647.
——, 1975, A stochastic-conceptual analysis of one-dimensional groundwater flow in nonuniform homogeneous media: Water Resources Research, v. 11, p. 725–741.
Freeze, R. A., and Cherry, J. A., 1979, Groundwater: Englewood Cliffs, New Jersey, Prentice-Hall, 604 p.
Freeze, R. A., and Witherspoon, P. A., 1967, Theoretical analysis of regional groundwater flow; 2, Effect of water-table configuration and subsurface permeability variation: Water Resources Research, v. 3, p. 623–634.
Gillham, R. W., 1984, The effect of the capillary fringe on water-table response: Journal of Hydrology, v. 67, p. 307–324.
Gilman, K., and Newson, M., 1980, Soil pipes and pipeflow; A hydrological study in upland Wales; British Geomorphological Research Group Research Monograph 1: Norwich, U.K., GeoBooks, 110 p.
Heede, B. H., 1971, Characteristics and processes of soil piping in gullies: U.S.D.A. Forest Service Research Paper RM-68, Rocky Mountain Forest and Range Experiment Station, 15 p.
Heinzen, R. T., and Arulanandan, K., 1977, Factors influencing dispersive clays and methods of identification, in Sherard, J. L., and Decker, R. S., eds., Dispersive clays, related piping, and erosion in geotechnical projects: American Society for Testing and Materials ASTM STP 623, p. 202–217.
Hewlett, J. D., and Hibbert, A. R., 1967, Factors affecting the response of small watersheds to precipitation in humid areas, in Sopper, W. E., and Lull, H. W., eds., Forest hydrology: Oxford, Pergamon, p. 275–290.
Higgins, C. G., 1982, Drainage systems developed by sapping on Earth and Mars: Geology, v. 10, p. 147–152.
——, 1984, Piping and sapping; Development of landforms by groundwater outflow, in Lafleur, R. G., ed., Groundwater as a geomorphic agent: Boston, Massachusetts, Allen and Unwin, Inc., p. 18–58.
Holmgren, G.G.S., and Flanagan, C. P., 1977, Factors affecting spontaneous dispersion of soil materials as evidenced by the crumb test, in Sherard, J. L., and Decker, R. S., eds., Dispersive clays, related piping, and erosion in geotechnical projects: American Society for Testing and Materials Special Technical Paper 623, p. 218–239.
Horton, J. H., and Hawkins, R. H., 1965, Flow path of rain from the soil surface to the water table: Soil Science, v. 100, p. 377–383.
Howard, A. D., 1986a, Groundwater sapping on Mars and Earth, in Howard, A. D., Kochel, R. C., and Holt, H. E., eds., Proceedings and Field Guide, NASA Groundwater Sapping Conference, Flagstaff, Arizona, November, 1985: National Aeronautics and Space Administration, p. vi–xiv.
——, 1986b, Groundwater sapping experiments and modelling at the University of Virginia, in Howard, A. D., Kochel, R. C., and Holt, H. E., eds., Proceedings and Field Guide, NASA Groundwater Sapping Conference, Flagstaff, Arizona, November, 1985: National Aeronautics and Space Admnistration, p. 71–83.
Howard, A. D. and McLane, C. F., III, 1988, Erosion of cohesionless sediment by groundwater seepage: Water Resources Research, v. 24, p. 1659–1674.

Hubbert, M. K., 1940, The theory of groundwater motion: Journal of Geology, v. 48, p. 785–944.
Hursh, C. R., 1944, Report of the sub-committee on subsurface flow: EOS Transactions of the American Geophysical Union, v. 25, p. 743–746.
Hutchinson, J. N., 1968, Mass movement, in Fairbridge, R. W., ed., The Encyclopedia of Geomorphology: Dowden, Hutchinson and Ross, Inc., p. 688–695.
Iida, T., 1984, A hydrological method of estimation of the topographic effect on the saturated throughflow: Transactions of the Japanese Geomorphological Union, v. 2, p. 67–72.
Imeson, A. C., 1986, Investigating volumetric changes in clayey soils related to subsurface water movement and piping: Zeitschrift für Geomorphologie, Supplement Band 60, p. 115–130.
Imeson, A. C., Kwaad, F.J.P.M., and Verstraten, J. M., 1982, The relationship of soil physical and chemical properties to the development of badlands in Morocco, in Bryan, R. B., and Yair, A., eds., Badland geomorphology and piping: Norwich, U.K., GeoBooks, p. 47–70.
Ineson, J., 1963, A hydrogeologic study of the permeability of the Chalk: Journal of the Institution of Water Engineers, v. 17, p. 449–463.
Iverson, R. M., and Major, J. J., 1986, Groundwater seepage vectors and the potential for hillslope failure and debris flow mobilization: Water Resources Research, v. 22, p. 1543–1548.
Jones, A., 1971, Soil piping and stream initiation: Water Resources Research, v. 7, p. 602–610.
Jones, J.A.A., 1981, The nature of soil piping; A review: British Geomorphological Research Group Monograph 3: Norwich, U.K., GeoBooks, 301 p.
——, 1982, Experimental studies of pipe hydrology, in Bryan, R. B., and Yair, A., eds., Badland geomorphology and piping: Norwich, Geo Books, Inc., p. 355–370.
——, 1987, The initiation of natural drainage networks: Progress in Physical Geography, v. 11, p. 207–245.
Kälin, M., 1977, Hydraulic piping; Theoretical and experimental findings: Canadian Geotechnical Journal, v. 14, p. 107–124.
Kirkby, M. J., 1978, Implications for sediment transport, in Kirkby, M. J., ed., Hillslope hydrology: Chichester, U.K., John Wiley and Sons, p. 325–364.
——, 1986, A two-dimensional simulation model for slope and stream evolution, in Abrahams, A. D., ed., Hillslope processes: Winchester, Massachusetts, Allen and Unwin, Inc., p. 203–222.
Kirkby, M. J., and Chorley, R. J., 1967, Throughflow, overland flow, and erosion: International Association of Scientific Hydrology Bulletin, v. 12, p. 5–21.
Kochel, R. C., 1986, Role of groundwater sapping in development of large valley networks on Hawaii, in Howard, A. D., Kockel, R. C., and Holt, H. E., eds., Proceedings and Field Guide, NASA Groundwater Sapping Conference, Flagstaff, Arizona, November, 1985: National Aeronautics and Space Administration, p. 193–195.
Kochel, R. C., and Piper, J. F., 1986, Morphology of large valleys on Hawaii; Implications for groundwater sapping and comparisons to Martian valleys: 17th Lunar and Planetary Science Conference, p. 424–425.
Kochel, R. C., Howard, A. D., and McLane, C. F., 1985, Channel networks developed by groundwater sapping in fine-grained sediments; Analogs to some Martian valleys, in Woldenberg, M. J., ed., Models in Geomorphology: Boston, Massachusetts, Allen and Unwin, p. 313–341.
Laity, J. E., 1983, Diagenetic controls on groundwater sapping and valley formation, Colorado Plateau, revealed by optical and electron microscopy: Physical Geography, v. 4, p. 103–125.
Laity, J. E., and Malin, M. C., 1985, Sapping processes and the development of theater-headed valley networks on the Colorado Plateau: Geological Society of America Bulletin, v. 96, p. 203–217.
Laity, J. E., and Pieri, D. C., 1986, An assessment of tributary development flanking Ius Chasma, Mars: Geological Society of America Abstracts with Programs, v. 18, p. 186.
Löffler, E., 1977, Geomorphology of Papua New Guinea: Canberra, Australian National University Press, 195 p.

Lohnes, R. A., and Handy, R. L., 1968, Slope angles in friable loess: Journal of Geology, v. 76, p. 247–258.

Lloyd, J. W., and Farag, M. H., 1978, Fossil groundwater gradients in arid regional sedimentary basins: Groundwater, v. 16, p. 388–393.

Maxwell, T. A., 1982, Erosional patterns of the Gilf Kebir plateau and implications for the origin of Martian canyonlands, in El-Baz, F., and Maxwell, T. A., eds., Desert landforms of southwest Egypt: National Aeronautics and Space Administration Contract Report 3611, p. 281–300.

Musgrave, G. W., and Holtan, H. N., 1964, Infiltration, in Chow, V. T., ed., Handbook of applied hydrology: New York, McGraw-Hill Book Co., p. 12-1–12-30.

Newell, W. L., 1970, Factors influencing the grain of the topography along the Willoughby Arch in northeastern Vermont: Geografiska Annaler, v. 52A, p. 103–112.

Parker, G. C., 1963, Piping, a geomorphic agent in landform development of the drylands: International Association of Scientific Hydrology Publication 65, p. 103–113.

Parker, G. C., Shown, L. M. and Ratzlaff, K. W., 1964, Officer's Cave, a pseudokarst feature in altered tuff and volcanic ash of the John Day Formation in eastern Oregon: Geological Society of America Bulletin, v. 75, p. 393–402.

Parthenaides, E., 1971, Erosion and deposition of cohesive materials, in Shen, H. W., ed., River mechanics, v. 2: Littleton, Colorado, Water Resources Publications, p. 25–91.

Pearce, A. J., Stewart, M. K., and Sklash, M. G., 1986, Storm runoff generation in humid headwater catchments; 1, Where does the water come from?: Water Resources Research, v. 22, p. 1263–1272.

Pierson, T. C., 1983, Soil pipes and slope stablity: Quarterly Journal of Engineering Geology, v. 16, p. 1–11.

Ragan, R. M., 1968, An experimental investigation of partial area contributions: International Association of Scientific Hydrology Publication 76, p. 241–251.

Roessel, B.W.P., 1950, Hydrological problems concerning runoff in headwater regions: EOS Transactions of the American Geophysical Union, v. 31, p. 431–442.

Rubey, W. W., 1928, Gullies in the Great Plains formed by sinking of the ground: American Journal of Science, v. 215, p. 417–422.

Rubin, J., 1966, Theory of rainfall uptake by soils initially drier than field capacity and its applications: Water Resources Research, v. 2, p. 739–749.

Rulon, J. J., Rodway, R., and Freeze, R. A., 1985, The development of multiple seepage faces on layered slopes: Water Resources Research, v. 21, p. 1625–1636.

Sakura, Y., 1983, Role of capillary water zone in groundwater recharge; Observation of rain infiltration by lysimeter: Japanese Journal of Limnology, v. 44, p. 311–320.

Sakura, Y., and Taniguchi, M., 1983, Experiments of rain infiltration on characteristics of soil water movement using a soil column: Geographical Review of Japan, v. 56, p. 81–93.

Sakura, Y., Mochizuki, M., and Kawasaki, I., 1987, Experimental studies on valley headward erosion due to groundwater flow: Geophysical Bulletin of Hokkaido University, v. 49, p. 229–239.

Schick, A. P., 1965, Some effects of lineative factors on stream courses in homogeneous bedrock: International Association of Hydrological Sciences Bulletin, v. X(3), p. 5–11.

Selby, M. J., 1982, Hillslope materials and processes: Oxford, Oxford University Press, 264 p.

Shainberg, I., and Letey, J., 1984, Response of soils to sodic and saline conditions: Hilgardia, v. 53, p. 1–57.

Sharp, J. M., 1976, Groundwater sapping by freeze-and-thaw: Zeitschrift für Geomorphologie, v. 20, p. 784–787.

Sherard, J. L., and Decker, R. S., eds., 1977, Dispersive clays, related piping, and erosion in geotechnical projects: American Society for Testing and Materials Special Technical Paper 623, 486 p.

Sklash, M. G., and Farvolden, R. N., 1979, The role of groundwater in storm runoff: Journal of Hydrology, v. 43, p. 45–65.

Sklash, M. G., Stewart, M. K., and Pearce, A. J., 1986, Storm runoff generation in humid headwater catchments; 2, A case study of hillslope and low-order stream response: Water Resources Research, v. 22, p. 1273–1282.

Stocking, M., 1981, A model of piping in soils: Transactions of the Japanese Geomorphological Union, v. 2, p. 263–278.

Swanson, M. L., 1983, Soil piping and gully erosion along coastal San Mateo County, California [M.S. thesis]: Santa Cruz, University of California, 141 p.

Swanson, M. L., Kondolf, G. M., and Boison, P. J., 1989, An example of rapid gully initiation and extension by subsurface erosion; Coastal San Mateo County, California: Geomorphology, v. 2, p. 393–404.

Tanaka, T., 1982, The role of subsurface water exfiltration in soil erosion processes: International Association of Hydrological Sciences Publication 137, p. 73–80.

Tanaka, T., Yashuhara, M., and Marui, A., 1982, Pulsating flow phenomenon in soil pipe: Annual Report of the Institute of Geoscience, University of Tsukuba, v. 8, p. 33–36.

Terzaghi, K., 1943, Theoretical soil mechanics: New York, Wiley and Sons, 510 p.

Toth, J., 1962, A theory of groundwater motion in small drainage basins in central Alberta: Journal of Geophysical Research, v. 67, p. 4375–4387.

—— , 1963, A theoretical analysis of groundwater motion in small drainage basins: Journal of Geophysical Research, v. 68, p. 4795–4812.

—— , 1966, Mapping and interpretation of field phenomena for groundwater reconnaissance in a prairie environment, Alberta, Canada: International Association of Scientific Hydrology Bulletin, v. 11, p. 1–49.

Wang, H. F., and Anderson, M. P., 1982, Introduction to groundwater modelling; Finite difference and finite element methods: San Francisco, California, W. H. Freeman Co., 237 p.

Whipkey, R. Z., and Kirkby, M. J., 1978, Flow within the soil, in Kirkby, M. J., ed., Hillslope hydrology: Chichester, U.K., John Wiley and Sons, p. 121–144.

Wilson, C. J., and Dietrich, W. E., 1987, The contribution of bedrock groundwater flow to storm runoff and high pore pressure development in hollows: International Association of Hydrological Sciences Publication 165, p. 49–59.

Yair, A., Bryan, R. B., Lavee, H., and Adar, E., 1980, Runoff and erosion processes and rates in the Zin Valley badlands, Northern Negev, Israel: Earth Surface Processes, v. 5, p. 205–225.

Youngs, E. G., 1958, Redistribution of moisture in porous materials after infiltration, II: Soil Science, v. 86, p. 202–207.

Zaslavsky, D., and Kassiff, G., 1965, Theoretical formulation of piping mechanism in cohesive soils: Geotechnique, v. 15, p. 305–316.

Zaslavsky, D., and Rogowski, A. S., 1972, Hydrologic and morphologic implications of anisotropy and infiltration in soil profile development: Soil Science Society of America Proceedings, v. 33, p. 594–599.

MANUSCRIPT ACCEPTED BY THE SOCIETY NOVEMBER 14, 1989

NOTE ADDED IN PROOF

Since this chapter was written, the field guide edited by Howard, Kochel, and Holt (1986) and referred to several times in the text has been published in modified form as:

Howard, A. D., Kochel, R. C., and Holt, H. E., 1988, Sapping features of the Colorado Plateau; A comparative planetary geology field guide: National Aeronautics and Space Administration, Washington, D.C., p. 108.

It contains a wealth of well-illustrated material on groundwater sapping.

Chapter 2

Weathering, soil development, and landforms

C. R. Twidale
Department of Geology and Geophysics, University of Adelaide, Box 498 G.P.O., Adelaide 5001, South Australia

INTRODUCTION

Only 0.614 percent of the Earth's water occurs in the continental subsurface, and less than half of this (0.306 percent of the total) is in the shallow subsurface, which is arbitrarily defined as occurring within 800 m of the surface. Only a minute part of the total water, 0.0018 percent, is found within the all-important root zone (Nace, 1960). Despite its small total volume, however, this underground water has widespread and significant geomorphologic effects.

Shallow subterranean water, especially that which occurs within the root zone, has especially pronounced and obvious results that are vital not only to the development of scenery but also to pedogenesis and, hence, to the entire biosphere (including man). It has a significance out of all proportion to its absolute and relative volume. Without shallow groundwater or soil moisture, life on Earth would be reduced both in variety and volume. Soil and vegetation cover would be reduced, and the land surface would in consequence be much less well armored (Russell, 1958). The epicycle of erosion and deposition due to human interference with the vegetational veneer provides only a partial indication of the stabilizing role of vegetation. Organic acids derived from plant litter, for instance, are important in general weathering (e.g., Keller and Frederickson, 1952; Duchaufor, 1982; Huang and Schnitzer, 1986) and in such specific activities as the dissolution of iron compounds (e.g., Bloomfield, 1957; Hingston, 1962). Bacterial activities are increasingly recognized as of major importance in rock weathering (e.g., Trudinger and Swaine, 1979).

The level of groundwater fluctuates episodically, seasonally, and in the longer term. These alternations of wetting and drying cause hydrophilic clays to expand and contract, resulting in the formation of a particular type of patterned ground that is known by various names in different parts of the world. Generally and increasingly, however, such soils are known as "gilgai soils" (Hallsworth, 1968), or as "vertisols" in the American Soil Taxonomy (Soil Survey Staff, 1975).

The mechanical effects of variations in water volume are interesting and important, but they are as nothing compared with the effects of water in the weathering of rocks at and near the Earth's surface. The entire zone of groundwater can be regarded as one in which rocks are being weathered and landforms initiated. Many of the changes wrought by this water are destructive, although constructive effects resulting from groundwater activities also have had an enduring impact on the evolution of scenery. Many of the geomorphologic effects of groundwater activity can be stated so briefly and simply that their importance in space and time can be easily underestimated or even overlooked. Others are neglected or not fully appreciated, even now.

WEATHERING AND THE REGOLITH

Subterranean waters are the cause of much, probably most, rock weathering. Solution, hydration, and hydrolysis together account for much of the alteration in evidence at and near the Earth's surface. In quantitative terms, hydrolysis is of major significance in the breakdown of silicates (e.g., Birkeland, 1974; Duchaufor, 1982). Solution is also a fundamental process, for not only are all of the common rock-forming minerals soluble to some degree but solution also prepares the way for other chemical reactions. Solution is indeed essential to chemical weathering (Loughnan, 1969, p. 61). Carbonates are readily affected by dissolution; this reaction is, of course, the basis of most karst. But even quartz, the most stable of the common rock-forming minerals, is slightly soluble, and is converted to amorphous forms that are more readily dissolved (e.g., Krauskopf, 1956; Siever, 1962; Morey and others, 1964). Given time and favorable conditions, silica solution can proceed to such an extent that underground voids or low-density zones are created, leading to subsidence and surface collapse (Twidale, 1987).

Many salts and gases, including oxygen and carbon dioxide, are circulated in groundwater so that oxidation, reduction, and carbonation, all common weathering processes, are directly attributable to the presence of circulating water in the near-surface zone of earth. Water also weakens the electrostatic bonding between minerals, causing disaggregation (e.g., slaking in sediments), either directly or indirectly, by freeing the system and allowing the penetration of solutions that effect other reactions.

Twidale, C. R., 1990, Weathering, soil development, and landforms, *in* Higgins, C. G., and Coates, D. R., eds., Groundwater geomorphology; The role of subsurface water in Earth-surface processes and landforms: Boulder, Colorado, Geological Society of America Special Paper 252.

Frost shattering is a much-cited "physical" weathering process, yet whether it is the result wholly, or even primarily, of mechanical forces has been questioned, and it has been suggested that the disruption may be caused in part by what Wilhelmy (1958, p. 52 and the following) called *Hydratationssprengung,* or disaggregation due to the formation of layered or ordered water on mineral and crystal surfaces (White, 1976; Walder and Hallett, 1986). Similarly the effectiveness of insolation appears to be enhanced by the presence of moisture (Griggs, 1936). Thus, groundwater may well play a significant role in mechanical as well as chemical processes.

Reactions between water and such minerals as the various micas and feldspars take place rapidly in warm, humid conditions. Alteration of recent volcanic rocks has been noted within a few decades of extrusion in the Antilles and Indonesia, for example, and in southeastern Brazil and Madagascar, freshly quarried granitic rocks show signs of change within a decade of exposure. Given circulating groundwater and an open system in the subsurface, weathering effects due to groundwater can be expected to be enormous. But the resulting weathered mantle or regolith is neither homogeneous nor of uniform thickness. It forms a discontinuous veneer, the thickness of which varies from place to place. Some weathering rinds, like those due to oxidation of the peripheral zones of erratic granite blocks and boulders in tills in the Sierra Nevada, California, are only a few millimeters thick. On the other hand, great thicknesses of altered bedrock have been reported and attest to the efficacy of underground water as a weathering agent. Regoliths of 50 m thickness are commonplace in the humid tropics, and weathered mantles of 100, 150, and even 200 m thickness are reported from several climatic zones, although some may be inherited features (e.g., Barbier, 1957; Ollier, 1965; Thomas, 1965). The regolith is not homogeneous, but commonly displays pronounced zonation or horizon development. Some is the result of the superimposition of one set of processes on a preexisting regolith, but frequently the sequence of horizons denotes the sequential progress of weathering, with the initial weathering effects represented in the basal zone and the most advanced in the near-surface horizon.

The regolith stands in marked physical contrast with the bedrock from which it is derived. The bedrock is basically cohesive, tough, and resistant to erosion, whereas the regolith is typically friable, weak, and readily evacuated. Without groundwater and the conversion of rock to regolith there would be little or no erosion. Without the breakdown of rocks, erosional agents such as waves, rivers, and the wind would have only limited and localized effectiveness.

Regoliths were also widely developed in past ages, and remnants of such ancient weathered mantles remain as prominent features of the landscape in many areas. In some bedrock the regolith and bedrock grade one into the other, but quite commonly the two meet in a zone so narrow that it can be regarded as a plane known as the weathering front (Mabbutt, 1961a). Over wide areas, the regolith has been stripped to expose the associated weathering front as etch forms and surfaces that include many familiar landscape features.

It is convenient to consider first the differentiation of the regolith and its geomorphologic consequences, and then to discuss the destructive effects of weathering.

DIFFERENTIATION OF THE REGOLITH

Weathering causes disintegration and alteration. Some of the products of such decay are evacuated from the system. Others, however, remain within the regolith to form pans and layers of varied compositions, positions and scales.

Pedogenesis

Weathering due to the physical and chemical effects of underground water of the near-surface zone are crucial to soil development. The profile development typical of many soils, the degree of differentiation, and the character of the various soil layers is related to the nature of the groundwater movement within the near-surface zone of the regolith. Soil consists of weathered rock or parent material plus an organic component that is usually concentrated near the land surface, and which is derived from the decay of plant and animal matter. Pedologists include in their soil categories and classifications skeletal soils that are merely thin layers of disintegrated bedrock lacking any profile differentiation, and peats that are entirely organic in origin, so that precise definition is difficult. Nevertheless, soils may reasonably be described as discrete bodies produced by the interaction of climate, organisms, and surficial geologic materials in landscapes that are, increasingly, influenced by human activities (Olson, 1981).

The soil profile, described and classified in terms of laboratory analysis as well as field tests, is the basic unit of soil classification or taxonomy. Soil taxonomy is designed to summarize soil characteristics that influence land use and management.

Early attempts at soil classification were based on soil characteristics and areal distributions that emphasized the interrelated roles of climate and vegetation as pedogenic factors. C. F. Marbut (e.g., 1927), who was familiar with Glinka's work, and through him Dukuchaiev's, stressed both dynamic (climatic and biologic) and passive (geologic and topographic) factors in time. He distinguished between the soil and the underlying parent material, and recognized that some soils (the "great soil groups") were mature and zonal in their distribution (e.g., Fig. 1A). Many familiar soil terms appear in such classifications.

Over the past 40 years or so, however, the need was recognized for a more comprehensive system of classification that would place similar soils, whether virgin, cultivated, or eroded, in the same groups. The task of formulating such a classification was undertaken by the Soil Survey Staff of the U.S. Department of Agriculture; after several approximations, the current Comprehensive Classification System has emerged (Soil Survey Staff,

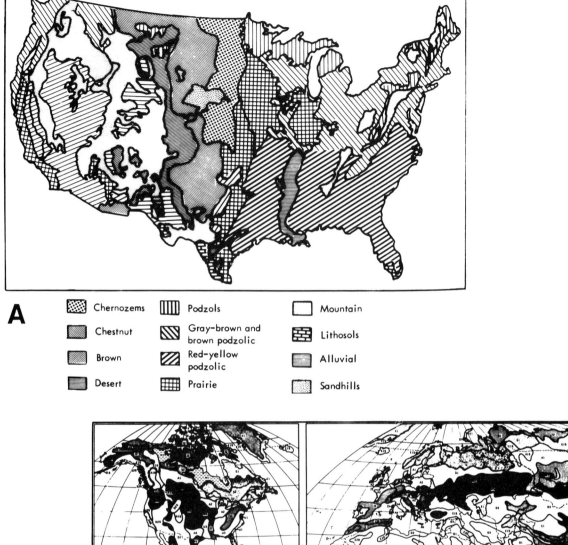
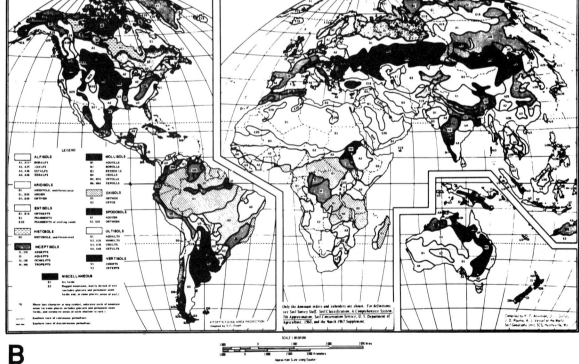

Figure 1. (A) Great soil groups of the United States according to the Marbut classification scheme (after Steila, 1976). (B) General soil map of the world showing orders and suborders according to *Soil Taxonomy* (after Olson, 1981; Soil Survey Section, 1975; Brady, 1974).

1975). It is more detailed (Fig. 1B) than the Great Soils Group method, and its terminology is not as instantly evocative to most workers because it is rooted in Latin and Greek (Table 1). But there is a direct relation between soil orders and degree of weathering (Fig. 2), and it allows the classification of a huge range of soil profiles.

Soil profiles are a function of climate, organisms, parent material, topography, time, and geologic history (e.g., Jenny, 1941; Stephens, 1947; Buol and others, 1973), although the relative significance of these factors varies from region to region and from site to site.

A soil profile typically displays several horizons disposed sensibly parallel with the land surface and differing from each other in mineralogy, chemistry, and various physical properties such as color, texture, and permeability. Soils that have very pronounced textural contrasts between layers are sometimes called texture differentiated or duplex soils. Soils may be buried or truncated, and many bear the imprint of different, even contrasted, processes.

Shallow surface waters contribute to pedogenesis in two ways. First, and as has already been indicated, water is responsible directly and indirectly for most of the weathering near the Earth's surface, for the alteration and disintegration of the parent material that is the basis of most soil formation. Organic acids and humic complexing agents play a central role, together with water. Second, water is the medium by which salts and fines are translocated both vertically and laterally within the regolith and within the soil. Illuviation and eluviation are due to groundwater movement. The water budget and rock permeability at a particular site determine whether leaching or precipitation occurs. An overall throughflow leads to leaching. On the other hand, an excess of evaporation over precipitation leads to salt accumulation at or near the top of the profile. Many factors, regional and local, influence soil movements, and feedback effects are commonplace. For example, fresh granite is of low porosity and permeability, but once water has penetrated by way of microfissures, etc., alteration occurs and porosity and permeability increase dramatically (Kessler and others, 1940). On the other

TABLE 1. SOIL ORDERS NAME DERIVATION, AREAL SIGNIFICANCE, AND MARBUT EQUIVALENTS

Soil Order	Derivation of Root Word	% of Total World Soils*	Rank (Total Area)	Marbut Equivalents
Entisols	Recent.	12.5	4	Azonal soils, some Low Humic Gley.
Vertisols	L: *verto* = to turn	2.1	9	Grumusols.
Inceptisols	L: *inceptum* = inception beginning	15.8	2	Ando, Sol Brun Acide, some Brown Forest, Low Humic Gley, Humic Gley.
Aridsols	L: *aridus* = dry	19.2	1	Desert, Reddish Desert, Serozem, Solonchak, some Brown and Reddish Brown soils, associated Solonetz.
Mollisols	L: *mollis* = soft	9.0	6	Chestnut, Chernozem, Brunizem, Rendzina, some Brown, Brown Forest, associated Humic Grey, and Solonetz.
Spodosols	Gr: *spodos* = wood ash	5.4	8	Podzols, Brown Podzolic, Groundwater Podzols.
Alfisols	Alf: combined from aluminum and iron	14.7	3	Gray-Brown Podzolic, Gray Wooded, Noncalcic Brown, Degraded Chernozem, associated Planosols and Half-Bog.
Ultisols	L: *ultimos* = ultimate	8.5	7	Red-Yellow Podzolic, Reddish-Brown Lateritic, associated Planosols, and some Half-Bogs.
Oxisols	Oxi: from oxide	9.2	5	Laterite soils, Latosols.
Histosols	Gr: *histos* = tissue	0.8	10	Bog soils.

*An additional 2.8 percent of the world total includes ice fields, unclassified lands, and others. (From Steila, 1976.)

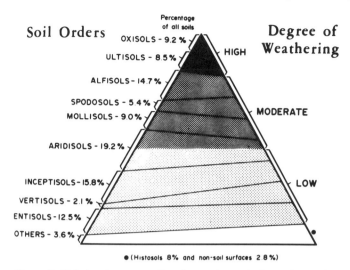

Figure 2. Relation between soil orders and weathering (after Steila, 1976).

hand, precipitation of carbonates or silica carried by downward percolating waters at the soil water table, and within the B-horizon, can impede flow and induce further precipitation.

Minor mineral accumulations

Horizon differentiation within soil profiles is, in many places, pronounced, so much so that adjacent horizons stand in marked physical contrast. External erosional agencies have exploited such differences to produce various erosional forms. Thus the friable A-horizon of soils is frequently stripped (sheet erosion). Duplex soils (with layers of contrasted permeability) and cambic and argillic soils may also allow tunneling to develop, for water readily percolates through the sandy permeable A-horizons, but runs along the top of the clayey B-horizons, eroding tunnels. In time the surface sands collapse into the tunnels, giving rise to either: (1) linear channels, to chains of alternating channels and tunnels, or (2) hollows as well as gullies that are due to soil collapse, as Rubey (1928) described from the Great Plains of the U.S.

Specific minerals such as lime or iron oxide commonly accumulate at shallow depths to form hardpans. Although thin, they impede subsurface infiltration of water. In extreme cases they can be a significant factor of causing floods. Thus, in 1952 an iron hardpan at shallow depth (ca. 30 cm) on Exmoor, southwestern England, greatly increased the magnitude and rate of run-off, giving rise to the disastrous Lynmouth flood of that year.

Some iron- and silica-rich skins or varnishes are thought to be due to solution and capillary migration of salts from within the near-surface zones of the host rock, although biological agents also have been invoked (e.g., Scheffer and others, 1963; Dorn and Oberlander, 1981).

Duricrusts

At a much larger scale, the most obvious expressions of mineral accumulation in the regolith are those that after desiccation, induration, and exposure, are known as duricrusts or, more recently, pedocretes (Netterberg, 1985).

Duricrusts of various and variable compositions have been described from many parts of the world, but the most commonplace both in space and time are laterite, ferricrete, silcrete, calcrete, and gypcrete.*

Laterite (the oxisols of modern soil classifications) consists of highly weathered bedrock, rich in secondary oxides of iron or alumina or both, and virtually devoid of bases and primary silicates, although quartz and kaolinite are commonly present (see e.g., Prescott and Pendleton, 1952; Maignien, 1966). Laterite consists of a friable A-horizon, commonly sandy and readily stripped, overlying a vesicular, pisolitic, ferruginous B-horizon, frequently with a rudimentary false bedding. This is underlain by mottled and pallid zones that give way to parent rock. Laterite hardens irreversibly after exposure and desiccation (Alexander and Cady, 1962). Intense desilication gives the alumina-rich bauxite. Both laterite and bauxite are forming today in the humid tropics, and optimally in monsoon lands such as southern India, Indonesia, Malaysia, and the Antilles, but they are widely preserved in relic form elsewhere (Fig. 3A).

Ferricrete is an encrustation rich in iron oxide, with no associated organized substrate of mottled and pallid materials.

Silcrete, also known as grey wethers, billy, and sarsenstone, consists of 95 percent or more SiO_2. It is widely preserved in central Australia and also in southern Africa, France, and southern England. It has a distinctive texture, with grains and shards of quartz set in a microcrystalline quartzitic matrix, or less commonly in an opaline matrix. The silica-rich layer varies in color from white to gray, brown-yellow, and red, and quite frequently includes materials of more than one of these colors. Columnar jointing is well developed; the sides of the columns display vertical grooves. It has a vitreous sheen and a well-developed conchoidal fracture. As far as can be determined, silcrete is not now forming, but the stratigraphic evidence suggests that it developed in the early Cenozoic under warm, humid conditions on surfaces of low relief. The extensive sheets of silcrete preserved in central Australia (Fig. 3B) are largely accumulational and alluvial, for rounded exotic pebbles and cobbles are characteristic of the material. Silcrete of curious mineralogy also formed in scarp-foot zones (e.g., Hutton and others, 1972). Over wide areas the silice-

---

*This terminology differs from that recommended by Goudie (1973) and others whose nomenclature is consistent and rational yet omits much of interest insofar as it fails, for example, to recall that laterite takes its name from the use of the material for brickmaking in southern India (Buchanan, 1807; Babington, 1821) and that bauxite takes its name from Les Baux in southern France. Also, it is useful to differentiate between iron encrustations or ferricrete and iron-rich horizons associated with mottled and pallid zones that together compose laterite profiles.

Figure 3. (A) Laterite-capped mesas in granite, northwestern Queensland (CSIRO). (B) Silcrete-capped mesa and butte, Rumbalara, Northern Territory (CSIRO).

ous zone is underlain by bleached country rock, but the two are not necessarily genetically related. In some parts of the Flinders Ranges, for example, silcrete overlies limestone.

Calcrete (e.g., Milnes and Hutton, 1983), also known as kankar, kunkar, nari, and caliche, is forming at the present time in semiarid and arid environments. Consisting of at least 80 percent $CaCO_3$ together with dolomite, quartz, and clay, calcrete is a pedogenic accumulation that occurs in nodular, honeycomb, and hardpan or sheet form. Travertine is mineralogically similar, but occurs in association with springs and dry river beds.

Gypcrete has accumulated as a pedogenic horizon on arid plains in, for example, central Australia and northern Iraq (Wopfner and Twidale, 1967; Tucker, 1978). It consists of crystalline gypsum preserved in layers a few meters thick.

Figure 4. Gravel-capped false cuesta separated from adjacent quartzitic hillslope by scarp-foot depression, western Cape Fold Belt, Republic of South Africa.

Most duricrust profiles are complex rather than simple (e.g. Milnes and others, 1985; Ollier and others, 1988). Thus, the plateau that occupies much of Kangaroo Island, South Australia, is capped by an ancient laterite (Daily and others, 1974). The pisolitic ironstone is overlain by a sandy A-horizon and is underlain by the mottled and pallid zone, but in some places at least, a siliceous accumulation zone occurs at the interface between a much-weathered zone above and a weathered but still relatively impermeable zone below. The band is horizontally disposed and cuts across the steeply dipping foliation of the weathered Cambrian schist.

On Kangaroo Island this deeper siliceous zone has no obvious topographic expression, but in the Daly (River) Basin of the Northern Territory a siliceous zone developed in the Late Cretaceous–middle Tertiary laterite of the region forms the protective capping of the Maranboy Surface, a plateau level located between the high laterite-capped Bradshaw Surface and the present river valleys and associated plains (Wright, 1963). The Maranboy Surface is due to the stripping of the higher zones of laterite, including some of the kaolinized, mottled, and pallid material. It has become stabilized on the intra-regolithic silica-rich zone.

Although they vary in composition and thickness, duricrusts are geomorphologically important for several reasons. They form the cappings of plateaus and ridges (Fig. 3). Ferruginous, bauxitic, and siliceous caprocks are especially durable (e.g., Daily and others, 1974). Scarp-foot silcrete forms the cappings of false cuestas in dissected piedmont zones (e.g., Twidale, 1978a; Twidale and Milnes, 1983); gibber and stone mantles have similar effects (Fig. 4). Calcrete and gypsum are less tough but, nevertheless, they commonly protect the land surface and form the cappings to prominent escarpments in arid and semiarid lands. For example, gypcrete overlying unconsolidated gypsiferous silts, caps the prominent escarpment that delimits Lake Eyre on its western side.

Duricrusts form morphostratigraphic markers on a regional scale. Thus, in Australia (see Twidale, 1983) the southern laterites and bauxites are of early or later Mesozoic age in South and Western Australia, respectively; the northern laterites are essentially of early-middle Tertiary age, as are the silcretes of central Australia; the gypcrete of the southwestern Lake Eyre Basin is a Pleistocene accumulation, as is the opaline silcrete that is co-existent with the gypcretes; the calcrete is a late Cenozoic deposit and is still forming. The various duricrusts are also good environmental indicators: laterite and silcrete form in warm, humid climates, and gypcrete in arid or semiarid climates. The duricrusts in sheet form have developed on or beneath surfaces of low relief. In some areas they have given rise to relief inversion at local and regional scales.

Lime in the form of travertine has been precipitated in valley floors and has been left high in the relief, e.g., in eastern Arabia (Miller, 1937), and in the Oakover and Ashburton valleys of the Pilbara of northwest Western Australia, although here the calcrete also contains pods and lenses of amorphous silica. Continued lime precipitation, either in the shallow subsurface or specifically in valley floors, has given rise to pseudoanticlines in

Figure 5. Thin slabs and flakes of rock (arkose) truncated by soil moisture attack, Ayers Rock, Northern Territory.

several regions (e.g., Price, 1925; Jennings and Sweeting, 1961; Mann and Horwitz, 1979).

The differentiation of the regolith is obviously of importance in various ways. Nevertheless, the main geomorphological significance of the regolith is that it forms a mantle that contains groundwater charged with salts, minerals, and biota. It thus effects the rotting of any fresh rock with which it comes into contact. Many familiar landforms, major and minor, originate at the weathering front through the interplay of groundwater and bedrock.

LANDFORM INITIATION IN THE SHALLOW SUBSURFACE

Several geomorphologic features point to weathering effects in the shallow subsurface being more pronounced than those at greater depths and those on exposed surfaces. A distinct zone of decay is frequently found just below the present or a former soil-rock contact around the bases of blocks, boulders, and cliffs. Aggressive moisture attack just below the soil surface has truncated flakes and slabs of rock developed in concentric zones at the margins of boulders and of inselbergs (Fig. 5). The thickness of the truncated layers varies between 1 mm and several centimeters, although several are commonly affected so that, *in toto*, the truncated zone may be almost 1 m thick. In many areas their exposure is due to recent soil erosion, in many instances related to anthropogenic activities.

The location of the optimal zone of weathering relative to the surface varies, as does the vertical extent of the zone of especially aggressive groundwater. Such effects are well demonstrated at Cunyarie Rocks, a complex of low domes located north of Kimba on northern Eyre Peninsula, South Australia. The most prominent dome stands only 2 m above the level of the adjacent gentle slope. The bare, broadly convex-crested area is dimpled and grooved through the development of rock basins and gutters. Several blocks, remnants of sheet structure, stand in orderly arrangement near the western downslope margin. Their flanks are flared or concave, and at some sites the flares merge laterally with tafoni.

Some of the rock basins are complex in that they consist of a smaller but deeper basin set in the floor of a larger shallow feature. Granite is exposed over all of the floor area in some basins. Others carry a fill of soil or granular detritus. The reason for the basin-in-basin morphology, as demonstrated by the steeper slope revealed by excavation and removal of the soil cover (Fig. 6), is that the detritus retains moisture sufficient to cause more rapid disintegration of the granite than on exposed faces. Most of the gutters have formed along fractures. Some were formerly occupied by moist detritus resulting in the flared sidewalls of the clefts (Fig. 7).

Flared slopes (Fig. 8) evolve in two stages (Fig. 9): the first involves subsurface moisture attack, and the second the exposure of the concave weathering front, usually by stream erosion (Twidale, 1962). The concavities typically flatten laterally and merge with rock platforms (Fig. 8).

Where the bedrock is not massive, basal fretting rather than regular flares results from the same sequence of weathering and surface lowering. Mushroom rocks also testify to the efficacy of subsurface moisture attack (Fig. 10A). Even more spectacular are the swamp slots developed at the soil-rock contact in limestone terrains (Fig. 10B).

THE BASE OF THE REGOLITH—THE WEATHERING FRONT

Meteoric waters penetrate unevenly into the near-surface bedrock. Fractures are prime avenues for infiltration and hence for weathering. Minerals that are particularly reactive, and fine-grained rocks, are rapidly altered and changed not only chemically but also physically. Every structural nuance finds expression in the form of the weathering front, which thus might be thought to be irregular both in gross and in detail. There is, however, a contrary tendency, in that irregular masses and projections, with a greater surface area exposed to moisture attack, are reduced more rapidly than are plane surfaces. Thus, over wide areas the weathering front is sensibly even. Elsewhere, however, the weathering intensive weathering, evacuation of solubles and fine detritus, volume reduction, settling, and surface lowering. At a quite different scale, Trendall (1962) has cogently argued that surface lowering on a regional scale may follow prolonged and intense weathering and consequent volume reduction.

The most important effect of groundwater is that hard, cohesive rock is transformed to friable material that is readily eroded by whatever agent is active in a given area. The change itself may be enough to induce stream incision, but climatic or base-level changes are also capable of initiating or triggering the stripping of the regolith and the partial or complete exposure of the weathering front in an etch surface (Wayland, 1934; Willis, 1936; see also Falconer, 1911; Jutson, 1914). A two-stage mechanism involving simultaneous planation of the land surface and the formation of a weathering front at the base of the regolith is the essential feature of Büdel's (1957) *doppelten Einebnungsflächen,* or double planation, because the weathering front, after the stripping of the regolith, is exposed as a second surface of low relief.

The landforms revealed as a result of stripping of the regolith and exposure of the weathering front vary both in type and scale. In granite and other massive rocks, for example (Figs. 8 and 13), they include several minor features ranging in size from crystal-size pitting to basins, gutters, clefts, flared slopes, enclosed marginal depressions, corestone boulders, and granite pinnacles (e.g., Clayton, 1956; Boyé and Fritsch, 1973; Twidale and Bourne, 1975a, 1976; Twidale, 1982a; Twidale and Campbell, 1984). Major forms of etch origin include plains of regional extent, isolated hills or inselbergs, and inselberg massifs.

The role of subsurface weathering, or two-stage development (Linton, 1955), was recognized almost two centuries ago by Hassenfratz (1791) when he described as *dégagés* boulders of the Aumont district of the southern Massif Central of France, which had originated as corestones in grus. Similarly, Logan (1849, 1851) realized that the grooves he observed on large residual granite boulders on Palo Ubin, near Singapore, continued below the soil and originated in the subsurface. The concept has never been entirely lost, although it has been misplaced over the intervening years (e.g., Twidale, 1978b).

Figure 6. Moisture retained in detritus accumulated in the floor of this rock basin at the granitic Cunyarie Rocks, northeastern Eyre Peninsula, South Australia, has rotted the rock with which it came into contact to produce a second basin set in the floor of the first.

front is characterized by hills or protuberances that project into the regolith. The latter commonly includes corestones, more or less rounded masses of various diameters that have resisted weathering and persist as isolated masses within the regolith, delimited and defined by discrete sectors of the weathering front (Fig. 11). The front also penetrates along fractures and fissures so that in places it has the form of an intricate rather than a plane surface (Fig. 12A). Once the bedrock is differentiated into fresh and weathered rock, groundwater runs along the interface between the two, eroding linear gutters.

Also, the weathering of the bedrock may cause volume reduction as the products of alteration are evacuated in solution, or as fines are removed by heavy throughflow. Thus, Ruxton (1958) has suggested that some scarp-foot depressions are due to

Boulders

The weathering of strongly foliated rocks gives rise to penitent rocks, monkstones, or tombstones (Ackermann, 1962; Twidale, 1971, p. 28–29), and on a larger scale the *Gefügerelief* of Turner (1952). More commonly, however, fields of boulders are developed on rocks characterized by evenly spaced orthogonal fracture systems. Such corestones and related boulders are characteristic of granite, basalt, dolerite, and to a lesser degree, limestone and quartzite. (Figs. 13C and 14). The basic unit is the fracture-defined block several centimeters or a few metres in diameter. Water infiltrates into fractures, altering the rock and eventually penetrating into the block through the weathered outer zones (Fig. 12). Weathering is more rapid at corners and edges than on plane faces, so that the cubic, rhomboidal, or quad-

Figure 7. Flared sidewalls of fracture-controlled clefts, Cunyarie Rocks, Eyre Peninsula, South Australia.

rangular block is converted into a spherical or ellipsoidal kernel or corestone (MacCulloch, 1814; de la Beche, 1839; Jones, 1859). These corestones are exposed as boulders with the stripping of the weathered rock (Fig. 12ii). Boulders are found in a wide range of climatic conditions and have manifestly been exposed by various agents (rivers, waves, periglacial wasting, etc.). It seems certain also that, in detail, different weathering processes have effected the initial differentiation of corestones and regolith.

Inselbergs

Inselbergs in granite and granitic gneiss (Figs. 15 and 16) develop in various ways (see Twidale, 1982a, b) but many, perhaps most, evolve in two stages. They are initiated as projections of more resistant (usually more massive) masses at the base of the regolith. Falconer (1911, p. 246) was clearly aware of this when he wrote:

A plane surface of granite and gneiss subjected to long-continued weathering at base level would be decomposed to unequal depths, mainly according to the composition and texture of the various rocks. When elevation and erosion ensued, the weathered crust would be removed, and an irregular surface would be produced from which the more resistant rocks would project.

That many inselbergs are of this derivation (Fig. 17) is suggested by several lines of evidence and argument (Twidale, 1982a, b), but particularly by the fact that they can be seen in various stages of exposure from beneath the regolith, as well as in nascent form, beneath the natural land surface (Fig. 18). As with boulders, the inselbergs are weathered and eroded only very slowly once they stand above the regolith and are not continually in contact with water (see below).

Apart from size, one difference between boulders and inselbergs is that the former are usually isolated and detached as a result of weathering from all sides of the fracture-defined block, whereas the latter tend to remain attached to the rock mass at their bases. On the other hand, some large masses of fresh rock near the base of the saprolith take the form of attached rounded pinnacles or projections (Fig. 13D; see also Twidale and Campbell, 1984), and presumably some, perhaps most, inselbergs are of the same nature. Also, where bosses or domical hills stand in ordered, fracture-determined arrangements within massifs (as in the Everard Ranges of central Australia and the Kamiesberge of Namaqualand), each boss, formed by preferential weathering along the fracture zones that define it and that determine the location and pattern of clefts and valleys, is but an upward projection of a mass, the bulk of which remains hidden.

There is, however, suggestion that some inselbergs, like boulders, are detached from the main mass of fresh rock. Brajnikov (1953) in particular has urged that the *morros* and sugarloafs of southeastern Brazil are really gigantic corestones, that they are not *enracinés,* or rooted in the bedrock.

Weathering, soil development, and landforms

Figure 8. Flared slope about 8 m high, with associated basal platform, Ucontitchie Hill, a granite inselberg on northwestern Eyre Peninsula, South Australia.

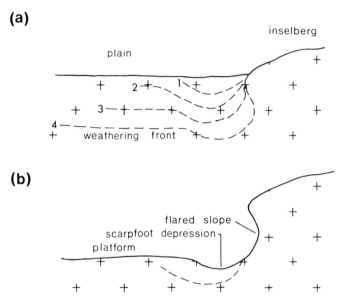

Figure 9. Stages in the development of flared slopes.

Figure 10. (A) Mushroom rock in level-bedded sandstone, southern Drakensberg, Republic of South Africa. (B) Swamp slot in limestone, near Ipoh, West Malaysia.

Figure 11. Corestones in granite, Rocky Mountains, west of Boulder, Colorado.

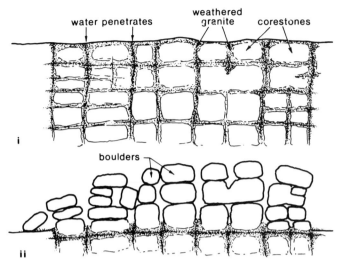

Figure 12. Two-stage development of boulders.

Figure 13. (A) Pitted surface on granite boulder, near Tampin, West Malaysia. The land has been cleared and the timber burned, resulting in spalling of part of the outer surface of the boulder. (B) Gutter (along fracture) and many shallow depressions developed on granite platform in the Kwaterski Rocks, northwestern Eyre Peninsula, South Australia. (C) Granite boulders, eastern Mt. Lofty Ranges, South Australia. (D) Granite pinnacles, Murphy Haystacks, northwestern Eyre Peninsula, South Australia.

Etch plains

Many plains are of etch type (Fig. 15). Rock platforms (or pediments; see Twidale, 1978c) commonly extend a few meters from the frequently flared bases of inselbergs, or they stand in isolation (Fig. 19). A few are several hundreds of meters across. Whatever their dimensions, however, they are grooved and dimpled by the development of gutters and basins. They commonly carry patches of the regolith that formerly covered the entire platform, and the boulders that rest on the platforms are interpreted as corestones released through the evacuation of fines.

Much of the Yilgarn Block of the southwest of Western Australia is occupied by a high plain above which stand numerous scattered remnants of a lateritized land surface preserved in mesas and buttes (Fig. 20). The laterite (and bauxitic equivalents) is well preserved in the Darling Range, where the duricrust is, according to Fairbridge and Finkl (1978) of late Mesozoic or earliest Tertiary (pre-Oligocene) age. Early this century, Jutson (1914) noted that the base of the lateritic regolith is essentially coincident with the New Plateau (see also Mabbutt, 1961b; Finkl and Churchward, 1973). He realized that the main high plain, or New Plateau, was due to the stripping of the lateritic regolith that had been developed beneath the Old Plateau (Fig. 20). Although he did not use the modern terminology, he clearly appreciated that the New Plateau is of etch character and is, in effect, an exposed weathering front.

Similar conclusions were reached concerning the origin of some of the high plains of east and central Africa. Subsequently, etch plains have been recognized in several other parts of the world (e.g., Mabbutt, 1965; Twidale and Bourne, 1975b; Thomas and Thorp, 1985). The lake-strewn high plain that occupies much of the Labrador Peninsula is surely of similar character—a weathering front that has been exposed largely by glacial erosion.

Figure 14. Corestones in basalt, southern Drakensberg, Republic of South Africa.

Shore platforms

Etch planation has been used to explain shore platforms that do not conform to the general observation (Hills, 1949, 1971) that, although they are well developed in a wide variety of lithologic environments, platforms are notably absent where granitic rocks are exposed on the coast. This is generally correct, for it seems that the well-developed jointing (orthogonal and sheeting) of such rocks have dominated development. Some quite wide platforms in granite and gneiss on the west coast of Eyre Peninsula extend as much as 200 m from the base of the backing cliffs, some

Figure 15. Inselberg landscape in granite and gneiss, Namaqualand, Republic of South Africa. The plain is cut in the fresh bedrock, and there is little development of regolith.

Figure 16. Inselberg massif in gneiss, southern Zimbabwe (L. A. Lister).

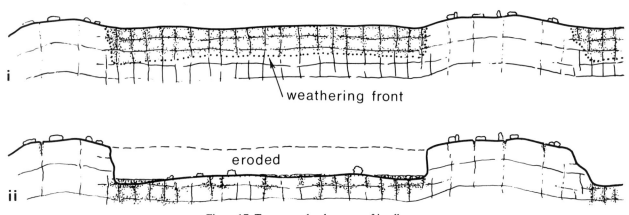

Figure 17. Two-stage development of inselbergs.

being irregular in detail, others quite smooth though strewn with granitic boulders (Fig. 21). Here the platforms are etch surfaces from which the Pliocene(?) regolith has been stripped by wave action, and the boulders are corestones released by the erosion of the matrix of grus (Twidale and others, 1977). Remnants of the former regolith, in some areas preserved by virtue of cementation by lime washed from the Pleistocene dune calcarenite, are preserved at various sites along the coast.

Some of the platforms that are cut in Pliocene Santa Cruz Formation exposed along the coast of central California (Bradley and Griggs, 1976) may be of similar character, for the weathering front there is closely coincident with the level of the platforms in fresh rock (Fig. 22).

Likewise the Carnatic, a coastal plain averaging some 80 km in width and located between the Bay of Bengal and the Eastern Ghats of southern India may be an etch feature initiated beneath the land surface, some remnants of which are preserved high in the Ghats. Cushing (1913) cited various morphological evidence pointing to a marine origin for the plain. The great width poses problems for marine planation because wave energy would be dissipated in crossing the fringing flats, unless of course the land were gradually sinking relative to the sea. On the other hand, as the platforms of the west coast of Eyre Peninsula demonstrate, waves can readily strip a regolith. The Carnatic may be of such an origin, although it must be stated that this suggestion merely eases, rather than resolves, the problem.

Nature of etch forms

All of these forms, large and small and of varied morphology, are of etch (subcutaneous) or intracutaneous type. They are developed on various rock types, although granitic and other massive, naturally impermeable, rocks are especially prone to the formation of etch forms because the weathering front is sharp and there is a marked contrast between regolith and the fresh bedrock on which it is developed. Etch forms are not exhumed, for they have never before seen the light of day. They are two-stage features, the first phase of development occurring below the land surface and the second involving the exposure of the weathering front by epigene agents. Etch forms may also be polygenetic in character because different climatic conditions might characterize each of the two stages. For example, the inselbergs of Encounter

Figure 18. Nascent inselberg (X) exposed in Vredefort Brick Pit, Orange Free State, Republic of South Africa.

Figure 19. Extensive platform in granite adjacent to Corrobinnie Hill, northwestern Eyre Peninsula, South Australia.

Figure 20. Laterite-capped mesa near Cue, central Western Australia. The laterite is due to the weathering of granite and gneiss. The surface of the mesa is a remnant of the Old Plateau of Jutson (1914) and is of Cretaceous-Eocene age. The surrounding plain is due to the stripping of the regolith and the exposure of the intrinsically fresh granite in the New Plateau.

Bay, south of Adelaide, South Australia, were initiated during a Mesozoic phase of deep lateritic weathering under torrid conditions that produced the laterite (of Kangaroo Island and adjacent areas) already referred to, but they have been exposed, in part at least, by marine agents during the later Cenozoic.

GROUNDWATER AND LANDSCAPE EVOLUTION

The regolith is a mass of weathered material soaked with groundwater that is charged with salts and biota. It is a rock mass that displays internal organization reflecting both stages in the decomposition of the original bedrock and groundwater movements within the regolith. Most importantly, it is a suppurating mass that gradually consumes any blocks enclosed within it, and is gradually gnawing away at both horizontal and vertical contacts with the bedrock. Where the groundwater is static or only slowly circulating, it may become saturated with the available salts, and the regolith may then become less aggressive and more protective. The formation of tough impermeable horizons within the regolith may have similar effects, and certainly duricrusts have a markedly stabilizing influence. In general, however, the regolith is a discontinuous, festering veneer that causes the alteration of any rock with which it comes into contact.

Where, for tectonic or structural reasons, the regolith and the groundwater it contains are unevenly distributed, several factors come into play that influence the course of landscape evolution. The efficacy of groundwater as a weathering agent determines the location of areas of weathered, unweathered, or little-weathered bedrock. Not only do those areas that for structural reasons are upstanding or low lying maintain their topographic status, but the contrasts tend to be emphasized through time as low-lying areas are weathered and eroded more and more.

Many workers, from de Saussure (1796) through Barton (1916) and Bain (1923) to Wahrhaftig (1965), have emphasized the importance of the contrast between wet and dry sites. Weak compartments are wetted, weathered, and eroded. Regions that are dry remain upstanding, and are changed only slowly while feedback effects take over. Boulders and inselbergs are comparatively stable once they are exposed to the air, and no longer in contact with moist detritus (e.g., Ruxton and Berry, 1957; Twidale, 1982a). It has been argued that castle koppies and nubbins result from the subsurface weathering of bornhardts with different degrees of exposure and burial (Twidale, 1981b). Wahrhaftig (1965) has shown that because weathering of exposed fresh rock is so relatively slow as to be negligible, zones of (low fracture density?) bare rock act as local baselevels, and regardless of the depth of weathering in the intervening regolith-covered areas, a stepped topography evolves (Fig. 23).

Vertical contacts between weathered and fresh compartments are developed. This can be seen at detailed and local scales at various sites; thus, the northern slope of Yarwondutta Rock (Twidale, 1982c; Twidale and Bourne, 1975c) is markedly stepped, arguably as a result of repeated phases of subsurface moisture attack (Fig. 24). On a regional scale also, there is a tendency not for reduction and planation but for the evolution of stepped forms and landscapes where the regolith charged with groundwater laps onto rock outcrops. Surfaces and forms are not necessarily eliminated. Upstanding areas tend to remain dry and upstanding while beneath the moist adjacent lowlands the weath-

Figure 21. Granite shore platform of etch type at Smooth Pool, near Streaky Bay, northwestern Eyre Peninsula, South Australia.

Figure 22. Shore platform in dipping Pliocene sediments, coincident with the base of intensive weathering preserved in the backing cliff, central California (W. C. Bradley).

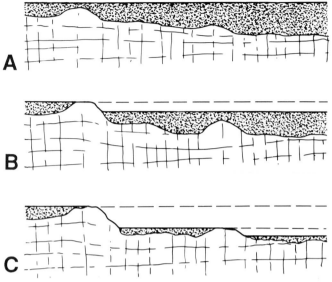

Figure 23. Development of stepped topography (after Wahrhaftig, 1965).

ering front is lowered. Uplands remain uplands and plains remain plains as reinforcement or positive feedback effects operate (Twidale and others, 1974).

Landscapes are not lowered *in toto*, but only in parts—the parts in contact with regolith and groundwater. Given tectonic stability or secular uplift (e.g., an isostatic response), relief amplitude will tend to increase rather than decrease (cf., Davis, 1909). The wearing back of the vertical contacts of upland and regolith implies a wearing back of scarps (cf., King, 1962), and the persistence of old forms and surfaces (Twidale, 1976). The scarp, however, is due initially and primarily to groundwater weathering, not to gullying or stream action, for stepped relief has developed on uplands too limited in area to generate significant runoff (e.g., Yarwondutta Rock and other granite inselbergs; see Twidale, 1982c).

Groundwater weathering also imposes a considerable measure of azonality or commonality on landscapes, superseding or reducing the effects of both climate and lithology. Groundwater exploits structural weaknesses at the weathering front in similar fashion regardless of atmospheric climate. The effects of air temperatures and humidity are very much cushioned and reduced. Hence, forms such as inselbergs and boulders, rock basins and gutters, plains, rock pediments, or platforms are found in a wide range of conventionally defined climatic regions. It is not a matter, as with pediments, of hydraulic and other factors transgressing climatic boundaries (Twidale, 1981b), but rather that groundwater has similar effects the world over. After exposure, the forms may be better and longer preserved in some climatic regions (e.g., arid) than others, but their initial development is similar in whatever part of the world the weathering front is located.

The forms may either be destroyed or differentiated by epigene processes after exposure. Thus, the pitted surfaces of microrelief developed at the weathering front in granitic and other rock types does not long survive exposure (Twidale and Bourne, 1976). On the other hand, rock basins of various forms have been developed—hemispherical pits, flat-floored pans, armchair-shaped hollows, and cylindrical depressions—through the interplay of moisture with the bedrock of various inclines and structural characteristics, but all began as shallow, saucer-shaped hollows in the weathering front.

Likewise, given a massive impermeable structure, different rock types are shaped in similar fashion under attack by groundwater: bornhardts and boulders formed in granite and gneiss are similar to those of sandstone, conglomerate, and even limestone; the clint and grike patterns, the gutters, and the basins of granite outcrops are emulated in sandstone, limestone, dacite, and conglomerate.

Figure 24. Stepped northwestern slope of Yarwondutta Rock, northwestern Eyre Peninsula, South Australia.

Figure 25. Granite clints developed (A) on a shore platform in granite, Point Brown, northwestern Eyre Peninsula, South Australia. (B) on granite at the Devil's Marbles, Northern Territory.

Thus, the grooved granitic pavements exposed in the coastal zone at Point Brown, on the west coast of Eyre Peninsula, in the Devil's Marbles complex of central Australia (Fig. 25), and in the Andorran Pyrenees, for example, are reminiscent of clint and grike pavements illustrated in Sweeting (1973, p. 96). The landforms are convergent in character with respect to both lithology and climate. Etch forms are azonal, and groundwater weathering is a unifying factor in landscape development.

Shallow groundwater has contributed markedly to the shaping of much of the Earth's scenery and continues to have a determining effect on the landscapes of the future. Groundwater is a potent and virtually ubiquitous solvent. It reacts with all of the common rock-forming minerals and through solution, hydration, and hydrolysis is responsible for a major part of the rock weathering evidenced at and near the Earth's surface. It carries bacteria, gases, and salts in solution and forms weak acids. It is the medium whereby salts and mineral fines are translocated both vertically within the profile and laterally beneath the land surface. Groundwater is a crucial constituent of rocks, but one that varies in volume both as a result of meteorologic and climatic fluctuations and variations in place and time.

Without groundwater, the biosphere and the surficial lithosphere would be different from those with which we are familiar. There would be little weathering and less erosion. Many familiar landforms and landscapes would not have evolved.

The precise character of the weathering effected wholly or partly by groundwater varies from place to place, as does the nature of the agents responsible for the exposure of the weathering front. The latter undoubtedly leave their characteristic mark. Running water is the principal erosional agency, and the nature of its work varies little regardless of climatic regime. But whatever weathering process is responsible for shaping the weathering front, structural controls are significant and the end products are similar. It matters little whether solution, hydration, hydrolysis, or oxidation is mainly responsible for the alteration and weakening of the bedrock in an area of compartments of contrasted fracture density. It matters little whether rivers or waves, for example, are responsible for stripping the regolith. The end product is an inselberg landscape. Hence, the development of inselbergs and associated plains is widespread in many climatic environments. Old shield lands are, however, favored because they are comparatively stable, and because they are old and there has been time for widespread deep weathering.

Shallow groundwater, then, is responsible for much direct weathering and, hence, erosion. In addition, differential weathering at the base of the regolith produces shapes that later become landforms, some minor, but others of regional extent (Twidale, 1986). These etch and intracutaneous forms are intrinsically azonal. Translocation and reprecipitation of salts and minerals within the regolith in due time give rise to another class of important features related to pans and duricrusts. Finally, groundwater not only initiates numerous landforms but, through gravity and the contrasted susceptibility of wet and dry rocks, strongly influences the character of landscape evolution.

ACKNOWLEDGMENTS

The writer thanks Professors H. Wopfner (Cologne), I. Douglas (Manchester), and A. Busacca (Pulman, Washington) for constructive comments on the first draft of this chapter.

REFERENCES CITED

Ackermann, E., 1962, Büssersteine-Zeugen vorzeitlicher Grundwasser-schwankungen: Zeitschrift für Geomorphologie, v. 6, p. 148–182.

Alexander, L. T., and Cady, J. G., 1962, Genesis and hardening of laterite in soils: U.S. Department of Agriculture Technical Bulletin 1282, 90 p.

Babington, B., 1821, Remarks on the geology of the country between Tellicherry and Madras: Transactions of the Geological Society of London, v. 5, p. 328–339.

Bain, A.D.N., 1923, The formation of inselberge: Geological Magazine, v. 60, p. 97–107.

Barbier, R., 1957, Amenagements hydroélectriques dans le Sud de Brésil: Compte Rendu Sommaire des Seances de la Societé Géologique de France, v. 6, p. 877–892.

Barton, D. C., 1916, Notes on the disintegration of granite in Egypt: Journal of Geology, v. 24, p. 382–393.

Beche, H. T. de la, 1839, Report on the geology of Cornwall, Devon, and West Somerset: Geological Survey of England and Wales, London, Longmans, Orme, Brown, Green and Longmans, 624 p.

Birkeland, P. W., 1974, Pedology, weathering, and geomorphological research: London, Oxford University Press, 285 p.

Bloomfield, C., 1957, The possible significance of polyphenols in soil formation: Journal of Science in Food Agriculture, v. 8, p. 389–392.

Boyé, M., and Fritsch, P., 1973, Dégagement artificiel d'un dôme crystallin au Sud-Cameroun: Travaux et Documents de Géographie Tropicale, v. 8, p. 69–96.

Bradley, W. C., and Griggs, G. B., 1976, Form, genesis, and deformation of central California wave-cut platforms: Geological Society of America Bulletin, v. 87, p. 433–449.

Brady, N. C., 1974, The nature and properties of soils: New York, MacMillan, 639 p.

Brajnikov, B., 1953, Les pains-de-sucre de Brésil; sont-ils enracinés?: Compte Rendu Sommaire des Seances de la Societé Géologique de France, v. 6, p. 267–269.

Buchanan, F., 1807, A journey from Madras through the countries of Mysore, Canara, and Malabar, etc., in 1800–01: London, Cadell and Davies, vol. 2, p. 436–460.

Büdel, J., 1957, Die 'doppelten Einebnungsflächen' in den feuchten Tropen: Zeitschrift für Geomorphologie, v. 1, p. 201–228.

Buol, S. W., Hole, F. D., and McCracken, R. J., 1973, Soil genesis and classification: Ames, Iowa State University Press, 360 p.

Clayton, R. W., 1956, Linear depressions (Bergfussniederungen) in savannah landscapes: Geographical Studies, v. 3, p. 102–126.

Cushing, S. W., 1913, The East Coast of India: American Geographical Society Bulletin, v. 45, p. 81–92.

Daily, B., Twidale, C. R., and Milnes, A. R., 1974, The age of the lateritized land surface on Kangaroo Island and adjacent areas of South Australia: Journal of the Geological Society of Australia, v. 21, p. 387–392.

Davis, W. M., 1909, Geographical essays: Boston, Massachusetts, Dover, 777 p.

Dorn, R. I., and Oberlander, T. M., 1981, Microbial origin of desert varnish: Science, v. 213, p. 1245–1247.

Duchaufor, P., 1982, Pedology, pedogenesis and classification, translated by Paton, T. R.: London, George Allen and Unwin, 448 p.

Fairbridge, R. W., and Finkl, C. W., 1978, Geomorphic analysis of the rifted cratonic margins of Western Australia: Zeitschrift für Geomorphologie, v. 22, p. 369–389.

Falconer, J. D., 1911, Geology and geography of northern Nigeria: London, MacMillan, 295 p.

Finkl, C. W., and Churchward, H. M., 1973, The etched landsurface of southwestern Western Australia: Journal of the Geological Society of Australia, v. 20, p. 295–307.

Goudie, A., 1973, Duricrusts in tropical and subtropical landscapes: Oxford, Oxford University Press, 174 p.

Griggs, D. T., 1936, The factor of fatigue in rock exfoliation: Journal of Geology, v. 44, p. 783–786.

Hallsworth, E. G., 1968, The gilgai phenomenon, in A handbook of Australian soils: Adelaide, Rellim, p. 415–420.

Hassenfratz, J.-H., 1791, Sur l'arrangement de plusieurs gros blocs de différentes pierres que l'on observe dans les montagnes: Annales de Chimie, v. 11, p. 95–107.

Hills, E. S., 1949, Shore platforms: Geological Magazine, v. 86, p. 137–152.

—— , 1971, A study of cliffy coastal profiles based on examples in Victoria:

Zeitschrift für Geomorphologie, v. 15, p. 137–180.

Hingston, F. J., 1962, Activity of polyphenolic constituents of leaves of *Eucalyptus* and other species in complexing and dissolving iron oxide: Australian Journal of Soil Research, v. 1, p. 63–73.

Huang, P. M., and Schnitzer, M., eds., 1986, Interactions of soil minerals with natural organics and microbea: Soil Science Society of America Special Publication 17, 606 p.

Hutton, J. T., Twidale, C. R., Milnes, A. R., and Rosser, H., 1972, Composition and genesis of silcretes and silcrete skins from the Beda Valley, southern Arcoona Plateau, South Australia: Journal of the Geological Society of Australia, v. 19, p. 31–39.

Jennings, J. N., and Sweeting, M. M., 1961, Caliche pseudo-anticlines in the Fitzroy Basin, Western Australia: American Journal of Science, v. 259, p. 635–639.

Jenny, H., 1941, Factors of soil formation: New York, McGraw-Hill, 281 p.

Jones, T. R., 1859, Notes on some granite tors: Geologist, v. 2, p. 301–312.

Jutson, J. T., 1914, An outline of the physiographical geology (geomorphology) of Western Australia: Geological Survey of Western Australia Bulletin, v. 61, 240 p.

Keller, W. D., and Frederickson, A. F., 1952, Role of plants and colloidal acids in the mechanism of weathering: American Journal of Science, v. 250, p. 594–608.

Kessler, D. W., Insley, H., and Sligh, W. H., 1940, Physical, mineralogical, and durability studies on the building and monumental granites of the United States: National Bureau of Standards Journal of Research, v. 24, p. 161–206.

King, L. C., 1962, Morphology of the Earth: Edinburgh, Oliver and Boyd, 699 p.

Krauskopf, K. B., 1956, Dissolution and precipitation of silica at low temperatures: Geochimica et Cosmochimica Acta, v. 10, p. 1–26.

Linton, D. L., 1955, The problem of tors: Geographical Journal, v. 121, p. 470–487.

Logan, J. R., 1849, The rocks of Palo Ubin: Genootschapp Kunsten Wetenschappen (Batavia), v. 22, p. 3–43.

—— , 1851, Notices of the geology of the Straits of Singapore: Geological Society of London Quarterly Journal, v. 7, p. 310–344.

Loughnan, F. C., 1969, Chemical weathering of the silicate minerals: Amsterdam, Elsevier, 154 p.

Mabbutt, J. A., 1961a, 'Basal surface' or 'weathering front': Proceedings of the Geologists' Association of London, v. 72, p. 357–358.

—— , 1961b, A stripped land surface in Western Australia: Transactions and Papers of the Institute of British Geographers, v. 29, p. 101–114.

—— , 1965, The weathered land surface in central Australia: Zeitschrift für Geomorphologie, v. 9, p. 82–114.

MacCulloch, J., 1814, On the granite tors of Cornwall: Transactions of the Geological Society of London, v. 2, p. 66–78.

Maignien, R., 1966, Review of research on laterites: UNESCO, Natural Resources Research, v. 4, 148 p.

Mann, A. W., and Horwitz, R. C., 1979, Groundwater calcrete deposits in Australia; Some observations from Western Australia: Journal of the Geological Society of Australia, v. 26, p. 293–303.

Marbut, C. F., 1927, A scheme for soil classification: Proceedings, 1st International Congress of Soil Science, v. 4, p. 1–31.

Miller, R. P., 1937, Drainage lines in bas-relief: Journal of Geology, v. 45, p. 432–438.

Milnes, A. R., and Hutton, J. T., 1983, Calcretes in Australia, *in* Soils; an Australian viewpoint: CSIRO/Academic Press, p. 119–162.

Milnes, A. R., Bourman, R. P., and Northcote, K. H., 1985, Field relationships of ferricretes and weathered zones in southern South Australia; A contribution to 'laterite' studies in Australia: Australian Journal of Soil Research, v. 23, p. 441–465.

Morey, G. W., Fournier, R. O., and Rowe, J. J., 1964, The solubility of amorphous silica at 25°C: Journal of Geophysical Research, v. 69, p. 1995–2002.

Nace, R. L., 1960, Water management, agriculture, and groundwater supplies: U.S. Geological Survey Circular 415, p. 1–11.

Netterberg, F., 1985, Pedocretes, *in* Brink, A.B.A., ed., Engineering geology of southern Africa; v. 4, Post-Gondwana deposits: Silverton, South Africa, Building Publications, p. 286–307.

Ollier, C. D., 1965, Some features of granite weathering in Australia: Zeitschrift für Geomorphologie, v. 9, p. 285–304.

Ollier, C. D., Chan, R. A., Craig, M. A., and Gibson, D. L., 1988, Aspects of landscape history and regolith in the Kalgoorlie region, Western Australia: Bureau of Mineral Resources Journal of Australian Geology and Geophysics, v. 10, p. 309–321.

Olson, G. W., 1981, Soils and the environment: New York, Chapman and Hall, 178 p.

Prescott, J. A., and Pendleton, R. L., 1952, Laterite and lateritic soils: Commonwealth Bureau of Soil Science Technical Communication, v. 47, 51 p.

Price, W. A., 1925, Caliche pseudo-anticlines: American Association of Petroleum Geologists Bulletin, v. 9, p. 1009–1017.

Rubey, W. W., 1928, Gullies in the Great Plains formed by sinking of the ground: American Journal of Science, v. 15, p. 417–422.

Russell, R. J., 1958, Geological geomorphology: Geological Society of America Bulletin, v. 69, p. 1–22.

Ruxton, B. P., 1958, Weathering and subsurface erosion in granite at the piedmont angle, Balos, Sudan: Geological Magazine, v. 45, p. 353–377.

Ruxton, B. P., and Berry, L. R., 1957, Weathering of granite and associated erosional features in Hong Kong: Geological Society of America Bulletin, v. 68, p. 1263–1292.

Saussure, H. B., de, 1796, Voyage dans les Alpes: Neuchatel, Fauche, vol. 1.

Scheffer, F., Meyer, B., and Kalk, E., 1963, Biologische Ursachan der Wüstenlackbildung: Zeitschrift für Geomorphologie, v. 7, p. 112–119.

Siever, R., 1962, Silica solubility, 0–200°C, and the diagenesis of siliceous sediments: Journal of Geology, v. 70, p. 127–150.

Soil Survey Staff, 1975, Soil taxonomy; A system of soil classification for making and interpreting soil surveys: U.S. Department of Agriculture and Soil Conservation Service Agriculture Handbook 436, 754 p.

Steila, D., 1976, The geography of soils: Englewood Cliffs, New Jersey, Prentice-Hall, 222 p.

Stephens, C. G., 1947, Functional synthesis in pedogenesis: Transactions of the Royal Society of South Australia, v. 71, p. 168–171.

Sweeting, M. M., 1973, Karst landforms: New York, Columbia University Press, 362 p.

Thomas, M. F., 1965, Some aspects of the geomorphology of domes and tors in Nigeria: Zeitschrift für Geomorphologie, v. 9, p. 63–81.

Thomas, M. F., and Thorp, M. B., 1985, Environmental change and episodic etch planation in the humid tropics of Sierra Leone; The Koidu etch plain, *in* Douglas, I., and Spencer, T., eds., Environmental change and tropical geomorphology: London, Allen and Unwin, p. 239–267.

Trendall, A. F., 1962, The formation of "apparent peneplains" by a process of combined lateritisation and surface wash: Zeitschrift für Geomorphologie, v. 6, p. 183–197.

Trudinger, P. A., and Swaine, D. J., eds., 1979, Biogeochemical cycling of mineral-forming elements: Amsterdam, Elsevier, 612 p.

Tucker, M. E., 1978, Gypsum crusts (gypcrete) and patterned ground from northern Iraq: Zeitschrift fur Geomorphologie, v. 22, p. 89–100.

Turner, F. J., 1952, Gefügerelief illustrated by 'schist tor' topography in central Otago, New Zealand: American Journal of Science, v. 250, p. 802–807.

Twidale, C. R., 1962, Steepened margins of inselbergs from north-western Eyre Peninsula, South Australia: Zeitschrift für Geomorphologie, v. 6, p. 51–69.

—— , 1971, Structural landforms: Canberra, Australian National University Press, 247 p.

—— , 1976, On the survival of palaeoforms: American Journal of Science, v. 276, p. 77–94.

—— , 1978a, On the origin of pediments in different structural settings: American Journal of Science, v. 278, p. 1138–1176.

—— , 1978b, Early explanations of granite boulders: Revue de Géomorphologie Dynamique, v. 27, p. 133–142.

—— , 1978c, Granite platforms and the pediment problem, *in* Davies, J. L., and Williams, M.A.J., Landform evolution in Australasia: Canberra, Australian

National University Press, p. 288–302.

——, 1981a, Granite inselbergs; Domed, block-strewn, and castellated: Geographical Journal, v. 147, p. 54–71.

——, 1981b, Origins and environments of pediments: Journal of the Geological Society of Australia, v. 28, p. 423–434.

——, 1982a, Granite landforms: Amsterdam, Elsevier, 372 p.

——, 1982b, The evolution of bornhardts: American Scientist, v. 70, p. 268–276.

——, 1982c, Les inselbergs à gradins et leur signification; l'exemple de l'Australie: Annales de Géographie, v. 91, p. 657–678.

——, 1983, Australian laterites and silcretes; Ages and significance: Revue de Géographie Physique et de Géologie Dynamique, v. 24, p. 35–45.

——, 1986, Granite landform evolution; Factors and implications: Geologische Rundschau, v. 75, p. 769–779.

——, 1987, Sinkholes (dolines) in lateritised sediments, western Sturt Plateau, Northern Territory, Australia: Geomorphology, v. 1, p. 33–52.

Twidale, C. R., and Bourne, J. A., 1975a, The subsurface initiation of some minor granite landforms: Journal of the Geological Society of Australia, v. 22, p. 477–484.

——, 1975b, Geomorphological evolution of part of the eastern Mt. Lofty Ranges, South Australia: Transactions of the Royal Society of South Australia, v. 99, p. 197–209.

——, 1975c, Episodic exposure of inselbergs: Geological Society of America Bulletin, v. 86, p. 1473–1481.

——, 1976, Origin and significance of pitting on granite rocks: Zeitschrift für Geomorphologie, v. 20, p. 405–416.

Twidale, C. R., and Campbell, E. M., 1984, Murphy Haystacks, Eyre Peninsula, South Australia: Transactions of the Royal Society of South Australia, v. 108, p. 95–103.

Twidale, C. R., and Milnes, A. R., 1983, Slope processes active in arid scarp retreat: Zeitschrift für Geomorphologie, v. 27, p. 343–361.

Twidale, C. R., Bourne, J. A., and Smith, D. M., 1974, Reinforcement and stabilisation mechanisms in landform development: Revue de Géomorphologie Dynamique, v. 23, p. 115–125.

Twidale, C. R., Bourne, J. A., and Twidale, N., 1977, Shore platforms and sealevel changes in the Gulfs region of South Australia: Transactions of the Royal Society of South Australia, v. 101, p. 63–74.

Wahrhaftig, C., 1965, Stepped topography of the southern Sierra Nevada, California: Geological Society of America Bulletin, v. 76, p. 1165–1190.

Walder, J. S., and Hallett, B., 1986, The physical basis of frost weathering; Toward a more fundamental and united perspective: Arctic and Alpine Research, v. 18, p. 27–32.

Wayland, E. J., 1934, Peneplains and some erosional landforms: Geological Survey of Uganda Annual Report Bulletin 1, p. 77–79.

White, S. E., 1976, Is frost action really only hydration shattering? A review: Arctic and Alpine Research, v. 8, p. 1–6.

Wilhelmy, H., 1958, Klimamorphologie der Massengesteine: Brunswick, Westermann, 238 p.

Willis, B., 1936, African plateaus and rift valleys, *in* Studies in comparative seismology: Carnegie Institute of Washington, D.C., Publication 470, 358 p.

Wopfner, H., and Twidale, C. R., 1967, Geomorphological history of the Lake Eyre basin, *in* Jennings, J. N., and Mabbutt, J. A., eds., Landform studies from Australia and New Guinea: Canberra, Australian National University Press, p. 118–143.

Wright, R. L., 1963, Deep weathering and erosion surfaces in the Daly River Basin, Northern Territory: Journal of the Geological Society of Australia, v. 10, p. 151–164.

Manuscript Accepted by the Society November 14, 1989

Chapter 3

The relation of subsurface water to downslope movement and failure

Donald R. Coates
Department of Geological Sciences, State University of New York at Binghamton, Binghamton, New York 13901

INTRODUCTION

There is a large literature on slope movement and the factors that contribute to hillside failure. In recent years the publication pace has increased, largely because of the realization that this field of investigation is necessary to provide societal safeguards and to minimize damage to installations. The list of pathfinders in these studies is too long to mention, with the exception of Karl Terzaghi, the founder of the disciplines of soil and rock mechanics. Also the published articles are now too numerous to single out those that are most noteworthy. However, background information on some of the leading books may prove helpful for the reader.

Slopes are a fundamental part of this chapter, and a general treatment of this topic can be found in books by Young (1972), Carson and Kirkby (1972), Schumm and Mosley (1973), and Abrahams (1986). Books that emphasize the destabilization of slopes include those by Coates and Vitek (1980), Brunsden and Prior (1984), Bromhead (1986), and Anderson and Richards (1987). An early major book devoted entirely to landslides and related features was written by Sharpe (1938). This was followed by other general landslide books by Eckel (1958), Zaruba and Mencl (1969), Coates (1977a), Schuster and Krizek (1978), and Crozier (1986). Books emphasizing specific landslide types include those by Voight (1978, 1979) which emphasizes rockslides, and Costa and Wieczorek (1987), which concentrates on debris flows and debris avalanches. Although many of these books mention the importance of subsurface water as a major factor in slope movements, there are no books that treat only this aspect of slope failure.

One of the earliest accounts of the importance of subsurface water in producing slope movements was given by Dana (1863, p. 649):

> The mass of earth on a side-hill, having over its surface, it may be, a growth of forest-trees, and, below, beds of gravel and stones, may become so weighted with the waters of a heavy rain, and so loosened below by the same means, as to slide down the slope by gravity.

A variety of terms are in use to describe hillslope degradation by gravity phenomena. These terms define factors such as the type of movement, rate of movement, nature of earth materials, direction of movement, amount of displaced material, and magnitude of the failure. The term "gravity movement" is the most inclusive and encompasses the full range of movement of material from a higher to lower elevation by gravitational forces. Such diverse changes as subsidence, collapse, boulder disintegration, frost heaving, creep, and landsliding that involve vertical or downslope displacements of material are all considered gravity movements. Other terms in frequent use include "mass movement," and "mass wasting."

Mass wasting is defined in the *Glossary of Geology and Related Sciences* as "A general term for the dislodgement and downslope transport of soil and rock material under the direct application of gravitational body stresses" (Bates and Jackson, 1980, p. 383). However, in the *Encyclopedia of Geomorphology,* mass wasting is described as including "free fall to subsidence" (Savage, 1968, p. 696) and is discussed as being a broader term than mass movement. Crozier (1986, p. 6) would expand the term even further in stating that it also refers to ". . . the mass reduction of the interfluves as opposed to the degradation by streams. In effect it must include the action of all non-linear erosional processes working on the slopes between streams."

In most geology textbooks, mass wasting and mass movement are used as synonymous terms. The *Glossary of Geology and Related Sciences* restricts mass movement to "A unit movement of a portion of the land surface" (Bates and Jackson, 1980, p. 383). Visser (1980, p. 14) also limits mass movement to "the gravitational transfer of loosened material where large quantities of debris move together as a coherent whole in close grain to grain contact." However, the confusion in terminology is shown by the expansion of mass movement in definitions by other authors. Hutchinson (1968, p. 688) states, "Mass movement comprises all . . . gravity-induced movements except those in which the material is carried directly by transporting media such as ice,

---

Coates, D. R., 1990, The relation of subsurface water to downslope movement and failure, *in* Higgins, C. G., and Coates, D. R., eds., Groundwater geomorphology; The role of subsurface water in Earth-surface processes and landforms: Boulder, Colorado, Geological Society of America Special Paper 252.

snow, water or air. . . . .A primary division is made between mass movements occurring on slopes and those involving only a sinking of the ground surface." These statements can be compared with those by Crozier (1986, p. 6): "Mass movement is distinguished. . . .by being outward or downward gravitational movement of earth material without the aid of running water as a transportational agent. This is a very precise and useful definition as it does not deny the importance of water in either its solid or liquid state as a destabilizing factor nor does it exclude subsidence and other movement on flat ground."

Because of these variations in terminology and disagreements as to what constitutes mass wasting and mass movement, in this chapter the type of terms advocated by Varnes (1978) are preferred, namely "slope movement" or "downslope movement." He excludes subsidence or other forms of ground sinking and instead restricts slope movements to five types: falls, topples, slides, spreads, and flows. Since this chapter is devoted to the role of subsurface water in the displacement of earth material on hillslopes, the range of features and processes associated with solifluction, earthflows, and mudslides must also be included. Thus, slope movement consists of downslope destabilization of earth materials (Fig. 1) with a spectrum of motion that can range from creep to the landslide end of the series.

The principal thrust of this chapter is that subsurface water can affect earth materials so as to exceed the threshold of stability, thereby producing failures on hillslopes and initiating motion of the material. Thus, a distinction is made between material transported by fluvial and sheet-flow water processes and water motion in the substrate. The subsurface water may be above or below the water table, and its reaction with the host material can produce an array of processes and results that include creep, landslides, piping, and sapping.

A variety of causes can initiate failures on hillslopes. Infiltration from storm rainfall and snowmelt are leading reasons for slope movements. In addition, such human-induced changes as highway construction and reservoir fluctuations have triggered many displacements worldwide.

HILLSLOPE STABILITY AND INSTABILITY

There are three primary factors that determine the stability of hillslopes:

1. The internal properties of the earth materials. These include the type of rock, regolith, sediments, and soil and their structural relations. All may have different stability properties and thresholds for slope movement. Their mass strength can be quantified using two components: their intrinsic cohesion and frictional resistance between grains or rock fractures.

2. The geomorphic setting and environment. Many features may influence the potential for slope displacements, such as total topographic relief, slope steepness and length, and shape of the land surface. In addition, other factors may also be crucial, including type of vegetation, slope orientation, climate, and antecedent moisture conditions.

3. Independent external events. These can provide the triggering mechanisms that result in increased stress or decreased strength and mobilize materials for movement. Prior to their displacement the earth materials are in a state of balanced stresses until a critical threshold is reached and the state of equilibrium is exceeded. This chapter focuses on the normal stresses in groundwater that produce slope failure when they exceed the shear resistance of the earth material. These relations for cohesive soil have been expressed by Terzaghi and Peck (1948), and for rock by Terzaghi (1962).

THE ROLE OF WATER IN HILLSLOPE FAILURE

Water is second only to gravity in producing slope instability. There are many varieties of subsurface water including water of hydration in mineral species, soil moisture, capillary water, vadose zone water, perched water, and groundwater. Water movement may include throughflow, interflow, percolation, and groundwater flow in confined, semi-confined, and unconfined systems.

The rate and amount of new water introduced into the subsurface system is a function of the type of earth materials, their structure, and the antecedent moisture conditions. When all other factors are equal, the magnitude, intensity, and duration of rainstorms determine the timing and size of hillslope failures. Sudden and heavy precipitation can weaken the substrate when it displaces air in pore spaces and fractures, thereby increasing the surcharge on potential failure surfaces, and providing pore-water pressure along such surfaces. The introduction of these positive pore-water pressures (piezometric or hydraulic pressures) and the buoyant effects of the water cause a decrease in effective normal stress and a loss in frictional resistance of the saturated material (Bromhead, 1987).

Hydration water in mineral and soil particles, absorbed water, and soil capillary moisture, although not strictly groundwater, can all contribute to enhancing potential slope movements. Increased moisture during wet weather reduces capillary tension because there is a rise in the water table that encompasses part of the capillary zone. In dry weather, evapotranspiration reduces capillary moisture and lowers the water table. The resulting increase in capillary tension also increases the effective strength of the soil. Any increase in absorbed moisture will decrease soil strength. In addition, an increase in water of hydration in minerals causes their expansion and decreases the bonding of soils, thereby lowering their strength. A rapid increase of moisture in dry soil can create pore-water pressure increases in trapped pore air, with local soil expansion and a decrease in soil stability. Increases in soil or rock moisture are therefore usually accompanied by a decrease in strength of earth materials (Sowers and Royster, 1978; Swanson and Swanston, 1977).

Antecedent moisture and water changes

Moisture in earth materials when coupled with the introduction of new water can provide the necessary conditions for slope failure. Coates (1977b, p. 7) has stated that:

Figure 1. Spanish Fork landslide near Thistle, Utah. This has been called the most costly landslide in U.S. history. It was also the fifth largest landslide in the conterminous U.S. during this century. Although creep had been occurring for many years, abnormal snowmelt and rains provided excessive infiltration during April 1983, which triggered the landsliding. Damage occurred to State Highways 6, 50, 89, and a railroad, and the town of Thistle was destroyed. The landslide originated in the upper left center of photograph and moved toward the center of the picture. Photograph courtesy of Earl Brabb.

Antecedent moisture conditions determine whether large amounts of rainfall will successfully trigger a landslide. If earth materials already contain significant moisture from prior rainfall, the severity of precipitation from a new storm can be less and still trigger landsliding. If other factors are equal, magnitude, intensity, and duration of the storm are all-important factors that can contribute to hillslope instability. Excessive precipitation weakens earth materials by displacing air in pore spaces and fractures, by increasing overburden pressure on potential slide surfaces, and by providing pore-water pressure along the shear surface.

Increases in water content can thereby decrease slope stability in such ways as:

1. Increasing the interstitial pore-water pressure. Positive pressures cause a reduction in frictional strength of earth materials.

2. Initiating cleft-water pressure within joints, voids, and other openings, which also reduces frictional resistance.

3. Producing seepage pressure wherein a type of drag stress contributes to shear stress, thereby causing subsurface erosion and/or reduction in underlying support of materials.

4. Conversion of clay-rich units into sediment slurries that provide avenues for sliding of overlying or adjacent material.

5. Addition of weight to the entire earth column. This water surcharge is especially effective in causing instability of sediments and those materials rich in clay minerals.

Many studies have addressed the relation of antecedent moisture and new rainfall to the initiation of landsliding (Okimura, 1981). In Hong Kong, landslide "disasters" occur when daily rainfall exceeds 100 mm after a 15-day antecedent rainfall in excess of 350 mm (Crozier, 1986). The six landslide episodes that occurred between 1933 and 1969 in the San Dimas Experimental Forest, California, were all triggered by 500 mm of rain over a 5-day period with 150 to 200 mm falling in a 24-hr storm (Rice, 1982). Campbell (1975) reported that rain of at least 6 mm/hr was capable of producing debris slides in the Santa Monica Mountains, California, if preceded by a 250-mm rainfall. In the Los Angeles area, California, debris slides are developed by 6.35-mm/hr rains when preceded by a rainfall of 250 mm (Cleveland, 1971).

Rainfall intensity and duration are also important in causing landsliding. In the Oregon Coast Range, Pierson (1977) found that debris slides were associated with 24-hr rainfalls of 130 to 140 mm, which included 6-hr bursts of at least 50 to 75 mm. Eyles and others (1978) showed that landslides on natural slopes occurred with 24-hr rains that exceeded 200 to 250 mm, whereas on artificially modified slopes the threshold for slides was reached with only 120 mm. Less than 100 mm of rainfall was needed to create sliding when soil moisture storage was full. In the Virginia mountains, Hack and Goodlet (1960) studied more than 100 landslides produced by a 9-in (230-mm) storm that lasted only a few hours, and Wolman and Gerson (1978, p. 202) concluded that intensive rainfall is the most important prerequisite for slope movement in humid montane environments. However, after studies in the Appalachian Plateau, Pomeroy (1984, p. 13) concluded "antecedent moisture is the determinant factor concerning how much rainfall in a single storm is required to initiate slope failure." His study of storm-induced slope movements in northwestern Pennsylvania showed that prior to a 120-mm rainfall, pre-storm rainfall totals throughout the area were substantially above the norms for 15-, 30-, 45-, 60-, and 75-day periods.

Various authors have quantified moisture and rainfall relations into equations that express thresholds for slope destabilization. Kohler and Linsley (1951) developed an "antecedent precipitation index," which has been used by Crozier and Eyles (1980) as an index of antecedent soil water content and applied to the 10 days prior to a rainfall that causes a landslide event. Subsequently, Crozier (1982) provided an "antecedent soil water status model" in order to define the threshold between triggering and nontriggering conditions of slope behavior and the potential for displacement. Success of the model depends on accurate determinations of the hydrologic properties of the terrain and the type of landslide involved. A value for soil moisture capacity is needed in respect to soil porosity and depth of the failure surface. A number that proved effective in his study area showed that an average value of 120 mm of water within the top 76 cm of the regolith was required prior to an important storm event for slope movement to occur (Crozier, 1986, p. 187). Another assumption in the model is that rainwater in excess of relative soil moisture saturation determines the degree of stability. Rate of water loss is important and varies with such conditions as evapotranspiration, soil type, permeability at the base of the regolith, and slope angle. Next, the probability of rainfall intensities within a 24-hr period is calculated, and from these data threshold lines are graphically portrayed that indicate the likelihood of landsliding events.

Mechanics of slope failure in sediments

Changes in hillslope characteristics and the water table can initiate substantial differences in pore-water pressures and seepage forces that destabilize the ground surface. These changes may be created naturally where streams undercut adjacent slopes or by abnormal rainstorms. Hillside excavations, as in road construction (Figs. 2 and 3), and water-level changes in reservoirs can also lead to slope movement. Terzaghi and Peck (1948) showed how percolating water exerts a seepage pressure on sediments due to its viscosity. Such pressure is in the direction of groundwater flow lines; its magnitude increases with the amount of hydraulic head, which determines seepage velocity. The seepage forces increase from higher terrain to the foot of slopes. Furthermore, these seepage forces are compounded by the weight of the water and earth material that acts along potential surfaces of discontinuities (Figs. 3 and 4). These relations have been shown by Terzaghi and Peck (1948) and Parizek (1971).

When P (total pressure per unit area) has the relation $P = \bar{P} + u$, the effective pressure, $\bar{P}$, transmitted grain-to-grain in a saturated soil is less by the amount that the displaced water tends to uplift soil grains (Parizek, 1971). For a soil or rock mass below the water table, $\bar{P}$ equals the submerged unit weight τ' times the thickness of the saturated column of soil. The submerged unit

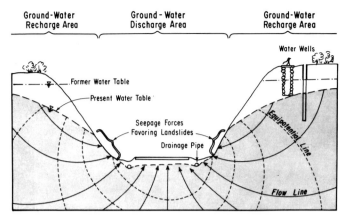

Figure 2. Beheading of water table by a highway excavation. Note changes produced in the water table and activation of seepage forces (Parizek, 1971).

weight τ' is equal to the difference between the unit weight of the saturated soil τ_{sat} and the unit weight of water, τ_ω:

$$\tau' = \tau_{sat} - \tau_w$$

The effective pressure, $\bar{P}$, on a potential failure surface may be further reduced in a groundwater discharge area by the amount that the excess (i.e., in excess of hydrostatic), pore-water pressure, U_ω, is acting upward against the base of that surface. The amount of the excess pore-water pressure depends on the hydraulic head, h, at the surface of failure with respect to the external water level times the unit weight of water, τ_ω, or:

$$U_\omega = h\tau\omega$$

Sapping and piping. Although the subsurface aqueous erosional processes that cause hillside sapping and piping are more fully discussed by other authors in this volume (see also Howard and others, 1988; Howard and McLane, 1988), this chapter would not be complete without some mention of these other causes of slope failure. Bradford and Piest (1980, p. 84) have shown that gully wall slumping is:

Mainly due to energy associated with water.... Slope failure due to the addition of water can be attributed to the weight of water added to the soil mass by surface infiltration and/or by a rise in the water table, and/or (2) increased seepage forces as water in the saturated zone exits through the bank as water table discharge or as drawdown of bank storage ... and/or (3) reduction in apparent cohesion of unsaturated soil, through reduction in capillary tension (or negative pore pressure).

They describe three different failure modes in the erosional development of valley-bottom gullies: deep-seated slides, slab failure, and base failure. Such movements are all associated with changes in soil moisture and subsurface water conditions.

Piping is the development of subsurface drainage conduits in nonlithified earth materials at depths above the nearby base level.

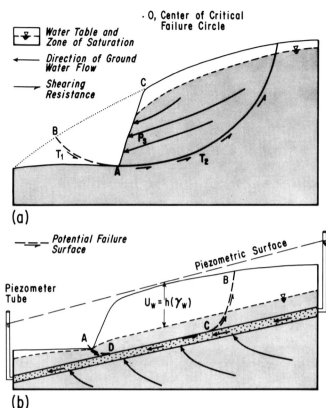

Figure 3. Potential slope failure due to (a) seepage force acting on a mass of saturated soil above a highway cut, and (b) uplift pressure originating within a thin, semiconfined aquifer below a cut. In (a), P_s equals the seepage force that is directed to the exposed cut in response to groundwater flow. T_2 is the sum of the shearing forces resisting failure, and T_1 is the amount by which the shearing resistance was reduced by the excavation. In (b), U_w is the excess hydrostatic head exerting uplift on the plane of failure ADCB, τ_w is the unit weight of water, and h the hydraulic head (Parizek, 1971).

It is both a natural phenomenon (Fig. 5) and a process that can be initiated and/or accelerated by human activities (Figs, 4, 6, and 7). The loss in slope stability occurs where the fine-grained sediments are selectively eroded from stratified materials (Fig. 4). The grain-by-grain removal of matrix material is caused by the discharge of subsurface water aided by seepage pressures acting outward from an unsupported hillslope. For piping to occur, several conditions must be fulfilled: (1) There must be a source of infiltrating water. (2) The percolation rate must exceed the absorption rate of some sediment horizon, and thus permit periodic saturation of overlying sediment. There must also be (3) an erodible layer above the retarding layer, (4) subsurface water capable of motion under a hydraulic gradient, and (5) an outlet for the flowing water, generally in the form of a discrete tube, tunnel, pipe, or conduit (Fig. 7). When all these requirements are met, piping may cause a slope instability that can become instrumental in producing landslides (Fig. 6).

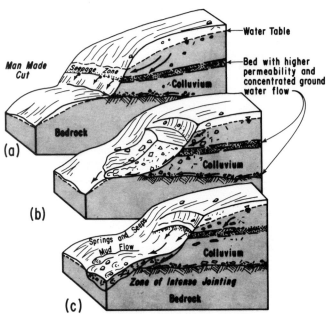

Figure 4. Slope failure by piping (a), (b), and (c). Erosion of fines from a thin aquifer is initiated by groundwater discharging to the new excavation (a). In (b) piping continues until the slope fails by slumping. With prolonged groundwater discharge and surface runoff (c), the slumped mass may give rise to mudflow, and the exposed slope is subjected to a new cycle (Parizek, 1971).

Mechanics of slope failure in bedrock

The many publications of Karl Terzaghi emphasize the theme of the importance of water in contributing to and causing destabilization of hillslopes, regardless of whether the underlying materials are surficial or bedrock. He referred to pore-water pressures in soil and cleft-water pressure in rock (Terzaghi 1950, 1962). Furthermore, he was particularly interested in making a distinction between primary and secondary permeability (openings in jointed or fractured rock). The water in the joints exerts a pressure on the walls equal to the unit weight of water times the height to which it would rise in a piezometer screened at the joint. Such a pressure is called the "cleft-water pressure," and like pore-water pressure it reduces frictional resistance along the joint walls.

The cleft-water pressure is zero at the water-table and it increases in a downward direction. Hence if a slope fails on account of cleft-water pressures, the failure will start at the foot of . . . [a] slope, whereby the rock will be displaced by the water pressure. As a result of the initial failure, the rock located above the seat of the failure is deprived of support and it will descend owing to its own weight.

The influence of the cleft-water pressures on the stability of steep rock slopes is well illustrated by the rock slide statistics prepared by the Norwegian Geotechnical Institute. . . .Within the area covered by the observations, the winters are severe, the snowfall abundant, and the heaviest rainfalls occur during the autumn months. . . .slide frequency was greatest in April, during the time of snow melt, and in October within the period of greatest rainfall (Terzaghi, 1962, p. 262).

Additional studies carried out by Bjerrum and Jorstad (1957) showed that Norwegian rock slides during the wet period between 1720 and 1760 were ten times greater than in the 1760–1810 period, a time of a drier climate. In a subsequent study, Bjerrum and Jorstad (1963) stated that the most important factor contributing to cliff falls is the combined action of gravitational shear forces and cleft-water pressures. They also suggested that the factor of fatigue is important with intermittent rise and fall of the cleft-water pressure.

CLASSIFICATION OF SLOPE MOVEMENT

There is a continuum of movement rates for surface materials on slopes. Creep is on one end of the spectrum and is generally restricted to slow and imperceptible motion. It results from gravitational force, and the rate is generally considered to be less than 0.3 m/yr. All slope movements faster than this fall into the landslide category, with velocities that can exceed 100 km/hr. There is a large variety of classification systems offered by different authors (see for example Hansen, 1984), but most agree that the most important diagnostic features include the type of movement, rate of motion, and nature of the earth materials. Unfortunately, slow solifluction phenomena have been omitted from many classification systems.

Slow movements

Most soil creep is seasonal and depends mainly on water. However, creep can occur in the absence of subsurface water with movement resulting from gravitational forces acting alone. Solifluction requires water and, as defined by Andersson (1906, p. 95–96), is "the process of slow flowage from higher to lower ground of masses of waste saturated with water." Although the original intention was to restrict usage to soils with a substratum of frozen ground, the term is now applied to other terrains; some authors have replaced it with the term "gelifluction" for those movements associated with frozen ground.

Ground displacements by the solifluction process can occur on low-angle slopes of 1 to 2° to steeper slopes of 14°. The rates of movement can range from 0.6 cm/yr on slopes that are dry much of the year to movements of 15 cm/yr on wetter slopes (Washburn, 1960; Rapp, 1960). In frozen ground terrane, gelifluction is initiated during the spring thaw when the upper soil is saturated. Downslope movement by gravitational forces under these conditions is facilitated by a fluid type of lubrication at the interface with the frozen zone sediment.

Solifluction processes can often form distinctive landforms, especially where such action is restricted to only a portion of the entire landscape. Solifluction lobes have been described in the cold-weather environments of Greenland (Washburn, 1969) and in Alaska (see Fig. 8). In the Great Bend region, 25 km southeast of Binghamton, New York, King and Coates (1973) and Coates and King (1973) describe a number of features that formed in a periglacial or immediately postglacial period. These landforms

Figure 5. Natural piping and landsliding in the Manti LaSal National Forest. The pipe openings were obliterated by the slope failure. Photograph courtesy of Earl Olson.

include patterned ground, boulder tongues, and solifluction lobes. The valley sides along the south-flowing Trowbridge Creek have several features that have been termed "concavo–convex landforms" (Fig. 9). They have been interpreted as solifluction lobes created after glacial ice had evacuated the area. The former tributaries to Trowbridge Creek were sites for deposition of thick till. Before vegetation could be established and help provide a stabilizing influence, the steeper masses of sediment, when water saturated, were mobilized and gradually moved downslope. Subsequent trenching at these sites has revealed subtle flow structures within the sediment and an imbrication of larger clasts.

Fast movements

The usual practice for designating faster movements of material on hillslopes is to call them landslides. Unfortunately the term "landslide" has generally been used in a dual sense—as a descriptor for the process of motion and as a landform created by the process. Earth materials are prone to dislodge when their moisture content is high. The subsurface water combines with the forces of gravity to produce stresses that are exceptionally difficult to withstand. For example, Sowers and Royster (1978, p. 83) state, "Water is a major factor in most landslides." Exceptions to this rule include the types of landslides that have been classified as rock avalanches. In such cases, earthquakes can provide the trigger for motion, and fluidization of the mass is by non-aqueous processes.

Hillslope stability is determined by the ratio of the shear strength of material resisting a slide and the shear stress that acts to produce movement. Slope stability or the safety factor (Gs) is represented by the following relation:

$$Gs = s/\tau$$

in which s is the shear strength along the base of sediment or rock column and τ is the shear stress. The shear strength, per unit area at the base of a column for a one-dimensional planar slope, is:

$$s = c + (p - h\tau_\omega) \tan \phi'$$

where s is the shearing resistance per unit of area, p is normal stress at a given point on a potential surface of sliding due to the

Figure 6. Slumping and landsliding of glaciolacustrine sediments in the Tioughnioga Valley. New York Route 79, immediately above the river in lower part of photograph, undercut the toe of the hillside, and drainage from the upper road was directed onto the sediment mass, creating a water surcharge. Both factors aided the instability and the development of extensive piping.

weight of the solids and water above that surface, h is the piezometric or hydraulic head at that point, τ_ω is the unit weight of water, $\tan \phi$ is the angle of internal shearing resistance, and c is cohesion. If the material is mostly granular, and therefore cohesionless, c becomes zero. An increase in h reduces shearing resistance. If the pressure $h\tau_\omega$ becomes equal to p, the total weight of the overburden is carried by the water, which has no shear strength and cannot support it, and liquefaction and failure take place.

When the resisting forces are overwhelmed by the driving forces, slope failure occurs, and—depending on environmental conditions, including terrain character, properties of the earth materials, and amount and character of subsurface water—a large range of different types of rapid slope movements may result. The following discussion concentrates on several different types of slope movement as examples of the importance of subsurface water in causing hillside failures, such as debris flows, debris avalanches, earth flows, and mudslides.

INFLUENCE OF TERRAIN AND MATERIALS ON SLOPE MOVEMENT

Although emphasis in this chapter is placed on the importance of subsurface water in producing slope failure, the character of the topography and earth materials must also be considered in the number and magnitude of hillside displacements.

There are several different ways in which terrain characteristics can influence slope movements. Hillslopes can be compared in terms of their length, steepness, and shape. Under appropriate conditions, each can be responsible for causing destabilization of earth materials. Another difference in slopes is their aspect or orientation, such as northeast-facing.

Topographic slope and colluvial hollows

Hack and Goodlet (1960) studied terrain characteristics in a 142-km^2 area of densely forested mountain land in Virginia. The region was in the headwaters of the Shenandoah River with an average relief of about 460 m. The purpose of the investigation was to evaluate the effects of a June 1949 storm that lasted only a few hours but produced 228 mm of rain. The term "chute" was used to describe the slope-movement scars, which were the result of debris avalanches.

The most spectacular remains of the 1949 flood are the many scars or chutes on the mountain slopes resulting from the avalanching of debris.
Many of the chutes are associated with small or incipient hollows on side slopes, and their location is favored by indentations or irregulari-

ties on the slopes that cause concentration of runoff. Some chutes occur adjacent to long grooves that appear to be the remains of an older chute now overgrown with trees and unaffected by the 1949 flood.

Most of the chutes occupy former depressions or groovelike areas, and the impression is inescapable that the chutes are indeed incipient hollows or channelways, that were partly obliterated during the passage of time by falling blocks and mass movements. They are, at rare intervals of time, flushed out and deepened by heavy runoff (Hack and Goodlet, 1960, p. 43–44).

Bogucki (1976, p. 188; 1977, p. 43) showed that major landslide scars in the Great Smoky Mountains, Tennessee, and the Adirondack Mountains, New York, originated at valley heads or were associated with small incipient hollows on side slopes. Studies in southern California have demonstrated the significant relation of colluvium-filled bedrock hollows as being the principal locus of debris-flow initiation (Dietrich and others, 1984, 1986; Reneau and Dietrich, 1987). The hollows are concave parts of upper hillslopes not occupied by channels. However, downslope is usually bedrock so that the channel is a seepage face draining groundwater from the colluvium-covered hollow.

In the hollow environment, high seepage forces caused by excessive pore pressure in the bedrock or colluvium develop channel heads and erode the colluvium. In steep basins with partial or complete saturation of the colluvium, such pressures are sufficient to produce debris flows. The instability depends on both the seepage forces and the strength of the colluvium. The pressure head is largely dependent on both slope length and gradient, and even hydraulic conductivity variations of two orders of magnitude are insufficient to alter the pressure values.

Dietrich and others (1986) explained the colluvial hollow-landslide characteristics by approximating the saturation flow depths and pressure heads as a function of basin geometry. These data were used with the Mohr-Coulomb failure criterion to predict the hollow length-gradient relations. They concluded that "the geometry of low gradient basins is controlled by the development of significant excessive pore pressures and that changes in length and commensurate drainage area are more important than gradient in controlling the pressure head at the downstream sections of basins" (p. 385). For slopes greater than $0.7 \tan \phi$, slope length can be short and still channel heads can be initiated.

Slope aspect

The influence of hillside exposure and orientation on slope movements has been addressed by many authors. In the northern hemisphere, slopes facing north and northeast receive less solar radiation than slopes facing other directions, and after rainfall the soils remain wet longer. East-facing slopes receive early morning sunlight, but the drying effect on such soils is less effective because of lower temperatures at that time of day. Conversely, higher evaporation rates occur on southerly slopes. In addition, slope asymmetry may occur more with steeper northeasterly slopes than elsewhere. Thus, the combination of longer moisture times and steeper slopes favors higher incidences of slope movements

Figure 7. View of an outlet conduit of piping in the glaciolacustrine sediments pictured in Figure 6. This site is in the oversteepened exposure immediately upslope from New York Route 79. Mattock is 45 cm long.

on northeast facing hillsides. These relations were found by Beaty (1956) for California, Harden (1976) and Colton and Holligan (1977) for Colorado, and Van Buskirk (1977) for northeastern Ohio. For example, Beaty (1956) reported that 70 percent of earthflows east of Berkeley were on north- and east-facing slopes. Hack and Goodlet (1960) observed that most of the 100 chutes they studied had northeast exposure. Studies of slope failures in western Pennsylvania by Pomeroy (1982b) showed that 75 percent of the features were on northwest-, north-, northeast-, and east-facing slopes (Fig. 10).

Multiple topographic factors that influence slope failure

In some terrains it may be possible to isolate a single cause for slope movement. Williams and Guy (1971, 1973) pointed out

Figure 8. Solifluction lobe in the Brooks Range, Alaska. This arcuate mass rises about 20 m above adjacent contours and is 260 m wide at the widest point. Photograph courtesy of William McMullen.

Figure 9. Topographic contour map of part of the Windsor, New York Quadrangle (U.S. Geological Survey 1:24,000 series). Solifluction processes have moved till from the valley reentrants on the west side of Trowbridge Creek. This type of landform has been called "concavo-convex hillslopes" (King and Coates, 1973). On the east side of the valley, the hill "1276" has been interpreted as a landslide.

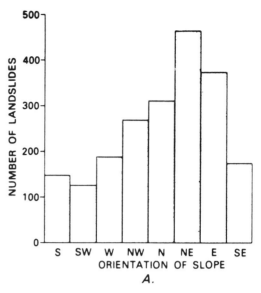

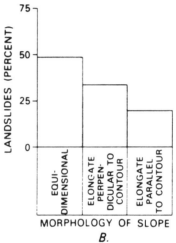

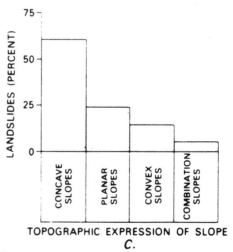

Figure 10. Statistical data comparing orientation, morphology, and topographic expression of slopes in areas of Washington County, Pennsylvania, where recent landslides have taken place. (A) Distribution of 2,032 landslides by orientation of slope; (B) percentage of 1,350 landslides classed by morphology; and (C) topographic expression of slope (from Pomeroy, 1982b).

Figure 11. (A) and (B) are typical scars of debris avalanches in Nelson County, Virginia. These landforms resulted from devastating rainstorms caused by Hurricane Camille in 1969 (from Williams and Guy, 1971).

that torrential rainfall was the primary cause of debris avalanches associated with Hurricane Camille in 1969. However, they also listed the possibility of multiple factors that can contribute to slope failure; these included: hillslope steepness, vegetation type and density, kind of bedrock, attitude of stratified or jointed bedrock, erosion or bombardment of the base of a slope by debris-laden streams, soil texture, orientation of hillslope, length of hillslope, soil depth, susceptibility of soil to infiltration, ability of soil to transmit water (hydraulic conductivity), initial presence of depressions or troughs along hillslopes, bolts of lightning, vibration of heavy thunder, and uprooting of trees due to storm winds.

A torrential rainfall with quantities as great as 500 mm in less than 24 hr was the trigger for the 186 debris avalanches in Nelson County, Virginia, studied by Williams and Guy (1971, 1973). Typical landslides were 200 to 800 ft in horizontal length, 25 to 75 ft wide, and 1 to 3 ft deep (Fig. 11). Some were as short as 20 ft, others as long as 1,000 ft and up to 200 ft wide. Hillside erosion amounted to 3.2 to 4.6 million ft^3/mi^2, or the equivalent of several thousand years of normal denudation. Terrain features determined the location of the majority of slope movements. "The presence of pre-existing depressions on the hillside and the aspect of hillside (N-, NE-, and E-facing slopes) were associated with the most intensive avalanching" (Williams and Guy, 1971, p. 25). Because 85 percent of the debris avalanches were sites of hillslope concavities and hollows, it was concluded that such locations offer the maximum position for collection and mobilization of earth materials by water surges and water-saturated regolith.

Studies in western Pennsylvania by Pomeroy (1980, 1982a and b, 1984) have shown there can be multiple causes for the different types of slope movements that occur throughout the region. Although antecedent moisture conditions and storm events were always considered to be principal reasons for the landsliding, topography and earth materials were also important in determining the specific location of the slope failure. In describing events near Johnstown, Pennsylvania, Pomeroy (1980, p. 1) stated, "Several hundred debris avalanches, debris slides, slumps, earthflows, and combinations of the various types took place as a result of 30 cm of rain that fell during a 9-hour period, July 19–20, 1977, in an area of about 60 to 70 km^2." Prior to the rainstorm the soil had been saturated by above-normal rainfall earlier in the month. Debris avalanches were the most conspicuous features, with maximum lengths of 300 m, widths of 25 m, and head scarps in colluvium to 4 m. They formed along mostly planar to gentle concave slopes of 20 to 40°. Earthflows began on more moderate slopes. The actual movement of the regolith in the debris avalanches took place in two phases. The first movement was planar or rotational sliding extending a short distance downhill from the head scarp, and the second motion was flow by a process referred to as "spontaneous liquefaction" (see Terzaghi, 1950, p. 107). According to Pomeroy (1980), this change in mobilization occurs because of the change in state from a supersaturated soil to a thick viscous fluid. Thus, the increase in pore-water pressure and concomitant decrease in shearing resistance of the colluvium produced transformation of material and terrain.

East Brady, Pennsylvania, another site in the Appalachian Plateau, suffered many slope movements from a 140-mm rainstorm on August 14, 1980 (Pomeroy, 1984). Furthermore, precipitation in a 2-week period prior to the event ranged from 90 to 115 mm. Such an abundance of moisture affected 2.5 percent of hillslope areas in the region with a density of 56 to 85 slope movements per km^2. Unlike the Johnstown, Pennsylvania, area, 90 percent of the debris avalanches occurred along previously existing hillside depressions or hollows. Such hollows seem to have developed originally because of an increased density of joints in the underlying bedrock (see also Woodruff, 1971; and Gray and Gardner, 1977).

... [I]ntensive ground-water movement over a period of time in the hollows (or coves) has produced a deeper regolith than that along adjacent slopes.

During the August 1980 storm, slope movements in the East Brady area took place chiefly on the bedrock-colluvium interface; less common were slippage surfaces within the colluvium. The base of the soil mantle overlying the weathered rock represents a weak plane on which creep has taken place. Surface water percolating through the thin soil (commonly less than 0.6–1.0 m thick) tends to accumulate on top of the rock surface (usually shale) or the thin soil above it. Observations indicate that shale or claystone is usually exposed along the head scarp of a slope movement where more permeable sandstone lies directly above and a concentration of water is present at the contact (Pomeroy, 1984, p. 13; see Fig. 12).

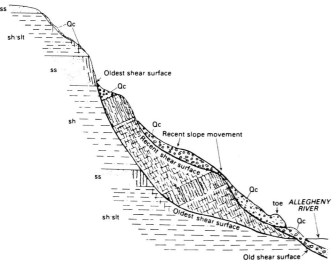

Figure 12. Schematic diagram showing rock slump and subsequent colluvial movement along the slope opposite East Brady, Pennsylvania. Note that the recent shear surface is probably the site of an older colluvial movement. Qc = Quaternary colluvial deposits (from Pomeroy, 1984).

LANDSLIDE TYPES

Although the classification of landslides contains many diverse types, this section discusses several that are representative of the role of subsurface water in their development. The terms "earthflow," "debris flow," and "mudslide" overlap, but this nomenclature is thoroughly embedded in the literature. However, it is possible for a single landslide to have two or more parts that show dissimilar failure modes of formation (Fig. 1). The somewhat associated slope movements of solifluction and debris avalanches are discussed elsewhere in this chapter.

Debris flows

These features have been studied by many researchers (Iverson and Major, 1986); they are the central theme in an entire book (Costa and Wieczorek, 1987), and they involve slope movements in which subsurface water is always a major component (Fig. 13). Pierson and Costa (1987) define debris flow in terms of a slurry flow that contains both inertial and viscous qualities.

Rodine (1974) and Johnson (1984) approached the low slurry strengths needed for mobilization of debris flows by considering water content, the principal determinant of slurry strength. They devised a mobility index (MI), defined by Johnson (1984) as the ratio of saturated water content of the inplace soil to the water content needed for flow of that soil down the available channel. Water content needed for flow was determined through innovative strength testing and measurement of channel form. Mobilization was considered likely where saturated water content of inplace soil was sufficient for debris flow down the available channel, and less likely where soil must take on additional water in order to flow. Rodine and Johnson found that soils involved in debris flow had an MI 0.85 (Seed and others, 1964, p. 77).

An approximation of the MI can be obtained by using the Atterberg liquid limit to represent the water content needed for flow. Thus, this approximate mobility index (AMI) is the ratio of saturated water content of inplace or undisturbed soil to its liquid limit. Qualitatively, the liquid limit seems suitable for this use because it is the water content at which soil behavior is marginally fluid under shallow conditions. Quantitatively, the liquid limit represents a shear strength of about 2kPa (Ellen and Fleming, 1987, p. 33–34).

In their work, Ellen and Fleming (1987) draw relations from the thousands of debris flows that mobilized from shallow slides in the San Francisco Bay region during the rainstorm of January 3–5, 1982. There are two fundamental ways for direct transformation from slide to flow: contractive soil behavior, which commonly results in liquefaction, and dilative soil behavior, which helps to mobilize the slide mass if additional water is incorporated. The concept of plug flow is used to show how this transition occurs (Johnson, 1970). Johnson has shown that in order for a sliding slab of soil to transform to a debris flow, its remolded thickness due to contraction must be less than or equal to the thickness of the sliding slab. Thus, mobilization requires a slurry strength low enough that such thickness is less than the slide thickness.

Wieczorek (1987) studied debris flows in the Santa Cruz Mountains, California, and developed an empirical model based on geology, hydrology, and topography to explain the triggering of debris flows with different combinations of intensity and duration of rainstorms once the antecedent and intensity-duration thresholds have been exceeded. He showed that debris flows did not occur unless at least 280 mm of rainfall had accumulated each season, thereby demonstrating the importance of pre-storm conditions.

Antecedent rainfall is important for establishing soil-moisture conditions conducive to rapid infiltration and build-up of high pore-water pressures during subsequent major storms. Prior to the development of positive pore-water pressures the infiltration of rainfall reduces intergranular capillary tension (alternately referred to as soil suction or negative pore-water pressure) in unsaturated or partly saturated soils. The reduction of capillary tension and the increase of positive pore-water pressure reduces soil strength and has been linked with triggering debris flows in many parts of the world (Wieczorek, 1987, p. 95).

Figure 13. Debris flow in Fishlake National Forest, Utah. This distal part of the flow shows a lobate character with imbrication of rock units. Photograph courtesy of Earl Olson.

Three different landform types result from rainstorm activity—deep slumps, shallow slides, and shallow slump/slides. The frequency and distribution of these categories of ground failure depend on the intensity and duration of individual storms. Deep slumps are related to long-duration, moderate-intensity storms. Shallow slides occur from short-duration, high-intensity storms, and shallow slump/slides from an intermediate position between storm extremes.

Nilsen and Turner (1975) studied the incidence of slope movements throughout Contra Costa County, adjacent to San Francisco, during the 1950-1971 period. They showed a strong relation between storms, wet and dry years, and new landslide activity. Furthermore, they found that most landslides had reactivated from old landslide material. During the dry years of 1966-1967, only 38 landslides were reported, whereas there were 66 during the wetter years of 1968-1969. Most landslides occur from storms with more than 180 mm of rain in which the ground had some moisture from earlier precipitation. They also found a corollary that more rainfall was required to produce slope movement at the beginning of a rainy season than at the end. Their data support the same types of conclusions reached by Cleveland (1971) and Rice (1982) in southern California; there, however, because of less vegetation and thicker regolith, the threshold value for slope destabilization is nearly 300 mm of rainfall.

Earthflows

Earthflows are characterized by a type of movement where the mass is no longer liquefied when remolded due to an increase in sediment. Thus, pore-water pressures are not in excess of hydrostatic pressures, so the full weight of the flowing mass is supported in grain-to-grain collisions or contact. Earthflows are one of the most common of all slope movements, especially in hilly and mountainous regions. Their landform identifying characteristics include dish-shaped scars, bulging toes, tongue or teardrop shapes, and spoon-like rupture surfaces (Figs. 14 and 15). They may contain complex internal features resulting from plastic deformation of a Bingham-body with finite shear strength.

Earthflows commonly contain a significant amount of entrained water, in the range of 27 to about 60 percent by volume. Their mobilization can be accompanied by an increase in water content and softening of the material. They are, therefore, softer and more fluid than the material in earth slumps or earth block slides as defined by Varnes (1978). They are also stiffer and less fluid than material in debris flows.

Case study: The Davilla Hill landslide. Keefer and Johnson (1983) studied slope movements for 4 years in this earthflow complex in central California. The features are in colluvial and residual soils that cover northeast-trending ridges. The topography on the ridges is rounded and hummocky because of the pervasive slope degradation by the earthflows. These landslides occur on slopes that range from 8 to 45°. The bedrock consists of poorly consolidated clastic rocks with a clay matrix. Piezometers were installed and portable tensiometers were used to quantify the relations of groundwater and pore-water pressures. There was free water in all piezometers of the active earthflows. Water levels ranged from 1.28 m below ground to slightly artesian. During a 3-yr period, 65 to 68 tensiometric measurements showed that pore-water pressures 61 cm below ground were all negative, thus confirming that groundwater levels were deeper than 61 cm when earthflows were inactive.

Mobilization and rate of earthflow movement were correlated with precipitation:

When the earth flows mobilized, the total precipitation after February 1 was between 8.8 and 17.3 cm, and the total annual precipitation between 29.9 and 38.4 cm.
Once the earth flows mobilized, their velocities correlated with the amount of daily precipitation; that is, they moved rapidly during times when precipitation was high and relatively slowly during times when precipitation was low (Keefer and Johnson, 1983, p. 29).

The active earthflows were always wet throughout, with water at or near the surface in cracks. This was just the opposite of inactive earthflows that contained no water in surface materials or cracks. In addition, there were sharp boundaries beneath the active earthflows in which the soil within a few millimeters of the basal shear surface had an elevated water content. The average water content for soil in the moving parts was 38.4 percent, and 31.4 percent for inactive parts.

Figure 15. Small rotational slump in the Laramie Formation (Upper Cretaceous) triggered by heavy rains May 5–6, 1973, Jefferson County, Colorado. The oversteepened roadcut contributed to the likelihood of slope instability at this site. Photograph by Wallace R. Hansen.

The conclusion of this study was that earthflows are mobilized by increases in pore-water pressure. To explain the full range of motion requires the integration of principles of equilibrium mechanics, the Mohr-Coulomb soil failure criterion for soils, and Terzaghi's theory of effective stress.

Case study: the Minor Creek landslide. The Minor Creek landslide of northwestern California is a compound, complex, earthflow-like landslide that covers about 10 ha. The feature is more than 50 years old, with a hummocky, stepped surface that slopes about 15°. The earthflow was studied for a 12-yr period, 1973 to 1985, using a variety of methods in attempting to understand the influence of rainfall recharge and spatially variable groundwater flow on slope movement (Iverson, 1986; Iverson and Major, 1987).

In the summer of 1982, 60 piezometers were installed, and 4 additional ones were emplaced in 1983. The wells were 1 to 17 m deep, and water levels were measured weekly after installation. Hydraulic heads were higher in shallow wells, showing there was a downward component of the hydraulic gradient and that deep portions of the groundwater flow field were continuously recharged by the surficial percolation. During rainy seasons the landslide is nearly saturated because water-level depths in shallow wells become less than 1 m, with many only a few decimeters deep. Such saturation during the wet season implies that groundwater flow is driven predominantly by elevation contrasts. Landslide motion commonly accelerates between November and March, and then moves for several months at a pace 10 to 100 times faster than its slow summer creep rate of 1 to 4 mm/month. The rapid movements persist into May or June, at which time the decelerated summer rate commences.

Figure 14. Earthflow developed from a slump near Berkeley, California. Photograph taken by G. K. Gilbert, U.S. Geological Survey No. 3007.

Temporal and spatial variations in groundwater flow influenced the effective stress distribution and motion. High groundwater head at the landslide base corresponded with times of rapid motion. Such rapid motion occurs when a critical water level is exceeded. This coincides with seasonal hydrograph peaks. Differences in head can trigger rapid motion in the same manner as the wet season head, when the landslide is almost completely saturated. Furthermore, landslide motion is closely regulated by the direction and magnitude of near-surface hydraulic gradients and by waves of pore pressure caused by intermittent rainfall. Single rainstorms produce short-period waves that attenuate before reaching the landslide base. In contrast, seasonal rainfall cycles produce long-period waves that modify basal pore pressure, but only after time lags that can range from weeks to months. These lags are related to antecedent moisture storage. Slope movement is triggered when the pore-pressure waves reach the landslide base. A hummocky land surface enhances local instability by creating additional hummocks and overall slope reduction (Iverson and Major, 1987).

Mudslides

There is a confusing literature and nomenclature regarding the entrainment of "mud" in erosional processes. In this chapter, the terminology of Sharpe (1938) and Brunsden (1984) will be used. Sharpe (1938) classifies mudflows into three principal categories: semiarid, alpine, and volcanic (sometimes referred to as lahars). Important diagnostic features of mudflows include their high proportion of water, thus showing their kinship more to surface-water phenomena than to subsurface-water conditions. Mudflows, according to Sharpe (1938, p. 55), "usually follow former stream courses and, unlike earthflows, commonly recut in the same channel." The highly fluid character of alpine mudflows is shown by their paired-levee morphology that merges into the bulbous front of the distal margin. Some authors have inappropriately used such terms as "slushflows" and "slush avalanches" to describe these features in arctic and alpine environments. However, these terms should be restricted to features that develop from ice, snow, and frozen and thawed ground.

Brunsden (1984, p. 363) defines mudslides as:

. . . a form of mass movement in which masses of softened, argillaceous, silty or very fine sandy debris advance chiefly by sliding on discrete boundary shear surfaces in relatively slow moving, lobate or elongate forms. . . .Mudslides usually occur in saturated clays of all types, from soft intact clays to stiff fissured clays which have been progressively softened, weathered or broken by movement. Such clays generally occur in the medium plasticity class. . . .The materials may be lightly or heavily overconsolidated.

Slope failure by mudslide activity can be of four different types, according to Takada (1968):

Type A—the slide called 'fluid type'; such a manner of the movement in which the rate of the slide movement of the soil mass is fastest at the ground surface and yet no movement is made in the vicinity of the bed rock.

Type B—the slide called 'layer slide'; the commonest type of the slide movement of the soil mass.

Type C—the slide called 'protrusion of an intermediate layer'; such a manner of the slide movement in which only the soil mass of one layer between the ground surface and the bed rock begins to move, which may be considered as the cause of a depression or sinking zone within a landslide area.

Type D—the slide called 'multi-slide layers'; such a manner of movements in which two or more different layers make respective slide movements at different times. . . .at different rates of movement.

The source area of mudslides commonly consists of various types of failure scars with disturbed slumps, rotational slumps, or even translational features. Mudflow tracks may contain steep, straight, or gently concave surfaces and channels through which the material passes to its downslope resting site. The accumulation zone comprises single to overlapping lobes of debris that form at the foot of the feeder channels. The lobe profile often contains a flatter tread of 1 to 5° followed by a convex curved toe with a 10 to 25° inclination at the outermost face.

The literature on mudslides is consistent in asserting that subsurface water supplies the additional stress needed to produce slope movement. For example, Hutchinson (1970) and Hutchinson and Bhandari (1971) have shown that mobilization of mudslides is related to the onset of seasonally wetter conditions, as reflected by a groundwater surplus and a general rise in the phreatic zone. They also showed that movements were controlled by the discharge of debris from feeder tracks that resulted from the generation of excess pore pressure with a corresponding reduction in shear strength of materials.

Case study: The Denmark coast. Several coastal regions of Denmark contain a wide variety of slope-movement phenomena, including mudslides of different ages that can be either deep-seated rotational landslides or shallow failures in the clay-rich sediments of Eocene age (Prior, 1973, 1976, Prior and Eve, 1975). The mudslides generally occur on 6 to 30° slopes and are the product of the complex interaction of such factors as steep slopes, high relief, marine trimming, clay-rich sediment, and weather. However, the saturated condition of subsurface water acting on the montmorillonitic clays is cited as the dominant cause of movement. "Pore-water pressure distribution within slope materials is fundamental to a consideration of their stability" (Prior and Eve, 1975, p. 19).

Four different areas were included in these studies, at Rojle, Klint, Rosnaes, and Helgenaes. The mechanics of landsliding is similar at all localities, and landslides differ only in terms of age and amount of basal trimming caused by marine erosion. Total length of landslides exceeds 300 m, with headwall excavation up to 10 m in height. Below these scars are deeply incised elongate channels bounded by very pronounced marginal shear zones. Transverse cracks may extent across the debris in the channels. The high clay content produces absorption of water from precipitation and groundwater. Such absorption produces a remolded plastic to viscous mud with very low strength. Thus, large pore-water pressures are generated within the sediment mass. Even artesian pressures occur in association with some of the steep feeder slopes above the smaller angles of accumulation. These pressures are related to the mechanism of undrained loading, wherein the material discharged from the feeder slopes creates loading at the proximal side of the accumulation slopes.

The mudslide movement rates correlate with seasonal weather conditions. In winter and spring the landslide areas become very wet, with areas of standing water and seepage water from the toe areas. By contrast, during summer the sites are hard and dry, with a compact surface crust. The high water availability during winter seasons results from both increase in rainfall and decrease in evapotranspiration. Thus, the maximum pore-water pressures coincide with the spring thaw, whereas low pore-water pressures are present in summer.

Sensitive clay slides

In geotechnical terms, sensitivity is the ratio of the undisturbed compressive strength of an unconsolidated soil to the remolded compressive strength measured with a constant moisture content. Other terms have also been used and associated with sensitive clays. For example, "The term quick clay was initially applied to clays with sensitivity values greater than 16; however, quick clay has been used more widely as a group name for all sensitive clays" (Bentley and Smalley, 1984, p. 457). In his discussion of quick clay, Kerr (1963) lists four physical properties that enhance their ability to flow: (1) more than 50 percent of solid matter has particles smaller than 2 μm in diameter, (2) the fine particles are not coagulated but are loosely dispersed through the mass, (3) the water content is often higher than 50 percent by weight, and (4) salt content is comparatively low, less than 5 g/l.

Sensitive clays have remolded strength 25 percent or less of their undisturbed strength, and some have ratios as high as 1 to 100. The liquidity index is generally greater than 1.0, indicating that the moisture content exceeds the liquid limit. Furthermore, the plastic limit is much lower than normal when judged by grain-size distribution (Seed, 1964). Such conditions are a product of the open, flocculated structure of the sediment with the "house of cards" fabric. Once the strength has been exceeded, there is a volumetric decrease with concomitant decrease in pore-water pressure. This causes the "cards" (loosely bonded, platy fragments of silty clay) to separate and transform the brittle mass to one that behaves like a fluid. This conversion of earth material, from a grain-to-grain support system to one with fluid support, sets in motion a flow-sliding event.

Landslides that result from disturbance of sensitive clay are one member of a complete gradation of features that "involve not only liquefaction of the subjacent material but also retrogressive failure and liquefaction of the entire slide mass" (Varnes, 1978, p. 19). Such landsliding is most common in glaciomarine and glacioestuarine sediments of Pleistocene age in Canada, Scandinavia, Greenland, and New Zealand. In these materials, freshwater leaching of the clay removes or weakens the cationic bonds that were responsible for creating the flocculated structure during deposition in a saline environment. For example, the infamous Leda clay of the St. Lawrence region has produced numerous landslide events in Canada, and several have led to major disasters (Gadd, 1975). This type of failure is referred to in the literature by a variety of terms, including "earthflows," "lateral spreads," "quick-clay slides," and "retrogressive flowslides."

The mechanism of flow and the resulting landforms produced by sensitive clay landslides are distinctive. Failure usually starts in the lower part of the slope. After the initial slip, the landslide develops by successive retrogressive slices, extending laterally as it progresses. The somewhat concentric pattern continues headward until stability is attained. In this process, long, narrow, transverse blocks of the overlying less sensitive strata subside, are stretched, and become rafted into the fluid mud. Thus, there is a gravitational remolding of the material, which transforms the clay into a viscous slurry. The resulting crater is usually somewhat pear-shaped, bulging at the toe with the larger and lower stable base having a slight inclination downslope. As pointed out by Bentley and Smalley (1984), there is a range in size and geometry of such landslides: pre-failure slope heights of 10 to 75 m and slope inclination of 0 to 65°; depth to failure surface 10 to 50 m, height of backscar 5 to 40 m, total retrogression 30 m to 6 km, and width 20 m to 1.5 km.

Mollard (1977) ascribes the pattern of transverse ribs (a distinctive fingerprint) to a failure mechanism in which plastic flow and extrusion occur in a soft layer that underlies a substantially stiffer upper layer capped by a saturated sand. Extrusion of the underlying, remolded, soft clay very rapidly transports successive and intact slices of the stiff layer away from the retreating backscarp. The elongate, crescent-shaped slices move into the softened clay as it extrudes in an outward direction.

. . . present-day saturation and high pore-water pressures in marine silt and clay is a principal factor in the landslide mechanism, and the groundwater regime is very much affected by the presence or absence of a surficial sand cover, which results in low runoff and a high level or recharge on a year-round basis (Mollard, 1977, p. 50).

In the St. Lawrence region, Gadd (1975) has demonstrated there can be a time-dependent relation in the character of sensitive-clay landsliding. Here the early and prehistoric landslides were broad, shallow, and extensively retrogressive and formed with a high degree of liquefaction. Later stages of erosion reached lower sections of the clay sediments, which were more consolidated. Such material had drained over a long period of time and, therefore, was subjected to diagenesis that involved desiccation of the crust, chemical degradation, and some cementation. Thus, the more recent slides are narrower, deeper, and possess fewer liquefaction features. These differences are attributed to relative changes in position of the water table and the progressive nature of in situ weathering during the past 10,000 yr.

Although there is no universal agreement on the geomechanical causes for failure in sensitive clays, the majority of specialists cite changes in the card-house structure of the sedimentary particles. With the onslaught of fresh water that moves through the saline sediments, the salt concentration of the pore water becomes reduced and the electrokinetic potential increases on particle surfaces. The leaching increases the double-layer repulsion between particles, thereby reducing the plasticity index to a level where the natural moisture content exceeds the liquid limit. The increased parallelism of the particle arrangement coincides with salt removal and the low osmotic swelling pressure, which ultimately lead to destabilization of the sediment and produce the fluidity responsible for slurry movement.

Case study: St. Jean Vianney landslide, Canada. Several accounts have been written about this classic landslide in southern Quebec, Canada: LaSalle and Chagnon (1968), Tavenas and others (1971), LaRochelle (1974), and Bentley and Smalley (1984). St. Jean Vianney is 2.4 km north of the Saguenay River, a part of Canada that was covered by an extension of the Champlain Sea during the waning stages of Pleistocene glaciation. Glaciomarine sediments of 30 m thickness were deposited, consisting of clays with some silt and sand. This site contains the largest slide in sensitive clays in eastern Canada. Here a 20-km^2 area was affected by a two-stage landslide that occurred first about 500 B.P. and again about 400 B.P., and involved an estimated 270×10^6 m^3 of material. The village of St. Jean Vianney was later situated within the ancient crater of this landform.

A steady thaw of the soil began April 21, 1971, and rains of April 21–24 produced 18.1 mm. On April 24, a small landslide occurred on the west bank of the Petit Bras River, which forms the northeast boundary of the site. Its dimensions were 60 m wide and 150 m long. However, renewed rains of May 3–4, 1971, produced 18.5 mm of water, which triggered the massive failure on May 4. In a 15-min period the slide retrogressed 150 m from a position near the April 24 scar. This new event produced a damming action on the river because the dislocated material contained trees and vegetation, which gave temporary stability to the dam. Such blockage for a time prevented the outflow of the inner liquefied sediment until a sufficient pressure head developed to overcome this initial resistance. Thereafter the dam was breached, and the remolded material flowed down the river with an 18-m-high wave front that traveled 26 km/hr. The wave front had sufficient energy to destroy a 42-m-long concrete-reinforced bridge and central pier that was 2.9 km downstream.

The flowing mass of liquefied clay contained a 5.4×10^6 m^3 volume that discharged into the Saguenay River and carried 34 houses, a bus, and several cars. The total retrogression distance was 550 m, and affected a 268,000-m^2 area to an average depth of 22.5 m. The final movements exposed a large mass of saturated sand in the headwall area. Perched water tables were present in sand pockets throughout the mass. The May 4 slide involved clay that had been previously remolded by the ancient landslides of 400 to 500 B.P. The absence of retrogression of the April 24 slide suggests that groundwater conditions had been normal and had not produced widespread failure conditions. Thus, a new surcharge of additional water on May 3–4 was required, plus the more pervasive thawing of the sediments, before massive landsliding could occur. The following averages provide the sensitive-clay characteristics at St. Jean Vianney: moisture content, 45 percent; liquid limit, 30 percent; plastic limit, 22 percent; liquidity index, 3; clay fraction, 55 percent; and silt fraction, 40 percent.

Other related slope movements

Several different landslide types occur on the islands of St. Lucia and Barbados; these include translational slab slides, and rotational and slump earthflows (Prior and Ho, 1972). The translational types are largely developed in nonswelling kaolinitic soils, whereas the rotational types occur in swelling montmorillonitic soils that also include some kaolinite, illite, and chlorite. The high concentration of divalent calcium and magnesium in sliding-surface layers has the effect of raising the liquid limit and plasticity of the kaolinite clays. The total effect of soil leaching is to concentrate clay and exchangeable ions within a discrete horizon at depth, thereby creating a discontinuity subject to decoupling and slope movement. On Barbados, landslides are concentrated on the northeastern part of the island. Catastrophic slides and flows occur on the steep slopes from buildup of groundwater pressure due to seepage through the overlying coral cap during periods of heavy rains. In 1966, a 2-day storm produced 406 mm of rain that caused massive slope movements. The association of perched water tables with the high pore-water pressure above the plastic soil horizons reduced soil strength to its failure threshold thereby creating its instability.

On hillslopes of the Cascade Mountains, Oregon, creep, slump, earthflows, debris avalanches, and debris torrents are the principal mechanisms that transport earth materials to streams (Swanston and Swanson, 1976; Swanson and Swanston, 1977). Creep motion is by quasi-viscous flow in deformation at grain boundaries within clay mineral structures. Both interstitial and absorbed water contribute to motion by opening structures within and between mineral boundaries, thereby reducing friction of the soil mass. This action permits a remolding of the clay fraction, transforming it into a slurry, which lubricates the remaining soil. Such creep is the most persistent of all slope processes and in many places operates over more than 90 percent of the landscape.

The influence of moisture on the Lookout Creek earthflow is apparent from readings at the site from August 1975 to June 1976. Early during the wet season, movement rates were low even with precipitation of 270 mm, because of very low antecedent moisture. However, during spring snowmelt, late in the wet season, movement rates several times faster became prevalent.

Debris avalanches and debris torrents form a type of continuum of hillslope degradation processes in the Pacific Northwest (Fig. 16). Some torrents can be triggered by debris avalanches smaller than 100 m^3, but ultimately can involve more than 10,000 m^3 of debris entrained along the track of the torrent (Swanston and Swanson, 1976, p. 213).

Figure 16. Debris avalanche/torrent landform in Oregon. Clearcutting of the hillslopes aided in the destabilization, and heavy rainfalls caused slope movements of the saturated regolith. 1975 photograph by Fred Swanson.

DOWNSLOPE MOVEMENT AND ENVIRONMENTAL GEOLOGY

Society is affected by hillslope failure from both natural and man-induced causes. Natural slope movements, especially in the landslide motion range, present serious threats to humans in many different topographic settings. Landslides are considered one of the geologic hazards and yearly cause worldwide damage of billions of dollars and kill hundreds of people. In addition, poorly designed construction can adversely affect slope stability and set in motion a chain of events causing slope failures that ultimately lead to major catastrophes.

Highway construction

Excavation of hillsides and the resulting groundwater changes were briefly discussed in the section "Mechanics of slope failure in sediments" (see also Figs. 2 and 3). Seepage forces and water-table changes can induce slope movements in many terrains when the projects are improperly engineered; unusual climatic factors have also triggered landsliding (Fig. 1).

The Antrim coast of Ireland is the site for several different types of slope movement that include mudslides, earthflows, and rockfalls. Prior (1975) made a detailed study of one mudslide that occurred November 24, 1974. The slope material is a mixture of Liassic clay with till that contains basalt and chalk rubble. The plastic and liquid limits of the slide debris averaged 17 and 43 percent, respectively. The natural water content sampled 18 hr after the slide occurred was nearly 40 percent, thus approaching the liquid limit. Although the failure was triggered by heavy rainfall that totaled 247 mm on the 24th, it had been preceded by 89 mm on the 22nd and 150 mm on the 23rd. Road widening at the site occurred 18 months prior to the event, when the toe of the hillslope was trimmed. This lag time in slope failure was attributed to the circumstance that, shortly after excavation, substantial drainage on the slope occurred, thereby actually decreasing pore-water pressure and increasing slope stability. Later the pore-water pressure gradually increased due to water infiltration until the natural seepage and drainage conditions had been reestablished. At this time the shear strength of the soil was at a minimum. Thus, there is a tendency in such excavations to reverse from stable to unstable conditions. Studies showed that the 1973 excavation occurred at a time of minimum rainfall, and the steady natural seepage condition was not fully achieved by the following winter. Summer conditions in 1974 again led to reduced pore-water pressure, which then dramatically rose by October and reached the threshold state in November 1984. The storm of the 24th represented the final development of positive pore-water pressure causing hillslope failure.

Mathewson and Clary (1977) studied multiple landsliding along roadcuts of Interstate 45 near Centerville, Texas. There the materials consist of a fluvial deposit containing natural-levee clay and crevasse-splay sands. The backslopes are composed of overconsolidated, low- to high-plasticity clay and silt with poorly graded sand. Analysis of the sites showed that failure occurred with undrained (or possibly partially drained) conditions from the excavation and not under conditions of long-term steady-state seepage. Instead, the slope movements resulted from high pore-water pressure at a time when residual strength of the sediment had been greatly reduced. The presence of springs and ponded water before construction, and occurrence of "weeping" sand after construction, along with high water levels in drill holes, support the pore-water pressure explanation. "Continued movement of the slide masses . . . has been in response to periods of heavy rainfall which has increased the pore-water pressure in a material at or near its residual shear strength" (Mathewson and Clary, 1977, p. 220).

In 1964, construction of the Trans-Canada Highway in the Fraser Canyon revealed extensive fractures upslope of the highway cut at Hell's Gate Bluffs (Fig. 17). The rock slope failures at this site were studied by Piteau and others (1979) for the purpose

Figure 17. Oblique photograph of rockslide area in Hell's Gate Bluffs area, Fraser Canyon, British Columbia, Canada. Note the undercutting of bedrock hillsides adjacent to the upper road. Photograph courtesy of D. R. Piteau.

of providing information that could be used to increase slope stability by engineering techniques. The remedial work required the scaling and removal of talus material to 6 m depth over an 8,500-m^2 surface and up to 70 m above the highway. Overhangs were trimmed, and about 1,800 rock bolts and 121 grouted 70-tonne anchors were also installed. Monitoring showed that movement of the slide continued to take place above those locations where scaling and other remedial work terminated.

The failure mechanism involved the outward overturning of steeply dipping tabular fault blocks oriented parallel to the rock cut.

... there is a direct correlation between movement and precipitation. The greater the precipitation, the greater the movement, except during periods of freezing. It is interesting to note that instead of aggravating the problem, freezing and snowfall appear to retard movement. Although the average monthly precipitation peaks in January, the rate of movement starts to decrease after December (Piteau and others, 1979, p. 551).

In order to quantify these relations, the authors explained their justification for them:

The cumulative sums technique was adopted to determine the amount of time, or time lag, that exists between slope movement and events. ....Experience has shown that there is a very close empirical relationship between the natural fluctuations of piezometric surfaces and the curve of the cumulative sums of precipitation. This relationship seems to apply to water surface fluctuations in bedrock as well as to water-bearing zones in unconsolidated material" (Piteau and others, 1979, p. 551–552).

To test the conclusions reached by the cumulative sums technique a mathematical simulation model of climate and slope movement was applied. The basic assumptions used as input to this model were in terms of flash floods over the surface area of slope materials, absorption characteristics of the rock faces, drainage through the rock mass, time lapse between precipitation and subsequent movement, the principle of accumulation (the result of interactions between precipitation, temperature, absorption, and drainage), and the relation of movement to moisture accumulation. The conclusion reached by this methodology was that the simulation analyses confirmed the relations established by empirical and cumulative sums techniques. Thus, a deterministic mathematical correlation does exist between slope-face movement of the rock materials and weather conditions.

Reservoirs and slope failure

When the water level in a reservoir rises, there is a corresponding adjustment in the groundwater table to the level of the impoundment. Water in the adjoining earth materials will develop pore-water pressures that about equal the hydrostatic pressure in the reservoir pool (Wahlstrom, 1974). The resultant pore pressures are then greater than before reservoir rise. The reduction in the effective shear strength of the material can induce slope movement. Repetitive changes in water level, with both rise and drawdown, frequently create slope failures in reservoirs. The magnitude of the effects is related to the amount of change and its rate, and to the permeability of the adjacent earth materials (Terzaghi, 1950; Erskine, 1973; Hopkins and others, 1975). Lawson (1985) provided a review of bank failures and discussed a range of causes that can lead to their destabilization.

Landslides in the Columbia River valley have been studied by Jones and others (1961) and Palmer (1977). Franklin D. Roosevelt Lake, the reservoir of Grand Coulee Dam, was shown to be especially responsible for new landslides created as a result of its water levels. Jones and others (1961) described 10 different types of landslides in their sample of 321, and most were related to the human factor: (1) recent slump earthflows, (2) recent slump earthflows limited by bedrock, (3) ancient slump earthflows, (4) slip-off slopes, (5) multiple alcoves, (6) landslides off bedrock, (7) talus slumps, (8) landslides in artificial slopes, (9) mudflows, and (1) dry earthflows. Seventy percent of the landslides were slumps in categories 1 and 2. They quantitied many of the geomorphic relations, such as the ratio of horizontal and vertical distances from the foot to the crown, and determined that groundwater was statistically significant in landslide formation:

The principal cause of the landslides in the area was the weakening of sediments by ground water. The sediments generally have a lower shear strength when saturated or partly saturated than when dry. During the filling of Franklin D. Roosevelt Lake the sediments bordering the lake became saturated and many landslides occurred. Apparently, buildup of ground water in certain areas downstream from Grand Coulee Dam has been the principal cause of landsliding there (Jones and others, 1961, p. 31).

The landslides along the lake were most frequent during the filling stage of the reservoir, whereas those downstream from the dam were most frequent following sharp reduction in riverflow.

On October 9, 1963, a landslide mass of 260 million m^3 catastrophically plunged into the reservoir behind Vaiont Dam in Italy. It created giant waves more than 100 m high that overtopped the dam, and the wall of water was still 70 m deep 1 km below the dam (Kiersch, 1965). More than 2,000 people lost their lives in this disaster, although the dam was not damaged. Most researchers now attribute the landslide to changes in reservoir water level, altered bank storage conditions, and subsurface water under the mobilized material. Other factors also may have facilitated the slope movement. The geologic structure is favorable, with sedimentary strata dipping valleyward. The rocks contain abundant arcuate jointing fractures that also have inclinations toward the reservoir. Heavy rains in August and September 1963 produced increased percolation rates and helped raise groundwater levels. Such recharge added extra weight to the rock mass as well as contributing to a reduction in shear strength of materials. Identification has also been made of montmorillonite clay-rich horizons along the slip zone (Pariseau and Voight, 1979).

The 2-km-broad and 1-km-long slide moved at 60 km/hr

with a motion described as "quasi-plastic flow" by Muller (1964, p. 210). A review of benchmark surveys showed there previously had been extensive creep amounting to 2 to 4 m. The surficial piezometric measurements had also shown a water table in equilibrium with the average reservoir water level. However, there was a correspondence between creep-rate motion and reservoir stage: velocity typically increased with rising stage and diminished with drawdown (Pariseau and Voight, 1979, p. 35 and 39). Thus, a rising stage raises the water table, producing a buoyant effect that decreases frictional resistance. The increased rainfall would contribute to an additional water-table rise, thereby increasing joint pressures with possible artesian magnitude. These circumstances have provided most researchers with the ruling theory for motion. However, alternate theories have been proposed, also based on reactions of subsurface water as potential mechanisms for such quick and massive slope movements.

Habib (1967, 1976) and Goguel (1978) have introduced the idea of "vaporization of pore fluid" as an appropriate driving force in the mobilization and rapid movement of large landslides.

In this case [Vaiont], we have an example of a mass which had been stable, undergoing only slow creep movements, suddenly gliding freely along (predominantly) a bedding plane, and reaching very high velocity. This series of events implied that almost all frictional resistance had somehow disappeared. How can this drastic reduction in frictional resistance be explained?

With the pressure of several hundred metres of rock, the dissipation of energy by friction must produce a tremendous amount of heat at the base of the moving masses. For the Vaiont case, for example, simple calculations suggest that the temperature must have been high enough for the water to be vaporized, even under high pressure. If this vapor can be confined along the gliding plane, its pressure may be large enough to support the major part of the weight of the moving mass; thus friction between solids is drastically reduced, to a level just sufficient to compensate for losses of heat and the escape of steam (Goguel, 1978, p. 697–698).

Although there are competing ideas on the actual mechanism involved in the slope movement, most authors invoke the presence of subsurface water in some form and would probably agree with Pariseau and Voight (1979, p. 35) when they say "although the last word on the subject has not been written, it seems clear that the rising reservoir condition was a major factor in the ensuing catastrophe."

Many other case histories could be cited where the data prove the impact of reservoir water levels on creation of slope failures. Some examples are provided by Pariseau and Voight (1979):

1. The Santa Rosa spillway rockslide in Mexico, where slides of millions of cubic meters were involved.

2. Bighorn Reservoir was impounded by the Yellowtail Dam, Montana. The filling created the first slide of 40 million m^3, and subsequent slides occurred with the re-filling period each spring. The slides commonly activated portions of older slide masses.

3. Filling of the reservoir behind the Gepatsch Dam, Austria, in August 1964 initiated slope failures of 20 million m^3 of material that moved as rock slabs mixed with morainic material over a length of about 1 km. It reached motion rates of 8 cm/day a month later. When the reservoir was emptied and then refilled, motion again commenced when water levels rose to the previous height; speeds of 34 cm/day were reached in a vertical component direction.

CONTROL AND PREVENTION OF SLOPE MOVEMENTS

Geological engineering decisions for the remedy of slope-movement problems require an accurate database to determine the most effective methods for hillside stabilization. The necessary information includes knowledge of the type of slope movement, magnitude and frequency of movement, characteristics of the regolith and bedrock, topographic and hydrologic setting, and the nature of human activity at the site. Costs and legal aspects also need to be considered before any engineering attempt is made to solve the problem. In many instances, mathematical relations must be calculated to determine whether an approved safety factor, 1.0, can be reached when the resistance forces are compared with the movement forces.

Although water-control methods are the principal subject of this section, other downslope-movement management measures include avoidance methods, excavation methods, and restraining structures methods. Water plays such an important role in many slope movements that techniques for its control and removal are the most universally applied methods to prevent landslides and stabilize hillslopes. As in all control systems, the basic principles are to decrease stress within the system and increase the shear resistance of the earth materials. Coates has summarized these strategies in the following way:

1. Surface water. Surficial water is prevented from entering the landslide area, or is removed from the ground, by techniques such as the following: (a) water from streams, seeps and springs, and sheet wash is diverted by pipes or lined open trenches, (b) water that collects in depressions is drained away by lined surface trenches. (c) Cracks and other openings are filled with grouting or sealant to prevent water penetration into the regolith or bedrock. Sealants include such impermeable materials as clay, concrete, or bitumens. (d) Slopes are regraded to allow more uniform drainage into ditches and water disposal systems. At times this may include paving the area to allow more rapid runoff and prevent percolation, as was done in the Ventura Avenue oil field of California.

2. Subsurface water. These techniques are used to dewater earth materials and to lower subsurface water levels when appropriate. (a) Galleries and tunnels can be effective in some situations where size and cost are not important factors. They are large laterals that can be subsequently filled with permeable materials to aid in diversion of waters. Some contain infiltration offshoots into more troublesome areas. (b) Horizontal drains and tiles are installed near the surface and are especially effective in soils that are granular, uncemented, and very permeable. (c) Interceptor trenches are usually less than 2 m deep, but depths of 6 m are used under certain conditions. Their use is becoming more rare because of the effectiveness of combinations of other methods, but in special circumstances it may be necessary to use this technique to reach

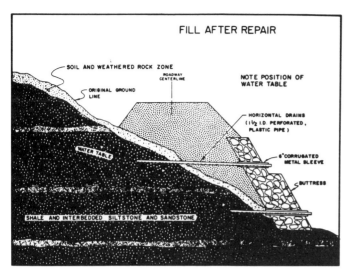

Figure 18. Typical slope stabilization methods used in highway construction projects in Tennessee. Note the placement of horizontal drain pipes to the water table (from Royster, 1977).

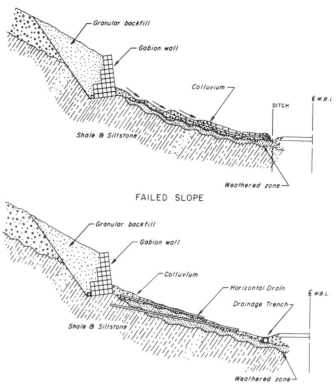

Figure 19. Schematic views of highway engineering project along Interstate 40 near Rockwood, Tennessee. Note that horizontal-type drains needed to be installed to help dewater earth materials and prevent further slope movement (from Royster, 1977).

deeper waters that have bypassed surficial or other catchment devices. (d) Vertical drains, holes, and wells take the form of various types of borings that become sumps or collector systems for water removal, commonly below the water table. Water can be transported away by pumping, continuous syphoning, or by subsurface pipes or movement through permeable strata when conditions permit (Coates, 1977b, p. 24).

Case study: Tennessee highways. (This section is from materials kindly supplied by David L. Royster, 1977.)

Water in all its forms . . . is the single most troublesome and perplexing substance that must be dealt with by transportation engineers. Of all these occurrences, subsurface water is probably the most perplexing because it is the least predictable, especially as it relates to the stability of cuts and embankments in geologically complex areas.

Subsurface water may act in any number of ways to reduce stability of cuts and embankments. These include subsurface erosion, lateral pressure in fractures and joints, decrease in cohesion, reduction in moisture tension, viscous drag due to seepage flow, and excess pore pressure (Royster, 1977, p. 1).

Although horizontal drains have been used in stabilization of landslides since about 1939 when California introduced their "Hydrauger" (Cedergren, 1967), the method did not begin to receive popular acceptance for use on a large scale by highway engineers in the eastern states until many years later. Tennessee first used this method and equipment in 1972 when a series of embankment failures occurred along Interstate 75 in Campbell County. Drains on this project totalled 18,288 m and were used in conjunction with rock buttresses (see Fig. 19). Some drains extend up to 183 m in length and initially produced flow up to 0.44 l/sec (7 gpm). During the 5-yr period, 1972–1977, a total of 46,600 m of horizontal drainage pipes were installed.

. . . the most complicated and frustrating landslide problems in the history of highway construction in the State of Tennessee have occurred along a 6.4 km (4-mile) section of Interstate 40 near Rockwood. . . .More than 30 slides had to be corrected before all four lanes could finally be opened to traffic in the late summer of 1974 (Royster, 1977, p. 28).

Bedrock includes limestone, shale, siltstone, sandstone, and conglomerates that contain folds, joints, and weathering. In places these lithologies are overlain by colluvium up to 15 m thick. Remedial measures included partial to total removal; minor grade and alignment changes; various restraint structures such as rock buttresses, gabion walls, and soil berms; and various dewatering systems such as French drains, vertical wells equipped with automatically actuating pumps, and horizontal drains (Figs. 18 and 19). The horizontal drains were installed principally to intercept subsurface water moving along the interface of the colluvium and bedrock surface, or to divert the water before reaching the interface. A total of 15,240 m of horizontal drains were installed, and although some drains proved to be unsuccessful, others had continuing flows that ranged from 0.001 to 0.13 l/sec (0.02 to 2 gpm). Four years after the drains had been installed, the system was apparently successful because there was no slope movement during this time.

CONCLUSIONS

The theme of this chapter is that subsurface waters are instrumental in producing a wide range of different types of downslope movement. The resulting downhill motion may be slow, as in creep, or fast, as in landsliding. Such displacements and failure may affect only selected portions of the landscape; elsewhere the movements may occur on the majority of hillsides. It has been shown that subsurface water is the chief reason for failure in some of the types of movement. These include solifluction, earthflows, debris flows, debris avalanches, sensitive-clay slides, and mudslides. However, water must be a significant component in many other types of downslope movement, such as slab slides, and mudflows.

Subsurface water has many facets, and all may contribute in some fashion to cause disruption of earth materials in the preparation process for downslope mobilization: water of hydration in mineral species, soil moisture, capillary water, perched water, vadose-zone water, and groundwater. Its movement also has different forms, such as throughflow, interflow, percolation, and groundwater flow in confined, semi-confined, and unconfined systems. Positive pore-water pressure is especially significant as a destabilizing force in soil and sediments, and cleft-water pressure provides a similar stress in fractured bedrock.

The amount of antecedent moisture in the substrate determines whether the amount, intensity, and duration of rainfall will initiate slope failure. There is universal agreement among specialists that subsurface water is the major cause for slope movement. However, other factors such as topography and type of material can also be important. Concave slopes and colluvium-filled hollows are favorate locations for debris flows. There is also a higher incidence of downslope movements on moister northeast-facing hillsides.

Because some slope movements, such as landsliding, cause property damage and loss of life, these hazards need careful study and, when necessary, must be environmentally managed. These investigations require the expertise of a wide range of earth science and engineering experts with an appreciation for the role that subsurface water plays in downslope movement.

ACKNOWLEDGMENTS

This chapter has greatly benefited from comments by David Varnes, Charles Higgins, and an anonymous geoscientist. All of their corrections and the great majority of their excellent suggestions for improvement have been incorporated into this final version of the manuscript. I especially appreciate the cooperation of Dr. Varnes, who waded through a second draft and provided additional helpful ideas. However, I shoulder the responsibility for any residual misunderstandings or inaccuracies.

REFERENCES CITED

Abrahams, A. D., ed., 1986, Hillslope processes: Boston, Massachusetts, Allen and Unwin, 416 p.

Anderson, M. G., and Richards, K. S., eds., 1987, Slope stability; Geotechnical engineering and geomorphology: New York, John Wiley and Sons, 656 p.

Andersson, J. G., 1906, Solifluction, a component of sub-aerial denudation: Journal of Geology, v. 14, p. 91–112.

Bates, R. L., and Jackson, J. A., eds., 1980, Glossary of geology and related sciences, 2nd ed.: American Geological Institute, 751 p.

Beaty, C. B., 1956, Landslides and slope exposure (California): Journal of Geology, v. 64, no. 1, p. 70–74.

Bentley, S. P., and Smalley, I. J., 1984, Landslips in sensitive clays, in Brunsden, D., and Prior, D. B., eds., Slope instability: New York, John Wiley and Sons, p. 457–490.

Bjerrum, L., and Jorstad, F., 1957, Rockfalls in Norway: Norway Geotechnical Institute Internal Report F-230.

—— , 1963, Correspondence on: Stability of steep slopes on hard unweathered rock, by K. Terzaghi: Geotechnique, v. 13, p. 171–173.

Bogucki, D. J., 1976, Debris slides in the Mt. LeConte area, Great Smoky Mountains National Park, U.S.A.: Geografiska Annaler, v. 58A, no. 3, p. 179–191.

—— , 1977, Debris slide hazards in the Adirondack province of New York State: Environmental Geology, v. 1, no. 6, p. 317–328.

Bradford, J. M., and Piest, R. F., 1980, Erosional development of valley-bottom gullies in the upper midwestern United States, in Coates, D. R., and Vitek, J. D., eds., Thresholds in geomorphology: London, George Allen and Unwin, p. 75–101.

Bromhead, E. N., 1986, The stability of slopes: New York, Chapman and Hall, 384 p.

—— , 1987, Groundwater and landslides; Principles and practice; Papers presented at the First Sino-British Geological Conference on Geotechnical Engineering and Hazard Assessment in Neotectonic Terrains, Taipei, Taiwan: Geological Society of China Memoir 9, p. 147–157.

Brunsden, D., 1984, Mudslides, in Brunsden, D., and Prior, D. B., eds., Slope instability: New York, John Wiley and Sons, p. 363–418.

Brunsden, D., and Prior, D. B., eds., 1984, Slope instability: New York, John Wiley and Sons, 620 p.

Campbell, R. H., 1975, Soil slips, debris flows, and rainstorms in the Santa Monica Mountains and vicinity, southern California: U.S. Geological Survey Professional Paper 851, 51 p.

Carson, M. A., and Kirkby, M. J., 1972, Hillslope form and process: Cambridge, Cambridge University Press, 475 p.

Cedergren, H. R., 1967, Seepage, drainage, and flow nets: New York, John Wiley and Sons, 489 p.

Cleveland, G. B., 1971, Regional landslide prediction: California Division of Mines and Geology Open-File Report, 33 p.

Coates, D. R., ed., 1977a, Landslides: Geological Society of America Reviews in Engineering Geology, v. 3, 278 p.

—— , 1977b, Landslide perspectives, in Coates, D. R., ed., Landslides: Geological Society of America Reviews in Engineering Geology, v. 3, p. 3–29.

Coates, D. R., and King, C.A.M., 1973, Glacial geology of Great Bend and adjacent region, in Coates, D. R., ed., Glacial geology of the Binghamton-western Catskill region: Binghamton, State University of New York Publication in Geomorphology Contribution 3, p. 3–30.

Coates, D. R., and Vitek, J. D., eds., 1980, Thresholds in geomorphology: London, George Allen and Unwin, 498 p.

Colton, R. B., and Holligan, J. A., 1977, Photo interpretative map showing areas underlain by landslide deposits and areas susceptible to landsliding in the Louisville Quadrangle, Boulder and Jefferson Counties, Colorado: U.S. Geological Survey Miscellaneous Field Studies Map MF-871, scale 1:24,000.

Costa, J. E., and Wieczorek, G. F., eds., 1987, Debris flows/avalanches; Process, recognition, and mitigation: Geological Society of America Reviews in Engineering Geology, v. 7, 239 p.

Crozier, M. J., 1982, A technique for predicting the probability of mudflow and rapid landslide occurrence, *in* Landslides and mudflows; Reports of Alma-Ata International Seminar, October 1981: UNESCO/UNEP, Centre of International Projects, GKNT, Moscow, p. 420–430.

——, 1986, Landslides; Causes, consequences, and environment: London, Croom Helm, 252 p.

Crozier, M. J., and Eyles, R. J., 1980, Assessing the probability of rapid mass movement; Third Australian–New Zealand Conference on Geomechanics, Wellington, 1980, Vol. 2: New Zealand Institution of Engineers, Proceedings of Technical Groups 6 (1G), p. 2.47–2.53.

Dana, J. D., 1863, Manual of geology: Philadelphia, Pennsylvania, Theodore Bliss and Co., 798 p.

Dietrich, W. E., Reneau, S. L., and Wilson, C. J., 1984, Importance of colluvium-filled bedrock hollows to debris flow studies: Geological Society of America Abstracts with Programs, v. 16, p. 488–489.

Dietrich, W. E., Wilson, C. J., and Reneau, S. L., 1986, Hollows, colluvium, and landslides in soil-mantled landscapes, *in* Abrahams, A. D., ed., Hillslope processes: Boston, Massachusetts, Allen and Unwin, p. 361–388.

Eckel, E. B., ed., 1958, Landslides and engineering practice: National Research Council Highway Research Board Special Report 29, 232 p.

Ellen, S. D., and Fleming, R. W., 1987, Mobilization of debris flows from soil slips, San Francisco Bay region, California, *in* Costa, J. E., and Wieczorek, G. F., eds., Debris flows/avalanches; Process, recognition, and mitigation: Geological Society of America Reviews in Engineering Geology, v. 7, p. 31–40.

Erskine, C. F., 1973, Landslides in the vicinity of the Fort Randall reservoir: U.S. Geological Survey Professional Paper 675, 64 p.

Eyles, R. J., Crozier, M. J., and Wheeler, R. H., 1978, Landslips in Wellington City: New Zealand Geographer, v. 34, no. 2, p. 58–74.

Gadd, N. R., 1975, Geology of Leda Clay, *in* Yatsu, E., Ward, A. J., and Adams, F., eds, Mass wasting: East Anglia, England, Geo Abstracts Ltd., p. 137–151.

Goguel, J., 1978, Scale-dependent rockslide mechanisms, with emphasis on the role of pore fluid vaporization, *in* Voight, B., ed., Rockslides and avalanches; 1, Natural phenomena: Amsterdam, Elsevier Scientific Publishing Company, p. 693–705.

Gray, R. E., and Gardner, G. D., 1977, Processes of colluvial slope development at McMechen, West Virginia: International Association Engineering Geology Bulletin 16, p. 29–32.

Habib, P., 1967, Sur un mode de glissement des massifs rocheux: Paris, Comptes-Rendus des seances de l'Academie des Sciences, v. 264, p. 151–153.

——, 1976, Production of gaseous pore pressure during rock slides: Rock Mechanics, v. 7, p. 193–197.

Hack, J. T., and Goodlet, J. C., 1960, Geomorphology and forest ecology of a mountain region in the central Appalachians: U.S. Geological Survey Professional Paper 347, 66 p.

Hansen, A., 1984, Landslide hazard analysis, *in* Brunsden, D., and Prior, D. B., eds., Slope instability: New York, John Wiley and Sons, p. 523–602.

Harden, C. P., 1976, Landslides near Aspen, Colorado: University of Colorado Institute of Arctic and Alpine Research Occasional Paper 20, 60 p.

Hopkins, T. C., Allen, D. L., and Deen, R. C., 1975, Effects of water on slope stability: Kentucky Bureau of Highways Division of Research Research Report 435, 41 p.

Howard, A. D., and McLane, C. F., 1988, Erosion of cohesionless sediment by groundwater seepage: Water Resources Research, v. 24, no. 10, p. 1659–1674.

Howard, A. D., Kochel, R. C., and Holt, H. E., eds., 1988, Sapping features of the Colorado Plateau: National Aeronautics and Space Administration, U.S. Government Printing Office, 108 p.

Hutchinson, J. N., 1968, Mass movement, *in* Fairbridge, R. W., ed., Encyclopedia of geomorphology: New York, Reinhold Book Corp., p. 688–696.

——, 1970, A coastal mudflow on the London Clay cliffs at Beltinge, North Kent: Geotechnique, v. 20, p. 412–438.

Hutchinson, J. N., and Bhandari, R., 1971, Undrained loading; A fundamental mechanism of mudflows and other mass movements: Geotechnique, v. 21, p. 353–358.

Iverson, R. M., 1986, Dynamics of slow landslides; A theory for time-dependent behavior, *in* Abrahams, A. D., ed., Hillslope processes: Boston, Massachusetts, Allen and Unwin, p. 297–317.

Iverson, R. M., and Major, J. J., 1986, Groundwater seepage vectors and the potential for hillslope failure and debris flow mobilization: Water Resources Research, v. 22, p. 1543–1548.

——, 1987, Rainfall, ground-water flow, and seasonal movement at Minor Creek Landslide, northwestern California; Physical interpretation of empirical relations: Geological Society of America Bulletin, v. 99, p. 579–594.

Johnson, A. M., 1970, Physical processes in geology: San Francisco, California, Freeman, Cooper and Co., 577 p.

Johnson, A. M., with contributions by Rodine, J. D., 1984, Debris flow, *in* Brunsden, D., and Prior, D. B., eds., Slope instability: New York, John Wiley and Sons, p. 257–361.

Jones, F. O., Embody, D. R., and Peterson, W. L., 1961, Landslides along the Columbia River Valley northwestern Washington: U.S. Geological Survey Professional Paper 367, 98 p.

Keefer, D. K., and Johnson, A. M., 1983, Earth flows; Morphology, mobilization, and movement: U.S. Geological Survey Professional Paper 1264, 56 p.

Kerr, P. F., 1963, Quick clay: Scientific American, v. 209, p. 132–142.

Kiersch, G. A., 1965, The Vaiont Reservoir disaster: California Division of Mines and Geology Mineral Information Service, v. 18, no. 7, p. 129–138.

King, C.A.M., and Coates, D. R., 1973, Glacio-periglacial landforms within the Susquehanna Great Bend area of New York and Pennsylvania: Quaternary Research, v. 3, p. 600–620.

Kohler, M. A., and Linsley, R. K., 1951, Predicting the runoff from storm rainfall: U.S. Weather Bureau Research Paper 34.

LaRochelle, P., 1974, Rapport de synthese des etudes de la Coulee d'argile de Saint-Jean Vianney: Gouvernement du Quebec, Ministere des Richesse Naturelles, v. 5–51, 75 p.

LaSalle, P., and Chagnon, J.-Y., 1968, An ancient landslide along the Saguenay River, Quebec: Canadian Journal of Earth Sciences, v. 5, p. 548–549.

Lawson, D. E., 1985, Erosion of northern reservoir shores: U.S. Army Corps of Engineers CRREL Monograph 85-1, 198 p.

Mathewson, C. C., and Clary, J. H., 1977, Engineering geology of multiple landsliding along I-45 road cut near Centerville, Texas, *in* Coates, D. R., ed., Landslides: Geological Society of America Reviews in Engineering Geology, v. 3, p. 213–223.

Mollard, J. D., 1977, Regional landslide types in Canada, *in* Coates, D. R., ed., Landslides: Geological Society of America Reviews in Engineering Geology, v. 3, p. 29–56.

Muller, L., 1964, The rock slide in the Vaiont Valley: Rock Mechanics and Engineering Geology, v. 2, p. 148–212.

Nilsen, T. H., and Turner, B. L., 1975, Influence of rainfall and ancient landslide deposits on recent landslides (1950–71) in urban areas of Contra Costa County, California: U.S. Geological Survey Bulletin 1388, 18 p.

Okimura, T., 1981, Analysis of mountainslope failures based on the simulated groundwater level in bore hole, *in* Proceedings Third Meeting of the IGU Commission on Field Experiments in Geomorphology, Kyoto, Japan: Japanese Geomorphological Union, p. 95–103.

Palmer, L., 1977, Large landslides of the Columbia River gorge, Oregon and Washington, *in* Coates, D. R., ed., Landslides: Geological Society of America Reviews in Engineering Geology, v. 3, p. 69–83.

Pariseau, W. G., and Voight, B., 1979, Rockslides and avalanches; Basic principles and perspectives in the realm of civil and mining operations, *in* Voight, B., ed., Rockslides and avalanches; 2, Engineering sites: Amsterdam, Elsevier Scientific Publishing Company, p. 1–92.

Parizek, R. R., 1971, Impact of highways on the hydrogeological environment, *in*

Coates, D. R., ed., Environmental geomorphology: Binghamton, State University of New York, p. 151-183.

Pierson, T. C., 1977, Factors controlling debris-flow initiation on forested hillslopes in the Oregon Coast Range [Ph.D. thesis]: Seattle, University of Washington, 166 p.

Pierson, T. C., and Costa, J. E., 1987, A rheologic classification of subaerial sediment-water flows, *in* Costa, J. E., and Wieczorek, G. F., eds., Debris flows/avalanches; Process, recognition, and mitigation: Geological Society of America Reviews in Engineering Geology, v. 7, p. 1-12.

Piteau, D. R., McLeod, B. C., Parkes, D. R., and Lou, J. K., 1979, Rock slope failure at Hell's Gate, British Columbia, Canada, *in* Voight, B., ed., Rockslides and avalanches; 2, Engineering sites: Amsterdam, Elsevier Publishing Company, p. 541-574.

Pomeroy, J. S., 1980, Storm-induced debris avalanching and related phenomena in the Johnstown area, Pennsylvania, with references to other studies in the Appalachians: U.S. Geological Survey Professional Paper 1191, 24 p.

—— , 1982a, Mass movement in two selected areas of western Washington County, Pennsylvania: U.S. Geological Survey Professional Paper 1170-B, 17 p.

—— , 1982b, Landslides in the greater Pittsburgh region, Pennsylvania: U.S. Geological Survey Professional Paper 1229, 48 p.

—— , 1984, Storm-induced slope movements at East Brady, northwestern Pennsylvania: U.S. Geological Survey Bulletin 1618, 16 p.

Prior, D. B., 1973, Coastal landslides and swelling clays at Rosnaes, Denmark: Geografisk Tidsskrift, v. 72, p. 45-58.

—— , 1975, A mudslide on the Antrim coast, 24th November 1974: Irish Geography, v. 8, p. 55-62.

—— , 1976, Coastal mudslide morphology and processes on Eocene clays in Denmark: Geografisk Tidsskrift, v. 76, p. 14-33.

Prior, D. B., and Eve, R. M., 1975, Coastal landslide morphology at Rosnaes, Denmark: Geografisk Tidsskrift, v. 74, p. 12-20.

Prior, D. B., and Ho, C., 1972, Coastal and mountain slope instability on the islands of St. Lucia and Barbados: Engineering Geology, v. 6, p. 1-18.

Rapp, A., 1960, Recent development of mountain slopes in Karkevagge and surroundings, northern Scandinavia: Geografiska Annaler, v. 42, no. 2-7, p. 65-200.

Reneau, S. L., and Dietrich, W. E., 1987, The importance of hollows in debris flows studies; Examples from Marin County, California, *in* Costa, J. E., and Wieczorek, G. F., eds., Debris flows/avalanches; Process, recognition, and mitigation: Geological Society of America Reviews in Engineering Geology, v. 7, p. 165-180.

Rice, R. M., 1982, Sedimentation in the chaparral; How do you handle unusual events?, *in* Swanson, F. J., Janda, R. J., Dunne, T., and Swanston, D. N., eds., Sediment budgets and routing in forested drainage basins: U.S. Department of Agriculture Forest Service General Technical Report PNW-141, p. 39-49.

Rodine, J. D., 1974, Analysis of the mobilization of debris flows [Ph.D. thesis]: Stanford, California, Stanford University, 226 p.

Royster, D. L., 1977, Landslide remedial measures; Presentation at the 37th Annual Southeastern Association of State Highway and Transportation Officials Convention, Nashville, Tennessee Publication Authority No. 1011, 62 p.

Savage, C. N., 1968, Mass wasting, *in* Fairbridge, R. W., ed., Encyclopedia of geomorphology: New York, Reinhold Book Corp., p. 697-700.

Schumm, S. A., and Mosley, M. P., eds., 1973, Slope morphology: Stroudsburg, Pennsylvania, Dowden, Hutchinson and Ross, 454 p.

Schuster, R. L., and Krizek, R. J., eds., 1978, Landslides analysis and control: Washington, D.C., National Academy of Sciences Special Report 176, 234 p.

Seed, H. B., Woodward, R. J., and Lundgren, R., 1964, Fundamental aspects of Atterberg limits: Journal of Soil Mechanics and Foundations Division, ASCE 90 (SM6) 4140, p. 75-105.

Sharpe, C.F.S., 1938, Landslides and related phenomena: New York, Cooper Square Pub., Inc., 137 p.

Sowers, G. F., and Royster, D. L., 1978, Field investigation, *in* Schuster, R. L., and Krizek, R. J., eds., Landslides analysis and control, Washington, D.C., National Academy of Sciences, p. 81-111.

Swanson, F. J., and Swanston, D. N., 1977, Complex mass-movement terrains in the western Cascade Range, Oregon, *in* Coates, D. R., ed., Landslides: Geological Society of America Reviews in Engineering Geology, v. 3, p. 113-124.

Swanston, D. N., and Swanson, F. J., 1976, Timber harvesting, mass erosion, and steep land forest geomorphology in the Pacific Northwest, *in* Coates, D. R., ed., Geomorphology and engineering: Stroudsburg, Pennsylvania, Dowden, Hutchinson and Ross, p. 199-221.

Takada, Y., 1968, A geophysical study of landslides (mechanisms of landslides): Disaster Prevention Research Institute Bulletin, v. 18, p. 59-77.

Tavenas, F. A., Chagnon, J. Y., and La Rochelle, P., 1971, The Saint-Jean Vianney landslide; Observations and eyewitnesses accounts: Canadian Geotechnical Journal, v. 8, p. 463-478.

Terzaghi, K., 1950, Mechanisms of landslides, *in* Paige, S., Chairman, Applications of geology in engineering practice: Geological Society of America Berkey Volume, p. 83-123.

—— , 1962, Stability of steep slopes on hard unweathered rock: Geotechnique, v. 12, p. 251-270.

Terzaghi, K., and Peck, R. B., 1948, Soil mechanics in engineering practice: New York, John Wiley and Sons, 566 p.

Van Buskirk, D. R., 1977, The relationship between aspect and slope conditions in northeastern Ohio: Geological Society of America Abstracts with Programs, v. 9, p. 327.

Varnes, D. J., 1978, Slope movement types and processes, *in* Schuster, R. L., and Krizek, R. J., eds., Landslides analysis and control: Washington, D.C., National Academy of Sciences Special Report 176, p. 12-33.

Voight, B., ed., 1978, Rockslides and avalanches; 1, Natural phenomena: Amsterdam, Elsevier Scientific Publishing Company, 833 p.

—— , 1979, Rockslides and avalanches; 2, Engineering sites: Amsterdam, Elsevier Scientific Publishing Company, 850 p.

Visser, W. A., ed., 1980, Geological nomenclature: The Hague, Royal Geological and Mining Society of the Netherlands, Martinus Nijhoff, 540 p.

Wahlstrom, E. E., 1974, Dams, dam foundations, and reservoir sites: Amsterdam, Elsevier Scientific Publishing Co., 278 p.

Washburn, A. L., 1969, Weathering, frost action, and patterned ground in the Mesters Vig district, northeast Greenland: Meddelelser om Groenland, v. 176, no. 4, 303 p.

Wieczorek, G. F., 1987, Effect of rainfall intensity and duration on debris flows in central Santa Cruz Mountains, California, *in* Costa, J. E., and Wieczorek, G. F., eds., Debris flows/avalanches; Process, recognition, and mitigation: Geological Society of America Reviews in Engineering Geology, v. 7, p. 93-104.

Williams, G. P., and Guy, H. P., 1971, Debris avalanches; A geomorphic hazard, *in* Coates, D. R., ed., Environmental geomorphology: Binghamton, State University of New York, p. 25-46.

—— , 1973, Erosional and depositional aspects of Hurricane Camille in Virginia, 1969: U.S. Geological Survey Professional Paper 804, 80 p.

Wolman, M. G., and Gerson, R., 1978, Relative scales of time and effectiveness of climate in watershed geomorphology, *in* Earth surface processes: New York, John Wiley and Sons, v. 3, p. 189-208.

Woodruff, J. F., 1971, Debris avalanches as an erosional agent in the Appalachian Mountains: Journal of Geography, v. 70, no. 7, p. 399-406.

Young, A., 1972, Slopes: Edinburgh, Oliver and Boyd, 288 p.

Zaruba, Q., and Mencl, V., 1969, Landslides and their control: Amsterdam, Elsevier, 205 p.

MANUSCRIPT ACCEPTED BY THE SOCIETY NOVEMBER 14, 1989

Chapter 4

Piping and pseudokarst in drylands

Garald G. Parker, Sr.
3414 Reynoldswood Drive, Tampa, Florida 33618 (U.S. Geological Survey retired.)
Charles G. Higgins
Department of Geology, University of California, Davis, California 95616
  with case studies by
Garald G. Parker, Sr. and
Warren W. Wood
U.S. Geological Survey, 431 National Center, Reston, Virginia 22092

INTRODUCTION

The surficial erosional processes that have sculptured the world's drylands are generally well known. For the most part, however, earth scientists have been concerned with processes operating on the land's surface, not beneath it. Thus, the effects of subsurface erosion generally have gone unnoticed and unappreciated. In certain parts of the Earth, especially in some arid and semiarid lands, subsurface erosion in essentially insoluble clastic rocks can be a major factor in the erosional process and thus in the shaping of landforms (Fig. 1).

One aspect of subsurface erosion is commonly called piping, a natural process by which percolating water produces tubular subsurface drainage channels in insoluble clastic rocks. In the introduction to *Badland geomorphology and piping,* Bryan and Yair (1982, p. 9) state that in some badlands, "subsurface flow provides the dominant denudational process." Piping appears in different physical forms and originates in different ways in different climatic, pedologic, and geologic settings. Thus, it is not strange that the process has engendered a variety of different names and ideas regarding its genesis, development, and role in erosion. The piping process and its results have been variously called: "seepage erosion" (Richthofen, 1877); "suffossion" (Pavlov, 1898); "natural tunneling" (Fuller, 1922); "sub-cutaneous erosion" (Guthrie-Smith, 1926); "gullies-by-sinking" (Rubey, 1928); "sink-hole erosion" (Thorp, 1936); "squirrel-hole erosion" (Sharpe, 1938); "tunnel erosion" (Bennett, 1939); "rodentless rodent erosion" (Bond, 1941); "pothole gullying" (Cole and others, 1943); "tunnel-gully erosion" (Gibbs, 1945); "sink-hole erosion" (Cockfield and Buckham, 1946); "tunneling erosion" (Downes, 1946); "soil piping" (Carroll, 1949); "collapse structure" (Hardy, 1950); "soil caves" (Funkhouser, 1951); "pothole erosion" (Kingsbury, 1952); and "piping" (Fletcher and others, 1948, 1954). The last term has since come into general use, and is adopted here (see "Nomenclature" below).

Before the 1940s, few geologists had written on piping, although some of its effects have long been known (e.g., Cussen, 1888; Haworth, 1897; Johnson, 1901). The subject had attracted more attention from soil scientists and engineers, but even from these sources reports were few, and there was no comprehensive report on the subject until the studies of Parker (1963), Parker and Jenne (1967), Bell (1968), and Heede (1971) in the drylands of the United States; and of Gilman and Newson (1980) and J.A.A. Jones (1981) in the humid regions of Great Britain. Piping in humid lands is discussed in the next chapter (Jones, this volume); the present chapter is concerned chiefly with drylands piping. Some excellent modern studies of piping in arid and semiarid regions are reported in Bryan and Yair (1982); others are cited by Jones (1981, 1987).

Piping develops in several different ways. The different types are, however, only variants of that type of subsurface erosion long known to civil engineers as responsible for failures of earthen dams and embankments due to seepage through, around, or under the structures. In the "heave" type of failure, which is common in such materials as quicksand, an open tube or pipe generally does not persist. Instead, overlying materials cave into the evacuated area, and "subsidence gullies" or "gullies-by-sinking" develop. In cohesive materials such as clay and loess the pipes remain open—at least until their roofs cave in, thus producing sink-like openings to the land surface. Eventually, when the entire roof over a pipe caves in, a steep-walled gully may result.

In coherent clastic rocks such as shale, the geomorphic effects of piping may simulate solutional erosion in calcareous rocks. Where closely and intensively developed, the pipes, caves,

---

Parker, G. G., Sr., and Higgins, C. G., 1990, Piping and pseudokarst in drylands, with case studies by Parker, G. G., Sr., and Wood, W. W., in Higgins, C. G., and Coates, D. R., eds., Groundwater geomorphology; The role of subsurface water in Earth-surface processes and landforms: Boulder, Colorado, Geological Society of America Special Paper 252.

Figure 1. Wide valley with vertical walls, box gullies, and isolated towers, part of an extensive landscape formed by combined surface and subsurface erosional processes in Pleistocene loess in north-central China, north of Tai-yuen-fu, Shansi Province. Engraving from Richthofen (1877, Fig. 13).

sinkholes, natural bridges, blind and dry hanging valleys, and subterranean drainage result in a karstlike topography (Fig. 2). The resulting landscape has been called "pseudokarst," a term that originated in Europe for "features of non-solutional origin which are analogous to those of areas of karstic geomorphology" (Halliday, 1960). According to Bates and Jackson (1987), the term was first used by von Knebel (1906).

Whereas karst features are essentially the product of the removal of solid soluble rocks on a molecule-by-molecule basis in solution in moving water, pseudokarst features are the product of the removal of solid clastic rocks on a grain-by-grain basis in suspension in moving water.

In some places, solution of soluble constituents in clastic rocks doubtless is involved in the development of pseudokarsts. However, solution is not at all essential to the piping process, which is basic to pseudokarst development. Whether the separation and removal of rock is accomplished in solution or in suspension is of no essential consequence. Either results in the development of landforms that are, to all appearances, of the same shape and general configuration. However, their duration as landforms is quite different. Karst forms change slowly and may maintain their general appearance and form over the centuries, whereas pseudokarst forms may develop or change radically as a result of one major storm event, as illustrated in the case study of Officer's Cave later in this chapter. The important point, geomorphologically, is that essentially insoluble clastic rocks can, under certain circumstances and conditions, undergo degradation that results in landforms largely duplicating those of carbonate rocks.

OCCURRENCE OF PIPING AND PSEUDOKARST; EARLY STUDIES

Piping is widespread over arid and semiarid areas of the Earth, and in some places has created serious erosion problems. Reports of its effects began with Richthofen's (1877) observations on landforms developed in the vast Pleistocene loess deposits in the drylands of north-central and western China. These cover an area of 777,000 km^2 with an average thickness of about 215 m. Sinkholes, caves, natural bridges, pipe-generated gullies, and subterranean drainage, mostly developing along high valley walls, are common there (Richthofen, 1877; Fuller, 1922; Thorp, 1936). Vertical columnar structure largely controls the location and development of these features in loess (Fig. 1). This, together with the loose, friable nature of the wind-deposited loess, allows rapid downward movement of rainfall or snowmelt. The downward seepage readily removes the loosely packed silt and clay particles, creating underground openings or channels, which commonly discharge at the base of the riverbank cliffs.

Fuller (1922) described "vertical pipes or wells" up to 3 m in diameter and as much as 9 m deep in loess. These led into "downward sloping tubes," 0.9 m wide and 3 m deep, associated with natural bridges and vertical-walled gullies, all developed

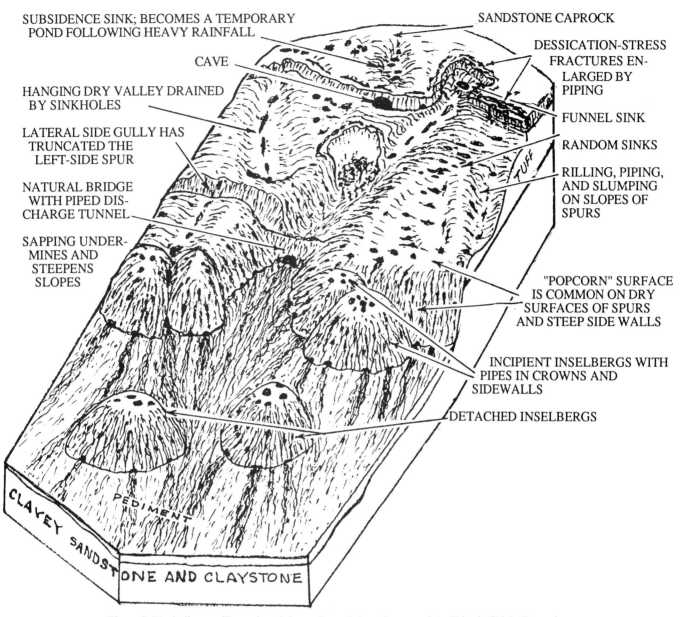

Figure 2. Block diagram illustrating piping and pseudokarst features of the Triassic Chinle Formation badlands in Petrified Forest National Park, Arizona. Vertical relief is about 30 m (100 ft). Adapted from Mears (1968).

along the loess plateau rims. Deep box gullies extending inward from the cliff faces were formed by collapse of large underground sloping-to-horizontal pipes. Open holes high in the cliff faces probably reflect pipe discharge at higher valley-floor levels during earlier phases of cave development. Some of these higher openings are windows in outer walls of vertical pipes. The larger ground-level caves, which gave shelter to prehistoric animals, were enlarged and used by prehistoric Man as permanent dwellings. As late as Thorp's visit, caves were occupied by human families or domesticated animals or used for storage (Fig. 3). Some may still be in use today.

The earliest report of piping in New Zealand is by Cussen (1888), in the Waikato River basin, although he ascribed the observed features to solution. Henderson and Grange (1926) observed piping effects in the Huntly-Kawhai Subdivision of the northern part of North Island, and Guthrie-Smith (1926) briefly described the development and evolution of subsurface tunnels in surficial tuff on his sheep station in east-central North Island. In a report that is still unequaled for clarity of description and directness of vision, Gibbs (1945) reconstructed the progression of piping-induced erosion on the Wither Hills in northeastern South Island, where recent devegetation, burning, and overgrazing had instigated what he termed "tunnel-gully erosion." Later, Ward (1966) described two types of vertical pipes developed on gentle

Figure 3. Peasant loess-dwelling in the Belgian mission station Si-wan-tsze near Kalgan in north-central China. Underground rooms have been formed in the Pleistocene loess by enlargement or modification of natural piping tunnels. Engraving from Richthofen (1877, Fig. 7).

slopes near Whangarei, in northern Auckland, and Blong (1965) observed piping phenomena in north-central North Island. Many of these early studies were reviewed by Visser (1969), who added his own observations on pipes and the piping process on a farm near Matakana, in central Auckland.

In Australia, Downes (1946) found pipes on hillslopes near Victoria, and several investigators studied the effects of "tunnel erosion" in New South Wales (Monteith, 1956; Newman and Phillips, 1957; Richie, 1963; Charman, 1969, 1970a, b).

On Molokai Island, Hawaii, Kingsbury (1952) noted sinkhole development, or "pothole erosion," forming holes as large as 2.5 m across and 4 m deep. In the continental United States, piping is most common in denuded areas of the dry West where gullying or arroyo cutting has developed steep-gradient subsurface drainage. Erasmus Haworth (1897) was the first American geologist to note that steep-walled gullies as deep as 2 m in the loess soils of western Kansas had been formed by processes now called piping. W. D. Johnson (1901), working in the Kansas High Plains, attributed the process to initial differential compaction producing depressions wherein water collects and seeps downward through the loess to erode both vertical and horizontal underground drainage channels. He noted that the collapse of weakened channel roofs in places creates new gullies and temporary natural bridges that diverted additional surface runoff underground. W. W. Rubey's (1928) study of "sinking of the ground" in the High Plains is one of the best early perceptions of this process. Rubey hypothesized that water percolating into the ground forms tunnels. Enlargement of the tunnels undermines their roofs, and the ground above them sinks, finally collapsing to form open gullies. As the gullies develop, soil creep opens cracks above the margins. Surface wash is then diverted into the cracks to initiate renewed tunneling.

Cockfield and Buckham (1946) and Buckham and Cockfield (1950) described a similar development of gullies by collapse of subsurface conduits in loess-like silt along the South Thompson River in south-central British Columbia. Their proposed mechanism was much like Rubey's except they attributed the level of piping to control by a temporary water table.

In the late 1940s, members of the U.S. Soil Conservation Service working in Arizona recognized that piping involves a complex set of problems in soil physics and soil structure, soil and water chemistry, and the behavior of irrigated or flooded soils, as well as problems of gullying and the role of weather, cultivation, and other factors. They were among the first to use the term "piping" (Fletcher and Carroll, 1948; Carroll, 1949; Fletcher and others, 1954; Peterson, 1954). In a related study of the effects of ripping and pitting on soils in New Mexico, Hickey and Dortignac discovered that runoff from some plots decreased or ceased altogether because "subterranean channels were being formed and runoff was occurring below the surface of the ground." They speculated that "it is possible that the mechanical treatments initiated or speeded up this soil piping" (Hickey and Dortignac, 1963, p. 22).

Fletcher and his colleagues' work and, later, Parker's 1963

paper stimulated interest and new research by others in subsurface erosion in arid and semiarid regions: Brown (1962) and Heede (1971) studied piping in Colorado, where the latter's observations of piping-induced gully development are among the most perceptive in the literature. Mears (1963, 1968) described piping in Arizona and Wyoming; Warn (1966) illustrated pseudokarst in the Imperial Valley of California; Niel Jones (1969) reported the effects of piping near Benson, Arizona; and Bell (1968) made a detailed study of piping and pseudokarst in the Paleocene Tongue River and Sentinel Butte Formations of the North Dakota Badlands. Parker and his U.S. Geological Survey research team covered much of the drylands of the American West from the 100th meridian to the Pacific Ocean (Parker, 1963; Parker and Jenne, 1966, 1967).

These studies prompted others to investigate similar cases of "tunnel erosion." Charman (1969) discussed the influence of Na salts in coastal areas of New South Wales, Australia; Dobrovolny (1962) noted effects of piping in landslide clays and silts in Bolivia; Bishop (1962) described the development of piping-generated gullies in Queen Elizabeth National Park, Uganda; Berry (1964, 1970) reported piping in semiarid parts of central Sudan and Tanzania; Downing (1968) described its effects in Natal; and Smith (1968) noted its relation to periglacial boulder concentrations.

In the 1970s, Otvos presented a review of the process and its terminology, and piping phenomena were reported from more drylands areas. For example, Bariss (1971) noted the piping origin of gullies in central Nebraska loess; Drew (1972) described the development of piping in the slopes of Big Muddy Valley in southern Saskatchewan; Rathjens (1973) reported pseudokarst in a savanna climate of Afghanistan and near Jaipur, India; Stocking (1976) commented on "tunnel erosion" in a sodic soil of Rhodesia; and Galarowski (1976) observed piping in the Bieszcady Mountains of the East Carpathians. Gile and others (1981) described the role of soils in the geomorphology of the Basin and Range in New Mexico, and Imeson and others (1982) described the effects of soil chemistry and physics in the development of badlands in Morocco.

Rooyani (1985) studied soil properties influencing piping in the subhumid climate of Lesotho, with warm, moist summers and cold, dry winters. Several studies in semiarid to subhumid areas in the American West were reported in a symposium, "Subsurface Processes in Steep Zero-Order Basins," at the 1986 Fall Meeting of the American Geophysical Union in San Francisco. Other recent studies of drylands piping are contained or summarized in Bryan and Yair (1982).

Meanwhile, there had been occasional reports of piping phenomena in humid areas. Travellers had long reported odd pitting and fluting of noncalcareous rock (e.g., Branner, 1913), but some began to describe more complex pseudokarst features. For example, Downes (1946) reported a mean annual precipitation of 585 mm for the piped area near Victoria, Australia, and Gibbs (1945) noted 710 mm in the Wither Hills of New Zealand. Later, Funkhouser (1951) wrote of soil caves in Ecuador; Feininger (1969) described pseudokarst developed by piping in saprolite on quartz diorite in the humid tropical climate of northern Colombia; Khobzi (1972) attributed huge shallow depressions, as much as 1 km in diameter, in the Andes of Colombia to piping processes; and Löffler (1974) reported widespread piping and pseudokarst in the humid tropical lowlands of Papua New Guinea, where the principal effect is the undercutting and headward growth of gullies.

The most thorough reports of piping in humid regions are by Gilman and Newson (1980) in upland Wales, and a definitive series of studies by Anthony Jones (1971, 1975, 1981, 1987). According to Bryan and others (1978), piping in humid lands differs from dryland piping in several respects, notably that dryland pipes tend to form in bedrock or alluvium rather than in soil because dryland soils are thin, and that the bedrock surrounding a flowing pipe may be dry, whereas in humid lands the soil is saturated. Aspects of humid-lands piping are discussed further in Jones' recent works (1987, and the following chapter in this volume). The present chapter is chiefly concerned with piping and pseudokarst in the world's drylands.

PROCESSES AND NOMENCLATURE

Meanings of "piping"

The term *piping* has long been used in the engineering literature to designate some erosional mechanisms and effects of water seepage under dams and foundations, but its meaning has become blurred with use. In his introductory chapter for this volume, Dunne warns that some of the later applications of "piping" in hydrology and geomorphology may be at odds with its meaning in civil and military engineering, and recommends adoption of a different word that does not carry such a weight of varying meanings. However, the term is now so well established that it seems more advisable to clarify its present usage and review its history.

Some of the misunderstanding about the meaning of "piping" stems from the usage by Terzaghi (1943, p. 258), who used the term only for what is known as "heave," "blowout," or "sand boils" of sediment and water at the surface. As he explained in an earlier work (Terzaghi, 1922), this phenomenon results from the dilation and fluidization of sediment when the hydraulic pressure exceeds the lithostatic pressure, or as he later put it in Terzaghi and Peck (1967, p. 170), when "the seepage pressure of the water that percolates upward through the soil beneath the toe [of a dam] becomes greater than the effective weight of the soil." In this process there is simply spontaneous failure, without the creation of subsurface conduits or tunnel-like voids—the features commonly called "soil pipes."

In 1967, Terzaghi and Peck extended the use of "piping" to include the "formation of a pipe-shaped discharge channel or tunnel" under the dam that "may be due to scour or subsurface erosion that starts at springs near the downstream toe and proceeds upstream" (Terzaghi and Peck, 1967, p. 169-170). Moreover, they commented (p. 618) that failure by heave is rare, and that almost all piping failures are by "subsurface erosion." Some authors have given the name "tunnel erosion" to the latter process or its results (see Dunne, this volume; also the *AGI glossary* [Bates and Jackson, 1987, p. 705]), but this may be confused with the engineering term "tunnelling," discussed below.

Despite the confusion about proper terminology, it seems appropriate to retain "piping" for soil-pipe formation, because the term has long been used in this sense in the literature of hydrology and geomorphology as well as in some early engineering reports. For example, in one of the first studies of confined flow of water through sand, conducted in May 1896, Colonel Clibborn (1909) clearly distinguished the enlargement of open conduits by "sand draw" or "piping" from the lifting and "blowing" (or, in modern terms, "heave") resulting from excessive upward hydraulic pressures. He used "piping" only for the former.

In these usages, some authors apply "piping" to both the process and also to the resultant soil pipes and tunnels. In engineering (e.g., Terzaghi and Peck, 1967; Cedergren, 1977) the term is often used for the resulting failure of a dam or levee. Because this chapter is little concerned with such failures, the term "piping" is hereafter restricted to the formation of subsurface pipes. The hydraulic process of "heave" and its results are not considered here, but are treated briefly in the preceding chapter (Coates, this volume).

Suffossion (suffosion)

One possible alternative to "piping" is the term "suffossion." According to Soleilhavoup and Cailleux (1978) it is derived from the Latin *suffossio* and its verb *suffodere*, "to excavate underneath," with a secondary meaning "to pierce upward." These authors surveyed more than a dozen glossaries and other works in which the word is used, commonly spelled *"suffosion,"* and their composite description of its effects closely matches that of features attributed to piping: surface sinks, hollows, and well-like shafts, with associated shallow subterranean drainage channels, all resulting from mechanical erosion by percolating waters in erosible, insoluble materials, especially clay, sand, or silt.

However, there are several reasons not to adopt "suffossion" as a replacement for "piping." First, it is rarely used, especially in English, where it does not appear in recent large dictionaries such as *Webster's*. In older ones such as the 1933 *Oxford English Dictionary* and the 1900 *Century Dictionary*, it is simply defined as "digging under" or "undermining," a meaning similar to that of "sapping" as used in a somewhat different sense elsewhere in this volume. Second, there is not uniform agreement about its meaning; some authors cited by Soleilhavoup and Cailleux (1978) extend it to include karst dissolution phenomena (see also Otvos, 1976) or restrict it to periglacial conditions, while the *AGI Glossary* defines it quite differently, as diapiric upwelling of sediment and water from thawed ground ice, a sense related to its secondary Latin derivation but not echoed in other sources. The 1964 *Grand Larousse* dictionary restricts "suffosion" (through a synonym, "inféroflux") to underflow and entrainment of fine particles beneath riverbeds. Third, in the discussion of his paper, Soleilhavoup opined that the term should be used only where the effects include the development of vertical shafts and sinks (Soleilhavoup and Cailleux, 1978, p. 223). This seems too restrictive for the range of features associated with the phenomenon here called "piping."

Seepage and piping

In engineering, the movement of water through substrate is commonly called "seepage," "water creep," or simply "creep." It is also known as "underflow" (Nelson and Nelson, 1967, p. 392), "percolating water" (*AGI Glossary*, p. 477), or, even, in a gross misuse of the term, as "piping" (Nelson and Nelson, 1967, p. 282; *AGI Glossary*, p. 731). In this chapter such water movement is termed "seepage" or "subsurface flow."

There is no generally accepted term for the erosion initiated by such seepage. Cedergren (1977, p. 5) calls it "migration of soil particles." Terzaghi and Peck (1967) refer to it vaguely as "subsurface erosion" or "scour" or "spring erosion." Hutchinson (1968, p. 694) uses "piping" as synonymous with "subsurface erosion" of fine materials, and uses "roofing" for one of its effects. *Webster's Dictionary* echoes this meaning of "piping" as an erosional process, "6: water erosion in a layer of subsoil or under or through a dam resulting in the formation of tunnels and caving" (Gove, 1966, p. 1722). However, as noted above, many authors seem to use "piping" to refer both to the erosional process and its results, or to the results alone, including both the development of soil pipes and their enlargement (e.g., *AGI Glossary*, p. 505; Nelson and Nelson, 1967, p. 282). This rather broad sense of the word is used in this chapter.

Note that most definitions refer to piping with respect to its effect on dams and foundations, reflecting the engineering origins of the term; none relate it to erosion of hillslopes and streambanks, which is the chief focus of this chapter and the next.

Processes responsible for piping

Few authors discuss the specific process, or processes, responsible for the erosional detachment, entrainment, and transport of soil particles that result in the opening and enlargement of subsurface soil pipes or tunnels. Doubtless this is because it is nearly impossible to study the process in operation. Also, the literature records at least three, and possibly four or more, processes responsible for the opening and enlarging of soil pipes.

Eluviation. Where a layer or stringer of sand or gravel is interbedded with finer sediment, and where grains of the fines are smaller than the interstices of the sand or gravel, seepage may entrain the fine grains and transport them through the sand. Fine cracks, joints, and faults may serve the same function as sand or gravel. So may animal burrows and holes left by rotting roots or buried wood (Cedergren, 1977, p. 8). Such selective erosion, which engineers try to control with filters and drainage blankets, is perhaps the predominant way in which soil pipes are initiated.

The process resembles elutriation, as defined in the *AGI Glossary* (p. 211):

(b) . . . removal of material from a mixture . . . by washing and decanting, leaving the heavier particles behind. (c) The washing away of the lighter-weight or finer particles in a soil by the splashing of raindrops.

However, the term "elutriation" is generally associated with laboratory procedures, not with subsurface drainage. The selective removal of fine particles more closely resembles the eluviation of clay particles during soil development. Most accounts of that process restrict it to simple downward transport by percolating soil water, whereas much piping is the result of lateral subsurface flow. However, since lateral transport *is* included in *Webster's* definition ("the transportation of dissolved or suspended soil material within the soil by the downward or lateral movement of water when rainfall exceeds evaporation"), this kind of erosion is here called "eluviation." Dunne (this volume) describes this and the following process in some detail and includes them both in "seepage erosion," a general process that includes erosional effects of seepage in both soils and consolidated rocks.

Seepage-face erosion. Where saturated susceptible soil is exposed on a sloping surface, concentrated seepage may entrain fine particles at the point of outflow and carry them away downstream, thus creating a niche or re-entrant that progressively cuts back into the slope to form an underground conduit. Concentration of the seepage may be caused by soil cracks, lithology, local relief, or other factors that converge the flow lines. Such soil-pipe development, not previously reported, has been observed on saline soils in Davis, California, by Aram Derewetzky (unpublished report discussed by Higgins in Chapter 6, this volume). The term "seepage-face erosion" is here proposed for this process.

Tunnel Scour. After a pipe or opening through the soil or erosible bedrock has been initiated—whether by eluviation, seepage-face erosion, rotting of roots, animal burrowing, or cracks and fractures—concentrated flow of water through it enlarges it by scour or plucking of detachable particles from the walls. The mechanics of this subterranean erosion are described in detail by Dunne (this volume). The process, commonly called tunnel erosion (e.g., Bennett, 1939), is here given Dunne's more descriptive name, tunnel scour.

Tunnelling. Soil-pipe development by either eluviation or seepage-face erosion is conceived to proceed from a free face inward as described above. A variant of this, called tunnelling, is thought to proceed in the opposite direction. Nelson and Nelson (1967) describe the process:

A form of failure. . . .It is caused by cracks developing in the bank under dry conditions. When suddenly brought into contact with water, the cracks collapse internally and create a tunnel. In dams and embankments, the tunnelling starts at the wet face or upstream side and proceeds downstream. It thus differs from *piping* which starts downstream and proceeds up to the wet face. Tunnelling is therefore more dangerous because it is harder to detect. (Nelson and Nelson, 1967, p. 390; this definition is paraphrased in the *AGI Glossary*, p. 705).

If this procedure is not a separate means of forming soil pipes, then it may be simply a variant of tunnel scour, in which a preexisting conduit is enlarged. Unfortunately, this term has been compromised by its use also for ordinary upstream soil-pipe development (e.g., Ritchie, 1963).

All the mechanisms listed above cause the creation and/or enlargement of subsurface conduits. In the following pages "piping" refers to such development and includes the several processes that promote it.

PIPING CHARACTERISTICS AND PIPE INITIATION

Piping can occur in a variety of ways. Fletcher and Carroll (1948) recognized three kinds, which can be termed seepage-face piping, soil-crack piping, and rodent-hole piping. However, there are other kinds as well.

Heave

The process of heave, as generally known to construction engineers, occurs chiefly during the dewatering of deep, pumped excavations or during the rise of impounded water behind levees and dams. "Boils" result where newly created, large hydraulic-head differentials cause channelized subsurface flow, with entrainment of water-saturated earth materials. The suspended sediment discharges into the excavation or on the down-gradient side of a dam or levee. Although this form of subsurface erosion rarely produces open pipes, it may remove enough material to collapse the overlying surface with sudden destruction of the superjacent structure.

Similar failures can develop naturally. For example, a landslide blocking a stream can impound a lake behind the landslide dam. Where there are permeable, erodible materials such as silt or sand in the slide mass or in the underlying valley fill, hydraulic forces from the rising lake level may force groundwater flow through permeable zones in or beneath the dam. The resulting entrainment of the sediment can produce boils that may undermine and eventually destroy the dam, although Costa and Schuster (1988, p. 1058) state that this mode of "failure of landslide dams from piping and seepage is uncommon."

Eluviation

Entrainment and transport of silt and clay grains through the interstices of unconsolidated sediments or regolith is a major factor in the development of pedogenic soil horizons. As noted above, this process can also initiate discrete crack-like or pipe-like openings in the soil; these may then be further enlarged by tunnel scour. Cedergren (1977) attributes the failure of California's St. Francis Dam in 1928 and Baldwin Hills Reservoir in 1963 to such piping.

Solution of carbonate cement of otherwise noncalcareous sediments can provide initial pathways for eluviation piping, as documented later in this chapter by Warran Wood's case study of the development of playa lake basins on the Staked Plains of Texas and New Mexico. By definition, however, solution is not the major factor in the piping process.

Seepage-face erosion

Pipes may also form from the surface inward where sufficient hydraulic head exists to move water through a stratum with enough velocity to entrain dispersed clay, fine sand, or even silt at the head of a gully, side of an embankment, or other steep slope. In such cases, outlets grow headward into the slope. Fletcher and Carroll (1948, p. 546) outlined this type of piping process:

The highly dispersed subsoil merely comes into suspension in the lateral flow and is washed out. The resultant rounded tunnel or pipe progresses back into the land toward the water source until caving and sloughing shuts off the water source and closes the pipe.

Some pipes continue to grow, however, and may even sap their way upward to the land surface, forming a sinkhole. With continued enlargement both in length and girth, the pipe roof partly caves in, forming a small gully, but leaving temporary natural bridges along the way. Finally these all cave in and the gully becomes a surface-water channel connected with the main channel.

Such gully-wall piping is widespread in dryland valley fills in the American West as a result of occasional heavy rainfalls or of over-irrigation, with the resulting subsurface flow to a gully through a permeable, erodible stratum (Harris and Fletcher, 1951). The hydraulics of a developing pipe beginning at a gully wall is depicted in Figure 4. The permeable bed is saturated with water that moves under hydraulic head to its discharge at a nearby gully wall or other steep slope. Water begins to seep from the bed at the point of highest permeability or lowest elevation in the face of the slope. As water leaves the face it carries away disaggregated and dispersed silt particles in suspension. This action initially creates a small orifice for the developing pipe in the cliff face (Fig. 4A).

Initially the pipe is narrow and short and its area of intake is small. Progressive erosion, either at the seepage face itself or by eluviation within the permeable layer, increases both the seepage flow and the width and depth of the pipe (Fig. 4B). After irrigation or a heavy rain the conduit may carry a steady flow of water as from a hydrant (Fig. 5). Selby (1982) notes that headward growth of pipes can be very rapid: rates of 45 m/hr are reported from pumice soils in New Zealand. Pipes developed in this manner tend to have visible outlets except where saturated ground causes slumping and temporary closure of pipe outlets.

Soil-crack development, tunnel scour, and pipe enlargement

Once connected subsurface openings have been formed, ephemeral subsurface flow can scour soil particles from the walls and enlarge the openings to form pipes. As noted above, such

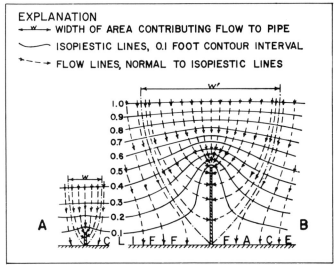

Figure 4. Flow nets illustrating increase of intake area as a pipe grows headward through seepage-face erosion. A, incipient state; B, after the pipe has extended considerably headward and increased greatly in diameter. Modified from Terzaghi and Peck (1967, Fig. 63.4).

openings may originate as animal burrows and holes left by rotting roots or buried wood, or they may be opened by eluviation. In cold regions, "piping is also promoted by surface crack systems left by ice wedges during freeze-thaw cycles.... Pipes developed in blocks of stagnant ice continue in the underlying silt at places" (Bell, 1968, p. 250). Commonly, however, passages for water are provided by soil cracks formed in other ways, such as soil compaction or dessication.

Soil-compaction cracking. Soil cracking can make an essentially nonpermeable material permeable, at least temporarily. Subsidence associated with soil compaction commonly leads to fracturing of alluvium (Knight, 1959). One kind of subsidence, generally associated with percolating irrigation water, results from the breakdown of secondary aggregates and collapse of voids following the saturating of previously unwetted, low-density earth materials. In drylands, mudflow deposits and loess are particularly susceptible to such hydrocompaction and subsidence when wetted (Bull, 1964). Subsidence of this type, which may produce a volume decrease of as much as 13 percent (Hardy, 1950) and a lowering of the ground surface by 3 to 4.6 m, commonly cracks the alluvium. Where such cracks reach the surface, they provide conduits for infiltration, and pipes may then develop at depth where the water escapes laterally to a nearby ditch, canal, gully, wadi, or arroyo.

Deep pumping of groundwater can also cause subsidence cracking, followed by piping that can create extensive gully systems, as discussed by Péwé (this volume). The subsidence greatly increases the intensity of piping in any earth material subject to piping by: (a) concentrating runoff. (b) causing stress cracks, and (c) developing partial lateral pipes by uneven subsidence and differential consolidation between different strata.

Figure 5. Excess irrigation water discharges into a desert wash from a soil pipe beneath a newly flooded field. Materials being piped are soils and alluvium from an irrigation project in Pakistan. Photograph by J. B. Culbertson, U.S. Geological Survey.

Desiccation cracking. Most piping in the western drylands originates from desiccation cracks, a characteristic feature of soils with swelling clays that dry and crack during hot, dry summers (Parker and Jenne, 1966). Aided by gravity, cracks along steep slopes widen. Runoff from sudden violent thundershowers of dryland summers may fill the cracks. Where an outlet exists for flow from the crack, a current of moving water develops that wets and saturates the crack walls. In silt or loess, dry crumb aggregates disintegrate rapidly, and the dispersed clays and silts are entrained in the flow. Hodges and Bryan attempted to study the breakdown of such aggregates in the field, and concluded that a major factor in the process is slaking, resulting from "air compression ahead of a wetting front" (Hodges and Bryan, 1982, p. 31), as first proposed by Yoder (1936).

Where the hydraulic gradient is sufficient, the fine-grained particles are transported in suspension and possibly by traction to an incipient pipe outlet in a gully wall, arroyo, or embankment. In this manner, cracks become widened into pipes. In some cases, wetting and swelling close the upper parts of the cracks, but an opening too large to close remains at depth to provide a tunnel for future subsurface flow. Widening at depth may also cause the crack above to close by gravitational settling (Heede, 1986). In some highly permeable and erosible materials the surface relief need not be steep, for some pipes are observed in little gullies where the overall relief is less than a meter. As Stocking (1981) found in central Zimbabwe, the hydraulic gradient is more important than the surface relief for pipe development.

Most dessication-stress cracks that become highly piped are near and parallel to gully walls in silty alluvium, loess, or other highly erodible materials. Cracks form most readily there because wetting and drying of the sediments, which diminishes downward and inward into the bank, is greatest at the surface and wall face. Here the major swelling movement is toward the unconfined face of the gully wall, and subsequent drying opens fractures that largely parallel the wall. Most of these gully-wall desiccation cracks form within about 0.6 to 1 m from the face of the gully wall, may roughly parallel the wall for distances of 10 m or more, and tend to be arcuate, closing at either end.

No single factor determines whether desiccation-crack piping will occur at a given site; whether a given material will or will not pipe depends on the interaction of several physicochemical-mineralogical and biological factors (Meileng and King, 1952; Parker and others, 1964). At least four minimum requirements exist: (1) there must be enough water to fill drainage cracks; (2) strata must contain swelling/shrinking clays; (3) strata must desiccate thoroughly, if only seasonally; and (4) there must be an outlet for drainage. Furthermore, as noted below, the intensity of piping is likely to be enhanced by high exchangeable sodium percentages, instability of crumb aggregates, minimal vegetative cover, and low slopes, which enhance infiltration. Where uneven and differential subsidence occurs, the minimum requirement for initiating piping would be a sufficient hydraulic gradient and a drainage outlet.

PROGRESSION OF PIPING AND DEVELOPMENT OF PSEUDOKARST

Initially, the flow through a permeable bed to a cliff face or to a growing pipe is laminar, but it becomes turbulent once the pipe is formed. As the subsurface drainage-flow net enlarges and soil pipes grow wider and longer (Fig. 4), the volume and rates of flow increase as corrasion enlarges the walls and roof. Some pipes, as in some examples cited below, become quite large. Heede was able to enter one on upper Alkali Creek, in western Colorado, and photographed and surveyed its entire length over a lateral distance of nearly 8 m (Heede, 1971, Fig. 2). As such a pipe grows, parts of the roof cave in at the more weakened areas along its course to produce a line of sinks, which then locally funnel additional surface runoff into the subterranean pipe (Leopold and Miller, 1956). Between the sinks, temporary natural bridges are the last remnants of the pipe's roof (Fig. 6). Other landscape elements may include blind and hanging dry valleys and detached haystack-shaped hills, or "inselbergs" (Fig. 2).

In limestone or dolomite terrane, such topography, formed by solution, is called "karst." To indicate a resemblance, the term "pseudokarst" is applied to the piped landscape. Although the landscapes may be similar, solution is not a major factor in the piping of clastic materials. Many of the elements of such landscapes are shown in Figure 2. Where relief is strong and pipe

Figure 6. Pseudokarst developing in Quaternary river-terrace alluvium near Benson, Arizona. These vertical sinks, 3.5 to 5 m in depth, discharge into a system of huge horizontal pipes that, in turn, debouch into the channel of the adjacent San Pedro River. The spaces between the sinkholes are earthen bridges over the pipes. Photograph by Lynn M. Shown, U.S.G.S.

outlets are far below the upland surface, large sink-and-cavern systems may develop, as at Officer's Cave in Oregon, described in a case study below.

Scale of pseudokarst

The piping conditions described above are for fairly level surfaces, e.g., on terraces or pediments, in which recent stream incision has made possible the development of relatively short, steep-gradient, subsurface drainage routes through easily erodible materials. Piping also occurs on hillside slopes and on the crowns and sides of badland hills, as shown in Figures 2 and 7. Piping tends to be deepest—generally as deep as the bed of the nearest channel adjacent to gullies or arroyos. Where relief is great and the local base level is low, large sinks and caverns may form (Fig. 7) and extensive karst-like landscapes may result (Fig. 1).

Miniature pseudokarst having intricate microrelief results where piping is limited by the shallow thickness of the permeable horizon. Bell (1968, p. 253) suggests that this type "be called 'retiform' after its net-like structure," and notes that it can develop rapidly—in two years—in new cut slopes as well as in natural ones. He describes its effect on highway cuts:

This class with its intricate system of hoppers or voids collects and conveys the meteoric water to the ditch grade or other base level by underground routes. The retiform effect is produced by jetting, headward piping, disintegration, and collapse of the sediments (Bell, 1968, p. 253).

This type of pseudokarst is illustrated in Figure 8, a view of shallow piping near Oceanside, and Figure 9, at Long Beach, California. Such examples commonly develop in duplex soils, where a relatively permeable, coarse-textured horizon is underlain by a less-permeable, finer-textured or argillic horizon (Rooyani, 1985).

Gully formation by piping

As piping and the pseudokarst landscape continue to develop, sinks merge laterally; finally, after all the roof of a pipe has caved in and the natural bridges have disappeared, only an open ragged gully remains. In some drylands this may be the major mode of gully development, and may in some cases occur so rapidly that evidence of the subsurface origin is not preserved. Much of the gully shown in Figure 10 developed by pipe collapse following a single rainstorm during the night of April 21–23, 1962. At about 6:00 pm that evening the ditch was in normal condition. Then, during the night, 2 in (50 mm) of rain fell, and by the next morning the entire section of pipe roof had collapsed, creating the open gully. The length of the caved-in pipe was about 25 m, and the average width of the resulting gully was about 1.5

m. At some dryland sites, following especially heavy rains, an incipient badlands pseudokarst may develop within a day or two where there was little or no previous indication of subsurface erosion.

Piping also aids the headward extension of gully headcuts, where the effects of piping in desiccation-stress cracks may exceed those of "over-the-brink" surface-water flow. Bishop (1962, p. 163) described such results:

... during heavy rainfall the water-table may rise rapidly and when supply exceeds drainage by a sufficient amount "perched water-tables" build up above layers of relatively impermeable strata. If the rainfall is sufficiently concentrated the surface of the water-table develops a slope toward the lines of drainage. Such a build up of ground-water results in rapid subterranean flow into gully heads, undercutting both unconsolidated beds and any surface cover of vegetation. The gullying proceeds in steps above each relatively impermeable band and temporary tunnels up to three feet [0.9 m] in diameter can be seen leading off from the heads of active gullies. With the collapse of tunnels gullies may advance headwards as much as 10 or 15 feet [3.3 or 4.6 m] in a single storm.

Bell witnessed similar effects in the North Dakota Badlands:

These intermittent side streams and gullies with near vertical walls are advancing headward by piping in increments of several feet as the arches and blocks collapse. Thus, piping progresses headward along the hidden lines of seepage in the formations and in great steps in the headward advancement of gullies (Bell, 1968, p. 252).

Stocking (1981) found gullies closely associated with piping in central Zimbabwe, and proposed that surface-stream entrenchment had acted as a trigger for pipe erosion by providing a drainage exit, thereby steepening the hydraulic gradient. The resulting pipes then aided in the extension of tributary gullies developed primarily by pipe collapse.

Figure 7. Pseudokarst developing on a spur of a sloping valley wall, east side of Chinle River, along U.S. Highway 191 near Round Rock, Arizona. Terrane is red, green, and white altered volcanic ash and tuff of the Triassic Chinle Shale. Except as noted, this and remaining photographs in this chapter are by G. G. Parker, Sr.

Figure 8. Intricate shallow piping producing miniature pseudokarst of the retiform type in a bed of silty sand 30 to 40 cm thick overlying a relatively impermeable layer. East side of U.S. Interstate Highway 5 near Oceanside, California. Notebook is 15 × 20 cm (6 × 8 in).

On July 7, 1961, the senior author witnessed gully-head recession caused by piping on a tributary to the Rio Puerco, west of Belem, New Mexico. Slashing rain had caused a flash flood during the night. By noon the following day the flood had peaked and receded, leaving the gully head about 10.5 m upstream from its position the day before. Three years later (July 15, 1964) on a tributary to Judito Wash near the Judito Trading Post, he observed the retreat of a gully head 4 m high as a result of a similar flash flood. Other aspects of gully-head recession caused in part by seepage-face erosion are discussed by Higgins (Chapter 6, this volume).

Bell (1968) believed that in the pseudokarst of the North Dakota Badlands, particularly in the miniature "retiform" type, a cyclic progression of gully development by piping begins and ends with surface drainage:

A cycle of piping is ... started by [stream incision or] regional uplift. The process begins when surface drainage is directed and diverted underground

Figure 9. Demoiselle-like miniature pseudokarst developed in a shallow, gravelly silt bed on the northeastern slope of Signal Hill, Long Beach, California. Most pipes have collapsed, leaving open channelways. The notebook is 15 × 20 cm (6 × 8 in).

Underground drainage routes then increase in number and spread in the area as the new base level of surface streams extend[s] headward in the valleys. . . .In this way, piping at the head of advancing levels (steps) in the stream channel help[s] produce the base level for piping along the sides of streams. Maturity is approached with maximum underground drainage.

Following pipe collapse and exposures of a surficial gully network,

The piped slope first acquires maximum channeling between sharp ridges that divide at cusps and thin spires. The sharp ridges and cusps disappear in the second year when the ridges are lowered and rounded. The retiform piped area is further subdued as the basal pipes are choked with the rising apron of sediments. Ideally, the lower part is buried in its own sediments and the upper part is obliterated as the surface is renewed by erosion, by landslides, or by the work of man, and a new cycle is started. . . .Thus there is a Davis type age cycle of pseudokarst topography (Bell, 1968, p. 253).

Such cyclic development may also occur in larger gully systems. Discharge from pipe outlets at the base of a gully wall commonly undercuts the wall and causes a slab to fall into the channel. Large accumulations of fallen debris may force the gully flow against the opposite wall. This is a common cause of widening of drylands gullies. Heede (1971) found that such slab collapse can lead to natural reclamation of the saline soils:

. . . it appears that these [collapsed] slopes become more stable. Growth of vegetation on the deposition slopes is favored by increased soil moisture. More water will infiltrate into the slopes due to the gentler slope gradient formed by the deposits. . . .As sodium is leached from the soil, the pH will decrease and calcium will be more soluble and available for plant growth. The greater calcium influence will also stimulate flocculation and will speed the leaching process. . . .Thus, gully side slope stabilization may progress rapidly, once plants become established.

The final stage of this development may resemble that of old age of a karst topography on a somewhat small scale (Heede, 1971, p. 13–14).

Piping and landsliding

A few authors have commented on the role of piping in slope instability and landsliding. In coastal Alaska, landslide initiation may be related to a rapid rise of perched groundwater owing to "rapid subsurface flow . . . through discontinuous networks of soil macropores" (Sidle and Swanston, 1986). Undermining of gully heads and walls by piping promotes tension cracks with subsequent collapse. The failed blocks may then themselves become affected where "open joint systems in the weakened blocks provide excellent avenues for meteoric water to attack and destroy the mass internally by maximum piping" (Bell, 1968, p. 255).

Pierson's (1983) model studies suggest that piping may also play a major part in some hillslope failures. Where pipes become blocked or their drainage is impaired, they may fill with water during storms. Pore pressures in the surrounding soil become even greater than those produced by total saturation of the soil, and locally "could trigger landslides at sites that would otherwise be stable" (Pierson, 1983, p. 1). He reports that features of landslide scars from many areas support this conclusion.

Examples of pseudokarst

Painted Desert, Arizona. Mears (1963) and Parker (1963) have described the effects of piping in the brick-hard red and green shales, siltstones, and white tuffs of the Triassic Chinle and Moenkopi Formations. There, the crowns and steep side slopes of desert inselbergs, bedrock spurs, and sloping valley walls are riddled by pipes, producing localized pseudokarsts (Fig. 2). Piping there occurs not in soil or alluvium but in tuffaceous and clayey bedrock. The diagram shows two spurs protruding from the sandstone-capped upland. They in turn have hanging dry

Figure 10. Gully resulting from overnight pipe collapse following heavy rain during the night of April 20–21, 1962. Gully is aligned in a drainage ditch along the north side of U.S. Highway 164, east of Spring Creek, about 32 km east of Rangely, Colorado. Materials piped are greenish gray alluvium derived from the Cretaceous Mancos Shale. Photographed April 21, 1962, after the storm had passed.

openings, that riddle the hilltops. A few minutes to a half-hour or so later this water reappears at discrete, small pipe orifices, or as small seeps or tiny wet-weather springs at or near the bases of the hills. Water from the small pipe orifices may shoot out as small jets under considerable hydraulic pressure, Silt, clay, and fine sand are thus discharged to the pediment surface and carried away both in shallow rills and by sheetflow. These flows are an effective force in the development and maintenance of the knickpoints, inselbergs, and pediments. Piping, rilling, slumping, and mass wasting of the slopes all act more or less concomitantly to produce this pseudokarst.

Other examples of hilltop piping are in the gray and tan ashy beds of the Miocene-Pliocene Furnace Creek Formation, a few kilometers south of Death Valley National Monument headquarters, California; in deformed Pleistocene saline lacustrine sediments farther south in Death Valley near the Saratoga Springs; and in the gray-to-white-weathered smectitic tuffaceous sediments of the late Oligocene and early Miocene John Day Formation in eastern Oregon, as described in a case study, below.

Near Benson, Arizona. Figure 6 shows a sinkhole pseudokarst developing on formerly irrigated farmland on the west side of the San Pedro River. There, piping in Quaternary river-terrace alluvium derived from Cretaceous Mancos Shale has created a system of caverns and sinks. The sinkhole pipes are raggedly irregular in shape, 2.5 to 3 m in depth, and connect with flat-bottomed, gently sloping, vault-roofed pipes that discharge at base level into the channel of the San Pedro River bed. Range animals find refuge in the vault-roofed caves that have developed along these huge horizontal pipes. Continued collapse of the soil bridges that connect the sinks will in time yield a steep-walled open-gully system, formed not by surface wash but by subsurface undermining.

Pakistan. Figure 11 shows effects of piping, sapping, and cave development along the bank of a Pakistan wadi. The valley wall contains clastic beds of differing texture, cementation, and permeability. However, vertical pipes are present in all beds; even the crumpled slump block is piped. Discharge along the plane of contact between beds 2 and 3 appears to be the reason for an alignment of pipes at this same height leading off to the right side of the picture. A cave developed in a brown silty clay is to the right of "1." Scenes similar to this are common in parts of Arizona, Utah, western Colorado, southern New Mexico, and Nevada.

Kutz Canyon, New Mexico. Figure 12 depicts two of the largest known pipes in the United States, about 17 m deep and with greatest diameters 6 to 8 m. They are developed in the north wall of Kutz Canyon, about 2.5 km south of Bloomfield, New Mexico. These pipes occur in hard gray sandstone and claystone of the Paleocene Nacimiento Formation, which is cemented with coatings and inter-grain fillings of a smectite clay. The piping appears to be guided in its development by a set of vertical cracks or joints.

The central dome-shaped feature between pipes 1 and 2 may well be the precursor of an inselberg that will be formed

valleys drained by sinks. Alignment of some sinks may reflect major joint sets in the substrate. The slopes of the spurs are scored by shoestring rills that may reflect surface runoff or be developed, in part or mainly, by piping. Inlets to slope pipes may occur at random places in the channels of the rills. Collapse of pipes hastens the rate of rill deepening and slope erosion.

Along the more deeply incised channelways, undermining of steep slopes causes slumping, forming rubble dams (shown in mid-drainage, Fig. 2). Continued slope undermining and erosion leads to the detachment of "inselbergs," or residual hills, which may themselves be laced with piping channels. Examples of such piped hills occur near Round Rock, on the east side of Chinle Valley (Fig. 7); near the headquarters of Petrified Forest National Monument and about 6.5 km north of Cameron. At such places, little of the rain that falls on top of the hills runs down their sides; instead, the precipitation vanishes into funnel-like sinks, or pipe

Figure 11. Piping along a wadi wall in western Pakistan. Lowermost unit is silty clay, overlain at the level of the cave roof by volcanic ash and silty sand. Slide block to right of cave. Piping occurs in all units. The height of the bank is estimated to be about 2.5 m. Photograph by James K. Culbertson, U.S.G.S.

once erosion has separated it from the remainder of the retreating valley wall. Mass wasting and slumping of the outer face of the valley wall is evidenced by the debris along the foot of the cliff, which is itself piped in places.

The rate of erosion here is slow. There have been essentially no discernible changes in the shape or sizes of these pseudokarst features in the past 44 years of observation.

Other large pipes, with diameters as great as 30.5 m, have been described on a river terrace in the White Silts at Kamloops, British Columbia, Canada. According to Buckham and Cockfield (1950), these huge pipes and related gullying developed from consolidation and sinking of the ground, which subsequently cracked and was then attacked by piping on a grand scale. The great pipes in Kutz Canyon appear to have developed by the coalescing of adjacent smaller pipes.

CONDITIONS FAVORING PIPING

"It appears that tunnel erosion is the result of a wide variety of conditions and is not restricted by either climate or particular soil conditions" (Stocking, 1976, p. 35). However, certain combinations of conditions do seem to promote piping in some regions. Although it is reported from humid and subhumid regions in many parts of the world, piping appears to be most pronounced in the drylands. This may be in part because the results are more visible there, where soils are thin and vegetation is sparce. Bell (1968, p. 250) tersely outlined the requirements:

"Piping develops as a short cut in the total drainage system where any physical and chemical advantages help intercept surface water." Selby is more specific:

Among the factors which dispose a soil to piping are: a seasonal or highly variable rainfall; a soil subject to cracking in dry periods; a reduction in vegetation cover; a relatively impermeable layer in the soil profile; the existence of a hydraulic gradient in the soil; and a dispersible soil layer (Selby, 1982, p. 110–111).

In a general way, piping seems to be favored by certain characteristics of earth materials, such as mineralogy, salt content, and permeability, and perhaps also by special requirements of climate and vegetation cover. These factors are briefly surveyed below.

Earth materials susceptible to piping

Piping has been observed to develop in a variety of materials, including clay-, silt-, and fine-sand-sized valley-fill alluvium and colluvium, silty soils, loess, volcanic tuff, and ashy sediments as well as weakly consolidated argillaceous sedimentary rocks such as siltstone, mudstone, and silty shales. The material may be sandy or even gravelly, as long as there is a substantial proportion of silt and clay. Sandstones, if clean and well cemented, rarely become piped, but where they are poorly cemented or contain silt or silty clay, they may develop pipes, some of spectacular size (Figs. 11 and 12). Even some volcanic agglomerates and breccias may become piped if they contain an ashy matrix, especially where they are weathered. For miles along the John Day River in eastern Oregon, north of Dayville, such breccias contain pipe openings in the cliff face 2 m or more in diameter.

However, if these were the sole requirements for piping, virtually all soils and sediments would show the effects. There appear to be additional lithologic requirements, particularly of mineralogy, salt content, and permeability.

Soil mineralogy. Parker and Jenne (1967) found that, in the western drylands, piping is facilitated where the materials contain at least 20 percent smectite clays. These swell and shrink when wetted and dried, thus promoting desiccation cracks that provide passages for infiltrating surface water. During the subsequent tunnel scour, swelling of the smectite tends to disrupt and disaggregate the sediment or soil. In addition to cracking when dry, the swelling clays may become slick, plastic, dispersed, and noncohesive when wet. This makes them especially vulnerable to erosion by moving water. Salt content also affects swelling (see below); Imeson and others (1982) found that in sodium-rich environments, illites and other clays will also swell.

In the dry state the mineral grains are principally stabilized into crumbs by flocculation and clay coatings. So long as these crumbs remain dry, or at least unsaturated, they can bear heavy loads and may be fairly permeable to infiltrating water. However, they rapidly lose this capacity upon wetting as the smectite clays swell. The higher the soluble salt content, especially of sodium ions (see below), the greater will be the volume change during swelling and the more the crumbs will deflocculate. The greater the Na to Ca + Ma ratio at a fixed total soluble salt content, the more readily the smectite clays will disperse. Dispersed clay platelets and silt-sized fragments will then wash away in suspension through fractures in the rock or through the interstices of the grains in beds of sand or gravel. This removal of clay in seepage water may be referred to as "colloidal erosion" or, preferred here, "eluviation." The larger the fractures or holes become, the more rapid and turbulent the water flow, and the erosion then becomes tunnel scour erosion. One must note, however, that where smectite predominates, swelling will reduce the pore space and decrease the permeability. Nearly pure clays, such as the bentonite beds in the North Dakota Badlands, effectively swell shut when wetted, becoming impermeable aquicludes and thereby promoting piping in the overlying beds (Bell, 1968). Similar swelling at the ground surface decreases the permeability and infiltration rate, thereby effectively increasing the amount of runoff available for interception by desiccation cracks and pipe inlets.

Bell pointed out another consequence of deflocculation where silt predominates and swelling clays are less than 25 percent of the sediment. There, clay aggregates the silt grains, but:

. . . wetting instantly disintegrates the aggregates and disperses the silt as if by melting. The shrinkage is at least 35 percent and the resulting collapse process aids the other processes that collectively produce the pipes and related features (Bell, 1968, p. 251).

Figure 12. Sink-and cavern system with giant pipe inlets developed by piping along a canyon wall in hard but weakly clay-cemented sandstone and siltstone of the Paleocene Nacimento Formation. Kutz Canyon, about 22.5 km south of Bloomfield, New Mexico.

Soil samples from 8 sites affected by piping in New Mexico and Arizona were analyzed by Lynn M. Shown and Karl W. Ratzlaff (unpublished U.S. Geological Survey report, 1963). X-ray diffraction showed that smectite was a major component of the clays at all sites, and was the only clay present at 5 of the sites. There were minor amounts of illite and/or kaolinite and/or mixed-layer illite-smectite at the other 3 sites. The smectite content of the soils ranged from 20 to 50 percent on a dry-weight basis. As pure smectite is capable of volume changes of as much as 1,600 percent in going from an air-dry state to maximum moisture sorption (Mielenz and King, 1952), the soils at the study sites could be expected to crack widely and deeply on thorough desiccation, such as occurs in the western drylands. Even in humid regions, such as the U.S. Gulf Coast, in areas of 120 to 150 cm/yr rainfall, smectite-rich vertisols have cracks as deep as 150 cm (T. C. Gustavson, personal communication, 1988), suggesting that smectite may be a significant factor in pipe development there too, when desiccation may be caused by summer droughts. In the subhumid climate of the northern Lesotho lowlands, Rooyani detected a change from less- to more-expandable clay minerals at the soil horizons where piping occurs; smectite was dominant in the piping-prone horizons, explaining the presence of desiccation cracks there (Rooyani, 1985).

In some sandy materials, clay coatings and occasional cements of iron and manganese oxides, calcite, and silica bind the mineral grains together to form crumbs or larger masses that are firm and stable in the dry state. The clay coatings tend to swell on wetting, especially where sodium is a dominant cation. This reduces the cohesive strength of the material and enhances erosion.

Salt content of soils or soil water. Damage from piping has long prompted studies in Australia, where an estimated 8.7 percent of small earth dams fail as a result of piping (Aitchison and Wood, 1965). N. H. Monteith (1954), in New South Wales, was one of the early investigators to note the importance of sodium in the exchange complex of soil clays and its role in promoting piping. Many other authors have also found an association between piping activity and soil chemistry, particularly the sodium content or exchangeable sodium percentage (ESP). Such studies have been especially promoted in New South Wales, where in three coastal areas Charman (1969, 1970a, b) found soils subject to piping at 37, 41, and 51 percent, respectively, of the sites studied.

Ritchie (1963) analyzed soils from all over New South Wales and developed a Dispersal Index that differentiates piping-prone soils from those that are relatively safe to use in dams and embankments. The index measures the degree to which the clay is dispersed in water. This in turn reflects the saturation of the exchange complex with sodium relative to other ions.

Aitchison (1960) postulated that deflocculation and dispersion of clays are responsible for piping failures of earth dams. He and his colleagues later defined the process of failure:

... homogeneous banks of clay soil constructed in relatively dry conditions would have the porosity characteristics of silty materials because of the aggregated condition of the clay. On wetting, in circumstances promoting deflocculation, these clay aggregates would break down into finer clay particles capable of moving in suspension, under seepage flow, through the inter-aggregate pores and finally out of the dam. An accelerating flow rate would result, culminating in macrofailure by piping (Aitcheson and others, 1963, summarized by Aitcheson and Wood, 1965).

The authors went on to identify the boundary conditions between the aggregated and dispersed states in terms of the salt content of the soil, especially the ESP, and the total cation concentration of the water moving through it.

Aitchison continued these studies and further defined the chemical limits of the deflocculation process (Aitcheson and Wood, 1965). For example, a clay soil with a sodium adsorption ratio (SAR; as defined by Richards, 1954, SAR = $[Na] \div \sqrt{0.5 [Ca+Mg]}$) more than 5 is susceptible to piping failure when in contact with seepage water containing less than 3 milliequivalents per liter (meq/l) total cation concentration. Further, they found that a safe dam can be constructed of piping-prone materials either by chemical treatment of the water used on compaction of the soils or by ensuring that the field permeability of the compacted soils is reduced to less than 10^{-5} to 10^{-7} cm/sec.

Later, McIntyre (1979) examined the relations between ESP, clay mineralogy, and tendencies of slaking, swelling, dispersion, and their effects on hydraulic conductivity. He confirmed earlier findings that soils with ESP>15 are subject to serious damage to soil physical conditions when exposed to seepage water, but concluded that soils with ESP as low as 5 may also be adversely affected, especially plastic soils with a high smectite content.

Additional early studies of the influence of sodium on the dispersion of soils were noted by Parker, 1983, p. 109 and later by Heede, 1971, p. 6, who also found that "piping soils had a significantly higher exchangeable sodium percentage (ESP) and sodium adsorption ratio (SAR) than nonpiping soil." These findings underscore the observations of almost all authors that soil dispersion and piping is greatly facilitated by high salt content, especially high exchangeable sodium, in the soil. Parker (1964) reported analyses of piping-prone soils at 8 sites in Arizona and New Mexico that show similar results (Figs. 13a and b). All of the soil samples were alkaline, with pH ranging from 7.6 (except one 7.2) to 8.5. There was a tendency for pH to increase from the surface downward. At 7 sites, total soluble cation content (TSC; Ca + Mg + Na) was greater than 20 meq/l and ranged as high as 800 meq/l, and at 4 sites it increased significantly from the surface downward to the zone of pipe development. These variations may reflect leaching or possibly cation exchange in the upper parts of the soil. More significantly, at all but one site the percentage of sodium (SSP; $[Na \times 100]/(Ca + Mg + Na)$) ranged from 70 to 96 percent in the piping zone. Some results of the analyses of samples from Cornfield Wash, New Mexico, are shown in Figure 13A. Samples were collected at 25- to 50-mm increments from the land surface down into the pipe floor, 865 mm below.

Similar results have been reported by other authors. For example, In Lesotho, Rooyani found that ESP is not only high (6.1 to 12.2) in the surface soils, but there are also changes with depth. Not only is there a pronounced change in soil texture and clay mineralogy at the level of piping, but there is "a significant increase in SAR values ... [and] the sum of Mg and Na concentration is larger than the Ca concentration at this zone" (Rooyani, 1985, p. 520).

In a five-year study of piping in central Zimbabwe, a subhumid region with seasonal precipitation, Stocking (1981) found five chemical parameters that differ significantly between piping and nonpiping soils: Na^+, ESP, SAR, pH, and conductivity, or specific conductance. Moreover, he reported that ESP tends to increase downward in the soil to a maximum just above pipe roofs. From this he concluded that a:

... subsurface dam wall effect is responsible for the initiation of piping. High ESP values render the massive soil virtually impermeable because of spontaneous deflocculation blocking pore spaces. This forces lateral movement of water below the impermeable horizon. Lateral movement is then the initial control to the low angle segment of pipe systems.... Once established, the pattern of pipes is reinforced by the massiveness and shear strength of the soils with high ESP. (Stocking, 1981, p. 115).

This relation deserves further study, although Parker's own field investigations do not corroborate an ESP maximum above pipe roofs. ESP increased downward at the Cornfield Wash site (Fig. 13A) and at 1 other site among 8 that were sampled in New Mexico, Arizona, and Nebraska, as indicated in Figure 13B. However, ESP was maximum *beneath* the pipe roof at Cornfield Wash as well as at the other 7 sites (Fig. 13).

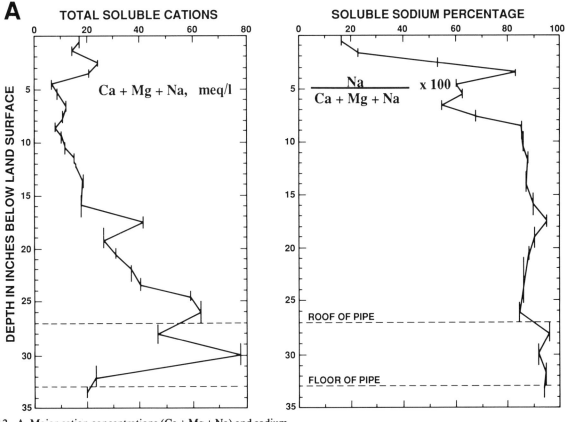

Figure 13. A. Major cation concentrations (Ca + Mg + Na) and sodium percentage (Na × 100/(Ca + Mg + Na)) in soil samples from Cornfield Wash, New Mexico. Vertical lines represent vertical range of samples. B. Vertical variations in soluble-sodium percentage at 8 study areas (Cornfield Wash, NM; Cuba, NM; Kutz Canyon, NM [2 sites]; Cameron, AZ; Chinle Valley, AZ; Benson, AZ [the low-percentage site]; and Harrison, NB). Analyses by Lynn M. Shown and Karl W. Ratzlaff (U.S.G.S unpublished report, 1963). After Parker (1964, Fig. 3).

Recently the overwhelming role of ESP in promoting piping has been questioned. For example, even though Rooyani (1985, p. 517–522) measured high ESP values in several piped soils in Lesotho, he also found that pipes may develop on duplex soils whether they are sodic (with ESP>15) or not. He also found that there is a significant increase in both the SAR and the ratio of (Mg + Na)/Ca, and he thinks that a result of a high proportion of Mg may be to enhance the effect of the Na. This corroborates McIntyre's findings that some soils swell and lose permeability with ESP as low as 5. Pipes near Davis, California, reported by Higgins (1984), also form in soils with low ESP but a relatively high Mg/Ca ratio, and Lynn Whittig (personal communication, 1988) has suggested that the high proportion of Mg may aid the dispersion of the clays there. Emerson (1977) has discussed the deleterious effects of magnesium on soil properties. As a result of such considerations, Rooyani proposes that SAR values "probably are more reliable criteria for explaining the physical deterioration of soils."

In the North Dakota Badlands, Bell found still other agents of dispersion. Subsurface water there gains organic acids from lignite. These "provide a kind of catalytic action through cation exchange by promoting a high degree of dispersion that may approach molecular subdivisions and aid piping by the disintegration of the sediments" (Bell, 1968, p. 246).

Aside from such technical considerations, the fact remains that many soils that have high cation concentrations, especially of Na, are prime candidates for piping erosion by subsurface water.

Permeability. An equally important factor is the relative permeability of strata or soil horizons. Some early investigators argued that zones of pipe development must be underlain by impermeable horizons that form perched water tables. Indeed, as described by Rooyani (1985) and others, shallow piping of the sort that leads to the development of small gullies tends to occur along the contact between relatively coarse-textured permeable horizons and underlying finer-textured, less permeable argillic ones in so-called duplex soils, as illustrated in Figures 8 and 9.

However, as noted above, Stocking (1981) has reported finding pipes *beneath* horizons with high ESP and impaired permeability.

Other authors, including Parker, have claimed that the ultimate level of piping is mainly controlled by the level of the pipe outlets, so that pipes at several levels, such as those in the wadi wall shown in Figure 11, may represent several stages of development during entrenchment of the local drainage. In the thick loess of northern China, pipe outlets are associated with prominent benches or terrace levels. Richthofen (1877) speculated on the origin of these levels, and they may in fact represent local base levels during the progressive deepening of the major streams. However, it is common in some gully walls to find some pipe orifices, generally the smaller ones, appearing well above the base of the cliff, commonly indicating perching by one or more horizons of higher permeability there. Similarly, in loess districts such as that in northern China, it is now generally recognized that thick accumulations of loess may contain one or more buried soils, or paleosols. The argillic B-horizons of these buried soils serve to impede downward percolation of water, and may aid development of soil pipes in the more permeable paleo–A horizons above them.

Similar variations of relative permeability in stratified sediments and sedimentary rock may also determine the level of pipe development. For example, in the alluvium cut by Town Dump Wash, near Bayfield, Colorado, Parker observed a large pipe outlet at a depth of about 2.5 m in a stratum slightly more sandy and permeable than the exposed strata above or below it. This zone of higher permeability serves as a drain during snow-melt runoff or rare prolonged rains, when the hydraulic conductivity is sufficient for the seepage flow to entrain and transport dispersed clay out to the gully. At all 8 of the sites selected for soil analysis the materials below the pipes were of low permeability.

Such controls by permeability are obviously very important. However, in massive sediments and rocks without aquicludes or aquitards, the level of piping is controlled chiefly by the depth of local drainageways or the water table, whichever is the higher. The frequent occurrence of pipe outlets at the level of gully floors reflects the importance of local base-level control where permeability is not a factor.

Climate

Piping in drylands occurs where, by definition, the climate is arid or semiarid, that is, where mean annual precipitation is less than potential mean annual evaporation. This varies with mean annual temperature, but many authors place the semiarid-subhumid precipitation boundary at about 600 to 700 mm. In the western drylands of the United States, piping is generally associated with regions where the mean annual precipitation is less than 400 mm.

However, as Jones (this volume) discusses, piping may also occur under subhumid and even humid conditions. Apparently of much greater importance than total rainfall is the intensity and duration of the precipitation and its seasonality. Where there are dry seasons, surface vegetation may be scanty and desiccation cracks may form in the soil. As noted above, such cracks are commonly associated with piping, where they serve to channel surface water down into the soil.

Jacobberger (1988) relates the initiation of piping and extensive gullying in central Mali to a decrease in precipitation that has promoted increased desiccation and decreased plant growth. The chief role of climate, then, may be its influence on the vegetation cover, as discussed below.

Paleoclimate and relict piping. Some sites of drylands piping seem virtually changeless over time. There the pseudokarst may be relict, reflecting different conditions that favored piping in the past. This is difficult to evaluate at very dry places, where there are only rare, widely separated opportunities to determine whether piping is currently active.

However, a few authors report instances of former pipe formation separated from the present by nonpiping intervals. In the Badlands of western North Dakota, Bell (1968) noted the presence of filled pipes in the Eocene Golden Valley Formation, below the Oligocene White River Formation, both now exposed high on Sentinel Butte, near the North Dakota–Montana border south of U.S. Interstate Highway 94. He proposed that piping was at a minimum during the late Pliocene and Pleistocene, and that the modern piping process began with the change to a semiarid climate with cold winters.

Vegetation cover

Piping seems to be most prevalent in drylands areas where vegetation is sparse, thinning, or denuded. Throughout the western U.S. drylands, the intensity of piping tends to be inversely related to the density of the vegetative cover. Reduction in plant cover may be related to changes in climate, including changes in the duration and intensity of precipitation. These factors have also been cited as initiators of the current cycle of gully and canyon cutting in the U.S. Southwest. Inasmuch as piping is so commonly related to gully development, it is not surprising that both might be causally related to a skimpy or reduced vegetation cover. As noted above, piping is also facilitated by generally high salt content, especially ESP and SAR; these salts also tend to decrease plant growth in soils where they are concentrated.

Most investigators of drylands piping have noted a relation between devegetation and the onset of piping. In many cases the devegetation resulted from interference: overgrazing (Bond, 1941), cutover and nonprotected slopes (Downes, 1946), recurrent burning (Thorp, 1936; Gibbs, 1945), and over-irrigation (Harris and Fletcher, 1951) are the chief causes, although overgrazing is by far the most common reported cause.

The effect of vegetative denudation, for any reason, is to decrease surface permeability and infiltration as a result of at least four factors: (1) a decrease in the surface litter and the organic-matter content of the soil; (2) an increased breakdown of soil aggregates and crumb structure due to rainbeat on the unprotected soil surface; (3) a raising of the surface and subsurface soil temperature, which aids in desiccation and cracking; and (4) a loss of the obstruction to runoff formerly provided by the vegetation cover.

PREVENTION AND CONTROL OF PIPING

Many authors have considered the problem of preventing piping under earth-fill dams and embankments. The usual solution is to install filters to prevent eluviation of soil particles or to compact the material sufficiently to reduce the hydraulic conductivity, or seepage velocity below the entrainment threshold. Aitcheson and Wood calculated that seepage should be limited to less than 10^{-5} cm/sec (better, 10^{-7} cm/sec) by compaction "at moisture conditions close to optimum moisture content" (Aitcheson and Wood, 1965, p. 445). They noted that some chemicals added during construction may effect a reduction in density and decrease in permeability. However, they also pointed out that filters "cannot prevent the passage of a deflocculated clay" in soils with appreciable exchangeable sodium, and cited the example of such a failure of a dam in Tasmania (p. 446). These authors suggested that in natural slopes and in dams already constructed of deflocculation-susceptible materials the risk of piping can be minimized by preventing deflocculation through chemical treatment.

Bell (1968, p. 255–256) went further. To avoid piping along canals and highways, he recommended excavating and replacing the piping-prone materials. He also warned that springs generated in excavations may "suddenly go below the surface along a desiccation crack or slump crack and start piping," but suggested no cure for this. Parker and Jenne (1967) recommended that highway routes across piping-prone areas should be relocated where possible. This would have saved expensive maintenance to many U.S. highways, including Highway 160 at Aztec Wash, near Cortez, Colorado. Where relocation is not feasible, they outlined several measures for drainage of runoff water from highways, such as asphalt curbs, lined drainageways, and carefully constructed drains and culverts. In some extreme cases, abutment slopes may have to be boxed with planking.

Other investigators have suggested other mechanical measures for controlling subsurface erosion. Gibbs thought that such control should focus on the "abnormal movement of the water" by encouraging dense and permanent vegetation or constructing barriers and channels (Gibbs, 1945, p. 146). Newman and Phillips (1957, p. 167) agreed, explaining that "A good vegetative cover builds fertility, alters the characteristics of the clay to some extent, encourages even infiltration and buffers the effect of rapid drying during the summer months by providing shade, roots and organic matter to prevent cracking." They recommended perennial, deep- and fibrous-rooted plants, but recognized that establishment of such a cover may be difficult, or even impossible on a tunnelled area with rainfall less than 432 mm (18 in) per year. They also noted that existing tunnels would have to be broken up to provide even infiltration of the surface. Commenting on control of piping in Zimbabwe, Stocking (1976, p. 35) concurs in the value of plants: "Vegetative control and good management are the only long-term solutions for tunnel-prone soils." He suggests the possible application of gypsum to the soil to lower the exchangeable sodium percentage, and agrees that existing and incipient tunnels must be destroyed by deep ripping or chisel-plowing.

No doubt some of the measures mentioned above work better than others, and some work in some piping-prone soils and not in others, but the most important aspect of prevention and control of piping is that it must be recognized as an important and destructive *subsurface* process that must be treated differently from *surface* erosional processes. Many conventional ways of addressing surface erosion, such as check dams and ditches, may actually exacerbate subsurface erosion by concentrating surface wash and channeling it into cracks and holes. Much progress has been made in understanding the workings and results of piping, and in learning how it may be prevented or treated, but these efforts will be in vain until subsurface erosion gains as much attention and concern as surface erosion has enjoyed in the past.

ABOUT THE FOLLOWING CASE STUDIES

As noted above, not only can piping aid in the formation of pseudokarst landscapes that resemble those developed by solution in soluble rocks, but enlargement of soil pipes can lead to the opening of large caves with integrated subterranean drainage systems. Such caves formed in insoluble materials are not uncommon in the world's drylands. They have served not only as shelters for animals and humans, but have had a variety of other uses. Artificially modified piping cavities in the thick Hwang-Ho (Yellow) River loess deposits of China's Kansu Province have been occupied by humans for as much as 1 m.y., and even recently, some of these were or still are in use as permanent homes for rural populations and their herd animals (Fuller, 1922; Thorpe, 1936). In Italy and Australia, ornate doors protect such caves, used as root cellars, and in Cappadocia, natural pipes opening on hillsides were once used as Coptic shrines and churches. The Qumran Caves in Jordan, once inhabited by the Essenes and where the Dead Sea Scrolls were found in the 1940s and 50s, are also in apparently noncalcareous, highly piped badlands, judging from photographs (DeVaux, 1967). Clausen (1970) has described badlands caves in Wyoming. However, none of these (Brainerd Mears, personal communication, 1988) is as extensive as Officer's Cave in the John Day Country of eastern Oregon. This feature is described in the following case study.

Eluviation can be partly responsible for the initiation of soil cracks and pipes that grow into large openings. More subtly, locally concentrated infiltration and percolation may not only aid in soil-profile development, but may transport sufficient sediment to lower depths to create shallow depressions and small lake basins. This process may have been partly responsible for an extensive landscape with hundreds of depressions in siliceous sandstone on the flat crest of the Chuska Mountains in northwest New Mexico, as described by Wright (1964). Another example is illustrated in the second case study below, by Warren Wood, of playa lake basins on the Staked Plains of Texas and New Mexico.

CASE STUDY: OFFICER'S CAVE, A PSEUDOKARST FEATURE IN THE JOHN DAY COUNTRY, GRANT COUNTY, EASTERN OREGON

Garald G. Parker, Sr.

Officer's Cave is the largest known pseudokarst cave in North America, developed entirely by piping, sapping, and collapse. It is in the Turtle Cove area of Big Basin, as shown in Figure 14. The cave is named after the Floyd Officer family, on whose former homestead the cave is located.

It is a well-known landmark of the region, and has been partially explored by many local citizens. In the summer of 1914, a party of geologists from the Oregon Bureau of Mines and Geology explored the cave and left the only published early description, quoted here in part:

Above the entrance to the cave is a cliff 20 feet high of approximately undisturbed John Day sediments. The entrance is a large room, approximately 15 feet high, 15 feet wide, and 100 feet long. Many large pieces have fallen from the roof, making progress difficult. Beyond this room the cave narrows to a single channel often 15 feet wide, but as low as two feet in height for some distance, to another large room somewhat like the first, though smaller. Along the bed of the passage is much drift material from the surface, but in the chamber near the entrance the walls and roof are blackened, presumably by the bats which are said to pass the winter here. After penetrating approximately 500 feet of this cave, the present survey party returned to the outside and made an examination of the surface. At a distance of 700 feet, more or less, S80°E from the entrance there is a large depression 20 feet in depth and 60 feet in diameter, into which several water channels run, and what was thought to be the upper end of the cave was found. This cave has many of the features of a limestone cavern but does not, of course, have either stalactites or stalagmites, and is confined, so far as was observed, to a single channel" (Collier, 1914, p. 13).

The cave and its setting

Officer's Cave underlies and essentially follows the trend of a narrow linear hill that trends about S80°E, which we call Officer's Cave Ridge. Paralleling the ridge on both north and south sides are dry, hanging valleys. These are drained at some places by large sinkholes, several as much as 5 m in diameter. Numerous minor sinkholes occur along the flanks of the ridge, and all drain directly into Officer's Cave Stream. These dry-valley sinks are integral parts of the master subterranean drainage system of the cave complex. Associated with the sinkholes are pipes, natural bridges, and several cave rooms developed like beads on a string. These, together with the subterranean drainage, give the area a karstlike development. This setting is shown in Figure 15.

Officer's Cave Ridge. Officer's Cave Ridge is one of a myriad of huge rotational landslide masses that characterize Big Basin. Where late Oligocene and early Miocene altered white volcanic ash and gray-tan tuff beds of the John Day Formation are overlain by the Picture Gorge Basalt, steep cliffs are common. When the cliff faces become too high and steep, or in places even overhanging, the softer John Day beds slump downward in huge

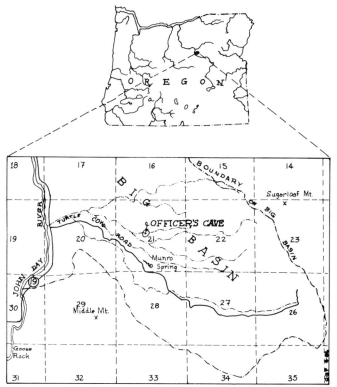

Figure 14. Index map of Officer's Cave area, near John Day, eastern Oregon.

masses, generally with a backward rotation. As this situation is repeated, the resultant landscape becomes an uneven, jumbled, distorted mass of hummocks, the oldest ones farthest from the retreating cliff face.

These erosional effects are actively changing the shape and bulk of Officer's Cave Ridge and the cave itself. Each runoff event, whether from snowmelt or thunderstorm flooding, results in surface runoff into the sinks that are aligned along the north and south dry valleys. Soon a torrent of cream-tan, roily water is discharged from the mouth of Officer's Cave Stream. Slabs of silty clay, parts of slumped rotational land slide blocks, partially block the channel. These and other smaller slump blocks from the walls of the sinks and their tributary area, plus steep wall-face erosion from higher up on the shoulder of Officer's Cave Ridge, all contribute to the load of sediment carried out of the ridge and on to discharge into the John Day River.

Officer's Cave, 1962–1964. In July 1962, the writer and an assistant made a five-day reconnaissance survey of the cave complex and its environs (Parker and others, 1964). Beginning at cave entrance IV, the main entrance to the cave (Fig. 16A), we explored and mapped the cave system a distance of about 67 m. Beyond, the passageway was choked by a rock fall from a large pipe. The debris included a few basalt cobbles, and leaves and stems of weeds and grass in a jumble of John Day rubble. This debris was partly from the near-vertical wall of the ridge adjacent

Figure 15. Officer's Cave Ridge at the cave site, June 1984. 1. West-facing, grassy, rubbly slope into Officers Cave Stream Valley. 2. Western remnant of the hanging dry valley that borders Officer's Cave Ridge on the north side. 3. Erosion remnant between small sinkhole (base of scar on right) and huge sinkhole below 4 at left. 4. Sinkhole. 5. Officer's Cave Ridge, north side covered by small slipettes. 6. Rubbly surface of tan–gray volcanic ash. 7. Smooth surface of overlying bed of volcanic ash. 8. Mat of vegetation growing on outer floor of the "front room." 9. Cave entrance (formerly IV). 10 through 13. Dry valleys and ridges south of Officer's Cave Ridge.

to the sinkhole; the remainder came from the wide funnel intake of surface-water runoff. We also explored cave entrances I, II, and III (Fig. 16A). Entrance I is along the bed of Officer's Cave Stream, an ephemeral creek that flows only after heavy snowmelt or intense rainfall. Entrances II and III, which apparently did not exist when Collier's study was made in 1914, were penetrated from mouth to streambed by clambering over and crawling through the clay-block landslide debris that covered their steeply pitched floors.

The cave stream channel slopes about 5°, or 8 percent, to the west and meanders through a chain of domed rooms of varying size, the largest being about 24 m (80 ft) in diameter. Its roof rises about 9 m (30 ft) above the floor, which is a huge jumble of fallen roof slabs absent only where the stream channel has been cleared of roof debris by flood flow. The channel meanders much like many surface stream channels; it is flat-bottomed and generally about 1 m wide. In several places the stream flows through tunnels only 0.8 to 0.9 m high.

In August 1964, we returned for five days to make a hydrogeological survey of Officer's Cave Ridge and the cave complex (Parker, 1966). This time we found the cave streambed relatively clear of fallen roof and wall debris, and we were able, with some excavation, to map the cave from the mouth at entry level I for a distance of 345 m (1,132 ft), where progress was halted by a large debris fall. We also discovered that a tubular chute had opened in the south side of the "front room" tight against the wall; it was not present in 1962. This pipe, or chimney, is almost vertical, following the cracking pattern of the clayey montmorillonitic bedrock of the ridge (Fig. 15). Being about 1.1 m in diameter, the pipe was large enough to allow us to descend through it by rope ladder directly to the cave streambed.

In 1962 and 1964, each of the cave entrances at II, III, and IV sloped backward (eastward) into the hill at angles up to about 45° (Fig. 16A), but the bedding planes dip southeastward at angles of less than 5°. In other words, even though several rotational slip blocks are present in the western end of Officer's Cave Ridge, block-slump rotation on the slip surfaces can account for only a minor part of the steep pitches of these entries. Apparently these slopes have been guided or controlled by fracture sets along which preferential erosion is dominant.

Fracturing also controls the shapes of the cave entrances, which tend to angularity rather than roundness (Figs. 16 and 17). Strong sets of vertical joints give the cave mouths near-vertical walls, a condition that Fuller (1922) and Thorp (1936) also noted in the piped forms in the loess of Kansu Province, China.

Enlargement of the cave is accomplished by spalling of montmorillonitic clays in the roof and walls. These materials swell when wetted and shrink when dried, and are most likely responsible for most of the fracturing of the weathered, altered tuff and ash in which the cavern system is developing. Spalled fragments then, in turn, are removed particle by particle in ephemeral runoff that is directed across the floor and discharged directly into Officer's Cave Stream. This runoff is derived from three sources: (1) snowmelt and storm water seeping through the

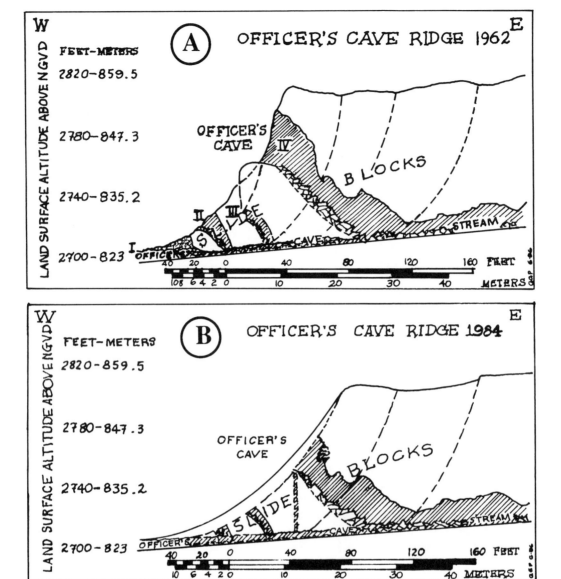

Figure 16. West–east section through the western end of Officer's Cave. A. Surveyed 1962 conditions. B. Inferred 1984 conditions.

roof and walls from the land surface above the cave by way of the numerous fractures in the bedrock; (2) water running into the sinks in the hanging valleys; and (3) water entering directly through the cave mouth either by wind-driven snow or rain, or by runoff over the cliff face and cave portals.

Base level of the cave stream is locally controlled by a hard layer in the surface channel about 18.7 m (20 yards) downstream from the outlet of the level I opening. As this layer erodes downward, the local base level for drainage will drop, and new and deeper caves probably will develop.

Little is known of the flow characteristics of Officer's Cave Stream. It flows ephemerally and probably with greatest volume and force after high-intensity rains from thunderstorms of late spring, summer, and early fall, and following rapid snowmelt runoff caused by the occasional warm, moist Pacific winds of the winter and early spring. At such times, flash floods would pass rapidly through the cave drainage system, their waters gathered through the numerous tributary sinks, and would remove from the stream channels the toes of the debris trains that lay on the sloping cave floors. Lacking support at the toes, the trains of debris would then move by slipping downslope and, somewhat depleted, come to rest again with toe support in the cave stream channel after the floods have passed.

Erosional removal of old slump blocks by the stream at the base of the cliff face triggers the breaking off of a new slump block and the establishment for some indefinite time of a young slide-block mass. These slump masses appear to weld their component parts together only to be attacked by piping and development of new cave entries, as seen in Figure 17. Eventually Officer's Cave Stream undercuts and saps the new slump mass, weakened by piping above, and then in one final big storm runoff event the slump mass disappears. The cycle then repeats itself and the ridge becomes shorter. Concomitantly, the cave lengthens headward (to the east).

Changes in the cave complex, 1962–1984. The appearance of Officer's Cave and Ridge is changing notably. When Coller (1914) described Officer's Cave, he noted that the entrance was about 4.5 m (15 ft) square, and that the clifftop was about 6 m (20 ft) above the cave roof. In 1962, the cave entrance was rectangular, and the clifftop only about 4.3 m (14 ft) above the ceiling of the cave. At the west end, where Collier reported the cave's front room to measure 4.6 m × 4.6 × 30.5 m (15 × 15 × 100 ft), in 1962 I found a nearly seven-fold enlargement, to 10.7 × 13.3 × 30.5 m (35 × 43.5 × 100 ft). Further, whereas Collier noted only one cave entrance, the writer found four: I at the cave stream outlet, II and III in what probably was a rotational slide block dropped from the western upper face of the cliff, and IV the enlarged cave "front room" of Collier (Figs. 16A and 17). Also, between 1962 and 1964, several new sinks had opened in the dry valleys that bound Officer's Cave Ridge (Parker, 1966).

I returned to the site on June 17, 1984, during a rainy, drizzly period that made it almost impossible to walk to the cave site. The John Day volcanic ash and tuff soil had become, when wetted, a greasy, slick clay, and it was not possible to examine the interior of the cave. However, one could examine the external conditions and changes at the cave mouth and at the western end of Officer's Cave Ridge (Figs. 15 and 18).

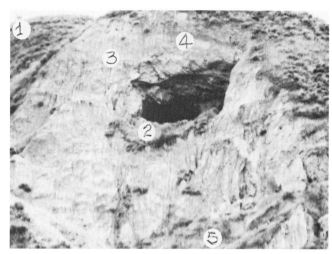

Figure 18. View S80°E of the western end of Officer's Cave Ridge and the main entrance of the cave, June 1984. 1. North-facing shoulder of Officers Cave Ridge. 2. Vegetation mat at the outer entrance to the "front room." 3. Contact of ash above the tuff. 4. Eastern edge of a set of vertical fractures that extend from the crest of the ridge. 5. Slide block carrying with it part of the floor mat from the cave above.

In 1984, although it was not possible to measure the change of location of the entry with respect to the top of Officer's Cave Ridge, it was apparent that considerable change had taken place, especially at the western end of the ridge. Most of the previous (1964) west face containing cave entries II and III had been eroded away by sapping and slumping into the cave stream. Another thick rotational (12 m) slumpblock had slipped diagonally down and to the south, carrying with it at least a third of the old "front room" (Fig. 16). In Figure 18 the western part of the old floor with its mat of grass and weeds is seen atop the rubble heap now resting at the foot of the west-face scarp.

Physical and chemical characteristics of the cave materials

Several factors are involved in the erosion of the materials of Officer's Cave Ridge. Montmorillonitic materials shrink and crack when dried after wetting. This allows water to move through drainable cracks until they are closed by swelling of the wetted wall materials. However, where coarser, permeable materials wash into an open crack, the crack thereafter has the capacity to allow water to seep through its permeable filling. Another factor is the strong tendency for dispersion and disaggregation of particles on all wetted surfaces. Being high in sodium, the clay and silt particles are dispersed and thus are easily detached in moving water, even that which seeps along fractures. The detached particles are then carried away in suspension, giving the water a whitish, chalky, turbid appearance.

In 1962, samples for laboratory analyses were collected from fresh-appearing solid bedrock in the north walls of cave entries II and IV (Fig. 16) about 6.2 m (20 ft) in from the portals. The materials are from an altered, montmorillonitic tuff that is

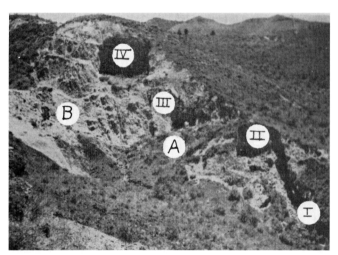

Figure 17. West end of Officer's Cave Ridge as it appeared in August, 1964. Cave entrances are marked I, II, III, and IV. The floor of No. I is the bed of Officer's Cave Stream. The vertically walled cut, about 1.5 m wide, connecting entrances I and II, developed as a product of sapping and slumping directly over the stream bed. Entry II appears to have been the front part of the old cave entrance, now numbered IV, that was carried as a slump block, A, in a downward slip to its 1964 position, possibly a decade or more earlier. The entire block was subsequently eroded away prior to June 21, 1984, when Parker revisited the cave after a 20-yr absence. At that time there were only two cave entrances, old numbers IV and I (see Fig. 18 for comparison). B marks the slip surface of recent slumping.

TABLE 1. ANALYSIS OF SELECTED SAMPLES FROM OFFICER'S CAVE*

Sample no.	Feet above cave floor	Soluble cations in meq/l†				SSP§	Texture	Saturation soil paste			
		Ca + Mg	Ca	Mg (by diff.)	Na			Relative cohesiveness	pH	Percent moisture	S.L.** percent
Cave level II											
4-2	4	0.4	0	0.4	12.0	96.8	clay	very sticky	8.13	67.4	25.5
5-2a	4	0.0	0	0.0	12.0	100.0	clay	extrem. sticky	8.00	83.6	24.5
6-2b	4	0.0	0	0.0	12.0	100.0	clay	very sticky	8.13	64.2	27.0
Cave level IV											
7-4	5	3.56	0	3.56	18.2	83.6	coarse sandy clay‡	sticky	7.98	53.8	30.7

*Analyses by Lynn M. Shown (after Parker and others, 1964)
†Milliequivalents per liter
§Soluble sodium percentage = $\frac{\text{meq/l Na}}{\text{meq/l (Ca+Mg+Na)}} \times 100$
**Shrinkage limit = Percent moisture at maximum shrinkage
‡This sample composed chiefly of sand-sized, resistant silt–clay aggregates

now a well-consolidated, pale-olive clay-siltstone. When dry it fractures into angular granules of sizes grading upward from approximately 0.002 mm. After soaking for 24 hr, many of the silt-size aggregates disintegrate into clay-size particles smaller than 0.002 mm. Comparison of the high saturation-moisture percentages with the much lower shrinkage limits (Table 1) shows that the material does not have to be nearly saturated for cracks to appear when it dries.

X-ray diffraction analysis revealed that the clay-size fraction is dominantly smectite, but there is some mixed-layer smectite and illite in which illite is only a minor component. The clay is saline and has a pH near 8 (Table 1). The high soluble-sodium percentages show that sodium is the main soluble cation, even though two samples contain some soluble magnesium. These magnesium concentrations probably are not high enough to repress the dispersion effect of the sodium, or perhaps the magnesium may even aid it. The saturation extract contained no soluble calcium, but digestion of sample cuts in dilute HCl disclosed the presence of a small amount of calcium carbonate. Microscopic analysis revealed the presence of a few small selenite crystals.

Case study summary

Officer's Cave and its related erosional features are developed in relatively insoluble clastic rocks in altered montmorillonitic tuff and volcanic ash. The appearance of the piped and caved area is that of a limestone terrain undergoing solutional degradation with resultant development of a karst topography. It has two hanging, dry valleys and contains sinkholes, natural bridges, and caved-in subterranean passages; drainage is internal to a cave-stream system. To this karstlike condition the term "pseudokarst" is applied. I have seen pseudokarst developed in clastic rocks elsewhere in the drylands of the western United States, but nowhere have I seen so large a karstlike cave.

The bedrock materials are not notably saline, calcareous, or gypsiferous; therefore, the development of the Officer's Cave pseudokarst is not primarily a result of solution. Rather, it comes about from subterranean erosion that is accomplished by water seeping and running through structural joints, dislocation cracks, and over the cave walls and floor, and by the erosive power of the ephemeral Officer's Cave Stream. Many of the bedrock fractures result from desiccation of the montmorillonitic clays that are so prominent in the John Day Formation. Water also enters the rocks through joints created by block slump movements of unstable topographic forms.

When dry, the John Day sediments are extremely hard, but when wet, the materials of the weathered surfaces and the walls of joints and cracks become plastic and slippery, are dispersed and readily disaggregated, and offer very little resistance to erosion by moving water. Likewise, movement along slip-plane surfaces is eased by the very slipperiness of the crack walls. The cave and its associated sinks and smaller caves are actively eroding and changing shape under current weather conditions. Apparently Officer's Cave has been undergoing rapid enlargement since about the turn of the century. Enlargement has been accomplished chiefly through slumping and collapse as a consequence of piping, spalling of the cave walls and roof, and particle-by-particle removal of the cave floor and debris blocks lying on it.

Figure 19. Playa lakes north of Lubbock, Texas. Diagonal line is U.S. Interstate Highway 27. Air photograph courtesy of High Plains Underground Water Conservation District.

The separated particles are taken away in suspension in Officer's Cave Stream, by means of which they reach the John Day River about 2.4 km (1.5 mi) away. Main episodes of enlargement occur at times of snowmelt, thunderstorm infiltration, and runoff.

CASE STUDY: SOLUTION AND PIPING DEVELOPMENT OF PLAYA LAKE BASINS ON THE LLANO ESTACADO OF TEXAS AND NEW MEXICO

Warren W. Wood

INTRODUCTION

The playa basins of the Llano Estacado (Southern High Plains; Figs. 8 and 9, Chapter 14, this volume), are one of the most interesting geomorphic features of the otherwise nearly characterless alluvial-eolian plateau of Texas and New Mexico. Approximately 30,000 of these flat, shallow basins, some of which are shown in Figure 19, occur above the zone of saturation, average less than 4 m in depth, tend to be circular in shape, and have flat floors generally less than 0.5 km in diameter at the lowest closed contour. Solutes in these ephemeral lakes are generally less than 150 mg/l, but suspended solids range from 400 to 500 mg/l, more than 50 percent of which is a smectite clay. There are approximately 40 larger, permanent saline lakes in the area, but many of these appear to have developed by a process of dissolution of underlying salt beds and subsequent collapse (Reeves and Temple, 1986; Gustavson and others, 1980) and are not considered in this study.

Reeves (1966) reviewed the numerous theories on the origin of playa basins of the Llano Estacado and concluded that most were probably eolian in origin. I agree that the initial origin was almost certainly eolian, but playa basin development is believed to occur largely by hydrologic processes (piping and solution) rather than eolian processes. This statement is based on: (1) a strong tendency for the playa basins to assume circular shapes, suggesting concentric expansion; (2) a pronounced development in some areas of series of playa basins in linear patterns that are not aligned with present or past prevailing wind directions and may reflect increased permeability due to fractures; (3) an apparently random location of most playa basins relative to topography; (4) a lack of lee-side dunes adjacent to many of the playa basins; (5) dunes, where they occur, have inadequate volume to explain the depressions; (6) very high clay content of dune material suggests that the dunes formed from clay aggregates formed in playa basins after these developments rather than original integral parts of playa formation; (7) the presence of playa basins as chains or series of shallow depressions along influent, occasionally incised stream channels (which suggests that significant internal removal of sediment has occurred by piping or other similar mechanisms, because fluvial transport and deposition of sediment there would otherwise quickly fill, then breach the stream-channel depressions during runoff events); and (8) the occurrence and documentation of solution and piping in the area (Osterkamp and Wood, 1987; Wood and Osterkamp, 1987).

Piping

Fenneman (1923, p. 126–128) and Rubey (1928, p. 418–422) document piping features related to gully systems of the High Plains. Gile and others (1981, p. 74, p. 101–102) discuss pipes near Las Cruces, New Mexico, that are typical of upper Pliocene argillic soils of the area. In a study of caliches of the Southern High Plains, McGrath (1984, p. 48–51) observed numerous pipes in the Miocene-Pliocene Ogallala Formation and overlying Pleistocene Blackwater Draw Formation. Piping of clastic material in the vicinity of stream channels and in soils around borrow pits or other high-slope features of the Southern High Plains is often pronounced. Parker (1963) has shown that piping commonly is a precursor of gullying.

Experiments of artificial groundwater recharge from a water-spreading basin north of Lubbock conducted by the author resulted in the opening of numerous pipes, which formed in the bottom of the spreading basin. The openings averaged nearly 0.1 m^2 in area and were at least 10 m deep. Another test, involving a well approximately 50 km north of Lubbock, also indicated that recharge occurred mostly through secondary openings (Brown and others, 1978, p. 27–29).

Diagnostic evidence of subsurface pipe formation and sediment transport from the voids was obtained by Dvoracek and Peterson (1970, p. 208–213) during recharge-well tests at Lubbock. Creation of cavities during the tests, at depths up to 15 m, resulted in voids as large as 2 m^3. Subsurface photography confirmed the measurements of void sizes and shapes. Wells drilled for other recharge studies near Hereford, Texas, about 50 km southwest of Amarillo (Keys and Brown, 1978), yielded information that strongly suggested the movement of water through secondary voids. Recovered core from a site near Stanton, Texas, showed poorly sorted, till-like material (Wood and Ehrlich, 1978), indicative of transport of fine sediment from the surface into pore spaces and solution openings. Additionally, this work documents the presence of solute openings by the rapid transport of particulate material in the subsurface.

Samples of soil and the underlying sediments from the edges of several playas generally show evidence of substantial downward movement and redeposition of kaolinite, illite, and other birefringent mixed-layer clays typical of vertisols. Samples of sediments underlying playas show extensive microfracturing, the fracture walls often lined with clay skins. The occurrence of disrupted aggregates of matrix material clogging the microfractures of some samples suggests possible plastic movement of silt and clay during wetting of recharge episodes. Microfractures are suggested as conduits for recharging water and movement of particulate matter. Thus, transport of particulate material in the unsaturated zone occurs over a size fraction that ranges from individual clay-size particles to aggregates of sand-size particles and may be an important process of mass transport in this system.

Dissolution

Dissolution of caliche and carbonate cement is believed to be critically important to the mechanisms of playa basin formation on the Llano Estacado, and is intimately involved with piping in this system. Several playa basins exposed by excavation and examined by the author have shown significant dissolution of caliche. Reeves (personal communication, 1987) has drilled over 20 basins and found significant removal of caliche in both the Black Water Draw and Ogallala. In the playa basins examined by the author, the amount of near-surface solution of carbonate can be shown to account for 30 to 40 percent of the volume of the basin. Wood and Petraitis (1984) have suggested that particulate organic material from the surface moves tens of meters into the unsaturated zone beneath the playa, where it is oxidized to CO_2 by aerobic bacteria and oxygen from the atmosphere. Because CO_2 is generated deep within the unsaturated zone, rather than in the shallow soil zone, concentration increases to very high values while establishing a gradient to the surface. Thus recharging water moving downward comes in contact with an increasing concentration of CO_2 and becomes thermodynamically subsaturated with respect to carbonate minerals, and dissolves them at depths of tens of meters below the soil zone. Dissolution of carbonate has two effects on playa basin development. First it creates part of the void space occupied by the basin and, secondly, by removing the ubiquitous carbonate cement it causes destabilization of the lithologic framework. Destabilization in turn promotes piping and further introduction of organic material from the surface, which on oxidation, promotes further dissolution.

Distribution of playa basins

Playa basin density generally increases from southwest to northeast in the Llano Estacado. This trend is believed to reflect both changes in precipitation and evaporation as well as soil types that result in an increased runoff toward the northeast. Increased runoff to the playa basins has the potential for increasing potential recharge flux, and thus for increasing the potential for moving particulate material into the unsaturated zone. Where infiltration rates are high or precipitation is low, runoff from the surface is reduced, resulting in insufficient recharge to cause extensive playa basin development.

Numerous sets of aligned playa basins and stream-channel segments of the Southern High Plains correlate with linear features (Finch and Wright, 1970; Finley and Gustavson, 1981). Reeves (1970, p. 63) asserts the solution along fractures in caliches of the Clovis, New Mexico, area results in surficial lineaments. In several places, northeast- and northwest-trending lineations defined by playa basins and stream channels continue as canyons off the escarpments, thereby suggesting structural control rather than eolian processes. It is proposed that fracturing or breaking of the caliches and other indurated horizons creates zones of weakness and increased permeability, and has promoted piping along the trace of the feature. It should be noted, however, that there are few documented fractures in the Ogallala and younger sediments, either because of a lack of structural activity or because the unconsolidated nature of some of the sediment does not preserve or transmit them.

Role of recharge in playa development

Early workers (Johnson, 1901) recognized that virtually all drainage on the Llano Estacado is internal toward the playa basins, but the important hydrologic question is whether the lakes evaporate to dryness or may serve as recharge sites. If recharge can be shown to occur through the playas, then the argument for solution-piping as their origin is significantly strengthened.

Theis (1937, p. 565), who otherwise contributed so much to the understanding of the hydrology of the region, devoted only one paragraph to the source of recharge and suggested that "recharge is favored in areas of sand dunes." He also implied that playa basins probably were not the source of recharge by stating that water in the playas "usually, but not in all localities remains through months of drought." Cronin (1964) felt that the low permeability of the playa lake sediments significantly inhibited recharge and that, if recharge does occur, it must "infiltrate through the sandy belt which commonly surrounds the perimeter above the clay and silt." However, other workers provide data to suggest that recharging does occur through the basins.

Investigations in the Texas segment of the Llano Estacado by White and others (1946) strongly suggest that significant recharge occurs through the playa basins. This conclusion was based on hydrographs of 5 wells adjacent to playa lakes, which showed rapid water level rises after the playas had filled with water. Similar data showing the rise of groundwater levels related to lake stage are given by Rettman and Leggat (1966). However, there were no corresponding hydrographs of wells in the adjacent outerbasin areas to serve as control in either study, and the data therefore suggest that recharge from playa basins occurred but are not conclusive. Shallow groundwater levels beneath the campus of Texas Tech University in Lubbock have been identified by Kier and others (1985) as resulting from the recharge of playa basins that have received increased storm runoff due to urban development.

Available data on solutes in lake water are consistent with a hypothesis that groundwater recharge occurs through playas. That is, if water were lost from the lake largely by evaporation, solutes should increase significantly over time. Data from 7 lakes near Amarillo (Lotspeich and others, 1969, Fig. 4) indicate that specific conductance, which is directly proportional to total dissolved solids, doubled over several months as the lakes went to dryness. This suggests that only about one-half of the water was lost to evaporation; the rest must have been recharged to the aquifer. Because no water-budget study on the lakes was conducted, no quantitative interpretation can be made. However, it is consistent with the hypothesis that a significant amount of the water collected in playa basins is recharged to the groundwater rather than evaporated.

The mineralogy and vegetation of the playa basins give no indication of saline condition. No evaporite minerals such as gypsum, halite, or mirabilite have been reported in the literature on these playas or found in playas I have investigated. Furthermore, the clay mineralogy of playa deposits is identical to that of upland source areas (Allen and others, 1972), which suggests the lack of diagenesis that is characteristic of lake sediments in a saline environment. The vegetation associated with the playa basins differs significantly in both genera and species from that of the surrounding plains (Reed, 1930). However, this is related to the presence of a heavy clay soil in most playa floors rather than to salinity. (Reed, 1930) states that halophilic plants, such as *Pseidoclapia arenaria,* grow around the larger saline lakes. These plants, however, are not associated with the playa basins described in this case study. This observation favors the view that a significant amount of water is recharged to the aquifer rather than being lost by evaporation.

The concept of concentrated local recharge from playas, rather than diffuse regional recharge over the entire basin, is consistent with tritium data. From 1958 through 1960, 62 water samples were collected from the High Plains aquifer in Lea County, New Mexico (unpublished U.S. Geological Survey analyses). Most of these samples had tritium value less than 5 TU (tritium units). Considering the analytical techniques and sample storage methods used at the time of sample collection, these values are probably within the error band of zero TU. However, approximately 20 percent of the samples had tritium values in excess of 5 TU, which indicates the presence of water recharged since 1952 (post-atom bomb). If recharge had been reasonably uniform throughout the area, all wells should show the presence of some tritium. However, if recharge occurred only in discrete zones, then some wells would show high values of tritium whereas others would show little or none. Wells yielding the higher values of tritium were not field checked, but the data are consistent with the concept that recharge occurs in discrete areas rather than uniformly throughout the area.

Another observation supporting the concept of recharge from the playa basins is based on chemical analyses of solutes from the unsaturated zone. Solute data from several depths in the unsaturated zone in an outerbasin area were obtained adjacent to the Lubbock, Texas, airport (Table 2). These data show that the concentrations of dissolved solids are at least twice that of the typical groundwater and imply low recharge flux through this area. Conversely, dissolved solutes from samples collected from the unsaturated zone at the edges of a playa lake in the city of Stanton, Texas (Table 2), have a lower concentration of solutes than typical groundwater in the area and are significantly lower than samples collected from the unsaturated zone in the outer-basin area. Although two sample sets do not constitute unequivocal evidence of recharge in the playa basin or the lack of it between basins, it is consistent with the recharge hypothesis outlined above. Stone (1985), using chloride concentrations obtained from soil water in the unsaturated zone, observed that, of the three settings he measured, the greatest recharge flux occurred beneath the center of a playa basin.

A water-budget study by Havens (1966) provided direct evidence that a significant amount of the water collected in 6 playa basins in New Mexico was recharged. In a similar water-budget approach, Myers (in Cronin, 1964) calculated that from 35 to 50 percent of the water collected in 1,348 lakes in a 4-county area of Texas was recharged to the aquifer. Myers' study showed that the larger the drainage area of the lake the greater the percentage of lake water recharged. More recently, the U.S. Bureau of Reclamation (1982) produced a detailed water-budget study of 29 playa basins in 13 counties covering a large part of the Llano Estacado. The study showed that, on the average, from 75 to 80 percent of the water entering the basins was recharged. The cumulative difference in volume of water infiltrated over

TABLE 2. CHEMICAL ANALYSES OF WATER FROM SATURATED AND UNSATURATED ZONES IN THE LLANO ESTACADO

	Lubbock Airport	Lubbock Airport	Stanton	Typical Ogallala
Latitude	33°40'00"N	33°40'00"N	32°07'30"N	33°44'20"N
Longitude	101°47'20"W	101°47'20"W	101°46'20"W	101°46'20"W
Date	4/14/72	4/22/84	3/10/72	6/26/83
Depth (M)	7.9	16.2	7.6	69
Silica (SiO_2) (mg/l)	74	68	36	41
Calcium (Ca) (mg/l)	200	170	30	50
Magnesium (Mg) (mg/l)	110	100	17	40
Sodium (Na) (mg/l)	360	130	5.8	56
Potassium (K) (mg/l)	10	19	7.4	13
Strontium (Sr) (µg/l)	3600	4100	500	1200
Lithium (Li) (µg/l)	180	190	10	100
Bicarbonate (HCO_3) (mg/l)	250	279	190	260
Carbonate (CO_3) (mg/l)	0	0	0	0
Sulfate (SO_4) (mg/l)	690	400	7.2	22
Chloride (Cl) (mg/l)	520	320	4.6	78
Fluoride (mg/l)	4.3	3.0	1.3	2.4
Nitrate (NO_3) (mg/l)	6.1	.41	0.8	4.0
Total Phosphorus (P) (µg/l)	40	1300	92	--
Boron (B) (µg/l)	610	52	90	290
Dissolved Solids (mg/l)	2090	1340	204	445
Specific Conductance (micromhos @ 25°C)	3200	2075	328	735
pH	7.79	8.30	7.50	7.48

water lost by evaporation in a typical playa during this study is illustrated in Figure 20. Certainly, the sand dune area of the Llano Estacado (Cronin, 1964) may be a locally important recharge area, as are the areas near the few ephemeral streams, but groundwater head distributions (Cronin, 1969) do not support the concept that a significant part of the total recharge to the High Plains aquifer in the Llano Estacado is from these sources.

Although the water-budget studies estimated a wide range in amounts of groundwater recharge, it is useful to compare these values with those calculated using the method of regional analyses outlined by Theis (1937). Luckey and others, 1986, using the Theis approach and more recent hydrologic data, found that a predevelopment recharge rate of approximately 0.3 cm/yr gave the best fit to a mathematical model of the aquifer system. This value is approximately 30 percent less than that calculated by applying the Bureau of Reclamation data to the amount of regional recharge estimated by Wood and Osterkamp (1984). Because none of the above-cited water-budget studies considered water losses from transpiration or replenishment of soil moisture, these studies tend to overestimate the amount of groundwater recharge and thus may account for the differences in calculated recharge between the two approaches.

Havens (1966), Cronin (1964), and the author have observed that the water level in many playa lakes drops rapidly after they are first filled and then declines slowly after the water level falls to a height of approximately 0.5 m above the basin floor. This observation supports the concept that recharge is occurring in the playa basin and suggests that water-level decline may reflect lower permeabilities of the floor deposits compared to the basin sides. Cronin (1964) and Havens (1966) mention a sandy area that typically surrounds the silt-clay filling of the playa basin floor. A hydrologic condition is envisioned in which groundwater recharge occurs in an annulus area adjacent to and surrounding the lake floor. If it can be assumed that most of the recharge to the aquifer occurs in this area, then the average recharge flux in the annulus area can be estimated. The playa drainage basins average approximately 2.6 km^2 in area. This value was determined by dividing the area not drained by streams (78,000 km^2) by 30,000 playa basins. The average playa floor is approximately 400 m in diameter. If we assume, based on topography, that the average annulus-shaped recharge area surrounding and adjacent to this 400-m-diameter floor is approximately 100 m wide, we then have 1.6×10^5 m^2 of recharge area per drainage basin, or a total of approximately 5×10^9 m^2 of potential recharge surface by the 30,000 basins. Assuming that all recharge, (0.3 cm/yr or 3.24×10^8 m^3/yr) occurs through the annulus area, then the average recharge flux through the annulus area is between 4 and 5 cm/yr.

Estimates of particulate flux through the basin

Quantitative estimates of the amount of eluviation and micropiping are difficult to obtain. A recent investigation of CO_2 gas flux in the unsaturated zone (Wood and Petraitis, 1984) provides an indirect method of estimating particulate flux from

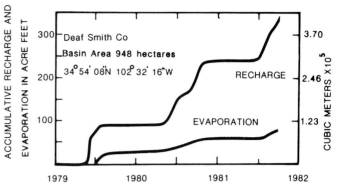

Figure 20. Cumulative recharge and evaporation from a typical playa lake basin on the Llano Estacado. Adapted from U.S. Bureau of Reclamation (1982).

the surface into the unsaturated zone. Wood and Petraitis suggest that observed CO_2 gas flux out of the unsaturated zone in this area is most likely the result of oxidation of organic material sorbed onto clay-sized particles that have been introduced from the surface. The concept of particulate transport in the unsaturated zone is supported by the work of Goss and others (1973), who demonstrated, using ^{137}Cs-tagged clay, that clay particles moved to a depth of at least 2.7 m below the surface during an artificial-recharge experiment. This observation is consistent, as are the numerous clay skins present throughout the upper part of the aquifer, with the suggestion that clay and organic particles move through the unsaturated zone in this environment.

The CO_2 flux out of the system for the two sites investigated by Wood and Petraitis (1984) averaged 28 (g/m^2)/yr. Calculations from groundwater solute data indicate that an additional 1 (g/m^2)/yr is converted to bicarbonate by dissolution of lithologic carbonates. Thus, assuming a steady-state carbon system, approximately 29 (g/m^2)/yr organic carbon enters the unsaturated zone in the recharge zone of the playa basins. Dissolved organic carbon (5 mg/l) introduced with the recharging water is insufficient to account for the observed CO_2 flux. The lithologic framework of the unsaturated zone contains little or no liable organic carbon, and the carbon isotopes of CO_2 preclude it from originating from a petroleum or deep-mantle source. Wood and Petraitis (1984) concluded that organic matter must be introduced from the surface as particulate material. Organic-carbon content of suspended solids in 9 water samples collected from 3 playa lakes averaged approximately 3.0 percent, most of which was sorbed on clay material (Ron Malcolm, oral communication, 1971). Assuming that the particulate matter of these samples is representative of particulate matter that recharged the aquifer, the carbon flux suggests that approximately 970 (g/m^2)/yr of inorganic particulate material, onto which the organic matter is sorbed, also enters the unsaturated zone. Assuming a dry bulk density of 1.7 g/cm^3, the amount of material introduced represents a volume of approximately 530 (cm^3/m^2)/yr, or a reduction in surface elevation of approximately 0.57 m/1,000 yr under present hydrologic conditions.

A sediment flux of 970 $(g/m^2)/yr$ into the unsaturated material would require a water flux of approximately 195 $(cm/yr)/m^2$, if 500 mg/l suspended is assumed as an average concentration. This is 50 times greater than what appears reasonable for this system. Thus, it is suggested that piping is responsible for much of the transport of the particulate material from the surface into the unsaturated zone.

Case study summary

Using the above observations, the following scenario is proposed for development of many playa basins of the Llano Estacado. Small proto-depressions, most of which were probably eolian in origin, concentrated surface runoff and increased recharge flux through them relative to the area between basins. Fine-grained organic and inorganic particles entered the unsaturated zone with recharging water. The organic material in the recharged water was oxidized to CO_2 by resident aerobic microbial communities. Some CO_2 escaped to the atmosphere and some formed carbonic acid, which dissolved carbonate and silicate minerals. Instability of the lithologic framework caused by dissolution, the presence of smectite clays, a thick unsaturated zone, and local concentrations of groundwater recharge promoted micropiping and eluviation of both the internal lithologic framework and material from the surface. The resulting depression increased the surficial catchment area. This increased groundwater flux and induced additional piping. As the catchment area increased exponentially with the basin radius, more material was brought into a basin than could be transported downward by recharging water, so that basin filling occurred simultaneously with its development. Where deposition occurs, infiltration capacities are reduced, causing recharge to be concentrated at the periphery of the basin floor. Thus, the playa floor tends to expand faster horizontally than vertically. Where well-developed caliches occur near the surface, horizontal expansion may be curtailed because of erosional resistance, and the basins become deeper. In these basins, surficial material is readily piped through small fractures, root tubes, and solution openings. Where resistant caliche beds are absent, basins gradually enlarge areally with each increment of sediment removed.

Arguments from several independent approaches support the hypothesis that the playa basins probably originated as small eolian depressions but have developed and enlarged largely by a combination of solution and micropiping. The mass-balance, chemical, mineralogical, hydrologic, and botanical evidence for groundwater recharge through the lake basin is persuasive and consistent with the solution-micropiping hypothesis, as is the geology and geologic history of the area.

CASE STUDY ACKNOWLEDGMENTS

Thanks go to Roger Lee and Craig Sprinkle of the U.S. Geological Survey for review and constructive comments. Many hours of discussion and days in the field with my colleague, Waite Osterkamp, have helped immensely in refining and clarifying my thinking on this study. Discussion (argument at extended length) with Dr. C. C. Reeves, Jr., Texas Tech University; Dr. John Hawley, New Mexico Bureau of Mines and Mineral Resources; Dr. Tom Gustavson, Texas Bureau of Economic Geology; and Dr. Vance Holliday, Texas A & M University, are also acknowledged with great pleasure. Clearly the opinions expressed are my own and the above-mentioned individuals are absolved from any endorsement of the theories expressed herein.

CHAPTER SUMMARY

Piping occurs widely in the Earth's drylands. As discussed here, piping refers to a form of subsurface erosion that by several processes produces subterranean drainage conduits or tunnels in relatively insoluble materials. It can affect hard, consolidated rock such as shale, claystone, and siltstone, and even some sandstones and volcanic breccias. It is even more effective in soft, unconsolidated materials such as clayey and silty alluvium and colluvium, loess, and volcanic ash and tuff. Silty sediments with 20 to 25 percent smectite clay and high exchangeable sodium content are especially susceptible to piping.

Piping sites range from valley bottoms and stream terraces to uplands. Rainfall, snowmelt runoff, or irrigation water sinks into the ground or is diverted into surface openings, most commonly desiccation cracks. The water then seeps downward through the soil or follows vertical openings until it reaches an especially permeable layer or is blocked by an impermeable horizon. It seeps laterally downgradient toward an outlet in a surface drainage feature such as a gully. Where the water emerges at the gully wall—as a seep at first, later as a more concentrated flow—grains of sediment are entrained, thus initiating an incipient pipe. With the passage of time and runoff events the tiny opening enlarges, extends headward upgradient, and finally grows so large that its roof caves in. This forms a new, small, ragged, lateral gully as a tributary to the parent gully. With time, more gullies are formed, and a large gully network develops.

One result of such piping erosion is the development of landforms much like those of typical limestone karst terrains. Such landscapes are called pseudokarsts, and their features may include sinks, natural bridges, hanging dry valleys, disappearing streams, and subterranean drainage. Pseudokarst develops chiefly by subsurface grain-by-grain erosion of largely insoluble materials, whereas true karst is formed by molecule-by-molecule removal of dissolved ions from soluble rocks such as limestone, dolomite, and gypsum. Whereas karst landforms may endure for centuries, pseudokarst forms, especially small ones, tend to be transitory, changing relatively rapidly and lasting a comparatively short time. Pseudokarst developed in soft valley fills may be altered radically by the effects of a single heavy rainstorm or snowmelt runoff. Gully heads have been known to retreat 10 m or more in such an event. However, pseudokarst developed in resistant rocks may last much longer. For instance, that near Round Rock, Arizona (Fig. 7), has changed only slightly in the 20-odd years since Parker first observed it in 1961.

Common causes of the initiation of piping in drylands areas are seasonal drought, land denudation, and entrenchment of surface drainage, aided by human activities such as deforestation and devegetation, recurrent burning, overgrazing, and over-irrigation. As with surface erosion, prevention of piping is better than control, but in most regions, little attention has been paid to instituting satisfactory means of preventing or controlling subsurface erosion. Such development will depend on a better understanding of the consequences and importance of the processes leading to piping in dryland environments.

CHAPTER ACKNOWLEDGMENTS

Parker acknowledges with pleasure the aid and assistance of his U.S. Geological Survey staff of the Denver Federal Center during the course of the field, office, and laboratory studies of piping in the western U.S. drylands. He is particularly indebted to Everette A. Jenne, Barbara J. Anderson, Lynn M. Shown and Karl W. Ratzlaff. Others who contributed significantly include Keith S. Essex, Richard F. Hadley, Reuben J. Miller, Alfred T. Myers, Stanley A. Schumm and John W. Shomaker. The authors also thank the reviewers of early drafts of this paper, Burchard Heede, Brainerd Mears, Jr., and Thomas Gustavson, whose comments were extremely helpful in improving this final version.

REFERENCES CITED

Aitchison, G. D., 1960, Discussion of Cole and Lewis, 1960: Proceedings 3rd Australia-New Zealand Conference on Soil Mechanics and Foundation Engineering, p. 230.

Aitchison, G. D., and Wood, C. C., 1965, Some interactions of compaction, permeability, and post-construction deflocculation affecting the probability of piping failure in small earth dams: Proceedings 6th International Conference on Soil Mechanics and Foundation Engineering, Montreal, v. 2, p. 442–446.

Aitchison, G. D., Ingles, O. G., and Wood, C. C., 1963, Post-construction deflocculation as a contributary factor in the failure of earth dams: Proceedings 4th Australia-New Zealand Conference on Soil Mechanics and Foundation Engineering, p. 275–276.

Allen, B. L., Harris, B. L., Davis, K. R., and Miller, G. B., 1972, The mineralogy and chemistry of High Plains playa lake soils and sediments: Lubbock, Texas Tech University Water Resources Center, 75 p.

Bariss, N., 1971, Gully formation in the loesses of central Nebraska: Rocky Mountain Social Science Journal, v. 8, no. 2, p. 47–59.

Bates, R. L., and Jackson, J. A., eds., 1987, Glossary of geology, 3rd ed.: American Geophysical Institute, 788 p.

Bell, G. L., 1968, Piping in the Badlands of North Dakota, in Proceedings 6th Annual Engineering Geology and Soils Engineering Symposium, Boise, Idaho: Boise, Idaho Department of Highways, p. 242–257.

Bennett, H. H., 1939, Soil conservation: New York, McGraw-Hill, 993 p.

Berry, L., 1964, Some erosional features of semi-arid Sudan: 20th International Geographical Congress Abstracts of Papers, supplement, p. 3.

—— , 1970, Some erosional features due to piping and sub-surface wash with special reference to Sudan: Geografiska Annaler, v. 52A, p. 113–119.

Bishop, W. W., 1962, Gully erosion in Queen Elizabeth National Park: Uganda Journal, v. 26, p. 161–165.

Blong, R. J., 1965, Subsurface water as a geomorphic agent with special reference to the Manakowhiriwhiri Catchment: Aukland Student Geographer, v. 2, p. 82–95.

Bond, R. M., 1941, Rodentless rodent erosion: Soil Conservation, v. 10, p. 269.

Branner, J. C., 1913, The fluting and pitting of granites in the tropics: Proceedings of the American Philosophical Society, v. 52, p. 163–174.

Brown, G. W., 1962, Piping erosion in Colorado: Journal of Soil and Water Conservation, v. 17, p. 220–222.

Brown, R. F., Signor, D. C., and Wood, W. W., 1978, Artificial ground-water recharge as a water-management technique on the Southern High Plains of Texas and New Mexico: Texas Department of Water Resources Report, 220, 32 p.

Bryan, R. B., and Yair, A., 1982, Badland geomorphology and piping: Norwich, England, GeoBooks, 408 p.

Bryan, R. B., Yair, A., and Hodges, W. K., 1978, Factors controlling the initiation of runoff and piping in Dinosaur Provincial Park, Alberta, Canada: Zeitschrift fur Geomorphologie, supplement to v. 29, p. 151–168.

Buckham, A. F., and Cockfield, W. E., 1950, Gullies formed by sinking of the ground: American Journal of Sciences, v. 248, p. 137–141.

Bull, W. B., 1964, Alluvian fans and near-surface subsidence in western Fresno County, California: U.S. Geological Survey Professional Paper 437-A, 70 p.

Carroll, P. H., 1949, Soil piping in southeastern Arizona: U.S. Soil Conservation Service, Region 5 Regional Bulletin 110, 21 p.

Cedergren, H. R., 1977, Seepage, drainage, and flow nets, 2nd ed.: New York, John Wiley and Sons, 534 p.

Charman, P.E.V., 1969, The influence of sodium slats on soils with reference to tunnel erosion in coastal areas; Part 1, Kempsey area: Soil Conservation Service of New South Wales Journal, v. 25, p. 331–347.

—— , 1970a, The influence of sodium salts on soils with reference to tunnel erosion in coastal areas; Part 2, Grafton area: Soul Conservation Service of New South Wales Journal, v. 26, p. 71–80.

—— , 1970b, The influence of sodium salts on soils with reference to tunnel erosion in coastal areas; Part 3, Taree area: Soil Conservation Service of New South Wales Journal, v. 26, p. 256–275.

Clausen, E. N., 1970, Badland caves in Wyoming: National Speleological Society Bulletin, v. 32, p. 59–69.

Clibborn, J., 1909, Experiments made on the passage of water through the sand of the Chenab River from the Khanki Weir Site, in Clibborn, J., Roorkee treatise on civil engineering; Irrigation work in India: Allahabad, United Provinces Government Press, Appendix D, p. xxv-xxxviii.

Cockfield, W. E., and Buckham, A. F., 1946, Sink-hole erosion in the white silts at Kamloops: Royal Society of Canada Transactions, 3rd series, v. 40, section 4, p. 1–10.

Cole, R. C., Koehler, L. F., Eggers, F. C., and Goff, A. M., 1943, The Tracy area, California: U.S. Department of Agriculture Soil Survey Series 1938, no. 5, 95 p.

Collier, A. J., 1914, The geology and mineral resources of the John Day region: Oregon Bureau of Mines and Mineral Resources, v. 1, no. 3, 47 p.

Costa, J. E., and Schuster, R. L., 1988, The formation and failure of natural dams: Geological Society of America Bulletin, v. 100, p. 1054–1068.

Cronin, J. G., 1964, A summary of the occurrence and development of ground water in the Southern High Plains of Texas: U.S. Geological Survey Water Supply Paper 1695, 88 p.

—— , 1969, Ground water in the Ogallala Formation in the Southern High Plains of Texas and New Mexico: U.S. Geological Survey Hydrologic Investigations Atlas HA-330, 9 p.

Cussen, L., 1888, Notes on the Waikato River basins: Transactions and Proceedings of the New Zealand Institute, v. 21, p. 406–416.

De Vaux, R., 1967, The men who hid the Deep Sea Scrolls, in Everyday life in

Bible times: National Geographic Society, p. 310–311.
Dobrovolny, E., 1962, Geologia del Valle de La Paz: La Paz, Bolivia, Departamento Nacional de Geologia Boletin 3, 153 p.
Downes, R. G., 1946, Tunneling erosion in northeastern Victoria: Australian Council for Scientific and Industrial Research Journal, v. 19, p. 283–292.
Downing, B. H., 1968, Subsurface erosion as a geomorphological agent in Natal: Geological Society of South Africa Transactions, v. 89, p. 131–138.
Drew, D. P., 1972, Geomorphology of the Big Muddy Valley area, southern Saskatchewan, with reference to the occurrence of piping, in Paul, A. H., Dale, E. H., and Schlichtmann, H., eds., Southern prairies field excursion background papers: Regina, University of Saskatchewan Department of Geography, p. 197–212.
Dvoracek, M. J., and Peterson, S. H., 1970, Recharging the Ogallala Formation using shallow holes, in Mattox, R. B., and Miller, W. D., eds., The Ogallala Aquifer: Lubbock, Texas Tech University International Center for Arid and Semi-Arid Land Studies Special Report 39, p. 205–218.
Emerson, W. W., 1977, Physical properties and structure, in Russell, J. S., and Greacen, E. L., eds., Soil factors in crop production in a semi-arid environment: St. Lucia, Queensland, University of Queensland Press, p. 78–104.
Feininger, T., 1969, Pseudokarst on quartz diorite, Columbia [sic]: Zeitschrift fur Geomorphologie, v. 13, p. 287–296.
Fenneman, N. M., 1923, Physiographic provinces and sections in western Oklahoma and adjacent parts of Texas: U.S. Geological Survey Bulletin 730, p. 115–134.
Finch, W. I., and Wright, J. C., 1970, Linear features and ground-water distribution in the Ogallala Formation on the Southern High Plains, in Mattox, R. B., and Miller, W. D., eds., The Ogallala Aquifer: Lubbock, Texas Tech University International Center for Arid and Semi-Arid Land Studies Special Report 39, p. 49–57.
Finley, R. J., and Gustavson, T. C., 1981, Lineament analysis based on Landsat imagery, Texas panhandle: Austin, University of Texas Bureau of Economic Geology Circular 81-5, 37 p.
Fletcher, J. E., and Carroll, P. H., 1948, Some properties of soils associated with piping in southeastern Arizona: Soil Science Society of America Proceedings, v. 13, p. 545–547.
Fletcher, J. E., Harris, K., Peterson, H. G., and Chandler, V. N., 1954, Piping: Transactions of the American Geophysical Union, v. 35, no. 2, p. 258–263.
Fuller, M. L., 1922, Some unusual erosion features in the loess of China: Geographical Review, v. 12, p. 570–584.
Funkhouser, J. W., 1951, Soil caves in tropical Ecuador: National Speleological Society News, v. 9, no. 5, p. 4.
Galarowski, T., 1976, New observations of the present-day suffosion (piping) processes in the Bereznica catchment basin in the Bieszcady mountains (the East Carpathians): Studia Geomorphologica Carpatho-Balcanica, v. 10, p. 115–124.
Gibbs, H. S., 1945, Tunnel-gully erosion on the Wither Hills, Marlborough: New Zealand Journal of Science and Technology, v. 27, sec. A(2), p. 135–146.
Gile, L. H., Hawley, J. W., and Grossman, R. B., 1981, Soils and geomorphology in the Basin and Range area of southern New Mexico: Guidebook to the Desert Project: New Mexico Bureau of Mines and Mineral Resources Memoir 39, 222 p.
Gilman, K., and Newson, M. D., 1980, Soil pipes and pipeflow; A hydrological study in upland Wales; British Geomorphological Research Group Research Monograph 1: Norwich, England GeoBooks, 110 p.
Goss, D. W., Smith, S. J., Stewart, B. A., and Jones, O. R., 1973, Fate of suspended sediment during basin recharge: Water Resources Research, v. 9, p. 668–675.
Gove, P. B., 1966, Webster's 3rd International Dictionary: Springfield, Massachusetts, G. and C. Merriam, 2662 p.
Gustavson, T. C., Finley, R. J., and Baumgardner, R. W., Jr., 1980, Preliminary rates of slope retreat and salt dissolution along the eastern caprock escarpment of the Southern High Plains and in the Canadian River Valley, in Geology and geohydrology of the Palo Duro Basin, Texas Panhandle: Austin, University of Texas Bureau of Economic Geology Geological Circular 80-7, p. 76–82.
Guthrie-Smith, H., 1926, Tutira, 2nd ed.: London and Edinburgh, W. Blackwood, 405 p.
Halladay, W. R., 1960, Pseudokarst in the United States: National Speleological Society Bulletin, v. 22, part 2, p. 109–113.
Hardy, R. M., 1950, Construction problems in silty soils: Engineering Institute of Canada Engineering Journal, September 1950, p. 3–8.
Harris, K., and Fletcher, J. E., 1951, The so-called soil piping, Yolo Ranch, Yuma County, Arizona: Phoenix U.S. Soil Conservation Service Arizona State Office unpublished report, 10 p.
Havens, J. S., 1966, Recharge studies on the High Plains in northern Lea County, New Mexico: U.S. Geological Survey Water Supply Paper 1819-F, p. F1–F52.
Haworth, E., 1897, Physiography of western Kansas: University Geological Survey Kansas, v. 2, p. 11–49.
Heede, B. H., 1971, Characteristics and processes of soil piping in gullies: U.S. Department of Agriculture Forest Service, Rocky Mountain Forest and Range Experiment Station Research Paper RM-68, 15 p.
—— , 1986, Soil piping processes in sodic soils: EOS Transactions of the American Geophysical Union, v. 67, p. 956.
Henderson, J., and Grange, L. I., 1926, The geology of the Huntly-Kawnia subdivision, Pirongia and Hauraki divisions: New Zealand Geological Survey Bulletin 28, n.s., 112 p.
Hickey, W. C., and Dortignac, E. J., 1963, An evaluation of soil ripping and soil pitting on runoff and erosion in the semiarid Southwest: International Association of Scientific Hydrology, v. 65, p. 22–23.
Higgins, C. G., 1984, Piping and sapping; Development of landforms by ground water outflow, in Ground water as a geomorphic agent: Boston, Allen and Unwin, p. 18–58.
Higgins, C. G., and 12 others, 1988, Landform development, in Back, W., Rosenshein, J. S., and Seaber, P. R., eds., Hydrogeology: Boulder, Colorado, Geological Society of America, The Geology of North America, v. O-2, p. 383–400.
Hodges, W. K., and Bryan, R. B., 1982, The influence of material behavior on runoff initiation in the Dinosaur Badlands, Canada, in Bryan, R. B., and Yair, A., eds., Badland geomorphology and piping: Norwich, England, GeoBooks, p. 13–46.
Hutchinson, J. N., 1968, Mass movement, in Fairbridge, R. W., ed., Encyclopedia of geomorphology: New York, Reinhold, p. 688–695.
Imeson, A. C., Kwaad, F.J.P.M., and Verstratten, J. M., 1982, The relationship of soil physical and chemical properties to the development of badlands in Morocco, in Bryan, R. B., and Yair, A., eds., 1982, Badland geomorphology and piping: Norwich, England, Geobooks, p. 47–70.
Jacobberger, P. A., 1988, Drought-related changes to geomorphologic processes in central Mali: Geological Society of America Bulletin, v. 100, p. 351–361.
Johnson, W. D., 1901, The High Plains and their utilization: U.S. Geological Survey 21st Annual Report 1890–1900, part 4, p. 601–741.
Jones, [J. A.] A., 1971, Soil piping and stream channel initiation: Water Resources Research, v. 7, p. 602–610.
Jones, J.A.A., 1975, Soil piping and the subsurface initiation of stream channel networks [Ph.D. thesis]: Cambridge University, 467 p.
—— , 1981, The nature of soil piping; A review of research; British Geomorphological Research Group Research Monograph 3: Norwich, England, GeoBooks, 301 p.
—— , 1987, The initiation of natural drainage networks: Progress in Physical Geography, v. 11, p. 207–245.
Jones, N. O., 1968, The development of piping erosion [Ph.D. thesis]: Tucson, University of Arizona, 163 p.
Keys, W. S., and Brown, R. F., 1978, The use of temperature logs to trace the movement of injected water: Groundwater, v. 16, p. 32–48.
Khobzi, J., 1972, Erosion chimique et méchanique dans la genèse de dépressions 'pseudo-karstiques' souvent endoréiques: Revue de Géomorphologie Dyna-

mique, v. 21, p. 57–70.

Kier, R. S., Stecher, S., and Brandes, R. J., 1985, Rising ground water levels, *in* Whetstone, G., ed., Ogallala Aquifer Symposium 2: Lubbock, Texas Tech University Water Resources Center, p. 416–439.

Kingsbury, J. W., 1952, Pothole erosion on the western part of Molokai Island, Territory of Hawaii: Journal of Soil and Water Conservation, v. 7, p. 197–198.

Knebel, W. von, 1906, Höhlekunde mit berucksichtigung der karstphänomene: Braunschweig, F. Vieweg und Sohn, 222 p.

Knight, K., 1959, The microscopic study of structure of collapsing sands, *in* Proceedings 2nd Southern African Regional Conference on Soil Mechanics and Foundation Engineering: Mozambique, Lourenco Marques, p. 69–72.

Leopold, L. B., and Miller, J. P., 1956, Ephemeral streams; Hydraulic factors and their relation to the drainage net: U.S. Geological Survey Professional Paper 282-A, 37 p.

Loffler, E., 1974, Piping and pseudokarst features in the tropical lowlands of New Guinea: Erdkunde, v. 28, p. 13–18.

Lotspeich, F. B., Hauser, V. L., and Lehman, O. R., 1969, Quality of water from playas on the Southern High Plains: Water Resources Research, v. 5, no. 1, p. 48–58.

Luckey, R. R., Gutentag, E. D., Heimes, F. J., and Weeks, J. B., 1986, Digital simulation of ground-water flow in the High Plains Aquifer in parts of Colorado, Kansas, Nebraska, New Mexico, Oklahoma, South Dakota, Texas, and Wyoming: U.S. Geological Survey Professional Paper 1400-D, 57 p.

McGrath, D. A., 1984, Morphological and mineralogical characteristics of indurated caliches of the Llano Estacado [M.S. thesis]: Lubbock, Texas Tech University, 206 p.

McIntyre, D. S., 1979, Exchangeable sodium, subplasticity, and hydraulic conductivity of some Australian soils: Australian Journal of Soil Research, v. 17, p. 115–120.

Mears, B., Jr., 1963, Karst-like features in badlands of the Arizona Petrified Forest: University of Wyoming Contributions to Geology, v. 2, p. 101–104.

—— , 1968, Piping, *in* Fairbridge, R. W., ed., Encyclopedia of geomorphology: New York, Reinhold, p. 849.

Mielenz, R. C., and King, M. E., 1952, Physical-chemical properties and engineering performance of clays: Denver, Colorado, U.S. Bureau of Reclamation Engineering Laboratory, not paged.

Monteith, N. H., 1954, Problems of some Hunter Valley soils: New South Wales Soil Conservation Service Journal, v. 10, p. 127.

Nelson, A., and Nelson, K. D., 1967, Dictionary of applied geology: London, George Newnes, 421 p. (also see Nelson, A., and Nelson, K. D., 1983, Dictionary of water and water engineering: Cleveland, Ohio, CRC Press, 271 p.)

Newman, J. C., and Phillips, J.R.H., 1957, Tunnel erosion in the Riverina: Soil Conservation Service of New South Wales Journal, v. 13, p. 159–169.

Osterkamp, W. R., and Wood, W. W., 1987, Playa-lake basins on the Southern High Plains of Texas and New Mexico; Part 1, Hydrologic, geomorphic, and geologic evidence for their development: Geological Society of America Bulletin, v. 99, p. 215–223.

Otvos, E. G., Jr., 1976, "Pseudokarst" and "pseudokarst terrains"; Problems of terminology: Geological Society of America Bulletin, v. 87, p. 1021–1027.

Parker, G. G., 1963, Piping, a geomorphic agent in landform development of the drylands: International Association of Scientific Hydrologists Publication 65, p. 103–113.

—— , 1966, Officer's Cave, eastern Oregon, revisited, *in* Abstracts for 1966: Geological Society of America Special Paper 101, p. 415.

Parker, G. G., and Jenne, E. A., 1966, Piping and collapse structure failures associated with western highways of the United States: Highway Research Abstracts, v. 36, no. 12, p. 115–116.

—— , 1967, Structural failure of western highways caused by piping: Highway Research Record, no. 203, p. 57–76.

Parker, G. G., Shown, L. M., and Ratzlaff, K. W., 1964, Officer's Cave, a pseudokarst feature in altered tuff and volcanic ash of the John Day Formation in eastern Oregon: Geological Society of America Bulletin, v. 75, p. 393–401.

Pavlov, A. P., 1898, Concerning the contour of plains and its change under the influence of subterranean and surface waters: Zhurn. Zemlevedenye, v. 3-4, p. 91–147.

Peterson, H. V., 1954, Discussion of piping: Transactions of the American Geophysical Union, v. 35, p. 263.

Pierson, T. C., 1983, Soil pipes and slope stability: London, Quarterly Journal of Engineering Geology, v. 16, p. 1–11.

Rathjens, C., 1973, Subterrane Abtragung (Piping): Zeitschrift fur Geomorphologie, Supplementband, v. 17, p. 168–176.

Reed, E. F., 1930, Vegetation of the playa lakes in the Staked Plains of Western Texas: Ecology, v. 11, no. 3, p. 597–600.

Reeves, C. C., Jr., 1966, Pluvial lake basins of West Texas: Journal of Geology, v. 74, no. 3, p. 269–291.

—— , 1970, Drainage pattern analysis, Southern High Plains, West Texas and eastern New Mexico, *in* Mattox, R. B., and Miller, W. D., eds., The Ogalalla Aquifer: Lubbock, Texas Tech University International Center for Arid and Semi-Arid Studies Special Report 39, p. 58–72.

Reeves, C. C., and Temple, J. M., 1986, Permian salt dissolution, alkaline lake basins, and nuclear-waste storage, Southern High Plains, Texas and New Mexico: Geology, v. 14, p. 939–942.

Rettman, P. L., and Leggat, E. R., 1966, Ground-water resources of Gains County, Texas: Texas Water Development Board Report 25, 185 p.

Richards, L. A., 1954, Diagnosis and improvement of saline and alkalai soils: U.S. Department of Agriculture Handbook 60, 224 p.

Richthofen, F. F., von, 1877, China; Ergebnisse eigener reisen und darauf gegründeter studien, Bd. 1, Einleittender theil: Berlin, D. Reimer, 758 p.

Ritchie, J. A., 1963, Earthwork tunnelling and the application of soil-testing procedure: Soil Conservation Service of New South Wales Journal, v. 19, p. 111–129.

Rooyani, F., 1985, A note on soil properties influencing piping at the contact zone between albic and argillic horizons of certain duplex soils (Aqualfs) in Lesotho, southern Africa: Soil Science, v. 139, p. 517–522.

Rubey, W. W., 1928, Gullies in the Great Plains formed by sinking of the ground: American Journal of Science, 5th series, v. 15, no. 85, p. 417–422.

Selby, M. J., 1982, Hillslope materials and processes: Oxford University Press, 264 p.

Sharpe, C.F.S., 1938, What is soil erosion?: U.S. Department of Agriculture Publication 286, 61 p.

Sidle, R. C., and Swanston, D. N., 1986, Groundwater dynamics in unstable zero-order basins of coastal Alaska: EOS Transactions of the American Geophysical Union, p. 956.

Smith, H.T.U., 1968, "Piping" in relation to periglacial boulder concentrations: Biuletyn Peryglacjalny, v. 17, p. 195–204.

Soleilhavoup, F., and Cailleux, A., 1978, Formes de suffossion actuelle pres de Tamanrasset (Hoggar, Sahara): Coloquio Estudo e Cartografia de Formaceos Superficiais e suas aplicacoes em Regioes Tropicais, Sao Paulo, v. 1, p. 215–224.

Stocking, M., 1976, Tunnel erosion: Zimbabwe Rhodesia Agricultural Journal, v. 73, no. 2, p. 35–39.

—— , 1981, A model of piping in soils: Proceedings of the 3rd Meeting of the International Geophysical Union Commission on Field Experiments in Geomorphology, p. 105–119.

Stone, W. J., 1985, Preliminary estimates of Ogallala Aquifer recharge using chloride in the unsaturated zone, Curry County, *in* Whetstone, G., ed., The Ogallala Aquifer Symposium 2: Lubbock, Texas Tech University Water Resources Center.

Terzaghi, K., 1922, Der Grundbruch an Stauwerken und seine Verhutung: Die Wasserkraft, v. 17, no. 24, p. 445–449.

—— , 1943, Theoretical soil mechanics: New York, John Wiley and Sons, 510 p.

Terzaghi, K., and Peck, R. B., 1967, Soil mechanics in engineering practice, 2nd ed.: New York, John Wiley and Sons, 729 p.

Theis, C. V., 1937, Amount of recharge in the Southern High Plains: American

Geophysical Union Transactions of the 18th Annual Meeting, p. 564–568.

Thorp, J., 1936, Geography of the soils of China: Nanking, National Geological Survey of China, 552 p.

U.S. Bureau of Reclamation, 1982, Llano Estacado playa lake water resources study: Amarillo, Texas, U.S. Department of the Interior, Bureau of Reclamation Southwest Regional Office.

Visser, S., 1969, Soil piping; A review with special references to Northland, New Zealand: Auckland Student Geographer, no. 6, p. 49–57.

Ward, A. J., 1966, Pipe/shaft phenomena in Northland: New Zealand Journal of Hydrology, v. 5, no. 2, p. 64–72.

Warn, F., 1966, Sinkhole development in the Imperial Valley, in Lung, R., and Proctor, R., eds., Engineering geology in Southern California: Los Angeles, California, Association of Engineering Geologists Special Publication, p. 144–145.

White, W. N., Broadhurst, W. L., and Lang, J. W., 1946, Ground water in the High Plains of Texas: U.S. Geological Survey Water Supply Paper 889-F, p. 381–421.

Wood, W. W., and Ehrlich, G. G., 1978, Use of baker's yeast to trace microbial movement in ground water: Ground Water, v. 16, no. 6, p. 398–403.

Wood, W. W., and Osterkamp, W. R., 1984, Recharge to the Ogallala Aquifer from playa lake basins on the Llano Estacado; An outrageous proposal?, in Whetstone, G., ed., The Ogallala Aquifer Symposium 2: Lubbock, Texas Tech University Water Resources Center.

—— , 1987, Playa-lake basins on the Southern High Plains of Texas and New Mexico; Part 2, A hydrologic model and mass-balancing arguments for their development: Geological Society of America Bulletin, v. 99, p. 224–230.

Wood, W. W., and Petraitis, M. J., 1984, Origin and distribution of carbon dioxide in the unsaturated zone of the Southern High Plains of Texas: Water Resources Research, v. 20, p. 1193–1208.

Wright, H. E., Jr., 1964, Origin of the lakes in the Chuska Mountains, northwestern New Mexico: Geological Society of America Bulletin, v. 75, p. 589–598.

Yoder, R. E., 1936, A direct method of aggregate analysis of soils and a study of the physical nature of erosion losses: American Society of Agronomy Journal, v. 28, p. 337–351.

MANUSCRIPT ACCEPTED BY THE SOCIETY NOVEMBER 14, 1989

Chapter 5

Piping effects in humid lands

J.A.A. Jones
Institute of Earth Studies, University College of Wales, Aberystwyth SY23 3DB, United Kingdom

INTRODUCTION

There is now a large body of literature on piping as a feature of drylands and badlands, as illustrated by the previous chapter. In fact, many have regarded piping as a process specific to such areas. However, a scan of the literature as a whole reveals that only one-third of all detailed references on piping in the natural landscape relates to arid or semiarid regions. Approximately 50 percent of references now relate to humid temperate climates, particularly in the United Kingdom and New Zealand, and a large portion of the remainder refer to the humid tropics or to humid mid-latitude continental areas.

DISTINCTIONS BETWEEN DRY-LAND AND HUMID-LAND PIPING

The widespread occurrence of piping across the climatic zones is clearly due to the variety of processes capable of initiating piping and the widespread occurrence of the minimum requirements for pipe initiation. The question therefore arises as to whether piping phenomena differ between climatic zones in terms of the initiating processes, or indeed in terms of their form and function in the hydropedogeomorphic system. In fact, the question must be extended beyond the dry land/humid land dichotomy to differences between humid climates, since piping has been reported from the tropical rain forest to periglacial regions (Jones, 1981).

The main evidence of differences to date relates to the processes of pipe initiation. Some initiating factors cut across climatic zones, particularly the presence of erodible soils, or soils susceptible to macropore development, overlying an impeding horizon. Other initiating factors are more specific to climatic regions, especially where a dry land regime has developed soils with high exchangeable sodium percentages (ESP), dispersive clays or highly expansive clays. Research in arid and semiarid regions has generally emphasized the role of soil chemistry and clay species in initiating piping through their effects on erodibility or macropore development. Highly expansive clays develop more contraction macropores during dry periods, allowing more water to enter the soil and providing rudimentary channelways that can be enlarged. Dispersive clays are particularly effective in aiding entrainment processes in the presence of freshly infiltrated rainwater with low electrolyte concentrations.

In archetypal humid regions the chemistry of dispersion is less important, and there is more emphasis on the role of high shear stresses generated by high throughflow discharges. The high throughflow stresses result from steep hydraulic gradients, normally associated with a strong positive water balance in the soil, periodic high-intensity storms, or drawdown at a free face. The region of annual water surplus in the UNESCO (1978) atlas of world water resources can be taken as a reasonable indication of the core area for these humid piping processes. Piping appears to be a particularly notable feature in higher rainfall areas such as the rain forest (Bremer, 1973; Loffler, 1974; Baillie, 1975; Franzle, 1976) and monsoon climates (e.g., Starkel, 1972). However, there is no simple relation between high rainfall or water surplus and the incidence of piping; soil characteristics such as permeability play an important role, as illustrated by the distribution of piping in Dominica (Fig. 9). Here, Walsh (1980) found piping less common in the high-rainfall areas. He attributed this to the high permeabilities of the topsoil in the heavier rainfall areas, but it may also relate to the species of clay minerals found in different parts of the island.

A number of areas of piping lie immediately outside the core region of water balance surplus, notably in Africa and Australia, which might be regarded as intergrade between humid and dry-land piping, at least in terms of initiating processes. These lie principally in the wet-and-dry tropics, in subtropical areas with a marked dry season (e.g., Rathjens, 1973; Loffler, 1974), and in the warm dry land margins of interior New South Wales and Victoria (e.g., Crouch and others, 1986; Downes, 1946). Many soils in these regions contain highly expansive and/or dispersive clays developed in a climatic environment that favors smectoid clay formation and neutral to alkaline soil conditions. These conditions spread over into semiarid regions where isolated heavy rains initiate the dispersive effect. This intergrade piping lies approximately between the zone of water surplus and the 600 mm annual isohyet, but it is the seasonal distribution of rainfall rather than the annual totals or water balances that tends to be more important for pipe development in these regions.

Jones, J.A.A., 1990, Piping effects in humid lands, *in* Higgins, C. G., and Coates, D. R., eds., Groundwater geomorphology; The role of subsurface water in Earth-surface processes and landforms: Boulder, Colorado, Geological Society of America Special Paper 252.

Even in very wet climates the temporal distribution of rainfall is liable to be a major factor in initiating piping, whether it be extreme rainfall or snowmelt events that generate the erosive stress or the extreme drought that creates cracks and loosens soil aggregates. An excellent example of the former situation is given by Starkel's (1972) observation of piping in the highlands of Sikkim initiated during a monsoon rainstorm that deposited a quarter of the annual 4,050 mm of rain at intensities of as much as 760 mm/hr. The effect of drought can be seen in the less dramatic example of pipes generated by cracking in the peaty soils of upland Wales under annual rainfalls of more than 2,200 mm (Gilman and Newson, 1980). In fact, for the proper study of the spatial distribution of piping processes there is an undeniable need for a "morphoclimatic" approach like that proposed by Ahnert (1987a), using magnitude-frequency distributions rather than standard climatic measures.

It is more difficult to make precise statements on differences in the form and function of piping between humid and dry climates, principally because of lack of research on these aspects in dryland regions. However, research in dryland areas of North America offers some basis for comparing pipe sizes and network form and is beginning to provide interesting comparisons in terms of hydrologic function and sediment yield. In the semiarid badlands of southern Alberta and Saskatchewan, piping has been found in multiple layers (Drew, 1972; Barendregt and Ongley, 1977; Bryan and Harvey, 1985) and even sponge-like networks (Bryan, personal communication, 1983) that are developed in shales and erosion glacis along canyon walls. In contrast, most observations in humid regions are of piping in relatively shallow soils with less extensive areas of free face, and here the networks are more typically developed on a single level (Jones, 1981, p. 136ff). Dry land pipes also tend to be larger, with typical diameters of 1 to 2 m compared with 0.3 m or or less for humid soil piping. The larger diameters are associated with very high, periodic yields of sediment from dryland pipes (Jones, 1968, p. 81; Barendregt and Ongley, 1977; Bryan and Campbell, 1986), and often rapid network extension or collapse (Jones, 1968, p. 42), in contrast to the lower sediment yields and more stable networks of most humid piping (Jones, 1981, 1987a). Recent work on the hydrological function of dryland piping in Alberta suggests that it may be a major drainage route, possibly directing up to 33 percent of the total basin drainage (Bryan and Harvey, 1985), and comparable to estimates in humid-temperate regions. In both the semiarid badlands of southern Alberta and humid Wales, investigations of the hydrological function of piping suggest highly variable responses from storm to storm, related to the extent of the networks activated and the wetness of the surrounding soils or mineral materials. Bryan and Harvey (1985) attribute reduced pipe flow contributions in wetter conditions to swelling of the shales and closure or constriction of shallow pipes. In contrast, Jones and Crane (1984) have attributed falling contributions in percentage terms in wetter conditions to increased yields from other sources such as saturation overland flow and riparian seepage, but it is conceivable that the peaty soils of Wales could swell and restrict flows in a similar way in the smaller pipes.

PROCESSES OF PIPE INITIATION IN HUMID LANDS

There are reasonable grounds, as we have noted, for making a distinction between the dry land processes discussed by Parker in chapter 4 and those found in the core area of humid piping, particularly as regards the role of dispersive properties. Nevertheless, dispersive processes may play an important role in the less humid "intergrade" areas. There have also been suggestions that climatic change may have confused the boundaries, in particular causing initiation processes more typical of drylands to be found in some humid areas.

General factors and processes

Piping is commonly associated with soils containing marked reductions in vertical permeability at some point in the soil profile and with horizons above that point, which have sufficient lateral permeability and hydraulic gradient to permit significant amounts of throughflow. The processes are aided by easily eroded horizons in this position and by soils with strong histic properties encouraging crack development either above or at this critical level. Cracking can increase infiltration, locally capturing more water for throughflow, and it can provide the rudiments of a subsurface channel, increasing flow velocities and therefore also boundary stress and erosion. Whichever factors are critical, be they permeability, erodibility, or volumetric change, the development of piping at a certain point in the profile is probably more a question of relative down-profile contrasts than of absolute values, as suggested by Jones (1978a; 1981, p. 67) and Imeson (1986). This is not to deny that certain minimum thresholds exist, as, for example, collated by Jones (1981, Table 27).

Hydraulic gradients and the permeability or macroporosity of the soil control the erosive stresses that generate piping by regulating pressures and flow velocities both within the soil matrix and within the macropores. The "heave" or seepage pressure process occurs when water emerging from a soil surface under pressure causes soil particles and aggregates to become weightless and to "boil" away (Terzaghi and Peck, 1966). Classically, Terzaghi saw the process as occurring when return flow exfiltrates through the soil surface, generating what would now be called "saturation overland flow." But the process may equally well occur on the interface between the soil matrix and macropores within a saturated soil, and thus contribute to the enlargement of macropores.

The second process is that of corrasion by which the boundaries of the pipes and macropores are eroded as a result of shear stress generated by channel flow. Figure 1 illustrates the forces associated with the two processes.

The boiling process only occurs beneath a phreatic surface

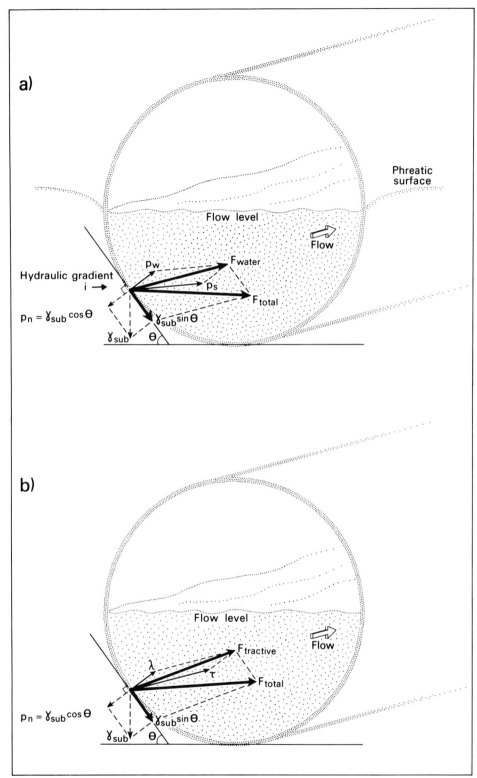

Figure 1. Forces involved in piping processes: (a) Terzaghian forces exerted by soil water on the wall of a pipe; (b) corrasion forces exerted by pipeflow on the wall of a pipe. p_s is the hydrodynamic pressure of throughflow, $p_s = \gamma_{water} \cdot i$; i the hydraulic gradient; γ_{sub} the submerged unit weight of a soil particle; p_w the static pore water pressure; F_{water} is the resultant of soil water pressure; F_{total} the overall resultant; τ is the drag force due to pipeflow, $\tau = \gamma_{water} \cdot u^2$, u being the velocity; λ the lift force; p_n normal stress per unit area; θ the angle of slope of the pipe wall; $F_{tractive}$ the resultant of tractive forces. Cohesion, c, reinforces p_n, so that erosive instability occurs when $F_{total} \geqslant p_n + c$.

and has been equated with the process of "sapping" often observed in gully walls, free faces, and spring heads (Higgins, 1984). Nevertheless, the critical phreatic level may be very temporary, occurring only in extreme storms, and it may take the form of a perched water table above a less permeable horizon within the soil profile. Once a significant-sized macropore or pipe has developed and pipeflow occurs, the boiling process will be supplemented by corrasion shear stresses. Indeed, the "boiling" may be accentuated by pipeflow as it draws water out into the channel by creating a suction force (the lift force in Fig. 1). While it is possible to equate Terzaghi's piping mechanism with the form of sapping in which no open void is created, wherever geomorphologically recognizable pipes are found, the mechanism must include an element of corrasion and sediment transport by channel-type flow. It is here that the distinction becomes blurred between the classic piping processes of the dam engineers and the so-called tunneling of the soil conservationists (compare Jones, 1981, p. 9, with Dunne, this volume). At the other end of the spectrum, piping developed in a vadose situation from the expansion and interconnection of structural or biotic voids in the soil may be created solely by boundary shear stresses. However, even here it is feasible that temporary saturation of the soil creates seepage pressures that aid erosion from time to time.

Pipe development is clearly related to soil erodibility. Jones (1971; 1981, p. 75ff) demonstrated that piping in the English Peak District is associated with layers of lower aggregate stability. An important corollary of this is that the piping has a more stable overlayer, which provides the roofing. An extreme example of this general layering sequence can be found in latosols, especially in vermiform laterites, which are very prone to piping beneath the duricrusts (e.g., Bowden, 1980).

A considerable amount has been written on the role of cohesive forces in restraining the onset of piping erosion (e.g., Zaslavsky and Kassiff, 1965; Perry, 1975). Particular attention has been given to the weakening of interparticle bonds in alkaline soils, where a zone of critical instability is now well established in the balance between the exchangeable sodium percentage (EPS) of the soil and the ionic concentration of the throughflow (e.g., Ingles, 1968; Sherard and Decker, 1977). The zone of instability represents a transition from stable flocculated clays at lower ESP and higher ionic concentrations to a stable deflocculated state in which the process of deflocculation has reduced permeability to the point where throughflow velocities are below the threshold for erosion (Fig. 2). The exact boundaries of the zone of instability vary according to the dominant clay mineral species. Montmorillonite exhibits the broadest zone of instability. It also has the highest coefficient of expansion, which means that it is particularly prone to cracking; this assists rapid infiltration and throughflow.

This is a process that is characteristic of semiarid and arid regions and of the intergrade subhumid regions, as seen in reports of field piping in interior New South Wales (e.g., Crouch and others, 1986; Ingles and Aitchison, 1970), in the wet-and-dry tropics of Africa (Stocking, 1979), and in detailed analyses of susceptible dam building material in America and Australia (e.g., Sherard and Decker, 1977). It is not likely to be important in perhumid regions where leaching tends to prevent significant accumulations of dispersive monovalent cations and to favor the formation of clay minerals with lower cation exchange capacities like kaolinite. These clay minerals also tend to have lower cracking potential than montmorillonite.

Nevertheless, it has been argued that cyclic salts provided by rainfall containing relatively high concentrations of sodium may be an important factor in stimulating piping erosion in coastal areas of New South Wales (Charman, 1969, 1970a, b). Charman argued that a delicate balance must exist between the rate of replenishment of sodium by rainfall and weathering of sedimentary rocks, on the one hand, and the flushing of sodium from the soil profile by leaching, on the other. This results in the critical degree of dispersion in the subsoil without significantly affecting the overall soil morphology, i.e., without creating true halomorphic soils.

Charman's soil tests indicated that between 37 and 50 percent of soils in the northern coastal districts of New South Wales are susceptible to pipe erosion, according to the established criteria of the NSW Soil Conservation Service (i.e., susceptible if Dispersal Index <3, and highly susceptible if the volumetric expansion is also >10 percent). ESP values are high in the dispersible subsoils. Nevertheless, he observed that piping is not found as

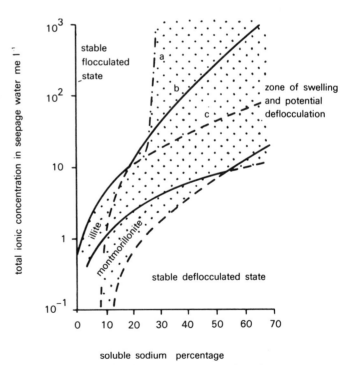

Figure 2. The onset of piping instability due to dispersion, in relation to exchangable sodium$_t$ percentage in the soils, ionic concentrations in throughflow, and clay mineral species. The critical range of instability is stippled. Montmorillonite deflocculates to the right of line b, illite right of line c (Ingles and Aitchison, 1970); line a marks the threshold in a "dominantly montmorillonitic" soil, according to Kassiff and Henkin (1967). See also Sherard and Decker (1977).

frequently in the field as the criteria would suggest (Charman, 1970a).

The discrepancy between the potential for piping and the actual occurrence seems to underline the need for other factors to be favorable for pipe development, particularly the hydraulic gradient and permeability or macroporosity, in addition to the susceptibility of the soil. It has to be admitted, however, that no single test of susceptibility is universally accepted or ideal, and refinement is an on-going process (Jones, 1981, Appendix I). Imeson (1986) also makes the point that the collection of samples of piped soils from the field may not reveal the true critical conditions. These may be ephemeral, and related to parameters that are not sampled, such as critical electrolyte concentrations in throughflow.

Effects of climatic change

Both Charman (1970a) and Downes (1954) speculated on the possibility of a lag effect resulting from climatic change in southeast Australia. Charman (1970a) pointed out that soil classification can be problematic in some of the podzolic soils in coastal New South Wales that appear to have been subjected to solodization in the past. In Victoria, Downes (1954) maintained that the historical effects are most pronounced north of Melbourne, where present-day rainfall ranges between 500 and 750 mm/yr. Downes traced solodization to the Recent arid period, up to 9000 B.P., when rainfall here was about half as much as now. Subsequent desalination has created a range of solonetzic, solodized solonetz, and solodic soils. Eluviation of sodium clays down-profile in the present climate and possibly breakdown of the clay complexes have created less permeable B horizons, which favor pipe development. At the same time, the strong columnar structure favorable to piping in the solonetz soils has been reduced but not entirely destroyed in the solodized solonetz and solod stages.

According to Downes (1956), the high level of cyclical salts brought in by rainfall tends to counteract leaching and reduce the degradation of the alkaline soils. Monteith (1954) calculated that present rainfall could add ca. 25 tonnes/ha in 10,000 yr in the Hunter Valley of New South Wales. However, stronger leaching in the areas of higher rainfall on the coast results in a more rapid destruction of solonetzic profiles than occurs inland (Charman, 1970a).

Research in southeast Australia suggests, therefore, that pipe-initiating processes more typical of semiarid regions are currently operating there in areas with annual rainfalls around 1,000 mm, and with annual surpluses of water, as a result of two anomalies: an arid interlude followed by recent climatic amelioration, and the current high levels of cyclical salts provided by coastal rainfall.

Climatic change has also been suggested as a long-term factor in the development of piping in a wet-and-dry tropical climate in Zimbabwe, where present-day rainfall is about 800 mm/yr (Stocking, 1976). Again, the piping is associated with high ESP levels (>50 percent), with gullying rather than piping

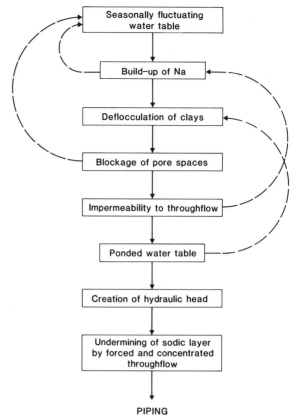

Figure 3. Stocking's model of pipe initiation in dispersive soils.

occurring wherever ESP levels are lower. Stocking believes that the sodium is derived from weathering of the local granite, but that its accumulation is the result of poor drainage conditions during a Pleistocene pluvial period, prior to the rejuvenation of the Umsweswe River. The sodic duplex soils have massive impermeable B horizons with dispersible material above and the area has high local relative relief as a result of stream incision. Stocking (1981) has presented a general model for pipe initiation in this environment in which soil dispersion is the key factor (Fig. 3).

Despite the emphasis on the roles of dispersion and climatic change, however, it is worth noting that the more common initiating factors of steep local hydraulic gradients and impermeable soil horizons are present in both the Zimbabwean and Australian examples. In fact, critical inspection of Charman's data suggests that he may have been overly influenced by the "dispersion hypothesis," which after all originated with the CSIRO in Australia (Wood and others, 1964). Many of the soils identified as susceptible by Charman appear to be true pedalfers, with the same structural features that are found sufficient in themselves to initiate piping in other humid regions.

Figure 4 has been constructed by combining Charman's tabulated data with Stephens' (1962) descriptions of type profiles. It shows two "peaks" of susceptibility, one in the podzolics and one in the solodic soils, each associated with moderate to strong

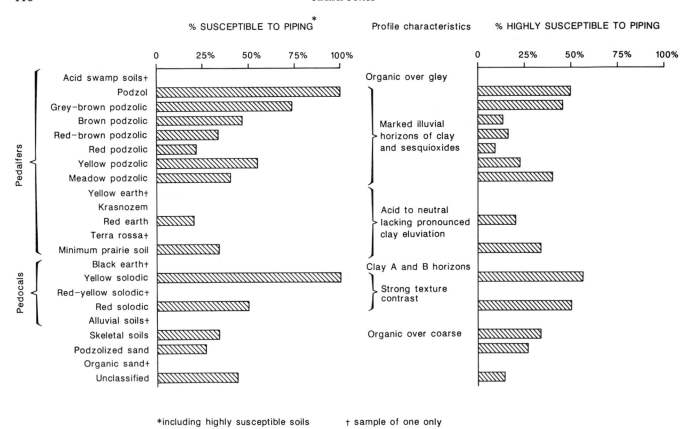

Figure 4. Susceptibility of soils to piping in New South Wales according to data collected by the Soil Conservation Service. Percentages have been calculated from the sample results presented by Charman (1969, 1970a, b). Susceptible soils have Dispersal Index <3; highly susceptible soils also have Volumetric Expansion >10 percent.

texture contrast between A and B horizons. The B horizons of susceptible soils are either less permeable or else prone to cracking and blocky structure, or both. Pedalfers that lack pronounced mobilization of clays down-profile, particularly the krasnozems, are markedly less susceptible than pedalfers that have illuvial B horizons, like the gray-brown podzolics, or even hardpans like the true podzols.

The only soils outside these two key groups that are susceptible are those with high rates of expansion and contraction, i.e., with a high cracking potential. This is shown in the second bar graph in Figure 4.

Geomorphic location of piping processes in a humid landscape

Analysis of reports published worldwide suggests that pipe development occurs preferentially on certain land surface units. Figure 5 shows the frequency of observed piping on each land surface unit of the NULM system (Conacher and Dalrymple, 1977) presented as probabilities of occurrence, i.e., probable pedogeomorphic locations given the presence of piping. This shows three peak locations: on unit 2, the seepage slope on the upper hillslope, units 5 and 6, the transportational mid-slope and colluvial toeslope, and unit 8, the flood plain. The distribution reflects the situation in the United Kingdom very well, although British references were by no means a majority in the reports used to compile the graph. Experience in the U.K. suggests that these three peaks are broadly associated with three different processes of pipe initiation. Desiccation cracking tends to be a more common initiating factor on the better drained slopes higher on the hillslope, including the seepage slope and the midslopes (compare Gilman and Newson, 1980; Jones and Crane, 1984). Mass movement tends to create usable linear voids, zones of higher permeability, and perhaps fractured soil aggregates on steeper slopes. Typically, mass-movement effects are found in midslope, although they may be found near the downslope end of the seepage slope, especially at the head of hillside hollows, and on river terraces (unit 2[2]) where rejuvenation scars provide the necessary topographic gradient. Mass-movement cracks may also be induced by lateral eluviation where this is sufficient to cause subsidence, especially where piping is already well developed, and these cracks may be used by a new phase of piping.

Terzaghi-style "boiling" is most likely near the pipe outfalls in the channel walls, rejuvenation scars, and the floodplains or terraces immediately behind them. Nevertheless, Jones (1978a) argued on the basis of a detailed inspection of 150 pipe outfalls in streambanks cut through lessivé brown earths in Derbyshire, England, that no more than 10 percent of these pipes seemed to have been initiated by Terzaghian mechanics, and that cracking in clay-enriched Bt horizons appeared to be a more common initiating process.

HYDROLOGIC EFFECTS OF HUMID PIPES

The hydrologic response of piping and its effects on stream runoff are essentially new areas of study. A certain amount of data has been collected on pipeflow in drylands by Drew (1972, 1982), Barendregt and Ongley (1977), Bryan and others (1978), Yair and others (1980), Bryan and Yair (1982), Bryan and Harvey (1985), and Bryan and Campbell (1986). But the bulk of information on pipe-flow response comes from the British Isles. Measurements here fall into three broad groups: (1) studies in which the hydrological role of pipeflow has been induced theoretically, such as those of McCaig (1983) and Wilson and Smart (1984); (2) studies based on point samples of pipe discharges, such as those of Weyman (1971), Stagg (1974), Waylen (1976), Finlayson (1977), and Cryer (1980); and (3) studies based on complete storm event hydrographs (Newson and Harrison, 1978; Gilman and Newson, 1980) or fully continuous monitoring (Jones, 1978a, 1982, 1986, 1987a and b, 1988; Jones and Crane, 1982, 1984).

The latter studies have been predominantly in upland peaty soils and podzols, i.e., histosols and aquods. Data on pipeflow in other soils in Britain are limited to a few measurements in brown earths taken by Waylen (1976) and Finlayson (1977), which have been analyzed in Jones (1981, p. 158ff).

Pipeflow has only been studied in four research areas in humid regions outside Britain: in the Schrondweilerbaach catchment in Luxembourg where pipeflow is seen to be an important source of quickflow, although it has not been studied separately from rillflow and overland flow, with which it interacts very closely (Bonell and others, 1984; Duijsings, 1985); on the tropical island of Dominica, where a brief program of spot discharge measurements was undertaken by Walsh and Howells (1988); on the subtropical island of Honshu, Japan, where complete hydrographs have been studied by Yasuhara (1980), Tanaka (1982), and Tsukamoto and others (1982); and in the boreal forest of the Laurentian hills in southern Quebec, where Roberge and Plamondon (1987) monitored pipeflow during the spring melt period. In contrast to the British sites, each of these areas is forested.

Quantitative importance of pipeflow

Monitoring programs in humid temperate regions have now confirmed that pipeflow is at least as important a process of hillslope drainage and stream flow generation as suggested in early unquantified speculations (Jones, 1971, 1979). Initial attempts to quantify pipeflow in Wales in the 1970s tended to point to contributions of the order of 20 to 25 percent of stream stormflow (Ward, 1975) and suggested that it was capable of initiating minor flood waves (Institute of Hydrology, 1974; Rodda and others, 1976). However, the bases for such estimates left much to be desired. They rested on observations of ephemerally flowing pipes during just three storms in the Institute of Hydrology's Upper Wye catchment, which were calculated to indicate that between 4 and 72 percent of rain falling upslope of the monitoring sites passed through the pipes (Gilman and Newson, 1980; Jones, 1981, p. 161).

Later work in the Upper Wye increased the number of storms to 15 and pipeflow discharges were measured as a percentage of flow into the flush at the base of a sideslope. This indicated proportions varying from less than 1 percent to as much as 61 percent (Gilman and Newson, 1980), but there was still no evidence of the importance of pipeflow to actual streamflow. Early continuous flow measurements covering 20 storms on the nearby Maesnant Basin did compare streamflow and pipeflow, and indicated an average contribution of about 25 percent (Jones,

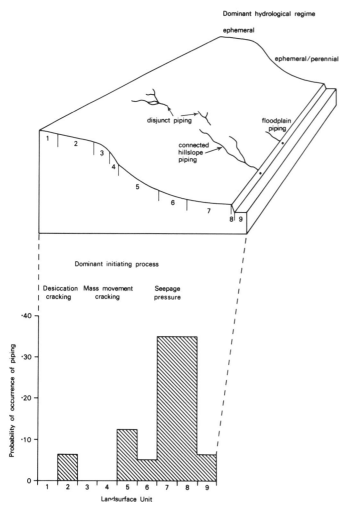

Figure 5. Distribution of piping in relation to Land-surface units. Probabilities indicate the likely location of piping when present, based on the frequency of published reports. Land-surface units are numbered according to the NULM scheme (Conacher and Dalrymple, 1977).

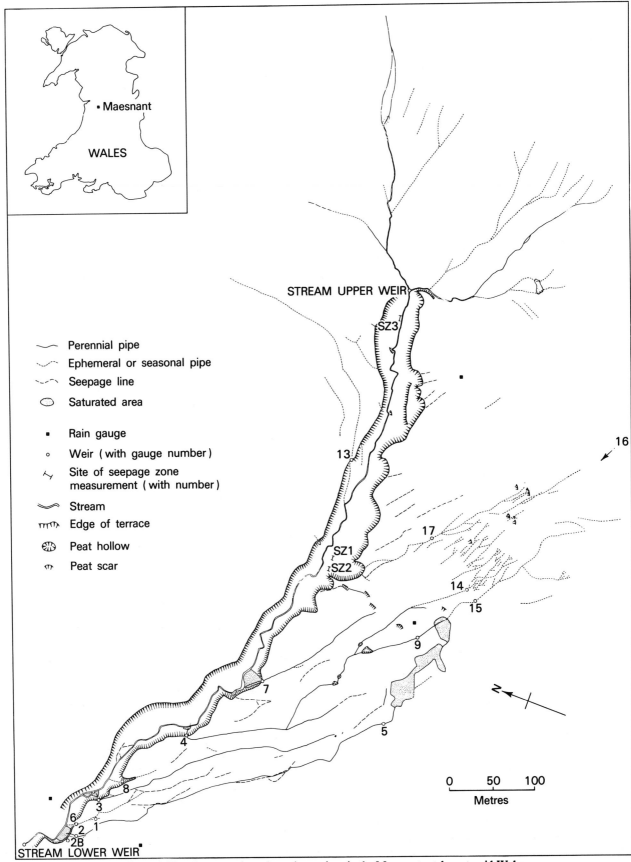

Figure 6. The pipe networks and gauging stations in the Maesnant catchment, mid-Wales.

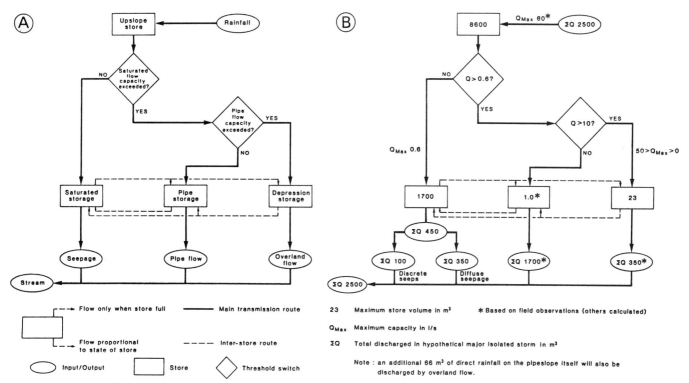

Figure 7. A flow diagram of hillslope hydrology in the Nant Llwch Basin, Brecon Beacons, South Wales, according to Wilson and Smart (1984). (A) Stores and thresholds, (B) estimated flows.

1978a), but on the basis of extrapolating measurements from a single pipe.

The subsequent intensive monitoring scheme begun at Maesnant in 1979 (Fig. 6) revealed the inadequacies of such an extrapolation. Two and one-half years of continuously logged data at up to 19 sites, covering 190 storms, indicated much higher yields. On average, pipes contributed 49 percent of stream stormflow and 46 percent of stream baseflow (Jones and Crane, 1984; Jones 1987a). As the importance of baseflow contributions indicates, most of this discharge reaches the stream by way of perennially flowing pipe sections.

Less direct evidence of the importance of pipeflow has been produced by Wilson and Smart (1984), based on theoretical calculations for a basin in the Brecon Beacons, South Wales. This suggested that 68 percent of streamflow is generated by ephemeral pipes during storms (Fig. 7). However, the methods of estimation are so indirect as to be somewhat open to question.

The only other calculations of pipeflow contributions in humid regions come from Japan, Quebec, and the West Indies. The Japanese research has yielded results comparable with the British data. Tanaka (1982) calculated that pipeflow issuing from the A horizon contribution 45 to 52 percent of total runoff in four monitored storms in the Hachioji basin the Tama Hills north of Tokyo. The pipeflow here issues from the base of slopes surrounding a hollow. It then flows across the flat floor of the streamhead hollow in unchanneled rivulets to reach the gauging weir at the head of the stream channel. In an adjacent basin, Tsukamoto and others (1982) monitored pipeflow within the soil profile in the bottom of a smaller, V-shaped, streamless sideslope hollow. They found that almost 100 percent of the throughflow passing through the top 600 mm of the Kanto loam soil was pipeflow, each pipe beginning to flow as the saturation level reached the floor of the pipe. The soil here is a silty clay loam with a very low saturated hydraulic conductivity of ca. 10^{-5} mm/s within the soil matrix (Fig. 8).

In the Lac Laflamme basin, 80 km north of Quebec City, Roberge and Plamondon (1987) estimate that pipeflow in the organic layer typically transmits 20 to 22 percent of rain-melt inputs, within groundwater flow through a sandy till transmitting the remainder. However, on occasions when groundwater levels are high, pipeflow can be the dominant process, accounting for up to 76 percent of inputs, especially at the end of the melt period. At other times of year, the pipes are thought to be dry.

The observations of Walsh and Howells (1988) on the tropical island of Dominica suggest a much less important hydrologic role for piping there than in the other environments. Their conclusions are, however, based on just two weeks of spot measurements in one basin, the Point Baptiste, in the ever-wet zone. Here, deep piping was estimated to contribute 16 percent of stream baseflow, and no significant stormflow reaction was observed in three storms of 25 to 43 mm. They observed that the main storm runoff process in the thick rain forest areas is diffuse throughflow within the highly permeable organic topsoil and above the silica pans. They speculate, however, that piping may be more impor-

Figure 8. Piping in the Kanto Loam monitored by Tsukamoto and others (1982).

tant where matrix permeabilities are more restricted, in the moderate to slow range below 25 mm/hr. As Figure 9 shows, these permeabilities are typically associated with the area of the island that experiences a marked dry season and where smectoid clays are the norm. Unfortunately, no pipeflow measurements were taken in this wet-and-dry tropical zone, and no survey was made of the frequency of piping, whereas one would suppose that this might be the most favorable environment for piping on the island. Clearly, it is impossible to extrapolate the Pointe Baptiste measurements even within Dominica, let alone the tropics as a whole. We must await considerably more data before any definitive statements can be made about pipeflow and runoff processes in the vast range of humid tropical environments.

Spatio-temporal variations in pipe response

Average pipeflow contributions give only a partial impression of the role of pipeflow drainage. The British observations have shown marked variations in pipeflow from storm to storm, depending largely on the antecedent wetness of the basin and the size of storm. Observations at Maesnant in particular have also revealed marked spatial variations in response across the hillslopes, both between and within the ephemeral and perennial pipe sections (Jones and Crane, 1984; Jones, 1987a, 1988).

Figure 10 summarizes the variation in the percentage contributed to the Maesnant stream by pipeflow and shows a peak contribution in events with moderately heavy rain falling on a moderately wet catchment. Although the measure of antecedent moisture taken here is the 7-day antecedent rainfall total, in fact the pattern is similar for 1-day, 4-day or geometric 7-day antecedent rainfall. When the catchment is dry and storms are light, the spatial extent of pipeflow is limited to the network of perennial pipes (Fig. 6). A large proportion of rainfall, and probably pipeflow as well, is lost to make up the moisture deficit in the soil matrix. Experiments by Newson and Harrison (1978), in which water was pumped into ephemeral pipes during dry weather, indicated large losses, amounting to 33 to 72 percent of the simulated pipeflow of 3 l/s. But losses are probably not quite of this order in natural pipeflow events, as indeed is suggested by the relative unimportance of the parameter "loss of moisture to the unsaturated zone" in Gilman and Newson's (1980) own computer simulations of pipeflow.

Pipeflow extends upslope into the ephemeral feeder network when the Maesnant catchment is moderately wet and storms moderately heavy (Fig. 11). At this time, substantial "channel precipitation" also enters the perennial pipe network directly from rain falling on the saturated bog areas around the heads of the main perennial pipes, and the bogs themselves appear to be fed by increased groundwater effluence from the slopes above. In these conditions, pipeflow can contribute substantially more than 50 percent of the water entering the stream.

When the catchment is very wet and/or the rainstorm is very heavy for the area, the percentage contributed by the pipes falls off (Fig. 10), even though the absolute quantity of pipeflow continues to increase and, indeed, many overflow pipes are only activated at this stage. In these wetter conditions, saturation overland flow and higher contributions from saturated areas near the stream begin to take over, and these other drainage processes combine to destroy the overall dominance of pipeflow. Albeit, a proportion of the yield from these alternative pathways has also passed through pipes at some stage at its journey down the hillslope.

Figure 12 is a schematic block diagram of the Maesnant basin illustrating the average quantities of flow across the hillslopes in near-optimal conditions for a higher percentage contribution from pipeflow, i.e., under average storm conditions of 3 mm 1-day antecedent rainfall and 30 mm of storm rainfall. The model plots the progress from the top of the side slopes (top of diagram) to the streambank and down the stream channel (from left to right).

Three key points are illustrated by this diagram. One is that the ephemerally flowing pipes are intimately linked with the

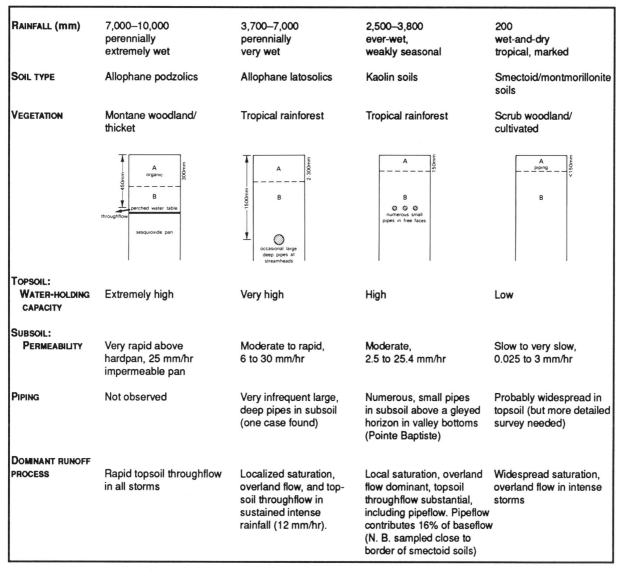

Figure 9. Distribution of piping in humid tropical soils in Dominica, based on surveys by Walsh (1980) and Walsh and Howells (1988), and personal communications from Walsh.

perennial network, whether as feeders lying generally upslope of the perennial pipes, or as overflows, which act as distributaries when the perennial pipes approach capacity flow. Certainly, in this catchment there are no grounds for studying only one part of the system, the perennial or the ephemeral pipes, in isolation as some have suggested and practiced (compare Gilman and Newson, 1980).

Secondly, the flows across the hillslopes involve complex interchanges between surface and subsurface waters and between channelized and diffuse flows. The pipe networks themselves do not operate in isolation from other hillslope processes; they cannot be studied separately from overland flow and seepage flow. This complex switching of flows has been noted by Wilson and Smart (1984) in the Brecon Beacons of South Wales. From their indirect estimates of pipeflow, they concluded that the ephemeral pipe networks are activated when the saturated flow capacity of 0.6 l/s in the upslope soil water store is exceeded (Fig. 7). In turn, saturation overland flow is generated after the average pipe capacity of 10 l/s has been exceeded. In Canada, Roberge and Plamondon (1987) noted the frequent exchange between surface and subsurface sections of channelized drainage during snowmelt under boreal forest. They also observed that while pipeflow hydrographs closely reflect the rain-melt input, there is a close correlation between groundwater levels and pipeflow activity.

The third crucial point that Figure 12 illustrates is the degree of spatial variability in pipeflow contributions, from 20 percent of receipts in the upper half of the stream to 90 percent in the lower. This variability makes it difficult to generalize from studying short sections of stream or small hillslope plots (see the large variation in pipe density in Fig. 21). Cross-slope variability is often enhanced by marked differences in response between adjacent pipes. These differences are most marked within the perennial network, reflecting differences in the size of network draining into them and of the microtopographic depressions which drain diffuse seepage and overland flow toward them. Figure 13 shows a range of an order of magnitude in maximum discharges between perennial outlets, and 350 percent in mean stormflow yields. This means that it is impossible to extrapolate measurements from one or two pipes to the whole system of pipes, at least in cases where there is marked range in the sizes of the constituent networks. The problem for the would-be measurer is to know whether such a range exists.

Experience at Maesant suggests that microtopography can be a useful guide to pipe discharges, and there is a reasonable correlation between depression size and mean stormflow discharges. However, the prediction of mean stormflow can be dramatically improved with a knowledge of the extent of the actual networks that maintain perennial or ephemeral flows (Jones, 1986). The ephemeral pipes show much smaller ranges in absolute terms, with maximum discharges all below 10 l/s. This reflects the much smaller areas of hillslope actually contributing flows to them during a storm, with mean calculated contributing areas of 0.18 ha, compared with 1.15 ha for the average perennial pipe. At least for ephemeral pipes, this would seem to vindicate the approach of Wilson and Smart (1984) of extrapolating from measurements on a few pipes, and implied in the sampling procedure used on the lower Nant Gerig by Gilman and Newson (1980). Certainly, this approach may well be adequate if the sole concern is the total contribution of ephemeral pipeflow to streamflow. Nevertheless, the variability in percentage terms between one ephemeral pipe and another is almost as great as for the

Figure 11. Pipeflow monitoring at the head of the ephemeral pipe network on Maesnant.

perennial pipes: 600 percent in peak discharges and 330 percent in mean stormflow yields.

One unfortunate characteristic of both types of pipe from the point of view of rapid assessment is that ranking in terms of *mean* stormflow is not always consistent with ranking in terms of *peak* discharges. Higher peak discharges appear to be associated with networks that have more ephemeral piping integrated with the trunk pipe, which can direct flows rapidly toward the trunk pipe during severe storms. In contrast, high average yields can be achieved by pipes that reach lower peak discharges but flow longer.

In addition to variability in flow volumes, the Maesnant program has revealed marked differences in the timing and length of stormflow response in the pipes. Most perennial pipes respond 30 to 100 min ahead of the stream, in effect shortening the response time of the stream and acting as a major trigger mechanism for stormflow in the stream (Jones, 1987a). The ephemeral pipes show a much wider range in response times, with average timing for commencement of flow varying from nearly 2 hr before the stream to almost 6 hr after, and drying up anywhere between 1 day before the end of stormflow in the stream and one-half day after. In contrast, perennial pipes normally cease stormflow within ±5 hr of the stream.

The perennial pipes trigger the start of stormflow in the stream because of their large integrated tributary networks, the rapid collection of "channel precipitation" in the bog at the head of their networks and the continual priming from effluent groundwater. However, once the ephemeral pipes are activated, their rise times are shorter, making up for "lost time," and their times of peak discharge are much more in accord with both the stream and the perennial pipes, thus adding their weight to peak runoff in the catchment.

Evidence from three other British catchments generally supports the broad response characteristics found in the Maesnant pipes, particularly in terms of the sensitivity of pipeflow to soil moisture status and the nonlinearity of response. Gilman and

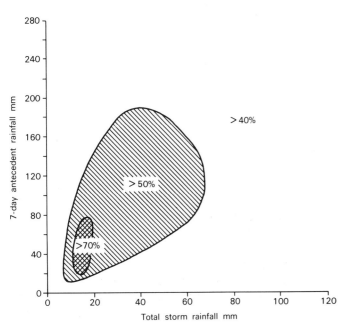

Figure 10. Proportional contributions of pipeflow to stormflow in the Maesnant stream, in relation to total weekly antecedent rainfall and total storm rainfall. Values are in percentages.

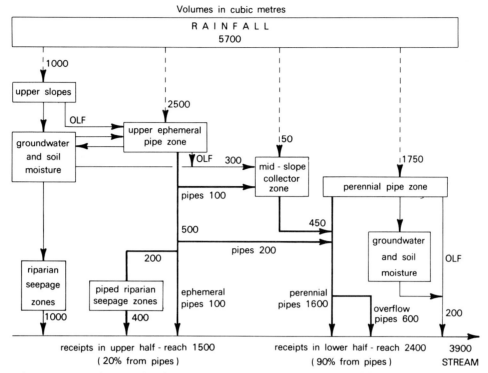

Figure 12. Flow diagram of drainage processes during the average storm of 30 mm in the Maesnant catchment.

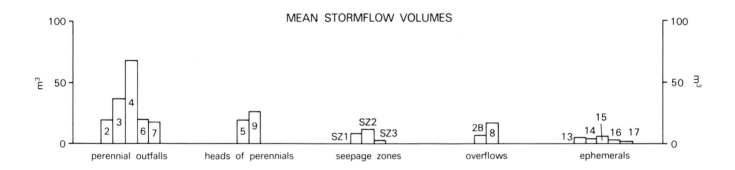

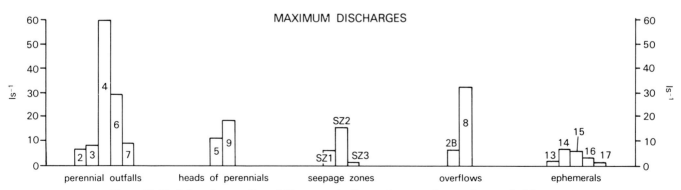

Figure 13. Variations in stormflow yield among pipeflow and seepage flow stations on the Maesnant. Numbers refer to gauging stations identified in Figure 6; SZ numbers are riparian seepage zones.

Newson's (1980) experiments with an optimized linear reservoir model on the ephemeral pipeflow data from Nant Gerig revealed two preeminent parameters controlling pipeflow: the soil moisture content at the start of the storm and the reservoir storage coefficient. More than 70 percent of the flow simulation experiments gave a satisfactory fit to the observed data using just these two parameters. This accords with recent statistical analyses of the Maesnant flow records, which reveal that antecedent rainfall or pipe discharge prior to a storm are broadly interchangeable statistically and second only to total storm rainfall as a predictor for stormflow volumes in all the major categories of pipe (Jones, 1987a).

The nonlinearity of response is emphasized by the observations of Wilson and Smart (1984) in the Nant Llwch, Brecon Beacons. They found that the rate of runoff from the piped subbasin was much quicker in events with 10-day antecedent precipitation indices above 50 mm, giving peak discharges averaging 3 times greater. Such results seem to have important implications for the applicability of models employing unit hydrograph theory in piped basins. McCaig (1983) inferred similar differences in Slithero Clough, Yorkshire, where he calculated that the proportion of streamflow derived from the pipe source area increased dramatically from less than 30 percent in a dry catchment to 70 percent in a catchment with 40 mm in storage. He concluded that there is a threshold at 20 mm of storage at which pipeflow is generated. This compares with a threshold of 75 mm calculated by Gilman and Newson (1980) in peaty soils in Wales and the somewhat lower thresholds suggested for flow in the ephemeral pipes on Maesnant with 7 mm of storm rainfall plus 3 mm 1-day antecedent rainfall (Jones, 1987a).

Effects on streamflow and contributing areas

Recent studies of the implications of piping for modeling streamflow have focused on the variations that the pipes create in dynamic contributing areas from one storm to another (McCaig, 1983; Wilson and Smart, 1984; Jones, 1986). McCaig's calculations of pipeflow contributions suggested that the contributing area for pipeflow was much larger than the area of surface saturation observed during 5 separate storm events, and that it increased more rapidly than the saturated area in more severe storms. On the other hand, Wilson and Smart could not prove any consistent differences between the DCAs in their paired piped and unpiped subbasins. Unfortunately, the lack of direct monitoring of pipeflow in these two cases must limit the reliance placed on them. McCaig (1983), for example, estimated the proportion of pipewater in the stream by comparing average total dissolved solids in samples of pipewater with the monitored TDS in the streamwater, yet recent monitoring of pipewater on Maesnant reveals marked differences in pipeflow solutes in individual storm events.

The calculations of contributing areas based on direct measurements from Maesnant suggest that while on average the pipes double the contributing area from 6 to 12 ha in a 23-ha subbasin, there are interesting variations between storms and between the ephemeral and perennial networks (Jones, 1986, 1987a). In the heavier storms, both ephemeral and perennial pipes tend to expand their own contributing areas to 2.5 to 3 times their average storm values. The expansion tends to be slightly greater among the perennial pipes, partly because they receive larger contributions from effluent groundwater from upslope. Although the proportional expansion may be comparable in the two parts of the network, quantitatively, of course, the perennials capture drainage from much larger expanses of hillside, with maximum DCAs averaging 3.5 ha compared with 0.4 ha for the ephemerals.

At maximum discharges the pipe network can extend the streamflow contributing area to more than 70 percent of the Maesnant subcatchment, linking sources on crests up to 750 m distant from the point of issue near the streambank. This substantially alters the general view of the nature of dynamic contributing areas as being adjacent to stream channels, expanding and contracting in contiguous bands, and recognizable by wetland vegetation and/or concavity in hillslope profiles or planform (cp. Hewlett and Nutter, 1970; Dunne and others, 1975; Kirkby, 1978). When the full pipe network is activated, i.e., in roughly 50 to 66 percent of storms on Maesnant, ephemeral sources on the upper slopes (e.g., unit 2 in Conacher and Dalrymple's [1975] classification) are linked to the stream and may transmit kinematic storm waves to the stream in under 2 hr (Jones, 1988). The water discharged from the upper slopes has filtered through different soils and has a different chemical signature from the water originating on the concave lower slopes (Jones and Hyett, 1987). Consequently, the relative balance between these source areas may have a significant effect on streamwater chemistry.

The only other attempt to estimate contributing areas comes from the work of Roberge and Plamondon (1987) in Quebec, based on the monitoring of a single pipe outlet during the 1984 and 1985 melt periods. In fact, they used this estimate as a basis for calculating the percentage of rain-melt passing through all the Lac Laflamme pipes. Interestingly, they estimated a maximum DCA for the pipe of 4,620 m^2 after 43 mm of input, which is similar to the estimates for ephemeral pipes on Maesnant, with an average maximum DCA of 4,498 m^2 (Jones, 1986, Table 1). Roberge and Plamondon also comment on the effects of pipeflow on the chemistry of the lake. As at Maesnant, pipeflow lowers pH in the surface waters by diverting seepage away from the mineral layers and reducing residence times (compare Jones and Hyett, 1987).

Effects on infiltration and permeability

The rapid expansion of streamflow contributing areas associated with piping and the evidence of marked cross-slope variability in pipe yields imply that piping has a major effect on infiltration rates and permeabilities.

In essence, we have an extension of the "minimum representative volume" problem highlighted by Beven and Germann (1982) in the context of identifying the right dimensions of sample for characterizing soil permeabilities or infiltration capacities.

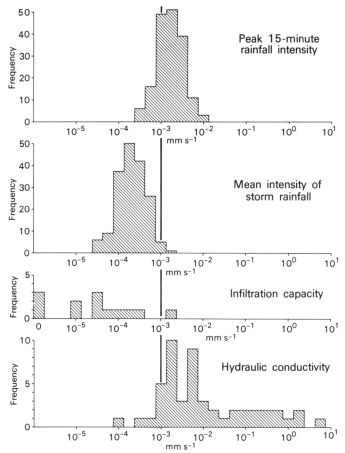

Figure 14. Rainfall intensities, infiltration capacities, and hydraulic conductivities on Maesnant.

The "minimum representative area" of hillslope needed to characterize hillslope-drainage processes is larger in the presence of piping, commensurate with the larger scale of variability in infiltration capacities, hydraulic conductivities, and throughflow yields across the slope.

Measurements taken within the main zone of piping at Macsnant reveal a number of interesting effects. First, comparison of rainfall intensities and infiltration capacities, as measured by a double-ring infiltrometer, suggest that, in terms of classical Hortonian hydrology, the main source of streamflow should be overland flow (Fig. 14). Sixty percent of storms had mean rainfall intensities above the mean infiltration capacity and all storms had peak 15-min intensities that exceeded it.

Putting a spatial interpretation on Figure 14, in 50 percent of storms we would expect Hortonian overland flow from 75 percent of the hillside. But this does not occur; overland flow is only observed covering short distances mainly in swales before entering cracks or blowholes. Such macropores are underrepresented in standard infiltration measurements; indeed, they are consciously avoided (compare Beven and Germann, 1982).

Secondly, saturated hydraulic conductivity measurements taken on undisturbed soil cores indicate that, once the rainwater has infiltrated, the soil is more capable of transmitting it. No storms had average intensities exceeding either the mean or the median conductivity, although 5 percent of storms briefly exceeded the median conductivity during peak intensity (Fig. 14). Even so, at 10^{-2} to 10^{-3} mm/s the conductivity is insufficient to provide water fast enough to contribute much stormflow to the stream. The mean conductivity suggests a 17-day travel time to the stream from mid-slope, whereas the stream tends to respond to rainfall in just 8.5 hr and to cease stormflow only 2 days after the end of the rain. Middle and upper slopes, therefore, could not contribute significant volumes of storm water without the presence of piping. In fact, spreading the calculated 6 ha of contributing area for diffuse, riparian flows evenly along the stream gives a 40-m-wide band, which the hydraulic conductivity measurements suggest could drain in 1.75 days. The conductivity measurements, therefore, concur reasonably well with estimated limits of contributing areas based on measured volumes of discharge.

Finally, intensive mapping of infiltration capacities and hydraulic conductivities within the zone of ephemeral piping indicated marked spatial patterns (Fig. 15). Bearing in mind that in taking both measurements the direct effects of pipes or major macropores were consciously avoided, these imply generally greater macropore development within the soil matrix in the central area of piping. This could be due to desiccation induced by pipe drainage. Equally, piping could have developed here precisely because of the greater permeability. No doubt elements of both lie behind the patterns.

PIPING EROSION AND THE DEVELOPMENT OF HUMID LANDSCAPES

Piping is still not a widely acknowledged erosion process in humid lands. Yet, when it is present, the hydrological effects discussed above may have important geomorphic consequences. There has been increasing recognition of the importance of understanding hillslope hydrologic pathways for modeling soil erosion and slope evolution, as noted by Burt (1987) in his review, but few attempts have been made to integrate piping into such schemes (compare Jones, 1987a, b). The purpose of this section is to explore the implications and, where possible, to adduce field evidence to support three main hypothetical propositions.

First, the provision of channelized flow within the hillslope domain extends channel erosion processes well beyond the streambank and the streamhead (Fig. 16). In this respect it is similar to rilling, but its effects are more varied. Piping may act as a source for the extension of integrated open-channel networks and as a means of linking and maintaining quasi-stable discontinuous open channels (Jones, 1987b). Being subsurface, piping is also able to operate under a complete vegetation cover. This makes piping a rather special form of erosion because it may well be as effective on well-vegetated humid hillslopes as on bare arid hillslopes, in defiance of the common relation between rainfall and rates of erosion, although quantitative evidence for this is currently lacking.

Secondly, the redistribution and acceleration of hillslope drainage processes by piping has implications for the pattern of hillslope evolution, especially through plan-form modification. Again, piping may take on a variety of roles here. It may have a direct effect as a form of channelized erosion, or indirect effects causing localized sinks and collapse features, landslides and slumping. Eventually, either individually or in combination, these effects may lead to the formation of hollows, dells, and even valleys (Tsukamoto and others, 1982; Jones, 1987b; compare Fig. 17).

Finally, the increased efficiency of hillslope drainage and the increases in contributing areas and runoff coefficients imply that storm discharges are increased within the stream-channel network. This in turn implies an increase in within-channel erosion rates. These in-channel effects have yet to be quantified and will not be discussed in any detail here. Nevertheless, Jones (1987a) has speculated that since piping provides 50 percent of storm runoff in the Maesnant stream, it is likely to substantially increase, perhaps even double, the mobilization of sediment within the stream channel. This is in addition to substantial quantities of fine sediments issuing into the stream from the pipe outfalls themselves.

Piping and slope development

Effects on the theory of slope evolution. Despite the emphasis in theoretical models of humid hillslope development upon evolution of the profile (compare Ahnert, 1987b; Armstrong, 1987), in reality adjacent profiles tend to differ. As Carson and Kirkby (1972) pointed out, spatial heterogeneity tends to predominate in hillslope profiles, with the result that plan-form development is an important aspect of slope evolution both per se and in its effects on profile development.

Piping is yet another process that increases cross-slope heterogeneity by altering the patterns and rates of erosion and deposition across the hillslopes. It is also a process that has so far been ignored in process-response models. The more efficient soil drainage created by piping reduces times of concentration for subsurface stormflow. Higher discharges of subsurface stormflow result both from these higher velociies and from capturing water that would otherwise drain down a different, adjacent section of the slope. Together these effects create greater erosion and erosion that is more spatially concentrated along the zones of piping. The positive effect of concentrated erosion is augmented by the negative effect of reduced overland flow frequencies and sheetwash on adjacent sections of slope. In this way, piping may create a microtopography on the surface of the slope, which could be the precursor of hollow and valley development in a rejuvenated environment. Any reduction in sheetwash effects on the surfaces within the piping zones tends to be countered either by more effective eluviation, subsidence and roof collapse there, or by the resurgence of pipeflow when stormflows exceed the capacity of the pipes.

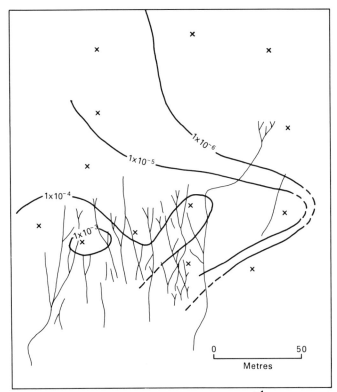
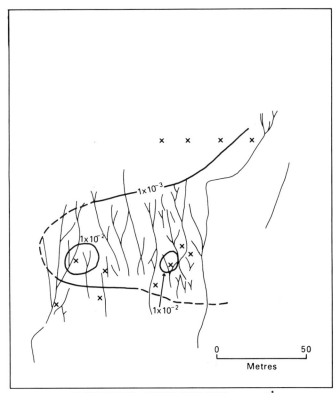

Figure 15. (A) Infiltration capacities and (B) saturated hydraulic conductivity in the main zone of ephemeral soil pipes on Maesnant. Measurement sites marked by crosses.

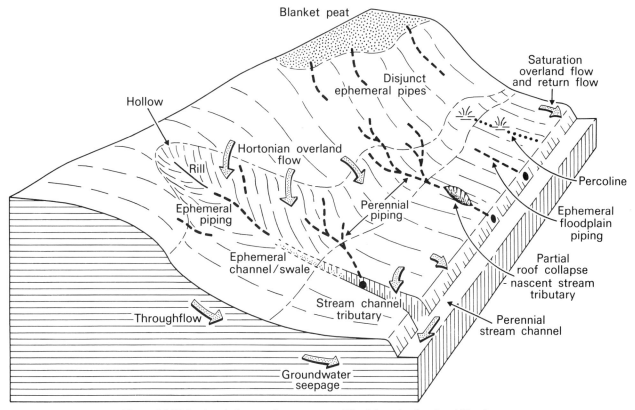

Figure 16. Piping in relation to other processes of fluvial erosion in a humid landscape.

In his formal model of hillslope evolution, Kirkby (1978) concluded that subsurface flow exerts a dominant control in humid, temperate climates and that the hillslope hydrology is crucial, particularly the relative proportions of water draining above or beneath the surface. Piping effects clearly corroborate Kirkby's basic conclusion, but they also force a reevaluation of some of his assumptions and results.

Piping affects two of Kirkby's fundamental assumptions regarding the speed of throughflow and the relative significance of solutional and mechanical erosion on humid hillslopes, which follows from this. Throughflow is assumed to be slower than overland flow (Whipkey and Kirkby, 1978, p. 131); therefore, throughflow is not generally a significant contributor to storm runoff, and because it is slow-moving, the water is more likely to reach chemical equilibrium with the soil, and solutional erosion is enhanced. So Kirkby (1978, p. 355) saw subsurface flow as significant in terms of solutional erosion but not mechanical erosion. He concluded that overall solutional erosion in humid regions is approximately proportional to the amount of subsurface flow. Moreover, on individual slopes, solutional erosion will be inversely proportional to a/s, the area drained per unit contour length divided by the slope, since, according to Kirkby, the proportion of rainwater infiltrating the soil is reduced downslope as the probability of overland flow increases.

In practice, field measurements of pipeflow velocities indicate average values two orders of magnitude faster than throughflow within the normal soil matrix and nearly an order of magnitude faster than overland flow (Fig. 18). Velocities of ca. 10^{-1} m/s in perennial pipes imply a marked potential for mechanical erosion. At the same time, solutional erosion may be reduced because of diminished opportunities for solute uptake in pipes. This is due partly to the shorter residence times caused by more rapid drainage of the slopes, giving less time for chemical equilibrium to develop between soil and drainage water. It is also partly due to lower areas of surface contact with the soil caused by the higher ratio of the cross-sectional area of flow to wetted perimeter in soil pipes compared with normal soil pores.

In the end, however, the relative effectiveness of piping as a solutional agent vis-à-vis diffuse throughflow depends on total ionic yields rather than concentrations, and total yields are the product of ionic concentrations and the volumes of total discharge. Pipes with high annual discharges may therefore be more effective solutional agents than matrix throughflow, despite adverse contrasts in solute concentrations. Pipes might even stimulate increased solutional erosion in the soil matrix by increasing the local hydraulic gradient and causing a more rapid turnover in soil water that has reached chemical saturation, thus allowing fresh, unsaturated rainwater to infiltrate the matrix. This latter effect would tend to work against Kirkby's inverse relation between solutional erosion and the a/s index, particularly where piping shows increased frequencies on the lower slopes and floodplain areas, which is often the case (Fig. 5). In such instances,

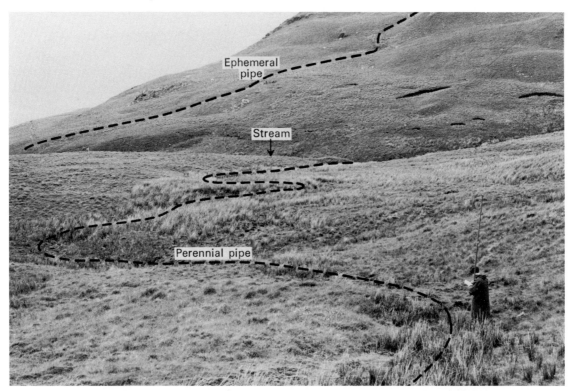

Figure 17. Piping in the headwater area of Maesnant. Perennial piping marked by the *Juncus* flush in the foreground occupies a well-defined valley and is essentially the source of the stream. The ephemeral pipe on the far slope is nearly 250 m long but has little surface expression at present except in the slight hollow around its source.

blowholes capturing overland flow would increase the proportion of rainwater infiltrating the soil downslope and the natural underdrainage will increase the rate of turnover in the soil water, and so reduce the likelihood of saturated solutions residing in the soil and retarding solute uptake.

Until we have more field measurements, of course, these theoretical implications must remain largely speculative, but they do suggest that piping could cause some interesting departures from currently accepted theory.

Measured erosion rates in humid piping. Looking at the field evidence, there are as yet very few measurements of erosion rates in either arid or humid areas (Jones, 1981, p. 212). Measurements that are available for humid areas suggest considerable variability, not only from one hydrologic region to another but even over short distances between pipes with different flow regimes.

Measurements of sediment yields in ephemeral pipes vary from 1 kg/yr of mainly organic material in mid-Wales (Newson, 1976) to 12 kg/yr of mineral material in the drier Drakensberg Mountains of South Africa (Humphreys, personal communication, 1983). In the perennial pipes of the Maesnant basin, yields of mineral sediment are much higher but equally variable, ranging from 15 kg/yr to 223 kg/yr at sites 3 and 4, respectively, in Figure 6. Moreover, there are marked seasonal variations, with yields at site 4 ranging from 6 kg/week in winter to 2 kg/week in summer, and generally coarser material being transported during the winter period (Fig. 19). The high variability, combined with the general lack of measurements, unfortunately makes it impossible to take a systematic global view of erosion rates at the present time.

The Maesnant is the only basin where any attempt has been made to monitor pipeflow sediments in anything approaching a comprehensive way. Here, the 2-year monitoring program indicated that pipeflow sediments were a much more significant source of stream sediment than expected; bed traps set behind the weirplates at the outfall of each perennial pipe in the subbasin collected an estimated 15 percent of the total sediment yield in the basin (Jones and Crane, 1984). Adding unmonitored pipe sources, especially in the streamhead area, could well raise this percentage to about 25 percent, making it a substantial erosion process for mineral material alone. The piping also appears to be a major source of peat erosion, the particles passing through the system largely in the form of suspended load. Erosion of the peaty O horizon by pipeflow may be a major factor in developing microtopographic hollows and in creating significantly thinner peat immediately above both ephemeral and perennial pipes.

Interesting links between sediment yield and microtopographic development have been provided in a recent statistical analysis of the Maesnant data by Jones (1987a). In this, weekly sediment yields were found to be strongly correlated with the size

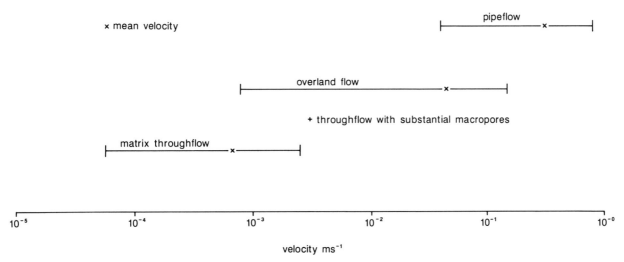

Figure 18. Pipeflow velocities in relation to rates of overland flow and throughflow. Based on values collated from available published measurements in a variety of environments.

of microtopographic hollow or "topographic contributing area" (TCA) centered on each major pipe (Table 1), suggesting that the pipes were responsible for creating the microtopography. A similar correlation was found with the total length of piping within the micro-hollows, suggesting that the larger networks are also the ones eroding most rapidly. This accords with the very close relation between sediment yield and 5 percent duration discharges, and likewise between these discharges and TCA or pipe lengths within the TCA (Table 1). Perhaps most interesting is the suggestion that the erosive potential of the perennial pipes owes more to the extent of the linked ephemeral network than to that of the tributary perennial network (Jones, 1987a). The correlations in columns 3 and 4 of Table 1 seem to imply that the more erosionally effective pipes are those with larger tributary ephemeral networks that can more rapidly collect and discharge stormflow.

It is the concentration of flow provided by the tributary network rather than actual sediment entrainment within the ephemeral reaches that seems to control sediment yields from the major pipes. Monitoring of hydraulic parameters during pipeflow events suggests that direct mechanical erosion is minimal within the ephemeral pipes. No observed events had Reynolds Numbers unequivocally within the turbulent range (Re<300 in all events) on Maesnant, comparable with values of 58<Re<425 calculated by Gilman and Newson (1980). In contrast, 4 percent of monitored flows in the perennial pipes had Reynolds Numbers clearly in the turbulent range (Re>2,500), and values were generally in excess of Re=1,000.

However, values of Re.f (Reynolds Number times Darcy-Weisbach friction coefficient) increased with discharge in all pipes, whether ephemeral or perennial, with the sole exception of the most highly developed pipe. This suggests that erosive events are too infrequent to have yet sculpted more efficient geometries at high flow. In fact, 70 percent of pipeflow sediments were

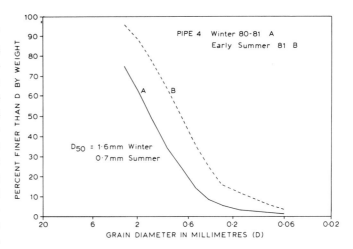

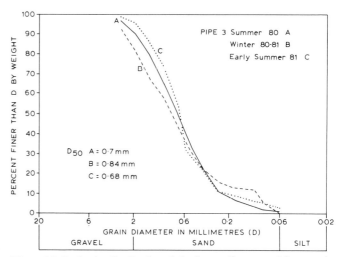

Figure 19. Grain-size distribution of pipeflow sediments on Maesnant in summer and winter flows. Seasonal differences are greater in pipe 4 which has a flashier regime and had a winter 5 percent duration discharge of 14.3 l/s compared with 5.6 l/s in pipe 3.

TABLE 1. CORRELATIONS BETWEEN PIPE SEDIMENT YIELD, 5 PERCENT DURATION DISCHARGE, AND PIPE NETWORK GEOMETRY

	5 percent discharge L/s	Mean stormflow m^3	Linked Pipe Lengths, m perennial	Linked Pipe Lengths, m ephemeral	Linked Pipe Lengths, m total	Proportion of perennial pipe %	TCA m^2	Pipe Lengths in TCA, m perennial	Pipe Lengths in TCA, m ephemeral	Pipe Lengths in TCA, m total	Pipe density in TCA m/m^2	Mean DCA m^2	Mean slope tan	Mean diameter m
Sediment yield kg/week	0.95	0.65	0.46	0.70	0.63	-0.70	0.70	0.65	0.71	0.70	-0.49	0.60	-0.47	0.43
5 percent duration discharge, L/s	1.00	0.70	0.51	0.71	0.66	-0.69	0.72	0.70	0.72	0.72	-0.55	0.68	-0.55	0.43

TCA = topographic contributing area, the area of microtopographic hollows centered on pipe.
DCA = dynamic contributing area, calculated as runoff volume/rainfall.
See Jones (1986) for detailed definitions.

derived from the single most highly developed pipe (pipe 4 on Fig. 6). In this pipe, the hydraulic geometry *is* better adjusted at high flows (Re = 2,433, f = 1, Manning's n = 0.07) than at low flows (Re = 193, f = 192, n = 0.85) (Jones, 1985). Pipe 4 has a 5 percent duration discharge of 10 l/s, compared with 5 l/s or less for other perennial pipes. It also has by far the largest tributary network, totaling almost 1 km of pipes, nearly half of it ephemeral (Jones, 1986, Table 3).

Turning to solutional erosion, once again the range of environments studied has been extremely circumscribed. Most measurements come from a limited range of soils in the British Isles, but even here there is a wide divergence in the relative importance of solute discharges in pipeflow. In ephemeral pipes developed in acid brown earths, both Oxley (1974) and McCaig (1979, 1983) found higher solute concentrations in pipeflow than in overland flow. At ca. 100 mg/l, McCaig's concentrations were 2 to 3 times those in saturation overland flow. In brown earths in the Mendip Hills, Finlayson (1977) measured total dissolved solids concentrations in pipeflow that were broadly comparable to those in matrix throughflow, at 50 to 100 mg/l, but considering the relative velocities of flow, of course, this would still make pipeflow the most efficient transporter of solutes in the immediate vicinity of the pipes. In contrast, in nearby podzols, pipeflow concentrations were comparable to the rainwater at just ca. 15 mg/l (Finlayson, 1977). In perennial pipes in peaty soils in mid-Wales, Cryer (1980) reported a complete reversal of roles, with specific conductances (at 350 to 400 mS/m, or approximately 25 mg/l TDS), less than half those in either overland flow or matrix water, and only slightly higher than in the rainwater (at 280 mS/m).

Even so, these results are of limited value and could be misleading, based as they are on either very brief monitoring or infrequent sampling. Recent work in the Maesnant basin is emphasizing the importance of frequent sampling, especially during storms, and of seasonal variations in pipeflow solutes. The research reveals higher solute concentrations during storm events in the perennial pipes and generally higher conductivities in the ephemeral pipes (which tend to flow only in the more major events) than in the perennials (Jones and Hyett, 1987). In fact, present indications are that conductivities in the ephemeral pipes (at about 600 mS/m or about 40 mg/l TDS) may be double those in perennial pipes and that those in perennial pipes may rise by 20 percent during storms. Particularly high concentrations occur during the first storms after a prolonged dry period, especially at the end of summer. Cryer's (1980) weekly sampling program clearly underestimated the importance of storm events.

The only information on pipeflow solutes in other humid climates comes from the humid tropics, from the Pointe Baptiste catchment in Dominica (Walsh and Howells, 1988). In this tropical environment, solute concentrations were high, suggesting both active solutional erosion and long residence times. The fact that the pipes flow perennially and show little response to daily rainfalls of up to 43 mm supports the view that the water has had a long residence time and is likely to have reached chemical equilibrium with the soil. Silica levels were about 75 mg/l, compared with 1-4 mg/l measured by Waylen (1976) at East Twins in southwestern England, and chlorine concentrations were 52 to 80 mg/l, compared with 7 to 10 mg/l in the English basin. The high solute levels in these tropical pipes resulted in specific conductances in the range 2,500 to 3,200 mS/m, 7 to 8 times higher than in Cryer's (1980) Welsh examples. Walsh speculates that pipeflow is an important mechanism for chemical denudation in the kaolin soils of the tropical rain forest in Dominica, and probably also in the strongly seasonal climate and smectoid soils of the west coast (Walsh and Howells, 1988; Walsh, personal communication, 1984).

Observed patterns of pipe erosion. Despite the large body of research papers reporting features of piping erosion from humid temperate regions, such as New Zealand (e.g., Gibbs, 1945; Laffan and Cutler, 1977), Tasmania (Colclough, 1978) and Poland (Galarowski, 1976), the warm dry-land margins of interior New South Wales (e.g., Newman and Phillips, 1957; Floyd,

1974) and to a lesser extent in the tropics (e.g., Loffler, 1974; Baillie, 1975), attention has been focused more on the factors responsible for pipe development than on the spatial patterns of erosion and the implications for geomorphic theory. It is only in very recent years that a number of process studies have begun to provide the field evidence to support incorporating piping processes in more general models of hillslope evolution.

The study of patterns of sheetwash erosion by McCaig (1984) in West Yorkshire, U.K., is an excellent demonstration of the extent to which the pattern and frequency of hillslope erosion in a piped catchment can be very different from that predicted on the basis of the factors that control surface erosion. Figure 20 illustrates the poor correspondence between surface erosion rates predicted on the basis of the ln (a/s) index and the measured rates. Rates were less than predicted below pipe inlets, where piping captures overland flow, and higher than predicted below pipe effluxes.

At Maesnant, the relation between the size of microtopographic depressions overlying the pipe systems and the volumes of stormflow and sediment yield within each network seems to offer a springboard for modeling. The larger depressions lying over the perennially flowing segments of the pipe networks tend to be of the order of 0.5 to 2.0 m in depth, 10 to 20 m across, and drained by a single master pipe system. Shallower depressions 0.1 to 0.5 m deep cover the zones of ephemeral piping farther upslope where the main piping zone is 100 m across. This suggests a downslope evolution from broad, shallow depressions toward narrower, deeper depressions as turbulent flows increase in magnitude and frequency and become more concentrated into a single integrated channelized system. There is, of course, something of a chicken-and-egg relation between the size of the depression and volume of pipeflow within it: on the one hand, the depressions will concentrate both surface and subsurface water, increasing flows; on the other, increased turbulent flow leads to enlargement of the depressions. Certainly, a number of the deeper depressions on the lower slopes have recognizable slump scars around the edges, indicating enlargement by subsurface erosion.

Although the size of microtopographic depression is proportional to the area drained per unit contour length—a in the a/s index—provided a is measured on a contour map of sufficient detail, the pattern of hillslope erosion generated by piping is clearly more linear than would be predicted by the index. The proviso of a sufficiently detailed contour map is crucial, because piping processes operate on a more localized scale than can be studied from even the largest scale published map, at least unless or until piping generates a major hollow or a permanent extension to the stream channel network.

Piping zones similar to these on Maesnant have been described in South Wales by Wilson and Smart (1984), where the topographic depressions are emphasized by vegetation boundaries and are visible on air photographs. Gilman and Newson (1980, p. 25) identified smaller-scale depressions overlying individual ephemeral pipes in the Institute of Hydrology research catchments in mid-Wales; these are also found within the broad zone of ephemeral pipes on Maesnant with depths commonly less than 0.1 m. Many of these depressions are remnants of the desiccation cracks that appear to have initiated the piping.

Piping has also been seen as an important factor in the extension of much larger hollows. Tsukamoto and others (1982) found that 94 percent of the landslide scars observed in the Tama Hills near Tokyo had pipes in them of the order of 140 mm in diameter. Most of the slides occurred on convergent slopes, and 85 percent of these were actually on the central thalweg of hollows, particularly at the point of change between the upper convex profile and the lower concave profile. They concluded that the landslides occur where the hydrodynamic pressure of throughflow, together with the pipeflow and occasionally aggravated by pipe blockages, is directed toward the surface, undermining the soil mass and lubricating the slip.

Many others have noted the association of piping and landslips. In the Romanian sub-Carpathians near Buzau, Balteanu (1986) has noted the role of a dense network of pipes, 0.5 to 1.0 m in diameter, in reducing slope stability and initiating landslides that tend to obstruct channels in zones of active gullying. The study of slope failure on the north shore of Lake Ontario by Bryan and Price (1980) is a clear example of mass movement initiated by concentrated hydraulic pressures generated by pipeflow. Since Terzaghi's (1931) initial description on the Mississippi, there have been numerous cases in which mass failures of alluvial streambanks have been attributed to piping within more permeable substrates (e.g., Brunsden and Kesel, 1973; Thorne and Tovey, 1981). Some of these attributions, such as that by

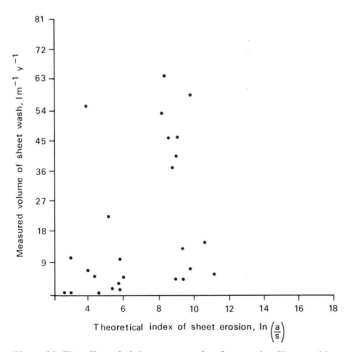

Figure 20. The effect of piping on rates of surface erosion illustrated by the relation between the ln (a/s) index and rates of surface wash erosion measured by McCaig (1984) in the Slithero Clough, Yorkshire. There is clearly no relation between observed and predicted wash rates.

Harrison and Petch (1984) explaining landslips on a meander scar of the River Irwell at Salford, England, have been quite speculative. On the other hand, the computerized shear failure analysis of bank materials along the Ohio River by Ullrich and others (1986) provides an impeccable example of the application of quantitative theory. It shows that piping is the cause of surface slab failures in the banks, sometimes weeks after an extreme flood event. The amount of piping could be predicted using the coefficient of permeability, the height of capillary rise and the slope of the underlying sand layers. The role of piping in riverbank failure has recently been reviewed by Jones (1989).

In many other cases, however, it is far from certain whether the piping preceded slope failure or was itself initiated by the concentration of throughflow toward the newly created free face (Jackson, 1966; Jones, 1981, p. 207ff). The description of the debris flow in the Ochil Hills, Scotland, by Jenkins and others (1988) seems to be a case in point. In fact, both sequences of events may coexist, for example, with piping perhaps initiated by mass-movement tension cracks allowing water in to reduce cohesion, increase the weight of the soil block, and initiate further movement (Ward, 1966; Balteanu, 1986).

In contrast, piping may stabilize hillslopes by providing a drain that prevents the build-up of excessive pore-water pressures within the soil: the exact opposite of the situation described by Tsukamoto and others (1982). It has been suggested that piping may inhibit bog bursts in this way (Jones, 1981, p. 210). Observations by Carling (1986) in the English Pennines do seem to support the view that slope failures occur where the absence of pipe networks has allowed critical pore pressures to develop. There are interesting implications here for the results of Freeze's (1986) simulation model of maximum stable slope angles. Freeze suggested that under low-intensity rainfall in a humid climate, where rainwater can infiltrate into less permeable soils, maximum stable angles will be almost halved. Conversely, where hydraulic conductivities are high enough to prevent soil saturation, steep slopes can be maintained. It seems likely that the addition of piping could transform slopes from the former category into the latter.

Piping and the development of humid drainage networks

The role of piping as a process initiating stream channel extension is well known and well documented from numerous climatic regions (Jones, 1971, 1981), but the significance of piping for drainage network development goes beyond this and exerts an influence even in areas not prone to active roof collapse and channel extension. It is clear that we need to consider the drainage network in general as including a variety of subsurface networks ranging from macropores through vughs or pseudopipes to ephemeral and perennial pipes (Jones, 1971, 1987b; Gregory and Walling, 1973; Gregory, 1979). The smaller-scale elements in this continuum are generally the least stable and may change significantly on a seasonal or even event-wise timescale. The larger elements, the major pipes and open channels, commonly require events of greater magnitude to cause any marked change, often well above the mean annual flood event.

The existence of these hillslope networks must exert an influence on the rest of the conventional drainage network. They have an effect equivalent to reducing the mean length of overland flow in Hortonian theory. One possible corollary of this is that a piped drainage basin requires a lower density of open-channel network to cope with a given water surplus. Some support for this comes from recent calculations by Jones (1987a) in which he applies Kirkby's (1978) formula for calculating drainage density to the Maesnant basin. The formula overestimates the open channel drainage density about threefold: 5.6 km/km^2 compared with the observed 2 km/km^2. Among the pipes themselves, however, the density reaches 153 km/km^2 in the ephemeral zone and 62 km/km^2 in the perennial zone (Fig. 21). Admittedly, the capacities of individual pipes are considerably lower than for the average stream channel, with mean diameters of 93 mm in the ephemerals and 240 mm in the perennials compared with a width of 2 m in the stream. If we compare the measured rates of storm discharge in the perennial pipes with those in the stream, we find that the stream discharges at a rate approximately 20 times faster than the average perennial pipe. Dividing the observed density of perennial pipes by this figure gives an "effective stream density" a little closer to that predicted by Kirkby's formula: 3.1 km/km^2.

In addition, piping offers a means of speeding up throughflow, so that it can become a significant source of stormflow in the open-channel network. In effect, it is a viable alternative mechanism to the other suggested mechanisms for the acceleration of throughflow, i.e., piston flow or translatory flow (Hewlett and Hibbert, 1967), and return flow in saturation overland flow or swale-flow (Hewlett and Nutter, 1970; Dunne and Black, 1970).

This could have two quite opposite results. On the one hand, the increased storm runoff could cause increased stream erosion, possibly increasing the rate of channel extension and even positively feeding incision so that stream rejuvenation might be both a cause and a consequence of piping. On the other hand, the pipe networks could stabilize the channel network, perhaps by diverting erosive overland flow or swale-flow below ground where roof erosion is a slow process and the pipes have more circumscribed capacities. At the very least, such a scenario would reduce the frequency of sheetwash events in piped hollows, and it could effectively stabilize the hillslopes against surface erosion.

However, there are generally very few case studies available to test some of these suggestions. Carling (1986) observed gullies actively headcutting along perennial pipes in the English Pennines. By comparison, in the peaty podzolic soils of mid-Wales, where most of the hydrological studies have concentrated, neither the stream networks nor the pipe networks change very quickly; even areas of roof collapse have changed little in more than 15 yr. Only extreme events with long return periods seem likely to alter these networks; no before-and-after studies have been possible, and it is very difficult to set up paired catchment studies of piped and unpiped basins. Newson (1975, 1980) does report the un-

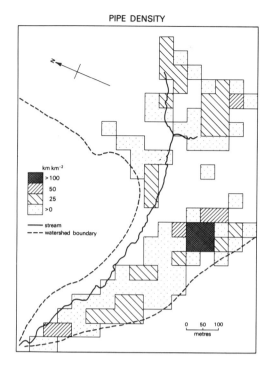 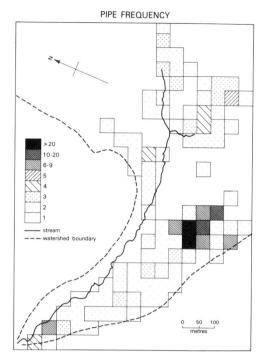

Figure 21. (A) Pipe densities and (B) pipe frequencies in the Maesnant catchment.

roofing of a perennial pipe flush here in one extreme rainstorm, and pipes apparently lubricating peat slides in stony areas. In contrast, he observed no sliding or collapse associated with typical ephemeral pipes on the mid-slopes. Extreme drought events, like summer 1976 in Wales, however, may have a more marked effect on the ephemeral networks (Table 2). After this drought, with an estimated return period of 200 yr, Gilman and Newson (1980) reported a marked increase in desiccation cracks, which initiated piping on upper and mid-slope areas.

The only areas in humid climates where the time scales of network development compare with those of arid badlands are areas of loess or loessial soils. These possess the important properties of low clay content, and therefore low cohesion, combined with easily shifted silt-size particles. The extension of gully networks by pipe collapse in loessic areas has been extensively described in Polish literature, for example, by Malicki (1946), Maruszczak (1965), Pinczes (1968), Walczowski (1971), and Zaborski (1972), but generally at a very qualitative level and offering no hydrologic or process measurements that might permit some degree of generalization and extrapolation. Better descriptions of gully extension have been provided from the slope loams of the Carpathians by Czeppe (1960) and Galarowski (1976). In Galarowski's case, discontinuous collapse occurred over a distance of 150 m within 9 months, a rate far exceeding the average of 5 m/yr observed in the badlands of the San Pedro valley, Arizona, by Jones (1968). Rapid development and collapse of piping, and subsequent gully extension has been described, again qualitatively, in the loessial soils of the Banks Peninsula, New Zealand, by Hosking (1967) and Hughes (1972).

None of these papers makes any substantial reference to the form of either the pipe networks or the extended channel networks (compare Jones, 1981, p. 136ff). This is clearly an important aspect that needs attention from both the hydrologic and geomorphic viewpoints. Collation of maps of pipe networks suggests at least three key properties of note. First, pipe networks tend to be more primitive than stream networks. They frequently anastomose, horizontally and even vertically, and the occasional triple junction may be found. Many networks are also highly fragmented, especially in their upper parts. These features result largely from the opportunistic nature of piping processes; they seek out weaknesses and the pipe network tends to reflect the spatial distribution of the initiating factors rather than the logic of

TABLE 2. FACTORS AFFECTING THE STABILITY OF PIPE NETWORKS

Predominant Type of Pipe	Predominant Location	Altering Process	Nature of Trigger Event
Ephemeral	Upper and upper midslopes	Desiccation cracking	Severe drought
Ephemeral/ perennial		mass movement)	
	Lower and lower midslopes	)	Heavy rain/ snowmelt
Perennial		roof collapse)	

hydraulic growth, especially when the networks are "young," ephemeral, and flow less frequently. There are analogies here with subsurface stream networks in karst, where structural features tend to be perpetuated in the network. Perhaps in both cases the ratio of erosive forces to the resistance of the materials is much lower than in the case of normal surface streams. Another possible analogy with cave systems is that, in many networks, there is less of the spatial competition for drainage water between adjacent pipes than there is between adjacent surface streams, particularly between pipes running in peaty horizons with very low hydraulic conductivity. In such soils, many sections of pipe networks have been considered to be primarily systems for water transmission rather than collection (Weyman, 1971; Gilman and Newson, 1980); this could be one explanation for the close clustered pattern of the networks.

The second key feature of pipe networks is the propensity for clustering, in favorable basins and in favorable sites within a given basin. A z-test of the presence/absence of piping based on the joins pattern in the grid squares of Figure 21 proves significant clustering throughout (z = 4.4 for the association of piped squares). Similar clustering has been found in the badland pipe networks of the San Pedro valley (Jones, 1978b). The only apparently contradictory data come from Nant Gerig, where the network appears to be more regularly spaced (Jones, 1981, p. 149). However, these measurements were taken only within the belt of piping and probably reflect regular desiccation cracks. The piping belt itself is markedly clustered on the hillside, as shown by the contrast between a pipe density of 180 km/km^2 in the piped area and 42 km/km^2 over the slope as a whole at Cerrig yr Wyn (Gilman and Newson, 1980). The fact that this clustering in favored areas is often reflected in the clustering of pipe outlets along the streambank led Jones (1971) to speculate that any evolution from pipe to open-channel networks must involve a low-density selection process, in which only the odd pipe in the cluster develops sufficiently to unroof.

A final characteristic of the mapped networks has been frequent discontinuities. In part, the disconnected networks may reflect "immature" development, or they may reflect different initiating processes such as the gaps in the Maesnant network between ephemeral pipes generated by desiccation cracking and the perennial pipes generated by "boiling" and mass-movement cracks. In other cases, the drainage networks are not as disconnected as a map of the pipe networks alone would suggest. In many cases the pipes act as part of an integrated network comprising pipes, swales, and rills, with the pipes sometimes in parallel and sometimes in series with the alternative routes (Jones, 1987b).

So far the discussion has concentrated on soil piping, but in a book on groundwater geomorphology it is important to note the similarity between the classical engineering process of piping as outlined, for example, by Terzaghi and Peck (1966) and Kalin (1977) and the classical geomorphic concept of spring sapping. The similarity was pointed out by Dunne (1980), who showed how measurements of piezometric head indicated concentrations of erosive forces around spring heads. Even so, we should remember that Terzaghi's theory of piping related to unconsolidated materials and the patterns of equipotential lines developing around bedrock springs are not so likely to follow the simple pattern of the original theory.

Higgins (1983, 1984) has pointed out some of the difficulties in identifying and quantifying the role of groundwater sapping. Significantly, the ideal examples of sapping processes he discusses are taken from unconsolidated beach materials (Higgins, 1982, 1984). Extending this concept to complete landscapes may be feasible in unusual circumstances, as in the drift-covered land surfaces of the Netherlands (De Vries, 1976), but in general it still presents unsolved and formidable problems for quantitative modeling at this scale, not least because sapping processes rarely operate in isolation.

There are also some purely semantic problems in comparing piping and sapping. As noted by Higgins (1984), and Higgins and others (1988), Terzaghi's use of the term "piping" subsumes "sapping," but the piping identified in the natural landscape by geomorphologists and soil scientists covers the much wider range of processes outlined earlier in this chapter and in Parker (this volume). It seems reasonable to confine the use of the term "sapping" to piping processes operating below the phreatic surface, whether it be the regional water table, a perched water table, or the surface of saturated throughflow within the solum. Often these sapping processes do not leave any lasting pipe-like void and fall into the category of "entrainment piping" in Parker and Jenne's (1967) schema.

One instance in which irregular linear voids are left by sapping or piping processes is in the "stone rivers" and boulder fields described by Czeppe (1965) in Spitsbergen, and Smith (1968) in the Falkland Islands, Pennsylvania, and Wisconsin. The essential feature in these cases is the coarse cobble and boulder matrix, which supports the structure after the fines have been winnowed away. Similar features, often grassed over, occur in the British uplands. Whether or not they owe their origins to a periglacial regime, as Smith and Czeppe argued, may be a moot point.

CONCLUSIONS

Piping is active throughout the humid regions of the world. It can be an important source of stream runoff and channel development, although it has a patchy distribution. Piping occurs basically in soils that: (1) are prone to developing saturated layers; (2) have moderate permeability in the saturated layers with sufficient macropore development to permit critical velocities of throughflow and preferably provide a primitive pipe structure that can be exploited; and (3) have erodible subsoils, but sufficient cohesion above to maintain a roof. Low cohesion, as in loess soils, can encourage rapid development, but this is usually accompanied by equally rapid destruction. Because of their high cohesion and low permeability, clays are generally not favorable unless cracking develops. Histic properties are ideal, and a moderate component of highly expansive clay minerals in the soil is particularly favorable.

Climate affects the type and severity of piping process, both through creating the pedogeomorphic environment and through controlling the frequency and severity of extreme events. In general, soil chemistry seems less important for piping processes in areas with a net annual water surplus than in dry lands. The seasonal distribution of water balance affects the relative role of desiccation cracking. Terzaghian mechanics and mass-movement cracking are probably more dominant in very wet climates. Nevertheless, dominant processes vary within a single climatic regime according to the geomorphic location, the local erosional environment, or land-surface unit. The occasional extreme event can also have a very persistent effect, either by altering the local erosional environment or by initiating pipe networks, which may then be maintained by far lesser flows.

The form of piping tends to be prescribed by the initiating processes during the early stages of development. As normal channel processes begin to dominate pipe development, so the networks tend to become more similar. Interestingly, what little information is available on the hydraulic geometries of pipes suggests that as pipes mature their geometries tend to develop toward those of the local stream channels in both dry and humid regions (Jones, 1981, p. 136ff).

The hydrologic and erosional functions of piping appear to differ on a basin-to-basin scale as well as with climatic regime. Most recent estimates of hydrologic contributions seem to converge on the range from 45 to 70 percent of streamflow, both in temperate and subtropical regimes. Limited evidence from a boreal forest catchment suggests a figure nearer 20 percent, while the only evidence from a basin in the ever-wet tropics suggests that piping there is mainly a source of base flow, contributing perhaps 16 percent of stream discharge. The erosional implications are extremely interesting and wide-ranging, but as yet there is a dearth of field evidence to support them. Large differences seem to occur between basins, which can partly be explained in terms of differing pipeflow regimes and the size of network collecting stormflow, e.g., in mid-Wales between the ephemeral pipes of Newson's (1976) basin and the perennial outfalls of Jones and Crane (1984). These contrasts extend right down to the level of adjacent pipes and make it dangerous to infer too much from limited sampling programs. Nevertheless, there is sufficient evidence to suggest that piping plays an important role in channeling drainage and erosion processes in many humid areas, with important implications for both slopes and drainage networks as well as for patterns of discharge.

ACKNOWLEDGMENTS

I thank Rory Walsh, Mike Stocking, and Peter Smart for offering material used in illustrations, and my reviewers for a number of valuable suggestions for improving the text. The Maesnant work was carried out under a Natural Environment Research Council Grant, GR3/3683, with the untiring assistance of Francis Crane; Ms. L. Follington assisted with the infiltrometer measurements. Contribution number 104 of the Institute of Earth Studies, University College of Wales, Aberystwyth, U.K.

REFERENCES CITED

Ahnert, F., 1987a, An approach to the identification of morphoclimates, in Gardiner, V., ed., International geomorphology, 1986, Part 2: Chichester, England, John Wiley and Sons, p. 159–187.

——, 1987b, Approaches to dynamic equilibrium in theoretical simulations of slope development: Earth Surface Processes and Landforms, v. 12, no. 1, p. 3–16.

Armstrong, A. C., 1987, Slopes, boundary conditions, and the development of convexo-concave forms; Some numerical experiments: Earth Surface Processes and Landforms, v. 12, no. 1, p. 17–30.

Baillie, I. C., 1975, Piping as an erosion process in the uplands of Sarawak: Journal of Tropical Geography, v. 41, p. 9–15.

Balteanu, D., 1986, The importance of mass movement in the Romanian sub-Carpathians: Zeitschrift für Geomorphologie, Supplement Band 58, p. 173–190.

Barendregt, R. W., and Ongley, E. D., 1977, Piping in the Milk River Canyon, southeastern Alberta; A contemporary dryland geomorphic process, in Erosion and solid material transport in inland waters; Proceedings of the Paris Symposium: International Association of Hydrological Sciences Publication 122, p. 233–243.

Beven, K., and Germann, P., 1982, Macropores and water flow in soils: Water Resources Research, v. 18, no. 5, p. 1311–1325.

Bonell, M., Hendricks, M. R., Imeson, A. C., and Hazlehoff, L., 1984, The generation of storm runoff in a forested clayey drainage basin in Luxembourg: Journal of Hydrology, v. 71, p. 53–77.

Bowden, D. J., 1980, Sub-laterite cave systems and other pseudokarst phenomena in the humid tropics; The example of Kasewe Hills, Sierra Leone: Zeitschrift für Geomorphologie, v. 24, no. 1, p. 77–80.

Bremer, H., 1973, Der Formungsmechanismus in tropischen Regenwald Amazoniens: Zeitschrift für Geomorphologie, Supplement Band 17, p. 195–222.

Brunsden, D., and Kesel, R. H., 1973, Slope development on a Mississippi River bluff in historic time: Journal of Geology, v. 81, p. 576–589.

Bryan, R. B., and Campbell, I. A., 1986, Runoff and sediment discharge in a semiarid ephemeral drainage basin: Zeitschrift für Geomorphologie, Supplement Band 58, p. 121–143.

Bryan, R. B., and Harvey, L. E., 1985, Observations on the geomorphic significance of tunnel erosion in a semiarid ephemeral drainage system: Geografiska Annaler, series A, v. 67, p. 257–273.

Bryan, R. B., and Price, A. G., 1980, Recession of the Scarborough Bluffs, Ontario, Canada: Zeitschrift für Geomorphologie, Supplement Band 34, p. 48–62.

Bryan, R. B., and Yair, A., eds., 1982, Badlands geomorphology and piping: Norwich, England, GeoBooks, 408 p.

Bryan, R. B., Yair, A., and Hodges, W. K., 1978, Factors controlling the initiation of runoff and piping in Dinosaur Provincial Park Badlands, Alberta, Canada: Zeitschrift für Geomorphologie, Supplement Band 29, p. 151–168.

Burt, T. P., 1987, Slopes and slope processes: Progress in Physical Geography, v. 11, p. 598–611.

Carling, P. A., 1986, Peat slides in Teesdale and Weardale, Northern Pennines, July 1983; Description and failure mechanism: Earth Surface Processes and Landforms, v. 11, p. 193–206.

Charman, P.E.V., 1969, The influence of sodium salts on soils with reference to tunnel erosion in coastal areas; Part 1, Kempsey area: New South Wales, Journal of Soil Conservation Service, v. 25, no. 4, p. 327–342.

——, 1970a, The influence of sodium salts on soils with reference to tunnel erosion in coastal areas; Part 2, Grafton area: New South Wales, Journal of Soil Conversation Service, v. 26, no. 1, p. 71–86.

——, 1970b, The influence of sodium salts on soils with reference to tunnel erosion in coastal areas; Part 3, Taree area: New South Wales, Journal of Soil Conservation, v. 26, no. 4, p. 256–275.

Colclough, J. D., 1978, Soil conservation in Tasmania: New South Wales, Journal of Soil Conservation Service, v. 34, no. 2, p. 63–72.

Conacher, A., and Dalrymple, J. B., 1977, The nine unit landsurface model; An approach to pedogeomorphic research: Geoderma, v. 18, no. 1/2, p. 1–154.

Crouch, R. J., McGarity, J. W., and Storrier, R. R., 1986, Tunnel formation processes in the Riverina area of N.S.W., Australia: Earth Surface Processes and Landforms, v. 11, p. 157–168.

Cryer, R., 1980, The chemical quality of some pipeflow waters in upland mid-Wales and its implications: Cambria, v. 6, no. 2, p. 1–19.

Czeppe, Z., 1960, Suffosional phenomena in slope loams of the Upper San drainage basin: Biuletyn Instytut Geologiczny, v. 9, p. 297–332.

——, 1965, Activity of running water in southwestern Spitsbergen: Geographia Polonica, v. 6, p. 141–150.

De Vries, J. J., 1976, The groundwater outcrop-erosion model; Evolution of the stream network in the Netherlands: Journal of Hydrology, v. 29, p. 43–50.

Downes, R. G., 1946, Tunnelling erosion in northeastern Victoria: Australia, Journal of the Council for Scientific and Industrial Research, v. 19, p. 283–292.

——, 1954, Cyclic salts as a dominant factor in the genesis of soils in southeastern Australia: Australian Journal of Agricultural Research, v. 5, p. 448.

——, 1956, Conservation problems on solodic soils in the State of Victoria (Australia): Journal of Soil and Water Conservation, v. 11, p. 228–232.

Drew, D. P., 1972, Geomorphology of the Big Muddy Valley area, southern Saskatchewan, with reference to the occurrence of piping, in Paul, A. H., Dale, E. H., and Schichtmann, H., eds., Southern Prairies field excursion background papers: Saskatoon, University of Saskatchewan Department of Geography, p. 197–212.

——, 1982, Piping in the Big Muddy badlands, southern Saskatchewan, Canada, in Bryan, R. B., and Yair, A., eds., Badland geomorphology and piping: Norwich, England, GeoBooks, p. 293–304.

Duijsings, J.J.H.M., 1985, Streambank contribution to the sediment budget of a forest stream: University of Amsterdam Laboratory for Physical Geography and Soil Science Publication 40, 190 p.

Dunne, T., 1980, Formation and controls of channel networks: Progress in Physical Geography, v. 4, no. 2, p. 211–239.

Dunne, T., and Black, R. D., 1970, Partial area contributions to storm runoff in a small New England watershed: Water Resources Research, v. 6, no. 5, p. 1296–1311.

Dunne, T., Moore, T. R., and Taylor, C. H., 1975, Recognition and prediction of runoff-producing zones in humid regions: Hydrological Sciences Bulletin, v. 20, no. 3, p. 305–327.

Finlayson, B. L., 1977, Runoff contributing areas and erosion: University of Oxford School of Geography Research Paper 18, 40 p.

Floyd, E. J., 1974, Tunnel erosion; A field study in the Riverina: New South Wales, Journal of Soil Conservation Service, v. 30, no. 3, p. 145–156.

Franzle, O., 1976, Ein morphodynamisches Grundmodell der Savannen, und Regenwaldgebeite: Zeitschrift für Geomorphologie, Supplement Band 24, p. 177–184.

Freeze, R. A., 1986, Modelling interrelationships between climate, hydrology, and hydrogeology, and the development of slopes, in Anderson, M. G., and Richards, K. S., eds., Slope stability: Chichester, England, John Wiley and Sons, p. 381–404.

Galarowski, T., 1976, New observations of the present-day suffosion (piping) processes in the Bereznica catchment basin in the Bieszczady Mountains (the East Carpathians): Studia Geomorphologica Carpatho-Balcanica, v. 10, p. 115–122.

Gibbs, H. S., 1945, Tunnel-gully erosion on the Wither Hills, Marlborough, New Zealand: New Zealand Journal of Science and Technology, v. 27, p. 135–146.

Gilman, K., and Newson, M. D., 1980, Soil pipes and pipeflows; A hydrological study in upland Wales; British Geomorphological Research Group Monograph 1: Norwich, England, GeoBooks, 114 p.

Gregory, K. J., 1979, Hydrogeomorphology; How applied should we become?: Progress in Physical Geography, v. 3, no. 1, p. 84–101.

Gregory, K. J., and Walling, D. E., 1973, Drainage basin form and process: London, Arnold, 456 p.

Harrison, C., and Petch, J. R., 1984, Ground movements in parts of Salford and Bury, Greater Manchester; Aspects of urban geomorphology: University of Salford Department of Geography Discussion Paper 27, 11 p.

Hewlett, J. D., and Hibbert, A. R., 1967, Factors affecting the response of small watersheds to precipitation in humid areas, in Proceedings of the International Symposium on Forest Hydrology, 1965, Pennsylvania State University: New York, Pergamon Press, p. 275–290.

Hewlett, J. D., and Nutter, W. L., 1970, The varying source area of streamflow from upland basins, in Proceedings of Symposium on Interdisciplinary Aspects of Watershed Management, Montana State University, Bozeman: New York, American Society of Civil Engineers, p. 65–83.

Higgins, C. G., 1982, Drainage systems developed by sapping on Earth and Mars: Geology, v. 10, p. 147–152.

——, 1983, Reply to Comment on 'Drainage systems developed by sapping on Earth and Mars': Geology, v. 11, p. 55–56.

——, 1984, Piping and sapping; Development of landforms by groundwater outflow, in La Fleur, R. G., ed., Groundwater as a geomorphic agent: Boston, Massachusetts, Allen and Unwin, p. 18–58.

Higgins, C. G., and 12 others, 1988, Landform development, in Back, W., Rosenshein, J. S., and Seaber, P. R., eds., Hydrogeology: Boulder, Colorado, Geological Society of America, The Geology of North America, v. O-2, p. 383–400.

Hosking, P. L., 1967, Tunneling erosion in New Zealand: Journal of Soil and Water Conservation, v. 22, no. 4, p. 149–151.

Hughes, P. J., 1972, Slope aspect and tunnel erosion in the loess of Banks Peninsula, New Zealand: New Zealand Journal of Hydrology, v. 11, no. 2, p. 94–98.

Imeson, A. C., 1986, Investigating volumetric changes in clayey soils related to subsurface water movement and piping: Zeitschrift für Geomorphologie, Supplement Band 60, p. 115–130.

Ingles, O. G., 1968, Soil chemistry relevant to the engineering behaviour of soils, in Lee, I. K., ed., Soil mechanics; Selected topics: London, Butterworth, p. 1–57.

Ingles, O. G., and Aitchison, G. D., 1970, Soil-water disequilibrium as a cause of subsidence in natural soils and earth embankments, in Proceedings, International Association of Hydrological Sciences-UNESCO International Symposium on Land Subsidence, 1969, Tokyo: International Association of Hydrological Sciences Publication 89, p. 342–353.

Institute of Hydrology, 1974, Research 1972–73: Wallingford, United Kingdom, Institute of Hydrology, Natural Environment Research Council, 66 p.

Jackson, R. J., 1966, Slips in relation to rainfall and soil characteristics: New Zealand Journal of Hydrology, v. 5, no. 2, p. 45–53.

Jenkins, A., Ashworth, P. J., Ferguson, R. I., Grieve, I. C., Rowling, P., and Stott, T. A., 1988, Slope failures in the Ochil Hills, Scotland, November 1984: Earth Surface Processes and Landforms, v. 13, p. 69–76.

Jones, J.A.A., 1971, Soil piping and stream channel initiation: Water Resources Research, v. 7, no. 3, p. 602–610.

——, 1978a, Soil pipe networks; Distribution and discharge; Cambria, v. 5, no. 1, p. 1–21.

——, 1978b, The spacing of streams in a random-walk model: Area, v. 10, no. 3, p. 190–197.

——, 1979, Extending the Hewlett model of stream runoff generation: Area, v. 11, no. 2, p. 110–114.

——, 1981, The nature of soil piping; A review of research; British Geomorphological Research Group Research Monograph 3: Norwich, England, Geo-Books, 301 p.

———, 1982, Experimental studies of pipe hydrology, in Bryan, R. B., and Yair, A., eds., Badland geomorphology and piping: Norwich, England, GeoBooks, p. 355–370.

———, 1985, Erosion by pipeflow, in Balteanu, D., Dragomirescu, S., and Muica, C., eds., Geomorphological research for land and water management: University of Bucarest Institute of Geography, p. 123–148.

———, 1986, Some limitations to the a/s index for predicting basin-wide patterns of soil water drainage: Zeitschrift für Geomorphologie, Supplement Band 60, p. 7–20.

———, 1987a, The effects of soil piping on contributing areas and erosion patterns: Earth Surface Processes and Landforms, v. 12, p. 229–248.

———, 1987b, The initiation of natural drainage networks: Progress in Physical Geography, v. 11, no. 2, p. 207–245.

———, 1988, Modelling pipeflow contributions to stream runoff: Hydrological Processes, v. 2, p. 1–17.

———, 1989, Bank erosion; A review of British research, in Ports, M. A., ed., Hydraulic Engineering, Proceedings of the 1989 National Conference on Hydraulic Engineering, New Orleans: New York, American Society of Civil Engineers, p. 283–288.

Jones, J.A.A., and Crane, F. G., 1982, New evidence of rapid interflow contributions to the streamflow hydrograph: Beitrage zur Hydrologie, v. 3, p. 219–232.

———, 1984, Pipeflow and pipe erosion in the Maesnant experimental catchment, in Burt, T. P., and Walling, D. E., eds., Field experiments in fluvial geomorphology: Norwich, England, GeoBooks, p. 55–72.

Jones, J.A.A., and Hyett, G. A., 1987, The effect of natural pipeflow solutes on the quality of upland streamwater in Wales: Vancouver, 19th General Assembly of the International Union of Geodesy and Geophysics, Abstracts, v. 3, p. 998.

Jones, N. O., 1968, The development of piping erosion [Ph.D. thesis]: Tucson, University of Arizona, 162 p.

Kalin, M., 1977, Hydraulic piping; Theoretical and experimental findings: Canadian Geotechnical Journal, v. 14, p. 107–124.

Kassiff, G., and Henkin, E. N., 1967, Engineering and physico-chemical properties affecting piping failure of low loess dams in the Negev, in Proceedings of 3rd Regional Conference on Soil Mechanics and Foundation Engineering, Israel: Jerusalem Academic Press, v. 1, p. 13–16.

Kirkby, M. J., 1978, Implications for sediment transport, in Kirkby, M. J., ed., Hillslope hydrology: London, John Wiley and Sons, p. 325–363.

Laffan, M. D., and Cutler, E.J.B., 1977, Landscape, soils, and erosion of a catchment in the Wither Hills, Marlborough; 2, Mechanisms of tunnel-gully erosion in Wither Hills soils from loessial drift and comparison with other loessial soils in the South Island: New Zealand Journal of Science, v. 20, p. 279–289.

Löffler, E., 1974, Piping and pseudokarst features in the tropical lowlands of New Guinea: Erdkunde, v. 28, no. 1, p. 13–18.

Malicki, A., 1946, Kras loessowy: Lublin, Poland, Annales Universitatis Mariae Curie-Sklodowska, section B, v. 1, no. 4, p. 131–155.

Maruszczak, H., 1965, Development conditions of the relief of loess areas in east-middle Europe: Geographia Polonica, v. 6, p. 93–104.

McCaig, M., 1979, The pipeflow streamhead; A type description: University of Leeds Department of Geography Working Paper 242, 15 p.

———, 1983, Contributions to storm quickflow in a small headwater catchment; The role of natural pipes and soil macropores: Earth Surface Processes and Landforms, v. 8, no. 3, p. 289–252.

———, 1984, The pattern of wash erosion around an upland stream head, in Burt, T. P., and Walling, D. E., eds., Catchment experiments in fluvial geomorphology: Norwich, England, GeoBooks, p. 87–114.

Monteith, N. H., 1954, Problems of some Hunter Valley soils: New South Wales, Journal of Soil Conservation Service, v. 10, p. 127–134.

Newman, J. C., and Phillips, J.R.H., 1957, Tunnel erosion in the Riverina: New South Wales, Journal of Soil Conservation Service, v. 13, p. 159–169.

Newson, M. D., 1975, The Plynlimon floods of August 15th/16th, 1973: Wallingford, United Kingdom, Institute of Hydrology Report 26, 58 p.

———, 1976, Soil piping in upland Wales; A call for more information: Cambria, v. 1, p. 33–39.

———, 1980, The geomorphological effectiveness of floods; A contribution stimulated by two recent events in mid-Wales: Earth Surface Processes, v. 5, p. 1–16.

Newson, M. D., and Harrison, J. G., 1978, Channel studies in the Plynlimon experimental catchments: Wallingford, England, Institute of Hydrology, Natural Environment Research Council Report 47, 61 p.

Oxley, N. C., 1974, Suspended sediment delivery rates and the solutes concentration of stream discharge in two Welsh catchments, in Gregory, K. J., and Walling, D. E., eds., Fluvial processes in instrumented watersheds: British Geomorphological Research Group Special Publication 6, p. 141–154.

Parker, G. G., and Jenne, E. A., 1967, Structural failure of western U.S. highways caused by piping: U.S. Geological Survey Water Resources Division, 27 p.

Perry, E. B., 1975, Piping in earth dams constructed of dispersive clay; Literature review and design of laboratory testing: Vicksburg, Mississippi, U.S. Army Engineer Waterways Experiment Station, U.S. Corps of Engineers Technical Report 5-75-15.

Pinczes, Z., 1968, Vonales erozio a Tokaji-hegy loszen: Foldrajzi Kozlemenyek, v. 16, no. 2, p. 159–171.

Rathjens, C., 1973, Suberrane Abtragung (piping): Zeitschrift für Geomorphologie, Supplement Band 17, p. 168–176.

Roberge, J., and Plamondon, A. P., 1987, Snowmelt runoff pathways in a boreal forest hillslope; The role of pipe throughflow: Journal of Hydrology, v. 95, p. 39–54.

Rodda, J. C., Downing, M. A., and Law, F. M., 1976, Systematic hydrology: London, Newnes-Butterworth, 399 p.

Sherard, J. L., and Decker, R. S., eds., 1977, Dispersive clays, related piping, and erosion in geotechnical projects: American Society for Testing Materials Special Technical Publication 623, 486 p.

Smith, H.T.U., 1968, "Piping" in relation to periglacial boulder concentrations: Biuletyn Peryglacialny, v. 17, p. 195–204.

Stagg, M. J., 1974, Storm runoff in a small catchment in the Mendip Hills [M.Sc. thesis]: University of Bristol, 163 p.

Starkel, L., 1972, The modeling of monsoon areas of India as related to catastrophic rainfall: Geographia Polonica, v. 23, p. 151–173.

Stephens, C. G., 1962, A manual of Australian soils, 3rd ed.: Melbourne, Commonwealth Scientific and Industrial Research Organisation, 61 p.

Stocking, M. A., 1976, Tunnel erosion: Rhodesia Agricultural Journal, v. 73, no. 2, p. 35–39.

———, 1979, A catena of sodium-rich soil: Journal of Soil Science, v. 30, no. 1, p. 139–146.

———, 1981, Model of piping in soils: Transactions of Japan Geomorphological Union, v. 2, no. 2, p. 263–278.

Tanaka, T., 1982, The role of subsurface water exfiltration in soil erosion processes, in Recent developments in the explanation and prediction of erosion and sediment yield; Proceedings of the Exeter Symposium: International Association of Hydrological Sciences Publication 137, p. 73–80.

Terzaghi, K., 1931, Earth slips and subsidences from underground erosion: Engineering New Record, v. 107, p. 90–92.

Terzaghi, K., and Peck, R. B., 1966, Soil mechanics in engineering practice: New York, John Wiley and Sons, 566 p.

Thorne, C. R., and Tovey, N. K., 1981, Stability of composite river banks: Earth Surface Processes and Landforms, v. 6, p. 469–484.

Tsukamoto, Y., Ohta, T., and Nogushi, H., 1982, Hydrological and geomorphological studies of debris slides on forested hillslopes in Japan, in Recent developments in the explanation and prediction of erosion and sediment yield; Proceedings of the Exeter Symposium: International Association of Hydrological Sciences Publication 137, p. 89–98.

Ullrich, C. R., Hagerty, D. J., and Holmberg, R. W., 1986, Surficial failures of alluvial streambanks: Canadian Geotechnical Journal, v. 23, p. 304–316.

UNESCO, 1978, World water balance and water resources of the earth: Paris,

UNESCO, 663 p.
Walczowski, A., 1971, Processy suffozji w okolicach Pacanowa: Biuletyn Instytut Geologiczny, v. 9, no. 242, p. 105–135.
Walsh, R.P.D., 1980, Runoff processes and models in the humid tropics: Zeitschrift für Geomorphologie, Supplement Band 36, p. 176–202.
Walsh, R.P.D., and Howells, K. A., 1988, Soil pipes and their role in runoff generation and chemical denudation in a humid tropical catchment in Dominica: Earth Surface Processes and Landforms, v. 13, no. 1, p. 9–18.
Ward, A. J., 1966, Pipe shaft phenomena in Northland: New Zealand Journal of Hydrology, v. 5, no. 2, p. 64–72.
Ward, R. C., 1975, Principles of hydrology, 2nd ed.: London, McGraw-Hill, 367 p.
Waylen, M. J., 1976, Some aspects of the hydrochemistry of a small drainage basin [Ph.D. thesis]: University of Bristol, 386 p.
Weyman, D. R., 1971, Surface and subsurface runoff in a small basin [Ph.D. thesis]: University of Bristol, 243 p.
Whipkey, R. Z., and Kirkby, M. J., 1978, Flow within the soil, *in* Kirkby, M. J., ed., Hillslope hydrology: London, John Wiley and Sons, p. 121–144.
Wilson, C. M., and Smart, P., 1984, Pipes and pipeflow processes in an upland catchment, Wales: Catena, v. 11, p. 145–158.
Wood, C. C., Aitchison, G. D., and Ingles, O. G., 1964, Physiochemical and engineering aspects of piping failures in small earth dams, *in* Colloquium on failure of small earth dams: Water Research Foundation of Australia, Ltd., and Soil Mechanics Section, C.S.I.R.O., paper 29, 13 p.
Yair, A., Bryan, R. B., Lavee, H., and Adar, E., 1980, Runoff and erosion processes and rates in the Zin Valley badlands, northern Negev, Israel: Earth Surface Processes, v. 5, p. 205–225.
Yasuhara, M., 1980, Streamflow generation in a small forested watershed [M.Sc. thesis]: Japan, University of Tsukuba, 55 p.
Zaborski, B., 1972, On the origins of gullies in loess: Acta Geographica Debrecina, v. 10, p. 109–111.
Zaslavsky, D., and Kassiff, G., 1965, Theoretical formulation of piping mechanism in cohesive soils: Geotechnique, v. 15, no. 3, p. 305–316.

MANUSCRIPT ACCEPTED BY THE SOCIETY NOVEMBER 14, 1989

Chapter 6

Gully development

Charles G. Higgins
Department of Geology, University of California, Davis, California 95616
 with a case study by
Barry R. Hill*
Department of Watershed Management, Humboldt State University, Arcata, California 95521
and André K. Lehre
Department of Geology, Humboldt State University, Arcata, California 95521

INTRODUCTION: RILLS VERSUS GULLIES

There has long been a tendency, among both laypersons and geomorphologists, to confuse *rills* and *gullies*. These terms have often been used interchangeably, and their definitions in most textbooks and dictionaries do little to aid understanding. From such sources one learns only that both are formed in soil or unconsolidated materials, that both carry only ephemeral streamflow, and that rills (commonly typified as "furrows") are smaller than gullies (which are sometimes characterized as "ravines"). Indeed, it is widely stated or implied that gullies are simply deepened, permanent rills that have become large enough that they cannot be eradicated by seasonal cultivation (cf., Bradford and Piest, 1980). Despite such vagaries of definition, many of these same sources contain hints of other differences between rills and gullies, and from the overall usage of these terms in the general literature, one can obtain a still clearer sense of distinction between them.

Many authors describe rills as being fairly straight and furrowlike, with flaring sides. They commonly have parallel or elongated dendritic patterns and tend to be associated with overland flow on bare slopes. There they generally head high on the hillslope, near its inflection point below the hill shoulder. Downslope, they may become wider and deeper but their sides still flare, and the channel at the bottom remains very narrow. These aspects of rills are illustrated in Figure 1.

In contrast, it is generally agreed that gullies have steep sides and, especially, steep headcuts or head scarps. They tend to have few, short branches. They are most commonly associated with vegetated terrain, where they develop on steep to fairly gentle slopes. They can even develop in nearly level cultivated fields. Wherever they form, they grow headward by erosional undercutting, or sapping, of the headcut. Although they may not become deeper downstream, many do tend to widen and to develop depositional flat bottoms that may have braided streamways. Many of these attributes of gullies are illustrated in Figure 1, which also depicts the common phenomenon of slumping or toppling of a block of soil along a steep gully wall as a consequence of sapping and oversteepening at its base.

Although it is widely stated that rills simply develop into gullies by erosional enlargement, differences in their morphology and occurrence suggest that they differ also in genesis and in behavior, as suggested by Imeson and Kwaad (1980). Indeed, one of the main characteristics of gullies is their ability to extend themselves headward by sapping at the base of their headcuts, a process known as "gullying." This would not be possible if the gullies had developed from rills, because rills, from their inception, head as far upslope as possible, near the lower limit of Horton's "belt of no erosion" (Horton, 1945, Fig. 14). Moreover, observations of rills on the kinds of bare slopes on which they typically develop, show that their erosional enlargement leads not to flat-bottomed gullies, but simply to larger valleys that retain the rill-like, characteristic v-shape with flaring sides.

From these and other differences, one may conclude that typical v-shaped rills are developed and enlarged chiefly by Horton overland flow, possibly much as Horton (1945) described, whereas typical flat-bottomed, steep-walled gullies are developed chiefly by sapping and migration of their headcuts, and may have no prior existence as rills. Figure 1 depicts a familiar situation of gullies heading in the tracks of preexisting rills. Such cases do suggest that the gullies are simply enlarged rills, but the abrupt change in morphology shows that the transformation is not so simple, and signals that there has been a change in process. This generally follows exposure of a water-bearing stratum or soil horizon as the rill is deepened. Seepage does the rest. The underlying stratum may also have different physical properties that promote headcut formation. Some authors believe that the u-shaped cross section of the typical gully depends on the nature

---

*Present address: 677 Ala Moana Boulevard, Suite 41, Gold Bond Building, Honolulu, Hawaii 96813.

Higgins, C. G., 1990, Gully development, with a case study by Hill, B. R., and Lehre, A. K., *in* Higgins, C. G., and Coates, D. R., eds., Groundwater geomorphology; The role of subsurface water in Earth-surface processes and landforms: Boulder, Colorado, Geological Society of America Special Paper 252.

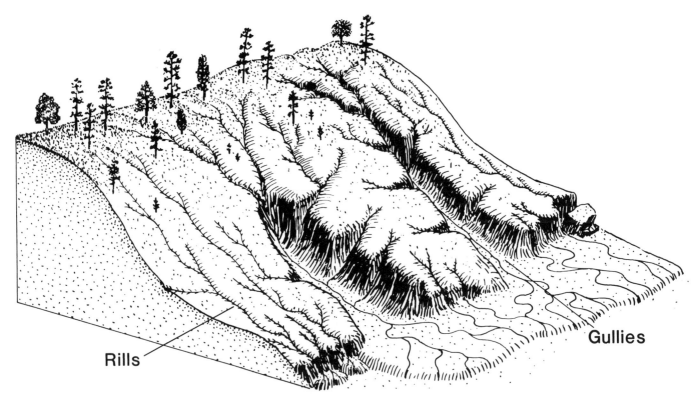

Figure 1. Rills versus gullies. Each has distinctively different morphology. After Stow and Hughes (1980, p. 26); reprinted by permission of the Geological Survey of Alabama.

and behavior of the substrate, such as slaking of dispersive soils. Imeson and Kwaad (1980) recognize four major types of gullies: one v-shaped (here regarded as a rill) and formed solely by overland flow, and three u-shaped, all associated with dispersive soils and formed by subsurface flow, with or without surface wash. In this chapter, the mechanisms of gully formation will be investigated following a brief discussion of gully types and sizes.

GULLY TYPES

Brice (1966) studied gullies developed in loess in south-central Nebraska, and classified them as *valley-bottom, valley-head,* and *valley-side gullies.* Those depicted in Figure 1 are valley-side gullies. Among these types, Brice recognized variations based on size, activity, morphology, and relation to local topography. *Discontinuous gullies,* discussed below, occur among all three major types.

He also recognized three kinds of gully scarps, all very steep, with inclinations of 45 to 90 degrees, and all promoted by erosion concentrated at their base. *Head scarps,* also known as *headcuts,* form at a gully head or channel head. *Channel scarps* are similar, but form breaks in the long profile of a well-defined channel. These are chiefly associated with valley-bottom gullies. *Side scarps* form the walls of active gullies. Brice's distinctions will be used in the following discussions.

Earlier, Ireland and others (1939, p. 43–44) suggested a different classification based on plan view: *linear, bulbous, dendritic, trellis, parallel,* and *compound.* They also classified the plan views of active gully heads as *pointed, rounded, notched* (like rounded but with a sharp notch in the semicircle), and *digitate.* Most gullies formed mainly by seepage erosion have rounded or digitate, sometimes notched, heads. These authors further characterized gully heads in profile as *inclined* (or markedly sloping), *vertical* or box head, *cave,* and *vegetated* (with an overhanging root mat or sod). They stated that the overhanging cave type "may be caused [either] by back-trickle and plunge-pool action or by seepage" (Ireland and others, 1939, p. 57). Most of the head scarps discussed in this chapter are of the overhanging type.

Heede (1970) distinguished two kinds of gullies: continuous and discontinuous. The latter are equivalent to the steep-headed features called discontinuous gullies in this chapter. However, Heede's "continuous gullies" include not only coalesced discontinuous ones, but also some well-branched drainages that begin high on a slope and are here termed rills.

GULLY SIZES

According to usage of the term, gullies range widely in size. Leopold and Miller (1956, p. 29) stated that "some gullies are narrow enough to step across but are deep enough to lose a giraffe in," and suggested that in New Mexico there is a continuum from ordinary small ones to large arroyos as much as 15 m deep and 200 km long. In Nebraska, Brice (1966) measured head scarps more than 10 m high in some steep valley heads, and found channel scarps that had retreated as much as 2.3 km in 15 years. Considering the lower limits of gully size in the light of observed rates of the retreat of gully heads in loess, he concluded that to be called a gully, a drainage channel must have "a width greater than about 1 foot, and a depth greater than about 2 feet" (Brice, 1966, p. 290). However, some drainage systems that resemble gullies in most other respects are not even that large. Some tiny ones superficially mimic rills, and this doubtless has contributed to some of the confusion about the origin of rills and gullies.

Small, rill-like gullies

In an unpublished report, Aram Derewetzky has described an area of small, rill-like, valley-side gullies on the northern outskirts of Davis, California. Some of these are pictured in Figure 2. Although their branching and elongate dendritic drainage make them resemble rill systems, other aspects of their morphology show that they are really miniature gullies. Instead of having v-shaped profiles with flaring sides, these little channels tend to be slot-like, with steep heads, nearly vertical walls, and very irregular long profiles. Their subsurface origin is obvious because some head at piping holes, and almost all have some underground segments that are still bridged-over, as can be seen in Figure 2. Significantly, they seem to be restricted to artificially steepened exposures of an alkali soil with high exchangeable sodium percentage, typically "greater than 20 per cent" (Andrews, 1972). Piping is commonly associated with such soils in drylands, as noted by Parker and Higgins (this volume), but the resulting gullies are not generally so small. In the Davis example, their small size may have depended in part on local disturbance and remolding of the soil.

The significant thing about these little features is their superficial resemblance to rill systems developed by surface wash and flow. Their true identity and origin become apparent only after close observation of their morphology and of their behavior immediately after rain storms. Clearly, one must use caution in drawing conclusions about the origins of such drainage nets; one cannot rely solely on their appearance in plan view, but must consider all aspects of their morphology. However, even this may fail to disclose their ultimate origin, because after completion of their development by piping, subsequent modification by surface wash may destroy the features characteristically formed by piping, replacing them with flaring slopes and other attributes of erosion by surface runoff.

HOW GULLIES ARE FORMED

Leopold and Miller (1956, p. 29) once wrote, "the mechanics of gully formation is very poorly understood," and this is still as true today as it was 30 years ago. One reason for this is that these features must be observed during the infrequent times when they are actively forming. Also, as Brice (1966, p. 295) has pointed out, "after a gully has become large, neither the exact way in which it was initiated nor the point of initiation can be

Figure 2. Rill-like gullies formed by piping on an embankment near Davis, California. Discontinuous segments are still underground. Hammer in lower left shows scale; height of slope about 4 m. Photograph by Aram Derewetzky, November 1984.

established." Despite these uncertainties, one can make *some* generalizations about gully development and growth.

The chief clues to the processes by which gullies are formed are their steep headcuts, or scarps. These and the steep walls can result from the development and collapse of soil pipes, as illustrated by the example cited above and as discussed in Chapters 4 and 5 in this volume. Stocking found a close relation between pipe development and gullying in central Zimbabwe, and observed that where there are sodium-rich soils, "gullies help to promote piping [by steepening the hydraulic gradient], and piping, in turn, aids the extension of the head cuts of gullies" by subsurface enlargement and collapse (Stocking, 1981, p. 116).

Some gullies may result from the widening of tension cracks by surface runoff (see Péwé's discussion of earth fissures, this volume). In other cases, near-vertical or even overhanging headcuts show that erosion must be concentrated at the base of the cut. An important mechanism by which this can be accomplished, and the one to be stressed in this chapter, is sapping by seepage erosion at the base of the scarp. This process is described in Dunne's introduction to this volume and by Jones (1987, p. 230–231). It entails grain-by-grain entrainment and removal of particles from the base of the headcut by the outflow of subsurface water and/or by saturation and reduction of strength of the soil. Either of these effects can undermine and weaken the scarp so that it collapses. Surface wash then removes the fallen debris, as described by Bradford and Piest (1980). In this manner the headcut is maintained near vertical as it retreats.

Describing gully growth, Emmett wrote:

Sapping at the base is one of the more effective processes in headward extension. Many gullies, particularly in relatively uniform material, advance headward, maintaining vertical cliffs at the gully headwall. In others, slumping at the base leaves a jumble of blocks at the vertical face. Sapping is intensified by the emergence of moisture moving under gravity to the base of the vertical wall (Emmett, 1968, p. 519).

Figure 3. Multiple headcut of a tiny gully near those shown in Figure 2. Collapse of the roof of the soil pipe that initiated the gully discloses active seepage erosion at the site of soilwater outflow. Photograph by Aram Derewetzky, November 1984.

Unquestionably, headcut sapping may also result from plunge-pool action by surface-water overwash, as argued by Ireland and others (1939) in their classic study of gullies in the South Carolina Piedmont. These authors minimized the role of spring sapping or seepage erosion in gully development, preferring to attribute headcut retreat to plunge-pool sapping at the base of overfalls. Doubtless the latter process is effective, even dominant, in the development of many gullies, especially where a carpet of turf or a clayey B-horizon of the soil forms a resistant rim and where the gully head is fed by surface drainageways. Indeed, surface wash and flow must affect all gully systems to some degree, no matter what their origin, and may even modify or obliterate any evidence of other origins. However, even in the South Carolina Piedmont, "many of the gullies . . . have been developed where formerly there was no channel drainage" (Ireland and others, 1939, p. 39), suggesting that surface wash there was absent or minimal.

Writing of the same region, Emmett commented:

In the gullies of the Piedmont of South Carolina, a stratum of relatively friable material occurs near the base of the headcut, but the top stratum is tough and resistant. This leads to *undercutting* by the emergence of groundwater, and large blocks of the upper material cave in, leading to rapid headward extension (Emmett, 1968, p. 519).

In such cases, where there is no local concentration of surface runoff above the gully head, sapping by piping or seepage outflow would seem to provide the only effective means by which erosion can be concentrated at the base of the headcut. In this chapter, seepage erosion is proposed as a major cause of the initiation and enlargement of many, if not most, gullies.

Active seepage and sapping of this kind have been observed only rarely, but are documented in Figure 3 as well as in the beach-face drainage system illustrated in Figure 1 of Chapter 14 (this volume). The site pictured in Figure 3 is the head scarp of a tiny gully near those shown in Figure 2. The roof of the soil pipe has collapsed, exposing an eroding face where active outflow is detaching the finer soil particles and transporting them in a braided wash down the channel. This example illustrates the action of seepage erosion and sapping; it also suggests that it is possible for soil pipes to form not only by lateral enlargement along a line of micropores or along a soil crack, but also by entrainment at a seepage face underground. Apparently this process can occur either underground or, as seen here and in Figure 1 of Chapter 14, at the surface. Some attempts to model head-scarp retreat in the laboratory are summarized below, in the discussion of valley-bottom gullies.

The seepage erosion and sapping process described here seems to operate over a wide range of scales, from tiny features like those described above to the stream valleys and canyons that Baker and colleagues describe (Chapter 11, this volume). Fenneman (1923) stated that the same mechanical principles that affect the growth and change of gullies and draws in parts of Texas and Oklahoma evidently also control the development of large valleys there. Hence, some aspects of the development of the gullies discussed in this chapter may be basically similar to the development of some valleys discussed in Chapter 11. The main differences, aside from size, are that the gullies are formed in unconsolidated rather than consolidated materials, and their flow is ephemeral or intermittent rather than perennial. Large valleys developed by sapping are generally served by water-table springs, but most gullies lie above the water table, much of the time at least, and are eroded only periodically by flow from groundwater or from temporarily saturated horizons in the vadose zone above the water table.

The examples of seepage-induced gullies cited below are presented in terms of the types identified by Brice (1966), and include gullies formed by recurrent outflow of groundwater as

Figure 4. Discontinuous valley-side gullies in coastal San Mateo County, California. A similarly eroded, more distant hillside is visible at the top of the picture. Photograph taken July 1983.

well as those formed mainly by seepage from perched water bodies.

Valley-side gullies

Figure 4 shows a severely gullied slope on the coastal hills between Pescadero and Pomponio State Beaches in San Mateo County, California. These are within an area studied by Swanson (1983), who attributed the erosion to outflow of water from a periodically saturated zone above a relatively impermeable argillic B soil horizon. There is no evidence of surface drainage at the heads of any of the gullies shown, except possibly at the one in the center of the picture. Figure 5 shows the turfed hillside above the head of a similar gully nearby. There, as at most of these valley-side gullies, one sees no convergence of surface-drainage lines nor any evidence of surface wash or overflow.

As seen in Figure 5, tumbled blocks of earth or sod, common at the heads of these gullies, show that the gullies are growing by undermining and collapse. However, although small natural bridges and short underground channels within the gullies show that soil pipes do develop there, one rarely sees open pipes at the gully heads. This lack of visible openings could result from pipe collapse at the head scarp after each episode of flow, but it could also indicate that the scarps are being undermined not by pipe collapse but by erosion at a seepage face, like that pictured in Figure 3, only on a larger scale.

Swanson (1983; Swanson and others, 1989) installed a grid of piezometers on a hillside like the one shown in Figure 4, and found high water levels in almost all of the tubes after rainstorms. In some, the water level even rose above the ground surface. Although Swanson thought that these high fluid pressures were related to subsurface pipe flow, they could also indicate upward-directed flow paths in a saturated zone perched above the clayey B horizon. Hill (1989) has found such flow associated with seepage outflow and sapping in a valley-bottom gully in Nevada, described in the case study in this chapter.

Valley-side gullies may also be fed intermittently by groundwater when the water table rises high enough to intersect the gully head. Dunne (1980) has documented such a condition in a small, recently deforested catchment in Vermont, where several first-order fingertip tributaries serve as examples of valley-side and valley-head gullies. There, overland flow occurs only under the snowpack, when the ground is frozen and erosion is negligible. During the rest of the year, groundwater moves toward the valley axis through major joints in the impermeable bedrock. During the snowmelt season, the water table reaches its highest level and intersects the channels of the fingertip tributaries. Small springs then appear at the heads of the tributaries, and where the hydraulic gradients are large enough, seepage erosion promotes sapping and headward retreat along the joints to form the channel network (Dunne, 1980). Dunne suggests that this is a leading cause of channel initiation and valley extension in northern Vermont "and areas of similar hydrology." Jones (1987, p. 224–226) explains in some detail the role played by hillslope hollows in concentrating subsurface throughflow and determining sites of spring sapping.

Unquestionably, closer observation of valley-side gullies, especially at the times when they are active, is needed to learn more about their initiation and development, but it seems likely that among the several processes that shape them, seepage erosion and sapping at the base of the scarp can be among the most important.

Discontinuous valley-side gullies. Most of the gullies shown in Figure 4 are discontinuous. That is, they have a distinctive steep head from which they become shallower downslope until they end in a small fan. This is common among valley-side gullies. Leopold and Miller's (1956) explanation for discontinuities in valley-bottom gullies (see below) may also apply to the hillside variety. However, some may simply result from the entrenchment of a typically concave channel into a convex or nearly rectilinear hillside that was shaped chiefly by creep under an earlier, different climatic regime. Such a relation appears to be common for gullies on rounded hillsides in California.

Valley-head gullies

Valley-head gullies differ from valley-side gullies only in their location and orientation in respect to the valley axis; Brice (1966) discusses both types together. As they do not seem to differ in origin and development, the above comments on valley-side gullies probably apply as well to valley-head gullies.

In a valley-head gully in loess in central Missouri, Roloff and others (1981) found that active extension owes chiefly to weakening of soil strength, followed by mass failure at the headcut. This occurs as a consequence of throughflow convergence controlled by a buried paleosol surface. Commenting generally on the effects of impeded circulation, they state, "Stratigraphy influences gullying processes in most landscapes" (Roloff and others, 1981, p. 119). Coelho Netto and others (1987) report that in eroded "rampas" of southeastern Brazil, advancing modern gullies are localized along earlier gully fills. These effects are similar to the convergence and concentration of outflow at valley-side gully sites in hillside hollows, which may be an important factor in determining gully-head initiation. Higgins (1984) has cited other examples of surface drainage control by buried topography.

A somewhat different example of hillside gully development owing to impaired percolation is described in the Rif Mountains of Morocco by van den Brink and Jungerius (1983). There, stony colluvium deposited in historic times overlies less permeable, weathered, older colluvium. The relatively high infiltration capacity of the surficial layer makes overland flow uncommon. However, where percolation is blocked at the base of the unit, subsurface flow occurs and facilitates gully growth. This is indicated by the field observation that "the gullies generated their own runoff and transported water before flow began on the surrounding fields" (van den Brink and Jungerius, 1983, p. 284).

Valley-bottom gullies

Brice (1966, Fig. 200) illustrates several varieties of valley-bottom gullies, all representing the entrenchment of a steep-headed, steep-sided channel into a broad valley floor. Some

Figure 5. Head of valley-side gully, one of several near Pompanio State Beach, California. Note undermined and collapsed soil blocks at head and lack of surface drainage above the scarp. Photograph taken July 1983.

examples are shown in Figure 6. In some valley-bottom gullies, the head scarp is at the head of the incised system, but in many cases it takes the form of an abrupt break along the long profile of the channel. Brice called these mid-channel nickpoints *channel scarps.* In south-central Nebraska, he measured valley-bottom scarps as high as 7.5 m and found that one channel scarp had retreated 2,300 m between 1937 and 1952.

It has long been maintained that such valley-bottom scarps develop and migrate mainly through surface-water overflow and plunge-pool action, and doubtless this is true in many cases. However, in other cases, much of the retreat may owe to sapping at the base of the scarp by water emerging from the valley alluvium. Numerous authors, including Leopold and Miller (1956), Schumm and Hadley (1957), Brice (1966), and Blong (1966), have reported that valley-bottom gullies tend to begin at sites where the valley floor is steepest. This has been taken to explain initiation by surface-water scour where the velocity is fastest.

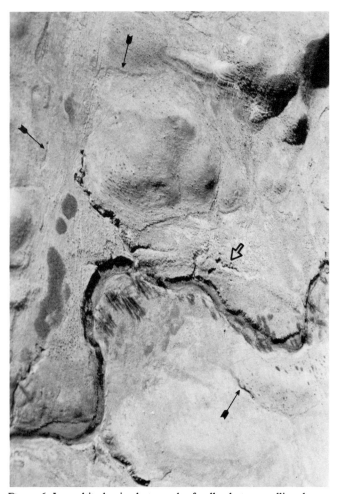

Figure 6. Low-altitude air photograph of valley-bottom gullies along a tributary to Dry Creek, 2 km east of Pierre Municipal Airport, South Dakota. Compare this scene with Brice's (1966, Fig. 200) drawing of varieties of valley-bottom gullies. The pronounced tributaries to the main incised channel are of the type here called "upland gullies." Some partially healed discontinuous gullies (indicated by arrows) occur along the edges of the main valley floor. Width of view about 320 m.

However, it could also explain initiation or maintenance by underground drainage where the subsurface flow gradient would also be greatest. Blong (1966) adds that incision may also be initiated where frictional resistance of the ground surface is lowest: for example, where vegetation is sparse or is removed along stock tracks. In such cases, the cut may be begun by surface wash, but may then be perpetuated by a combination of overflow and seepage. Cooke and Reeves (1976) report that two arroyos at San Xavier, Arizona, each now several kilometers long, were initiated by the excavation of trenches dug to the water table when local springs failed. Subsequent seepage at the heads of these infiltration galleries aided their extension.

Leopold and others (1964, p. 445), writing about nickpoints and head scarps in general, stated, "Field observations indicate that retreat of the face may be brought about by seepage or by pressure from water seeping out of the face following rains that raise the water table above the foot of the face." Expanding upon this theme, they added:

Sapping is intensified and in many cases probably caused by the emergence of moisture that infiltrated the unchanneled swale above the headcut and moved under gravity to the base of the vertical wall. The moisture produces a pore pressure tending to slough the supporting base of the gully head; perhaps more importantly, moisture in materials containing clay weakens the supporting foundation at the base of the vertical cut, both in tensile and compressive stress (Leopold and others, 1964, p. 446).

To be sure, surface wash must also affect these channels at times. Brice writes, "when a large volume of water spills over the scarp, as in Nebraska after a thunderstorm, the effect is rapid and impressive—the roar of the fall can be heard from a distance" (Brice, personal communication, 1987). However, Bradford and Piest (1980, p. 84) found that in valley-bottom gullies in the loess hills of western Iowa, such floods were ineffective in scouring the bank material, and "little soil is eroded from the standing banks by tractive forces of flowing water." Instead, gully growth owes chiefly to mass wasting concentrated near the headcut as a result, in part, of higher water-table levels there. At one site, the water table was more than 2 m above the base of the cut, and in some cases, seepage zones were higher than the maximum height of flowing water, suggesting that the source of the water was from infiltration on the valley floor or unchanneled swale, as well as from soaking of the channel walls during floods. Although large amounts of soil debris were transported annually through these gullies during a 13-year study, Bradford and Piest found that this transport chiefly represented cleanout of loose debris that had slumped from the head and sidewalls as a consequence of seepage. They also found that such cleanout was essential to maintain continued gully enlargement, so that the rate of gully growth depends on the supply of debris created by seepage undermining and slope failures as well as on the ability of surface wash to remove this debris from the base of the headcut and sidewalls.

Similarly, in the valley-head gully studied by Roloff and others, "overland flow was responsible for the removal of failure debris but not for bank erosion" (Roloff and others, 1981, p. 119). Heede (1971), too, found that enlargement of gullies by piping ceases and the gullies become stabilized where accumulation of sidewall debris buries the base of the scarp and creates a slope. These studies suggest that surface overflow is important in gully development but that in some cases it plays a secondary rather than a primary role.

Fenneman (1923) suggested that occasional surface wash not only helps maintain a gully's characteristic steep sides and flat bottom, but that the "inability of the stream to transport all of the debris supplied to it" is one factor that determines and limits "the depth of the draw or the height of its walls." Another factor is "nearly horizontal resistant beds that may act as local base-levels" (Fenneman, 1923, p. 128).

Subsurface strata may also control gullies in other ways. Development and localization of gullies by irregularities on a buried surface of impaired permeability not only affects valley-side gullies, as noted above, but in valley-bottom gullies, too, outflow through valley alluvium may be particularly strong where subsurface water is perched above a less permeable soil horizon, or above bedrock lying at or just below the base of the scarp. This seems to be true of the valley-bottom gully described by Hill later in this chapter. It may also be the case in many channels of ephemeral streams and gullies in the western United States, such as those discussed in the studies of Leopold and others (1966), Malde and Scott (1977), and Patton and Schumm (1981). All of these channels are described as having actively advancing, abrupt channel scarps or head scarps as high as 7 m, and significantly, most or all have impaired drainage at shallow depth. Leopold and others (1966) stated that the modern channel floor of the Arroyo de los Frijoles, near Santa Fe, is near the alluvium-bedrock contact. Similarly, Malde and Scott (1977) reported "impermeable volcanic rocks along the valley floor" below the head scarp of the Cañada de la Cueva, also near Santa Fe. In a nearby valley they found local cementation of the alluvium by caliche at about the depth of the channel scarps. In Sand Creek, Nebraska, Patton and Schumm (1981) reported that the retreat of a channel scarp had removed the alluvial fill down to the bedrock. In a more humid setting, at three separate localities in New South Wales, Blong and others (1982) reported occasional exposures of bedrock along the floors of gullies that they attributed both to throughflow and surface flow.

In all of these ephemeral streams the impedance of infiltration at the base of the permeable alluvium would promote active subsurface flow. However, it must be stressed that no one has observed the retreat of the scarps, and seepage erosion and sapping have not been demonstrated there. On the other hand, the resulting channels with their retreating headcuts resemble the flat-bottomed, steep-walled features attributed to spring sapping more than they resemble the rill-like features generally associated with overland flow. They also resemble the tiny gully-like features formed by seepage outflow on some beaches (see Fig. 1, Chapter 14, this volume; Higgins, 1984).

Association with resistant layers. Most valley-bottom gullies develop where some kind of resistant material caps the alluvium (Schumm, 1961; Gardner, 1984). This may be a clayey horizon, either depositional or pedogenic, or a mat of rooted vegetation. At the New Zealand sites that Blong (1966) studied, the steepness of the headcuts owes in part to a buried resistant soil horizon. The presence of a resistant cap and the accompanying steep headcut are commonly interpreted to signify erosion by plunge-pool action by surface-water overflow. However, as noted above, the chief role of surface wash may be to remove debris weakened by seepage. In many of the gullies studied by Blong, "it is difficult to ascertain whether surface water or subsurface water is the agent more active in promoting gully development" (Blong, 1966, p. 95). Moreover, some gullies lack a resistant veneer. On some beaches, little gullies can be seen to form by seepage erosion in uniformly erosible moist sand (Higgins, 1984). This suggests that seepage erosion and sapping can promote gullying in any material cohesive enough to form a vertical scarp and permeable enough to permit subsurface throughflow. Brice (1966) reported head scarps in massive loess, where he attributed the maintenance of the vertical face to sloughing along vertical joints. These examples suggest that the common relation between a steep headcut and a resistant cap does not necessarily mean that a head scarp or channel scarp has been developed entirely by surface wash, but may simply reflect the requirement for sufficient cohesiveness to support a vertical face.

Experimental studies of nickpoint development with and without resistant layers. Nickpoint retreat has been studied experimentally in models, ranging from a mathematical simulation by Pickup (1977) to laboratory analog models of Brush and Wolman (1960), Holland and Pickup (1976), and Gardner (1983). In three of these studies the channel was formed in homogeneous materials—noncohesive sands in Brush and Wolman's model, a sand-clay mixture in Gardner's, and an undefined erosible semicohesive material in Pickup's mathematical simulation. In the latter study, the nickpoint "maintains its height and steepness [as] it undergoes parallel retreat" (Pickup, 1977, p. 54). Similarly, in Holland and Pickup's (1976) model study, in which stream erosion is simulated in stratified sediment with alternating beds of cohesive and less cohesive materials, small channel scarps with "features analogous to their counterparts in the entrenchment of sedimentary rocks . . . tended to maintain their form as they retreated, provided that their plunge pools continued to function" (Holland and Pickup, 1976, p. 79). Conversely, where plunge pools became clogged with coarse debris from the scarp "the knickpoints tended to rotate and lose their identity" (p. 80). This is reminiscent of Bradford and Piest's (1980) findings that cleanout of sloughed debris by surface wash is essential for continued gully development. In both of the other model studies, however, upstream parallel retreat failed to occur in homogeneous materials, and both Brush and Wolman (1960) and Gardner (1983) concluded that resistant layers must be present for such retreat to occur (see Gardner, 1983, Fig. 13).

All of the above studies considered only the effects of surface-water flow; none incorporated any provision for possible effects of subsurface throughflow. This lack is corrected in a series of studies conducted jointly and separately by A. D. Howard and R. C. Kochel, and their colleagues at the University of Virginia and Southern Illinois University, respectively.

In the earliest of these studies, Howard and McLane (1988) observed the evolution of sapping erosion and headcut retreat in two dimensions in a tank filled with noncohesive sand and subjected to lateral groundwater flow. They then developed a two-dimensional simulation model that closely predicted the behavior of the analog model. Later, Kochel and others (1985) expanded these studies with three-dimensional analog models that provide subsurface flow in both layered and homogeneous materials. Some of the studies allow for variations in internal stratigraphy, cohesion, the slope of both the bedding and the surface, and even the presence of joints. In most model runs, the formation and headward retreat of head scarps was clearly reproduced. Some of the results are reported in Howard and McLane (1988), Howard (1988), and Kochel and others (1988), and are briefly reviewed in Howard's case study (Chapter 11, this volume).

These latter studies confirm my own observations of gully-head retreat in miniature channels produced solely by groundwater outflow in moist sand on a beach in San Diego (Higgins, 1984, Fig. 2.11), and suggest that the effects of subsurface throughflow may be important in the development and headward retreat of nickpoints and head scarps in many natural streams and valley-bottom gullies, with or without assistance from a resistant cap or from surface wash.

Discontinuous valley-bottom gullies. Valley-bottom gullies are commonly discontinuous. This occurs where the gradient of the incised channel is gentler than the valley floor. As characterized by Leopold and others, discontinuous gullies have:

... a vertical headcut, a channel immediately below the headcut which often is slightly deeper than it is wide, and a decreasing depth of the channel downstream. Where the plane of the gully floor intersects the more steeply sloping plane of the original valley floor, the gully walls have decreased to zero in height and a fan occurs (Leopold and others, 1964, p. 448).

Earlier, Leopold and Miller (1956) developed an explanation for this difference in gradients, based on presumed hydraulic relations. In their analysis, mutual accommodation among discharge, width, depth, and roughness causes the initial slope to be gentler than that of the ungullied valley floor. However, as the head scarp retreats and the gully becomes longer, it may overtake the toe of a second gully upstream; the two then coalesce. The effect of this growth in length is to increase both the width and the gradient, so that eventually "the gradual increase in width has required such an increase in slope that the gully bed becomes almost parallel to the original valley floor" (Leopold and Miller, 1956, p. 32). A similar progression in gully development has been reported by Heede (1960) in the Colorado Front Range and by Blong (1966, 1970) for discontinuous gullies on both valley floors and lower colluvial hillslopes in central North Island, New Zealand.

Not all discontinuous gullies follow this evolution. In some, deposition in the channel and at the toe may keep pace with head-scarp retreat, so that the whole system retreats upstream. In others, decrease or loss of surface overwash and/or subsurface outflow may reduce or end gully development, and revegetation then begins to soften or hide its outlines. Some partially healed gullies of this sort may be visible near the edges of the floodplain in Figure 6.

Upland gullies. A very common kind of gully begins at the bank of a stream channel or some other steep local slope and is eroded headward into a nearly level field or plain. These are similar in form to some of Brice's (1966) valley-bottom gullies. Gullies tributary to the main channel shown in Figure 6 are of this type. There does not seem to be an accepted name for such gullies, although the term "upland" has been used in aposition to "valley-bottom." Many are developed chiefly by direct surface runoff, but some can be shown to be formed mainly by seepage erosion and sapping.

Schumm and Phillips (1986) have recently described some valleys or channels in New Zealand that were initiated at a sea-cliff and extend into the Canterbury Plain. Some of these are of gully size and are actively forming at the present, but the largest are as much as 1 km long. The authors suggest that the larger channels were initially formed by overland flow, but were then widened and "modified by seepage and consequent sapping along zones of maximum transmissibility in the gravels of the Canterbury Plain" (Schumm and Phillips, 1986, p. 328). However, the authors suggest that the smaller channels are formed almost entirely by seepage erosion:

The smallest channels develop when seepage weakens the gravels 2 to 3 m below the cliff top and slumping results. Enlargement of this zone of removal causes collapse of the overlying loess and dry gravel, and a chute is formed....The process continues until a steep gully has developed (Schumm and Phillips, 1986, p. 328).

Continued development mantles the lower slopes with an armor of coarse gravel, which halts the enlargement of the gully when the gravel reaches the top of the seepage zone. However, "if there is sufficient seepage, or seepage supplemented by overland flow, which can remove the coarse sediment, then the gully can continue to enlarge, and a large, wide, flat-floored channel will develop" (Schumm and Phillips, 1986, p. 328).

Unquestionably, the kind of development described above, where upland gully systems are initiated and extended by outflow of subsurface water, is fairly common and is particularly indicated where the gullies show some or all of the features characteristic of valleys formed by sapping. As noted by Kochel and others (1985), Higgins (1984), and others, these features include: (1) u-shaped profiles with blunt, steep head scarps and steep side scarps, and with occasional slumped masses or blocks that leave alcoves in the valley walls; (2) low drainage density, with long main valleys and few, stubby tributaries; (3) aggraded valley floors that tend to form abrupt angles with the valley sides; (4) minor increase or even decrease in downstream width and depth; (5) alignment of the channels in preferred directions, controlled by directional permeability or structure in the material in which they are formed; (6) absence of evidence of contemporary surface runoff at their heads; and (7) presence of seepage outflow at still-active valley heads.

Where the sediments are coarse, like those of the Canterbury Plain, or where surface wash is incapable of removing debris sloughed from the walls, gully development may be self-limiting; but where the sediments are sands and silts that can be carried out of the system, gullies may be extended far headward into plains and cultivated fields. At some ill-defined point, they become ephemeral stream valleys like some of those described in Chapter 11 (this volume).

GULLY CONTROL

The traditional approach to gully prevention and control has been to divert or contain surface runoff, and to prevent it from washing over the head scarp. A common conservation measure is to create spreading works or dry wells so as to enhance infiltration and thus reduce overland flow. These procedures are all based on the assumption that overland flow does all the damage. However, it is clear that where piping and seepage erosion are important or dominant factors in head-scarp sapping, efforts should be directed toward reducing soilwater throughflow and groundwater outflow. At such sites, enhancement of infiltration can aggravate the very problems that the conservation measures were meant to solve (Morgan, 1980). In Nigeria, ditches, sumps, diversion channels, and small dams were constructed in the late 1940s in order to control gullying in the Agulu-Nanka region. However, these works seem to have had the opposite effect: "Instead of checking the gullying, the ditches and sumps increased groundwater percolation and flow rates, and hence the gullying" (Egboka and Okpoko, 1984, p. 343).

Little attention has been paid to the requirements for controlling gullies that grow in part by subsurface flow. Palmer (1965, p. 17) has stated the fundamental principle that must be followed: "Gullying may be controlled by modifying the factors that caused it." Specifically, "if water is observed in the gully or the soil depth is shallow to bedrock, seepage can be anticipated. Under these conditions, the gully should be filled and a toe drain provided" (Palmer, 1965, p. 27). Bradford and Piest (1980) suggest that mulch tillage or level terraces can greatly reduce gully growth by preventing debris cleanout and plunge-pool deepening by runoff, thereby slowing bank sloughing and allowing the healing process to begin. Morgan (1980, p. 290) concurs: "control of subsurface erosion is best achieved by establishing a healthy grass cover." However, Nir and Klein (1974) found that the introduction of such conservation measures in 1948 at a semiarid site in loess in the coastal plain of Israel led to rejuvenation of gully growth. There, fields were contour plowed and areas near the gullies were left unplowed. The authors attributed the renewed gullying to increased infiltration and throughflow as well as to increased scour by surface wash as a consequence of decreased availability of erosible soil and the resulting lower sediment load.

Another gully-control measure is to plant thirsty vegetation, such as willows, that help to dewater the slope above a gully head, although Thornes (1984, p. 139) cautions that "regional afforestation of upland slopes will not provide the substantive controls of soil erosion." Dewatering by continuous pumping and removal of surface runoff through concrete channels has been suggested as possible control measures for the Agulu-Nanka gullies in Nigeria (Egboka and Okpoko, 1984). However, Blong and others (1982, p. 385) have cautioned that "construction of gully head control structures alone cannot [entirely solve the problems] as the bulk of the sediment may come from the sidewalls." Morgan (1980) suggests other treatments: ripping up soil pipes before sowing grass, or chemically treating the soil to counteract effects of high sodium content. All such measures, though, may be beyond the resources of many small farmers, especially in developing nations. Stocking (1980, p. 518–519) considered the factors, including piping, that appear to control gully growth in central Zimbabwe, and concluded, "It must be questioned whether conservation projects are feasible or, indeed, desirable under such circumstances."

DEVELOPMENT OF GULLIES IN HUMID REGIONS

Most of the above examples of gullies formed wholly or in part by seepage and sapping are in drylands. Examples of gullies in humid regions, although less common in the literature, suggest no significant differences from dryland gullies. In humid regions, dense vegetation tends to inhibit the formation and growth of gullies. However, where vegetation is removed, gully development may be pervasive, as at Ducktown, Tennessee (Burt, 1956). As noted above, some of the gullies studied by Ireland and others (1939) in humid South Carolina owe in some part to seepage erosion. In the Agulu-Nanka region of Nigeria, where mean annual rainfall may exceed 2,000 mm, Egboka and Okpoko (1984) attribute much of the gully development to high pore-water pressures in the underlying sands and shales and to the acidity of the effluent groundwater, which aids the dissolution of cementing materials. Also, near Durham, New Hampshire, where a gully was developing from subsurface flow, Palmer (1965) observed crescentic scarps and ground subsidence much like the features described by Rubey (1928) in northern Texas. In both cases the effects were attributed to removal of fines by subsurface flow.

Although he concluded that "surface runoff is the major cause of gullying in New England," Palmer (1965, p. 32) also found that "seepage may produce gullies by removing soil material below the surface, especially where the depth to bedrock is shallow." As an example, Palmer cited a gully near Claremont, New Hampshire, developed in very fine sandy loam with unusually shallow depth to bedrock. Infiltration capacity of the surface was 1.6 in/hr after 60 min. During three spring seasons, no surface overflow was observed above the gully head "even when considerable gully damage was occurring below it. . . .[yet] Between 65 and 560 cubic yards of earth were carried off by eroding water each year" (Palmer, 1965, p. 9). Palmer surveyed the site every year from 1957 until 1960, and installed a weir with a water-stage recorder and an array of thermocouples and piezometers. From his data he concluded that "surface runoff, seepage, and freezing and thawing appear to be major causes of erosion at this site" (Palmer, 1965, p. 12) and that "deep frost penetration retards soil infiltration and inhibits erosion; conversely, shallow frost penetration of the soil, warm temperatures, and ample precipitation promote soil erosion during the spring thaw period" (Palmer, 1965, p. 9). These findings are similar to those reported from Nevada by Hill and Lehre in the following case study.

CASE STUDY: GULLY PROCESSES AT MAHOGANY MEADOW, NEVADA

Barry R. Hill and André K. Lehre

INTRODUCTION

Design of control measures and evaluation of the effects of resource management programs have been hindered by the lack of information concerning the processes of gully initiation and development. To identify the processes by which water enters and erodes an actively developing gully channel, we selected a small (0.024 km^2) meadow on the north slope of Badger Mountain in Humboldt County, within the Sheldon National Wildlife Refuge in northwestern Nevada.

STUDY AREA

The study site is known locally as Mahogany Meadow. Mahogany Meadow is roughly 400 m in length and ranges from approximately 30 to 80 m in width, with width decreasing toward the lower end of the meadow. The meadow has a gentle (2.5 percent) south-north topographic gradient, and is surrounded by gentle to moderately steep (12 to 40 percent) hillslopes. The meadow terminates at its lower end in a narrow bedrock canyon. Mahogany Spring, situated 0.5 km south of the meadow, flows intermittently onto the upper meadow.

The watershed surrounding the meadow has an area of 3.55 km^2. Elevations in the watershed range from 1,950 to 2,190 m above mean sea level. The climate of northwestern Nevada is semiarid and continental. Mean annual precipitation in the vicinity of the meadow is 30 cm, most of which falls as snow during the winter (U.S. Weather Bureau, 1960). Streamflow records for a tributary of Badger Creek, 15 km west of the meadow, indicate that all annual peak flows are snowmelt generated and occur between January and June (Butler and others, 1966; U.S. Geological Survey 1974). Vegetation of Mahogany Meadow consists primarily of grasses, sedges, and forbs, and the vegetation of hillslopes surrounding the meadow consists primarily of shrubs and perennial grasses.

Bedrock of the uplands surrounding the meadow consists of Tertiary rhyolite (Greene, 1984) that is extensively exposed on hillslopes and is highly weathered and fractured. Meadow stratigraphy consists of three alluvial units with little evidence of soil profile development. The uppermost stratum is a dark incipient A horizon that forms the present sod-like meadow topsoil. Texturally, this horizon is a silty clay loam. A gravelly clay loam designated the C1 unit lies below the A horizon. Near the meadow margins, the C1 is underlain by bedrock, but throughout most of the central portion of the meadow it is underlain by a distinct gravelly gleyed layer designated C2. Texturally, the C2 is a very gravelly loam.

The lower portion of the meadow is incised by a gully 50 m in length, 0.6 to 1.5 m deep, and 1.2 to 5.5 m wide (Fig. 7). Both

Figure 7. View upstream of gully at Mahogany Meadow, April 1985. Most of the water visible in the channel is groundwater discharge; a slight amount of flow from Mahogany Spring is passing over the headcut.

width and depth increase upstream, as is typical of valley-bottom gullies (Brice, 1966). The gully is incised to bedrock along portions of its length. Gully banks are nearly vertical in the root-bound A horizon, but are much more gentle and have a lobate appearance in the C1. Bed material consists largely of angular gravel and sand. A small tributary headcut has developed along a wheel rut on the east bank (Fig. 8). The gully has developed since 1963, as it does not appear in aerial photographs taken that year. Headcut retreat between 1978 and 1984 was approximately 3 m (D. L. Franzen, U.S. Fish and Wildlife Service, personal communication, 1984).

Gully erosion of wet meadows on the refuge may be due to formerly unrestricted livestock grazing (Swanson and others, 1987). The area encompassed by the present refuge boundaries has been used for the grazing of livestock for more than 100 years (U.S. Fish and Wildlife Service, 1980). Mahogany Meadow was fenced in 1984, but until that time was heavily used by cattle and was subject to occasional motor vehicle traffic.

No direct observations on streamflow generation have been reported in previous studies of the refuge or adjacent areas. Permeability of the bedrock has been discussed and the lack of surface runoff has been noted (Waring, 1908; Sinclair, 1963); nevertheless, infiltration-excess overland flow has generally been accepted as the process responsible for generation of peak flows (Snyder, 1951; Sinclair, 1963).

METHODS OF STUDY

Runoff and groundwater processes were monitored by means of piezometers, wells, crest-stage gages, water-level recorders, and overland flow troughs installed during the summer of 1984. Locations of instrument stations are shown in Figure 8. Both wells and piezometers consisted of plastic pipe, 2.5 cm in

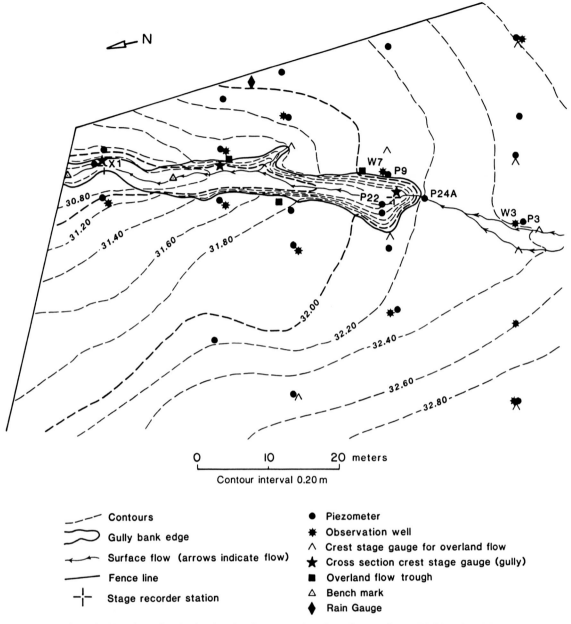

Figure 8. Plan view of study site showing instrument locations. Contour interval 0.20 m; local datum.

diameter, placed in auger holes dug to bedrock in meadow alluvium. Crest-stage gages were placed at various points on the meadow to measure the depth of overland flow. Three troughs for the direct measurement of overland flow at the gully banks were constructed of roof gutter and aluminum flashing. Two stage recorders and three crest-stage gages were installed in the gully, but were not used due to low flow conditions. Field instrumentation and study methods are described in detail in Hill (1987; 1990).

Field observations were conducted during the fall of 1984 and from March 28 to May 10, 1985. Observations were made on snow depth and water equivalent, extent of soil frost, rates of overland flow and stream (gully) flow, and groundwater levels in wells and piezometers. Surface-water discharge was measured volumetrically at site X1 (Fig. 8) and at the headcuts. Infiltration tests were made at four sites on the meadow using a crude single-ring infiltrometer.

OBSERVATIONS

On March 30, 1985, Mahogany Meadow was covered with a fairly uniform snowpack that completely filled the gully and covered all vegetation. Average snowpack depth was 38 cm, and average snowpack water equivalent was 10 cm (Hill, 1987). A period of warm weather began on March 31. By April 1, areas of the meadow were uncovered, and the study site was snow free by April 5, although drifts persisted at higher elevations until early May. Weather turned cold again on April 15. Frequent low-intensity snow and rainstorms passed over Badger Mountain, dropping a total of 2 cm of precipitation at the meadow.

During the snowmelt period, soil frost was extensive. Frost could not be penetrated with a soil auger, and appeared to be of the "concrete" type (Post and Dreibelbis, 1942). Soil frost attained a depth of roughly 40 cm at the west gully bank on March 31. Meltwater puddles were observed at times above frozen soil.

Except for small amounts of wind-blown snow that had entered through vents, all wells and piezometers were dry at the beginning of snowmelt on March 30. Water first appeared in well W3 above the gully head and in a piezometer near the north fence on April 8. Subsequently, water entered all wells and piezometers except for several wells along the meadow margins and several shallow piezometers. Daily water levels for selected stations during the final two weeks of observation are plotted in Figure 9. Groundwater flow at the meadow is described in detail in Hill (1990).

Results of volumetric measurements of flow over the main and tributary headcuts in early April, during the snowmelt period, are listed in Table 1. Crest-stage gages detected overland flow at only a few sites, and none of these showed consistent flow throughout the study interval. During the times at which snowmelt-generated overland flow was observed, the gully remained filled with snow, and no discharge from the gully at site

TABLE 1. VOLUMETRIC MEASUREMENTS FOR OVERLAND FLOW ENTERING GULLY HEADCUTS OVER FROZEN SOIL

Location	Date	Time	Discharge (ml/s)
Main headcut	4/01	15:30	58
Tributary headcut	4/05	13:00	30
Tributary headcut	4/05	14:00	63
Tributary headcut	4/05	15:00	65

X1 (Fig. 8) was observed. No overland flow was collected in any of the troughs at any time.

Flow began in the lower portion of the gully on April 15. At this time, no water was flowing over either headcut. Some flow over the tributary headcut was observed on April 17, most likely resulting from rainfall on that date. With this exception, no flow was observed at either headcut until April 28, when water flowing from Mahogany Spring onto the upper meadow reached the main headcut. Thereafter, flow over the headcut was sporadic, generally exhibiting a diurnal cycle but failing to reach the headcut at all on some days. Flow became continuous throughout the gully on April 22. Groundwater discharge from the gully was computed as total discharge from the gully at site X1 (Fig. 8) less the flow over the headcuts. Results are shown in Figure 10.

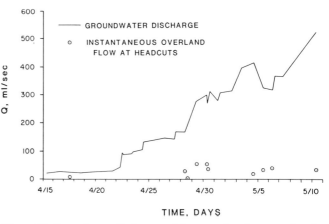

Figure 10. Groundwater discharge and flow over headcuts entering the gully between April 15 and May 10, 1985. All measurements were made volumetrically at X1 and at headcuts.

DISCUSSION

Weather conditions in the northwestern Great Basin during the winter of 1984–1985 were neither extremely dry nor very wet. Stream forecasts for March predicted maximum 24-hr peak flows to be 100 percent of historic averages (USDA Soil Conservation Service, 1985). Results of this study, therefore, may be considered applicable only to conditions of moderate precipitation and runoff, and not to extreme events, which might be of greater importance for erosional processes.

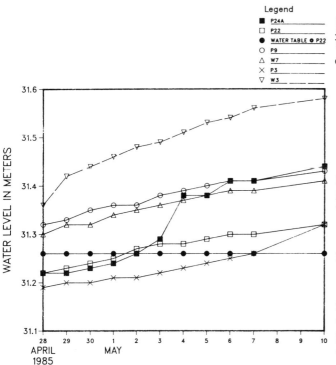

Figure 9. Water levels in wells and piezometers near the headcut during a two-week period in the spring of 1985. The water table at P22 was at the surface during this period.

Figure 11. "Boiling" of particles is occurring at the point marked by the screwdriver on the gully bank. Eroded material is being deposited in the gully bed at the point marked by the pencil.

Streamflow generation

Overland flow resulting from snowmelt in excess of infiltration was observed on only two days during the early snowmelt period (Table 1). As water levels in wells were more than 0.5 m below the meadow surface (Hill, 1987), this flow can be considered to be true infiltration-excess (Horton) overland flow. The maximum measured flow of this type was 65 ml/s. Infiltration-excess overland flow occurred only over frozen soil. The fate of the water after it had passed over the headcut was difficult to determine, because the gully remained filled with snow. Overland-flow crest-stage data (Hill, 1987) indicate that overland flow was restricted to disturbed areas and to a narrow portion of the central meadow carrying discharge from Mahogany Spring.

On the basis of our limited data, the peak snowmelt rate was lower than the infiltration capacity of unfrozen meadow soil by roughly an order of magnitude. Infiltration-excess overland flow during snowmelt is thus likely only when the soil is frozen. Because thawing of the soil was rapid after meltwater ponded on the meadow surface, it appears unlikely that soil frost can persist long enough to generate substantial overland flow except in cases of exceptionally deep frost penetration or extremely rapid snowmelt, as might occur during a rain-on-snow event.

Groundwater discharge constituted by far the greatest proportion of streamflow. Groundwater discharge resulted from groundwater flow through alluvium and through fractures in the bedrock underlying the meadow (Hill, 1990). Groundwater discharge was positively correlated with water levels in wells and piezometers. Water levels in paired wells and piezometers indicated vertically upward hydraulic gradients near the headcut in late April and early May (Hill, 1990) when groundwater discharge was greatest.

Erosion of gully banks

Erosion of gully banks during spring of 1985 was very slight. The headcut did not retreat, but adjacent portions of the bank retreated as much as 2 cm. Bank erosion was associated with outflow of groundwater and freeze-thaw cycles. All material eroded was of the C1 unit. Groundwater erosion was not uniform but occurred persistently in association with upward seepage at several locations along the banks of the gully, primarily near the headcut. At some points, erosion resulted from diffuse seepage from relatively large areas; elsewhere, discrete sources of upward flow, as evidenced by "boiling" of particles, could be identified (Figs. 11 and 12). All eroded bank material was deposited in the gully channel. Except for removal of a few sand grains during the initial few minutes of flow over the main headcut, overland flow was ineffective in transporting sediment or eroding gully banks. Apparently, gully discharge greater than any observed in 1985 is required to remove eroded bank material from the channel.

Frost heave was observed to contribute directly to mass wasting of bank material. Frost may also indirectly affect seepage erosion and sapping. On April 26, thawing of heavy frost near the headcut was accompanied by unusually rapid outflow of groundwater and sloughing of bank material. Within an hour, both groundwater outflow and bank erosion had ceased. Similar high rates of groundwater outflow and bank sloughing were observed only during thawing of frost along gully banks, although less dramatic seepage erosion occurred continuously after April 24.

CASE STUDY CONCLUSIONS

During the spring of 1985, groundwater discharge was the major source of water entering a small, actively eroding gully at

Figure 12. Diffuse seepage and sloughing along the east bank of the gully.

Mahogany Meadow. Surface flow originating as discharge from a nearby spring was a secondary and much smaller source of streamflow. Very small amounts of infiltration-excess overland flow entered the gully over frozen ground. Erosion of the gully banks was slight, and was due almost entirely to seepage erosion and frost action. Surface flow was not effective in transporting eroded sediment from the gully.

CASE STUDY ACKNOWLEDGMENTS

Field and laboratory work were completed while the senior author was a graduate student at Humboldt State University, Arcata, California. The suggestions of thesis advisors Bob Ziemer and Doug Jaeger are appreciated, as are those of Dick Iverson, Mike Lohrey, Judy Wait, Scott Lundstrom, and Sherm Swanson. Assistance in the field was provided by John Oldemeyer, Steve Woodis, Allan and Lorin Hill, and Rich Menchinella. We thank the staff of the Sheldon National Wildlife Refuge for their support; the assistance of Dave Franzen was particularly helpful.

CHAPTER CONCLUSIONS

Leopold and others (1964, p. 447) wrote, "In our experience in semiarid regions, the erosive action of water flowing over the vertical face of a gully head is not among the most active agents of headward progression." Earlier, Fenneman (1923, p. 128) had commented that "sapping at the base of the scarp . . . is the dominant factor in determining the form of the flat-bottomed, scarp-bordered gully or 'draw' of arid regions." Years later, these observations have yet to be universally accepted. In this chapter, both piping and seepage pressure are proposed as important, even dominant, agents in undermining head scarps and gully walls.

As summarized here, both field observations and model studies such as those by Kochel and Howard and their colleagues suggest that where gullies are formed at least in part by subsurface flow processes, their development, morphology, and rate of recession depend on several factors:

1. The thickness, differential permeability, and strength of subsurface strata or horizons.
2. The slope of both the land surface and the subsurface layers.
3. The orientation of fractures or other linear zones of weakness in the stronger or more cohesive horizons.

Exposure of a water-conducting layer at a seepage face is essential for head-scarp retreat by sapping. This may be initiated either by surface scour or by upward-directed outflow of subsurface water. Once the headcut is established, there must be some means of disaggregating and removing the debris that is undermined and falls into the channel. Commonly such removal is provided by surface wash. This may be the main function of the plunge pool at the gully head, where surface overflow plays a vital but secondary part. These relations have significant consequences for erosion control: to the extent that subsurface processes are effective in gully development, conservation measures must be modified to deal with problems of infiltration rather than surface runoff.

Past studies of gully development, gully control, and nickpoint retreat have focused almost exclusively on the mechanisms and effects of overwash and streamflow. This is understandable in view of the domination of the fluvial doctrine in geomorphic thought for the last 150 years. However, it is hoped that the studies reported here will foster an increased awareness of the possible surface effects of subsurface water so that, in future studies of gullies and their processes, observers will pay particular attention to the often hidden influence of piping, seepage erosion, and the outflow of subsurface water.

REFERENCES CITED

Andrews, W. F., 1972, Soil survey of Yolo County, California: U.S. Department of Agriculture Soil Conservation Service, 102 p.

Blong, R. J., 1966, Discontinuous gullies on the Volcanic Plateau: New Zealand Journal of Hydrology, v. 5, p. 87–99.

——, 1970, The development of discontinuous gullies in a pumice catchment: American Journal of Science, v. 268, p. 369–383.

Blong, R. J., Graham, O. P., and Veness, J. A., 1982, The role of sidewall processes in gully development; Some N.S.W. examples: Earth Surface Processes and Landforms, v. 7, p. 381–385.

Bradford, J. M., and Piest, R. F., 1980, Erosional development of valley-bottom gullies in the upper midwestern United States, in Coates, D. R., and Vitek, J. D., eds., Thresholds in geomorphology: London, George Allen and Unwin, p. 75–101.

Brice, J. C., 1966, Erosion and deposition in the loess-mantled Great Plains, Medicine Creek drainage basin, Nebraska: U.S. Geological Survey Professional Paper 352-H, p. 255–339.

Brush, L. M., Jr., and Woman, M. G., 1960, Knickpoint behavior in noncohesive material; A laboratory study: Geological Society of America Bulletin, v. 71, p. 59–74.

Burt, J. C., 1956, Desert in the Appalachians: Nature Magazine, v. 49, p. 486–488, 499.

Butler, E., Reid, J. K., and Benwick, V. K., 1966, Magnitude and frequency of floods in the United States; Part 10, The great Basin: U.S. Geological Survey Water Supply Paper 1684, 256 p.

Coelho Netto, A. L., Fernandes, N. F., and Moura, J. R., 1987, Drainage network expansion by gully advance through RAMPA deposits in SE Brazilian Plateau; Bananal, SP [abs.]: EOS Transactions of the American Geophysical Union, v. 68, no. 44, p. 1273.

Cooke, R. U., and Reeves, R. W., 1976, Arroyos and environmental change in the American South-west: Oxford, Clarendon Press, 213 p.

Dunne, T., 1980, Formation and controls of channel networks: Progress in Physical Geography, v. 4, p. 211–239.

Egboka, B.C.E., and Okpoko, E. I., 1984, Gully erosion in the Agulu-Nanka region of Anambra State, Nigeria, in Walling, D. E., Foster, S.S.D., and Wruzel, P., eds., Challenges in African hydrology and water resources; Proceedings of the Harare Symposium, July 1984: International Association of Scientific Hydrological Sciences Publication 144, p. 335–347.

Emmett, W. W., 1968, Gully erosion, in Fairbridge, R. W., ed., Encyclopedia of geomorphology: New York, Reinhold, p. 517–519.

Fenneman, N. M., 1923, Physiographic provinces and sections in western Oklahoma and adjacent parts of Texas: U.S. Geological Survey Bulletin 730, p. 115–134.

Gardner, T. W., 1983, Experimental study of knickpoint and longitudinal profile evolution in cohesive, homogeneous material: Geological Society of America Buletin, v. 94, p. 664–672.

——, 1984, Reply, 'Experimental study of knickpoint and longitudinal profile evolution in cohesive, homogeneous material': Geological Society of America Bulletin, v. 95, p. 123.

Greene, R. C., 1984, Geologic appraisal of the Charles Sheldon Wilderness Study Area, Nevada and Oregon: U.S. Geological Survey Bulletin 1538, p. 13–34.

Heede, B. H., 1960, A study of early gully-control structures in the Colorado Front Range: U.S. Forest Service Rocky Mountain Forest and Range Experiment Station Paper 55, 42 p.

——, 1970, Morphology of gullies in the Colorado Rocky Mountains: International Association of Scientific Hydrology Bulletin, v. 15, no. 2, p. 79–89.

——, 1971, Characteristics and processes of soil piping in gullies: Rocky Mountain Forest and Range Experiment Station Forest Service Research Paper RM-68, 15 p.

Higgins, C. G., 1984, Piping and sapping; Development of landforms by groundwater outflow, in LaFleur, R. G., ed., Groundwater as a geomorphic agent: Boston, Allen and Unwin, p. 18–58.

Hill, B. R., 1987, Snowmelt-generated runoff processes affecting wet-meadow gully erosion, northwest Nevada [M.S. thesis]: Arcata, California, Humboldt State University, 86 p.

——, 1990, Groundwater discharge to a headwater valley, northwestern Nevada, U.S.A.: Journal of Hydrology, v. 113, p. 265–283.

Holland, W. N., and Pickup, G., 1976, Flume study of knickpoint development in stratified sediment: Geological Society of America Bulletin, v. 87, p. 76–82.

Horton, R. E., 1945, Erosional development of streams and their drainage basins: Geological Society of America Bulletin, v. 56, p. 275–370.

Howard, A. D., 1988, Groundwater sapping experiments and modelling, in Howard, A. D., Kochel, R. C., and Holt, H. E., eds., Sapping features of the Colorado Plateau: National Aeronautics and Space Administration SP-491, p. 71–83.

Howard, A. D., and McLane, C. F., III, 1988, Erosion of cohesionless sediment by groundwater seepage: Water Resources Research, v. 24, p. 1659–1674.

Imeson, A. C., and Kwaad, F.J.P.M., 1980, Gully types and gully prediction: Geografisch Tijdschrift, n.r., v. 14, p. 430–441.

Ireland, H. A., Sharpe, C.F.S., and Eargle, D. H., 1939, Principles of gully erosion in the Piedmont of South Carolina: U.S. Department of Agriculture Technical Bulletin 633, 142 p.

Jones, J.A.A., 1987, The initiation of natural drainage networks: Progress in Physical Geography, v. 11, p. 207–245.

Kochel, R. C., Howard, A. D., and McLane, C. F., 1985, Channel networks developed by groundwater sapping in fine-grained sediments; Analogs to some Martian valleys, in Woldenberg, M. J., ed., Models in geomorphology: Boston, Massachusetts, Allen and Unwin, p. 313–341.

Kochel, R. C., Simmons, D. W., and Piper, J. F., 1988, Groundwater sapping experiments in weakly consolidated layered sediments; A qualitative summary, in Howard, A. D., Kochel, R. C., and Holt, H. E., eds., Sapping features of the Colorado Plateau: National Aeronautics and Space Administration SP-491, p. 84–93.

Leopold, L. B., and Miller, J. P., 1956, Ephemeral streams; Hydraulic factors and their relation to the drainage net: U.S. Geological Survey Professional Paper 282-A, p. 1–37.

Leopold, L. B., Wolman, M. G., and Miller, J. P., 1964, Fluvial processes in geomorphology: San Francisco, California, Freeman, 522 p.

Leopold, L. B., Emmett, W. W., and Myrick, R. M., 1966, Channel and hillslope processes in a semiarid area, New Mexico: U.S. Geological Survey Professional Paper 352-G, p. 193–253.

Malde, H. E., and Scott, A. G., 1977, Observations of contemporary arroyo cutting near Santa Fe, New Mexico, U.S.A.: Earth Surface Processes, v. 2, p. 39–54.

Morgan, R.P.C., 1980, Implications, in Kirkby, M. J., and Morgan, R.P.C., eds., Soil erosion: Chichester, John Wiley and Sons, p. 253–301.

Nir, D., and Klein, M., 1974, Gully erosion induced in land use in a semi-arid terrain (Nahal Shiqma, Israel): Zeitschrift für Geomorphologie, Supplement, v. 21, p. 191–201.

Palmer, R. S., 1965, Causes, control, and prevention of gullies at various New England locations: U.S. Department of Agriculture Agricultural Research Service Conservation Research Report 2, 34 p.

Patton, P. C., and Schumm, S. A., 1981, Ephemeral-stream processes; Implications for studies of Quaternary valley fills: Quaternary Research, v. 15, p. 24–43.

Pickup, G., 1977, Simulation modelling of river channel erosion, in Gregory, K. J., ed., River channel changes: Chichester, John Wiley and Sons, p. 47–60.

Post, F. A., and Dreibelbis, F.R., 1942, Some influences of frost penetration and microclimate on the water relationships of woodland, pristine and cultivated soils: Soil Science Society of America Proceedings, v. 7, p. 95–104.

Roloff, G., Bradford, J. M., and Scrivner, C. L., 1981, Gully development in the Deep Loess Hills region of central Missouri: Soil Science Society of America

Journal, v. 15, p. 119–123.

Rubey, W. W., 1928, Gullies in the Great Plains formed by sinking of the ground: American Journal of Science, v. 15, no. 89, p. 417–422.

Schumm, S. A., 1961, Effect of sediment characteristics on erosion and deposition in ephemeral-stream channels: U.S. Geological Survey Professional Paper 352-C, p. 31–70.

Schumm, S. A., and Hadley, R. F., 1957, Arroyos and the semiarid cycle of erosion: American Journal of Science, v. 255, p. 161–174.

Schumm, S. A., and Phillips, L., 1986, Composite channels of the Canterbury Plain, New Zealand; A Martian analog?: Geology, v. 14, p. 326–329.

Sinclair, W. C., 1963, Ground-water appraisal of the Long Valley-Massacre Lake region, Washoe County, Nevada: Nevada Department of Conservation and Natural Resources Ground-water Reconnaissance Series Report 15, 23 p.

Snyder, C. T., 1951, Prospects for stock water development in the Massacre Lake grazing area, Washoe County, Nevada: Salt Lake City, Utah, U.S. Geological Survey Water Resources Division, Technical Coordination Branch, unpublished report, 32 p.

Stocking, M. A., 1980, Examination of the factors controlling gully growth, *in* de Boodt, M., and Gabriels, D., eds., Assessment of erosion: Chichester, Wiley-Interscience, p. 505–520.

—— , 1981, A model of piping in soils, *in* Proceedings of the 3rd Meeting of the IGU Commission on Field Experiments in Geomorphology, Kyoto, Japan: Japanese Geomorphological Union, p. 105–119.

Stow, S. H., and Hughes, T. H., 1980, Geology and the urban environment, Cottondale Quadrangle, Tuscaloosa County, Alabama: Geological Survey of Alabama Atlas Series 9, 78 p.

Swanson, M. L., 1983, Soil piping and gully erosion along coastal San Mateo County, California [M.S. thesis]: Santa Cruz, University of California, 141 p.

Swanson, M. L., Kondolf, G. M., and Boison, P. J., 1989, An example of rapid gully initiation and extension by subsurface erosion; Coastal San Mateo County, California: Geomorphology, v. 2, p. 393–403.

Swanson, S., Franzen, D., and Manning, M., 1987, Rodero Creek; Rising water on the high desert: Journal of Soil and Water Conservation, v. 42, no. 6, p. 405–407.

Thornes, J. B., 1984, Gully growth and bifurcation, *in* Erosion Control; Man and nature; Proceedings of Conference 15, International Erosion Control Association, February 23-24, 1984, Denver, Colorado: Freedom, California, International Erosion Control Association, p. 131–140.

USDA Soil Conservation Service, 1985, Nevada water supply outlook; March 1, 1985: Reno, U.S. Department of Agriculture Soil Conservation Service, 17 p.

U.S. Fish and Wildlife Service, 1980, Summary document; Renewable natural resources, management plan, draft environmental impact statement, Sheldon National Wildlife Refuge, Washoe and Humboldt Counties, Nevada: Portland, U.S. Fish and Wildlife Service Region 1, 14 p.

U.S. Geological Survey, 1974, Surface water supply of the United States 1966-1970: U.S. Geological Survey Water Supply Paper 2127, 1143 p.

U.S. Weather Bureau, 1960, Climates of the states; Nevada: U.S. Weather Bureau Climatology of the United States 60-26, 15 p.

van den Brink, J. W., and Jungerius, P. D., 1983, The deposition of stony colluvium on clay soil as a cause of gully formation in the Rif Mountains, Morocco: Earth Surface Processes and Landforms, v. 8, p. 281–285.

Waring, G. A., 1908, Geology and water resources of south-central Oregon: U.S. Geological Survey Water Supply Paper 220, 86 p.

MANUSCRIPT ACCEPTED BY THE SOCIETY NOVEMBER 14, 1989

Printed in U.S.A.

Chapter 7

Surface and near-surface karst landforms

William B. White
Department of Geosciences and Materials Research Laboratory, Pennsylvania State University, University Park, Pennsylvania 16802

INTRODUCTION

Karst landscapes are the classic examples of the interface between groundwater hydrology and geomorphology. Karst landscapes are those in which dissolution predominates over other processes of mass wasting. Differential dissolution of the bedrock produces a characteristic sculpturing, diversion of drainage to subsurface routes, and closed depression features on many size scales. Karst features intrude themselves into human affairs. Highways are difficult to route through karst. Dams in karst often do not hold water. Solution, subsidence, and soil piping create sinkholes, undermine buildings and highways, and disrupt water and sewer lines. Groundwater in karst is erratic in supply and always threatened with pollution because of rapid and effective transport between surface water and groundwater.

Much has been written about karst landforms, especially their description, classification, and geographic distribution. See the books by Sweeting (1972), Jakucs (1977), and Jennings (1985) for descriptions of karst; Bogli (1980) for a speleological perspective; Milanovic (1981) for a hydrogeological discussion with engineering overtones; and White (1988) and Ford and Williams (1989) for overall textbooks. There is also an important series of papers reviewing karst country by country (Herak and Stringfield, 1972) and a recent volume on paleokarst (James and Choquette, 1988).

Karst is often classified in terms of the rock type on which it occurs. For example, there is limestone karst, dolomite karst, gypsum karst, salt karst, karst on silicic rocks, karst-like forms on ice (see Chapter 9, this volume), and suffosional pseudokarst in unconsolidated sediments (Chapters 4 and 5, this volume). Karst has been classified also in terms of its cover, for which the most careful, critical, and detailed analysis is that of Quinlan (1967, 1978). Distinctions were made between karst developed on exposed bare bedrock (naked karst), karst developed beneath a soil mantle (covered karst), karst developed within older bedrock (interstratal karst), and several other types of cover. There has been also a great deal of interest, particularly by European geomorphologists, in classifying karst in terms of climate. It was thought for some time that characteristic suites of landforms could be associated with arctic karst, alpine karst, humid temperate karst, tropical karst, and arid karst.

It appears to the present author that it is more useful to either arrange discussion of karst in descriptive terms based on the dominant landform or in terms of the environment in which the karst processes are operating. Thus, one might speak of doline karst, pavement karst, or cone and tower karst by referring to the dominant landforms. Alternatively, a term such as "alpine karst" includes not only the expected alpine climate but also high relief—a folded, faulted, and fractured bedrock perhaps still undergoing active tectonism—and the effects of present or past alpine glaciation.

To get beyond the stage of description and classification, karst landscapes have been modeled in terms of the chemistry of dissolution and the physics of fluid flow, and by various statistical and morphometric approaches. The geochemistry/hydrodynamics approach has been most frequently applied to caves, conduit drainage systems, and other underground aspects of karst hydrology (see Chapter 8, this volume). Mostly the approach to surface landforms has been statistical and morphometric (e.g., Williams, 1971, 1972a, b; White and White, 1979; Kemmerly, 1982; Troester and others, 1984).

The objective of the present chapter is to extend geochemical and flow-hydraulic reasoning to surface landforms. The problem is made more difficult because of the direct influence of the soil mantle, plant cover, human land use, and the much more direct impact of changing seasons and climatic regimes.

The discussion in this chapter deals entirely with surface and vadose zone karst features on limestone and dolomite. Specifically excluded are gypsum and salt karst, interstratal karst, paleokarst, various karst-like forms on ice and unconsolidated sediments, and the pseudokarstic features of glacial and volcanic terrains. The chapter deals with the geomorphic effects of surface waters, infiltration waters, and waters in the vadose zone. The geomorphic implications of the water table and lateral flow above and below the water table are dealt with in Chapter 8.

KARST LANDSCAPES

The sections that follow review briefly the main types of karst landscapes, beginning with the transition of fluvial land-

Figure 1. Holokarst. Limestone mountains north of Dubrovnik, Yugoslavia. All drainage is internal through closed depressions or open fractures.

scape morphology to karst landscape morphology. Only the most common types of landforms are included here.

Holokarst and fluviokarst

Jovan Cvijic, whom many would claim as the father of karst geomorphology, was much impressed with the karst landscapes of Yugoslavia. Ten thousand meters of carbonate rock in a tectonically active region has yielded closed depressions, underground drainage, and other karst landforms on a dramatic scale rarely exceeded on the entire planet. The Dinaric Karst (Fig. 1) is a chaotic landscape of sculptured limestone and dolomite, with many closed depressions on a vast array of size scales but with little evidence for a previous fluvial history. This, Cvijic called holokarst. Most karst landscapes, however, are constructed on carbonate rocks from 100 to 1,000 m thick, usually less than the topographic or structural relief of the drainage basin. In these more typical settings, the outcrop area of carbonate rock does not span the entire drainage basin, and there is always some influence of surface water flow and some evidence of the superposition of a fluvial landscape. Fluviokarst is the norm.

Some areas are fluviokarst on a basin scale but are completely dominated by karst drainage and landforms on a local scale. Other areas of fluviokarst are a landscape of ordinary valleys and surface drainage channels in which the downcutting valleys have intersected carbonate rocks with resulting diversion of some of the drainage underground. At topographic scale, the landscape retains it valley form, but in detail many dry valleys, blind valleys, and valley sinkholes are evident.

Sculptured bedrock surfaces

Sometimes called pavement karst, areas of exposed carbonate bedrock occur where glacial activity or some other process has stripped the mantling soil or rock debris. These areas vary in size from a few hundred meters on a side to many square kilometers. In general, patches of pavement karst are smaller than karst areas dominated by other landforms. Pavement karsts are characterized by solutionally widened joints and fractures (Figs. 2 and 3). In glaciated regions the bedrock is often planed nearly flat but

Figure 2. Pavement karst, Malham Cove, Yorkshire, England. Limestone was exposed by glacial scour and has been dissolved into a network of solutionally widened fractures (kluftkarren) and intermediate sculptured blocks. The blocks in turn are sculptured with small solution runnels that drain into the fractures.

Figure 3. Pavement karst, Pyrenees Mountains on Spanish/French border north of Val Roncal. The pavement is cut by solutionly widened fractures with vertical grooving.

Doline or sinkhole plain karst

Sinkholes are the most common landform type in soil-mantled, temperate-climate karst regions. The karst areas of southern Indiana, central Kentucky, parts of Missouri, the Highland Rim of Tennessee, some parts of the Appalachian Plateaus and Appalachian Valleys, north Florida, and many others are made up of a rolling sequence of closed depressions (Fig. 5). There has been a traditional picture of temperate-climate sinkholes as an otherwise flat plain with an occasional bowl-shaped depression indented into it. Such occur in a valley relation where the internal drainage through sinkholes is superimposed on antecedent fluvial drainage. However, in the sinkhole plain relation there is often no longer a readily discernible fluvial record. Although there may be flat areas—or areas that on topographic

etched with an array of small landforms usually termed "karren" (Bogli, 1960, 1980), which have been examined in some detail in terms of structural and hydrodynamic controls.

On tropical limestone surfaces, the bedrock is frequently sculptured into sharp, jagged pinnacle surfaces referred to as "spitzkarren." Pinnacle surfaces occur in a range of sizes, from minor jagged surfaces with a few centimeters of relief, through examples in Puerto Rico of pinnacles a few meters in height (Fig. 4), to examples in south China and Malaysia of pinnacles tens of meters in height. The most extreme example is near the summit of Mount Kaijende, New Guinea (Jennings and Bik, 1962), where pinnacles reach a height of more than 100 m.

The dominant chemical process in these terrains is the direct solution of the carbonate rock by rainfall. Sheet wash and runoff from the bare bedrock surface into the vertical fractures dissolve pathways through the vadose zone. These continue to enlarge until all that is left is the core of the original bedrock block, now standing in relief as a pinnacle.

Figure 5. Shallow dolines (sinkholes), Burkes Garden, Virginia.

maps are not bounded by closed depression contours— examination of these areas on the ground shows that in almost all cases any overland flow from these interdepression areas would, in fact, be internal runoff into some adjacent sinkhole. The speckled pattern of closed depressions is to a large extent an artifact of the relatively low relief compared with the usual contour interval of topographic maps. Many shallow swales and other closed depressions exist with depths that do not exceed one contour interval.

Doline karsts lend themselves more easily to morphometric analysis than do most other karst landscapes. Closed depressions often occur in large numbers but on a scale where they can be easily counted or measured on topographic maps or air photographs. Cramer (1941) long ago introduced the idea of a depression density, simply a count of the number of closed depressions per unit area. Williams (1971, 1972a, b) argued that the divides between large closed depressions in tropical regions were more significant than the depressions themselves and, thus, proposed the concept of polygonal karst as a basis for discussing patterns of closed depressions. White and White (1979) used the catchment

Figure 4. Pinnacle karst (spitzkarren) near Mayaguez, Puerto Rico.

Figure 6. Isolated limestone tower, Guajataca River Basin, Puerto Rico.

Figure 7. Contrasts in tropical karst development, northern Puerto Rico karst belt. Foreground is underlain with chalky Cibao Formation; towers in the background are developed in the massive Lares Limestone.

Figure 8. Cone karst on Lijiang River near Yangshuo town, Guangxi, south China.

area draining into closed depressions as a geomorphic measure that could be related to hydrologic characteristics of Appalachian karst drainage basins. In a wide variety of settings, the number of dolines of a given depth falls off exponentially with increasing depth, although the parameters of the exponential relation vary somewhat from region to region (Troester and others, 1984).

Cone and tower karst

The classic tropical karst occurs on massive and thick limestones such as occur on the Caribbean islands, parts of the South Pacific, south China, and other areas. There the limestones have been dissected by deep solution to produce a rugged and chaotic topography of deep, nonintegrated gorges separating residual hills, which may take on a cone or tower shape (Fig. 6). The distinction is sometimes made between "cockpit" karst, in which there are deep sinkholes with residual pyramidal hills at the junction of three or more sinkholes, and cone karst, in which arrays of conical or tower-like hills are separated by more or less linear gorges. It is apparent from examination of a number of areas that the distinction between cockpit karst and cone and tower karst is one of degree, and that many gradational forms exist. In certain areas such as the northern karst belt of Puerto Rico the deep ravines separating the karst towers are rugged, with few or no intermediate flat areas. In other areas of tower karst, such as south China and the mogote karst of Cuba, there are intermediate alluvial plains with the limestone towers rising above them. Whether these plains represent the influence of old base levels, former positions of the water table, levels of sedimentary infilling, or some other process has been rather widely discussed. Chinese geologists (Yuan, 1985) distinguish between "peak cluster" (fengcong) for the continuous towers with, at best, narrow gorges, and "peak forest" (fenglin) for the isolated towers separated by flat plains.

Some evidence for the effect of rock lithology on differential solution is found in the north Puerto Rico karst belt (Monroe, 1976). There is a well-developed cone and tower karst on the massive and relatively pure Lares and Aymamon Limestones (Fig. 7). The stratigraphically intermediate Cibao Formation is impure and chalky. The erosion of the Cibao leaves a substantial soil mantle, and a doline karst is developed on the Cibao rather than a cone and tower karst. The landscape on the Cibao Formation is not dramatically different from the doline karsts found in temperate climates.

The cone and tower karst of south China is instructive in comparison with the cone and tower karst in the young limestones of the Caribbean. The karst is formed in folded, faulted, and fractured limestones of Devonian and Mississippian age (Song, 1986; Smart and others, 1986). The geometric forms, by and large, ignore the details of the geologic structure, thus supporting the view that the conical form, particularly, is a direct result of the weathering process and not of rock structure. Geometrically ideal cones (Fig. 8) and towers of various shapes (Fig.

Figure 9. Tower karst on Lijiang River near Xingping town, Guangxi, south China.

Figure 10. Alpine karst. Pyrenees Mountains north of Pamplona, Spain.

Figure 11. Coastal karst on San Salvador Island, Bahamas, showing effects of organisms on corrosion of limestone.

9) exist together; it is apparent that the distinction between cones and towers is one of local detail and not regional significance.

Alpine karst

High alpine regions are usually characterized by sparse vegetation and extremes of seasonal weather conditions. The dissolution of the bedrock often takes place beneath a melting snowpack. The high relief and the seasonal inputs of water during periods of snowmelt result in shafts, sometimes more than 1 km deep, within the mountain mass. Surface landforms consist of a variety of closed depressions, and karren-covered bare bedrock (Fig. 10). Alpine karst occurs in most mountainous regions, including the Alps and the Pyrenees in Europe, the Rockies in Canada (Ford, 1979), and in limited areas in many of the mountains of the western United States, including the Teton Range (Medville and others, 1979), and the Uinta and Wasatch Ranges in Utah (White, 1979).

Coastal karst

Karst development on limestone coastal areas involves processes in addition to inorganic dissolution of carbonate minerals. Wave and wind action serve to keep the bedrock surface wet. In warm and tropical regions, the intertidal and active spray zones are excellent habitat for organisms of many kinds. In large part, the ubiquitous rough pinnacle surfaces on most coastal karst are due to direct attack by organisms (Fig. 11) or are caused by organic acid and carbon dioxide-rich solutions formed through biological processes (Trudgill, 1985). Plant growth, particularly algae, encrusts the carbonate rocks, and larger plants are encrusted by deposited carbonate minerals. The result is a rugged surface of plant-rock composite known as "phytokarst" (see e.g., Folk and others, 1973).

Fresh groundwater mixes with salt water within the karst system in coastal karst. Because of nonlinear relations between ionic strength and activity coefficients, the mixed waters are undersaturated with respect to calcite even if both the groundwater and seawater are saturated. The result is accelerated dissolution in the mixing zone. Back and others (1979) have claimed that crescent beaches and collapse zones near points of groundwater discharge result from mixing solution. There is now some evidence (Mylroie, 1988) that there are two classes of small caves on the young limestone islands of the Bahamas: water-table caves formed near the top of the freshwater lens, and halocline caves formed at the freshwater-saltwater interface.

GEOCHEMISTRY OF KARST PROCESSES

Equilibrium

The thermodynamics and solution chemistry of calcite and dolomite is thoroughly studied and has been reviewed in many places (e.g., White, 1988). For convenience, the usual reactions for carbonate dissolution are listed below.

$$CO_2(gas) \rightleftharpoons CO_2 \text{ (aqueous)}$$
$$CO_2(\text{aqueous}) + H_2O \rightleftharpoons H_2CO_3 \quad \Bigg\} \quad K_{CO_2}$$

$$H_2CO_3 \rightleftharpoons H^+ + HCO_3^- \quad K_1$$

$$HCO_3^- \rightleftharpoons H^+ + CO_3^{2-} \quad K_2$$

$$CaCO_3 \rightleftharpoons Ca^{2+} + CO_3^{2-} \quad K_c.$$

K in the above equations signifies the equilibrium constants for the reactions. For the best numerical values as a function of temperature, see Plummer and Busenberg (1982). Conventionally, the dissolution and hydration reactions for carbon dioxide are combined into a single equilibrium constant. So long as the only source of hydrogen ions is from the carbonic acid, the controlling variable in this sequence of reactions is the partial pressure of carbon dioxide in the environment. P_{CO_2}, in turn, is controlled by such environmental factors as soil type and thickness, plant cover, length of growing season, precipitation distribution, and temperature.

Kinetics

The dissolution of limestone or dolomite in carbon dioxide–rich water is a relatively slow process. The time scale for dissolution is long with respect to the time scale for water movement in the open-conduit permeability of the vadose or phreatic zone of a carbonate aquifer. Water will travel substantial distances before it becomes saturated. We must, therefore, consider kinetic factors as well as the mass transport of solutions in chemical equilibrium in modeling the development of karst landforms.

The atomic-scale mechanism of calcite and dolomite dissolution is surprisingly complex. This is because these minerals have a solubility intermediate between easily soluble rocks such as halite and gypsum, where the rate-controlling mechanism is the diffusion of ions to and away from the dissolving surface, and insoluble rocks such as silicates, where chemical reactions on the mineral surface are rate controlling. There have been many laboratory investigations of carbonate dissolution kinetics, of which the modern investigations begin with the work of Berner and Morse (1974). Variables investigated include temperature, carbon dioxide partial pressure, acidity, ionic strength, and flow velocity. Materials include both calcite and dolomite as powders, bulk rocks, and single crystals. See for example papers by Plummer and Wigley (1976), Sjöberg (1976), Plummer and others (1978), Keir (1980), Massard and Desplan (1980), Busenberg and Plummer (1982), Girou and others (1982), Compton and Daly (1984), Herman and White (1985), Baumann and others (1985), Buhmann and Dreybrodt, (1985a, b), and Rickard and Sjöberg (1983).

Many different rate equations have been proposed but the one most widely used in karst investigations is from Plummer and others (1978). Their equation is:

$$\text{Rate} = k_1 a_{H^+} + k_2 a_{H_2CO_3} + k_3 a_{H_2O} - k_4 a_{Ca^{2+}} a_{HCO_3^-}$$

where k_1 through k_4 are reaction rate constants and a signifies activities of the ions indicated. The equation identifies three dissolution mechanisms, each with its own forward rate constant. The back reaction rate constant can be derived from the forward rates:

$$k_4 = K_2/K_c \{k_1' + 1/a_{H^+}(s) [k_2 a_{H_2CO_3} + k_3 a_{H_2O}]\}.$$

$a_{H^+}(s)$ is the hydrogen ion activity at the reaction surface. k_1' is the effective forward rate constant for the direct attack of hydrogen ions at the surface and is taken by Plummer and others (1978) to be $k_1' = 10\,k_1$.

In common with most other rate equations, the Plummer and others equation does not work particularly well near equilibrium. It does describe precipitation reactions on the supersaturation side of the calcite solubility curve (Reddy and others, 1981). For further comparisons of rate process and comparisons with laboratory data, see Plummer and others (1979) and Dreybrodt (1988).

The Plummer and others rate equation can be dissected by calculating the individual rate processes separately (Fig. 12). The system is assumed to be open to carbon dioxide of specified partial pressure. The values chosen (log P_{CO_2} = –1.5, –2.5, and –3.5) are typical of soils in tropical and temperate climates with and without heavy rain, and bare limestone bedrock, respectively. (Tropical and temperate soil measurements mostly fall in the range 10^{-2} to 10^{-3}; in either environment, P_{CO_2} can be boosted to $10^{-1.5}$ after heavy rain.)

Carbonic acid was assumed to be the only source of hydrogen ions. Because the rates are plotted as a function of saturation index rather than pH, there is a different R_1 curve for each carbon dioxide partial pressure. Because the reactions occur at fixed P_{CO_2}, R_2 is independent of saturation index and there is a horizontal line that is a function only of P_{CO_2}. The third forward rate term is a constant because the activity of water can be taken as unity in groundwaters. The back reaction, R_4, has only a weak dependence on carbon dioxide partial pressure. The net rate curve, given by the heavy line in Figure 12, is obtained by adding the individual rate terms:

$$\text{Rate} = R_1 + R_2 + R_3 - R_4.$$

Somewhat surprisingly, the dissolution rate of calcite is almost independent of carbon dioxide partial pressure over the range of undersaturation typical of vadose zone karst processes. Also running counter to intuition, the forward rate is driven dominantly by R_3. Only at carbon dioxide pressures higher than those generally found in karst systems does the R_2 term have a greater value than R_3. The numerical value of the carbon dioxide pressure determines the saturation concentration of dissolved limestone (and dolomite) but has little influence on the dissolution kinetics.

Near equilibrium the dissolution rate decreases sharply. The

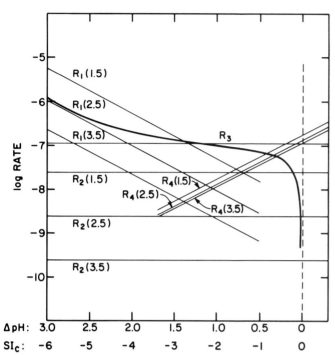

Figure 12. Dissolution rates as a function of saturation index, SI_c, and the shift in pH from the equilibrium value. Temperature = 10°C. R_1, R_2, R_3, and R_4 are calculated from the Plummer and others equation. Numbers in parenthesis are $-\log P_{CO_2}$. Rate is given in millimoles/cm²/sec.

inflection on the net rate curve occurs at about $SI_c = -0.3$, which is equivalent to approximately 50 percent saturation. This may be taken as a boundary between a regime of fast kinetics in undersaturated water and a rapidly decreasing rate in nearly saturated waters.

Kinetics and process time scales

Water flows over the land surface, infiltrates into the soil, migrates through the subcutaneous zone, and flows down joints, open fractures, solution chimneys, and vertical shafts to the water table. Each process has a characteristic time scale. What makes carbonate-dissolution kinetics important to the development of karst landforms is that the various chemical reactions also have characteristic time scales, which are often of the same magnitude as the time scales for fluid flow. If the kinetic time scale for a particular reaction is short compared with the residence time, then reactions will approach equilibrium and mass transport can be described in terms of equilibrium models. If the kinetic time scale is long with respect to residence time, then only the forward reaction terms need be considered and again the calculations are simplified. When the time scales for kinetics and mass transport are of the same magnitude, detailed analysis becomes complicated.

The dissolution of calcite can be expressed as a characteristic response time most easily by examining actual dissolution data (Fig. 13). When water saturated with carbon dioxide comes into contact with calcite, the uptake of calcium in solution at first proceeds rapidly, and then progressively more and more slowly as the solution becomes saturated. The calcium and bicarbonate ion contents of the solutions can be expressed as a saturation index, and it can be seen that about 100 hr is required for the solution to reach $SI_c = -0.3$. Groundwater flow processes that are rapid on this time scale will contain undersaturated water.

Most dissolved carbon dioxide is present as molecular CO_2. For chemical reactions to proceed, the molecular CO_2 must first be hydrated to carbonic acid:

$$CO_2 \text{ (aqueous)} + H_2O \rightleftharpoons H_2CO_3.$$

The rate of this reaction is known (Kern, 1960) and is described by the equation:

$$d[CO_2]/dt = k_{CO_2} [CO_2].$$

The rate constant $k_{CO_2} = 0.03$ sec^{-1} at 25°C. Thus the characteristic reaction time for the hydration of carbon dioxide is 30 sec. This time is too short to be an important rate factor for most groundwater movement in the phreatic zone but it is long compared to the travel time of water films moving down the sides of vertical shafts or in sinkhole drains and open fractures. Likewise, the flow of runoff over limestone pavement or down the sides of cone and tower karst is rapid compared with the CO_2 hydration rate. It is easy to find circumstances in the surface environment and in the vadose zone where the hydration of carbon dioxide may be rate controlling.

Mass transport kinetics and the uptake of carbon dioxide

The subcutaneous or epikarstic zone is, in fact, an extremely complex chemical system. The carbon dioxide is generated in situ

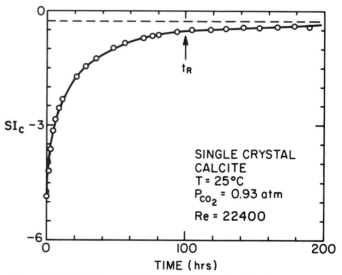

Figure 13. Dissolution of single crystal calcite showing time to 30 percent saturation. Data from Herman (1982).

in the soil mainly by decay of organic material and from plant roots, but in different amounts at different depths. When rainfall occurs, infiltrating groundwater percolates vertically downward through the unsaturated soil guided by the form of Darcy's Law appropriate to unsaturated vertical flow. Meanwhile the CO_2 generated in the soil atmosphere is migrating vertically upward, and a substantial amount of the soil CO_2 is lost to the surface atmosphere. Within the subcutaneous zone are three interlocked rate processes: two of mass transport and one of chemistry.

The vertical infiltration of stormwater is described by Darcy's Law for unsaturated flow (Philip, 1969),

$$\bar{v} = -K_s \nabla \phi$$

where $\bar{v}$ is the flow velocity, K_s is the hydraulic conductivity of the soil, and $\nabla \phi$ is the potential gradient. The controlling parameter is the hydraulic conductivity of the soil.

The infiltrating stormwater dissolves CO_2 from the soil atmosphere at a rate given by (House and others, 1984):

$$d[\text{dissolved carbon}]/dt = A/V \, k_T \{[CO_2]_s - [CO_2]_b\}.$$

[dissolved carbon] refers to the sum of all aqueous carbon species. A is the gas/liquid surface area, and V is liquid volume. k_T is the transfer velocity for CO_2 transfer across the liquid/gas interface. $[CO_2]_s$ and $[CO_2]_b$ are the concentrations of dissolved carbon dioxide at the liquid/gas interface and in the bulk solution, respectively.

The second mass transfer process is the upward migration of CO_2 to the atmosphere. It is assumed that this is a diffusive process; that is, we neglect CO_2 transport by convection or channelled gas flow in open cracks or channels in the soil. The transport rate should then be described by Fick's laws, which for plane parallel boundaries can be written as:

$$Q = -D_s \, d[CO_2]/dz.$$

The CO_2 transport is controlled by the diffusivity, D_s, which in turn depends on the permeability of the soil (deJong and Schappert, 1972).

The transport processes can be documented by measurements on the CO_2 partial pressures in the gas phase, on calculations of CO_2 partial pressures in soil water collected in soil lysimeters, and on calculations of CO_2 partial pressures in waters collected in cave streams and karst springs. The partial pressure of CO_2 in the soil water and deeper in the karst system is lower than that measured in the gas phase in the soil. In mass balance terms, CO_2 is partitioned between the upwardly migrating gas stream and the downwardly migrating infiltrating stormwater. The carbon dioxide concentration in the soil and its respiration into the atmosphere goes through a seasonal cycle, with high values in the summer and low values in winter (Edwards, 1975) in temperate climates. However, in warm climates P_{CO_2} is usually maximized in the wet season and minimized in the dry. Thus, in Mediterranean climates, karst P_{CO_2} is probably at a maximum in winter (Derek Ford, personal communication, 1989). A similar cycle is seen in cave atmospheres (Troester and White, 1984).

KARST ON A REGIONAL SCALE: DENUDATION RATES

Measurements of karst denudation

Because dissolution is the most important process for landscape sculpturing in karst, measurements of dissolved load combined with water balance calculations permit an estimate of the overall loss of bedrock, a quantity generally known as "karst denudation" (Jennings, 1983). Karst denudation is usually expressed in terms of a lowering of the land surface in units of mm/kyr (Fig. 14).

Figure 14. Sketch of land-surface lowering that defines the karst denudation concept.

Karst denudation can be calculated by delineating a drainage basin that can be demonstrated to discharge through a known outlet point, measuring the runoff at the outlet point, and through chemical analyses of the water, determining the dissolved load. Other approaches also use chemical analyses of the runoff water to measure dissolved load but use precipitation measurements, corrected for evapotranspiration losses, to determine water throughput. The concentration of dissolved carbonates varies from place to place, and a good deal of effort has gone into explanations for these variations in terms of geologic and climatic variables.

Most systematic of the early studies of karst denudation was that of the French geomorphologist, Jean Corbel, who traveled extensively in Europe and North America analyzing water samples and trying to fabricate a global climatic model for karst landscape development (Corbel, 1957, 1959). By applying what in hindsight seems extremely naive geochemical reasoning, Corbel concluded that the highest rate of chemical denudation takes place in cold climates—arctic and alpine environments—where low temperatures promote higher concentrations of dissolved carbon dioxide. The low-relief pavement karst with minor karren that is the characteristic landscape of arctic and alpine environments was argued to be due to the fact that development of these landforms had taken place since the melting of the last ice sheet little more than 10,000 years ago. In contrast, he thought that the dramatic and high-relief cone and tower karst characteristic of many tropical environments had been developing since the limestones emerged from the sea in early Cenozoic or even Mesozoic time.

Corbel's interpretations sparked considerable controversy, triggered an immense amount of work by others, and also gave the entire subject more of a climatic focus than it probably deserved. The voluminous literature is reviewed in most of the textbooks cited at the beginning of this chapter.

Smith and Atkinson (1976) and Atkinson and Smith (1976) critically evaluated the denudation data available at that time. They found that denudation rates could be grouped within various climatic regions but that runoff was the most important variable. Their results were expressed in the three regression equations:

$$D = 0.063 R + 5.7 \quad \text{Tropical}$$
$$D = 0.055 R + 7.9 \quad \text{Temperate}$$
$$D = 0.036 R + 7.4 \quad \text{Arctic/Alpine}$$

where D is denudation rate in mm/kyr and R is annual runoff in mm/yr. The distinction between the cold climate denudation rate and temperate and tropical climate denudation rates appears to be real. Although the denudation rate in tropical climates is higher than that in temperate climates, the small difference is likely not significant, given the large scatter in the data. This does not take us very far in providing a climatic basis for karst geomorphology and is even less helpful in explaining the factors controlling the development of the distinctive karst landforms.

The equilibrium model

Much of the climatic influence on the average denudation rate can be explained with a rather simplistic geochemical model (White, 1984). The solutional lowering of carbonate rock terrains is taken to be mostly the result of chemical reaction between calcite (or dolomite) and carbonic acid. If it is assumed that the carbonate reactions proceed to equilibrium, then a combination of mass balance and equilibrium chemical arguments lead to a theoretical expression for the denudation rate:

$$D = \frac{100}{\rho \sqrt[3]{4}} \left(\frac{K_c K_1 K_{CO_2}}{K_2} \right)^{1/3} P_{CO_2}^{1/3} (P - E)$$

where ρ is the density of the bedrock in g/cm^3, D is denudation rate in mm/kyr, and (P – E) is precipitation minus evapotranspiration in mm/yr. K signifies the equilibrium constants defined above.

Calculations of saturation indices show that much karst runoff is not saturated, that many karst springs have saturation indices in the range of –0.2 to –0.1, which would correspond to waters that are 60 to 80 percent saturated. For those karst drainage basins in which groundwater flow is through open conduits, the equilibrium equation overestimates the denudation rate by these percentages. However, agreement between the equilibrium equation, which has no adjustable parameters, and observed denudation rates was found to be rather good.

The equilibrium equation also accounts for many of the field observations on denudation rate. Denudation rate is predicted to vary linearly with runoff (or precipitation minus evapotranspiration), as exhibited by Smith and Atkinson's (1976) linear regression lines. However, denudation rate is predicted to vary only with the cube root of the carbon dioxide partial pressure. Although P_{CO_2} varies by about a factor of 100 from the atmosphere above bare bedrock pavement karst ($10^{-3.5}$) to maxima in the range of $10^{-1.5}$ in warm humid soils, the corresponding increase in denudation rate is only about a factor of 5. The contrast in CO_2 pressure from temperate to tropical soils is much smaller, and there is a corresponding lack of contrast between the two environments in overall denudation rate.

The factor of temperature on which Corbel placed such great emphasis turns out to be the least important climatic variable in the equilibrium calculation. Temperature appears only in the temperature dependence of the equilibrium constants that enter the equations. It is not merely the solubility of carbon dioxide that is important, but the product of a series of equilibrium constants, each with its own temperature dependence.

Environmental controls

The slope coefficients in the linear regression equations can be set equal to the coefficient of the equilibrium denudation equation and, with an assumption of temperature, the carbon dioxide pressure can be calculated (Table 1). The results are in remarkable agreement with observed carbon dioxide pressures. As regional averages, the arctic/alpine value is just a little above the atmospheric background, as would be expected from terrane consisting mostly of bare bedrock with some patches of moss, soil, and vegetation. The temperate-climate values are close to what has been observed in well waters and seepage waters (Langmuir, 1971; Drake and Harmon, 1973). The calculated carbon dioxide partial pressure for tropical areas is in the range of the sparse measurements that have been made.

Although the various climatic factors enter into the equilibrium model, and although the chemistry of the subcutaneous zone is extremely complex in detail, these factors are subdued by the mathematical relation and also by a tendency for variations to average out on basin or regional scales. In the final analysis, the rate of regional lowering of the karst land surface depends primarily on runoff, as Drake (1984) demonstrated from various field data.

Runoff itself is highly variable in most regions. It is well established that dissolved load is diluted by high discharges. However, it is the product of discharge and dissolved load, the flux of dissolved bedrock, that determines the overall mass wast-

TABLE 1. CARBON DIOXIDE PRESSURES CALCULATED FROM DENUDATION RATES

	Arctic/Alpine	Temperate	Tropical
Assumed mean temperature	0°C	10°C	20°C
Regression coefficient*	0.036	0.055	0.063
Calculated P_{CO_2} (atm)	$10^{-3.07}$	$10^{-2.31}$	$10^{-1.93}$

*From Smith and Atkinson (1976, Fig. 13.5).

ing of a karstic drainage basin. The flux of transported limestone bedrock in units of cm³/sec is given by:

$$F = Q_i H_i / \rho A$$

where Q_i and H_i refer to "instantaneous" discharge and hardness.

Most data in the literature give only changes in dissolved carbonate concentration through seasonal variations in runoff or through changes in discharge in the course of a storm event. Data are rarely available to determine the flux in response to different runoff conditions. However, by a stroke of good fortune, the small karstic basins of Thompson Spring, a diffuse flow system, and Rock Spring, an open-conduit drainage system, were instrumented for discharge and regularly sampled for chemical analysis during an 8-month period that included the Hurricane Agnes storm, which swept across Pennsylvania in June 1972. From these data (Jacobson, 1973; Jacobson and Langmuir, 1974), it is possible to calculate the flux of dissolved carbonate rock being removed from these two small basins (Fig. 15). What is immediately obvious from Figure 15 is that, although the concentration of dissolved carbonate is diluted during storms, the decrease in concentration is more than compensated by the increased discharge. The net result is a large increase in the rate of denudation during high-runoff events. Estimates of denudation rates based on spot measurements of uninstrumented drainage basins are likely to be somewhat low, because data are generally not collected during extreme storm runoff.

Karst surfaces

A much-discussed (and maligned) landscape feature is the erosion surface. This is generally taken to be the record of an old stable base level, which allowed a stream valley to widen, receive a veneer of alluvial material, and then be preserved when valley deepening resumed. Terrace levels and upland erosion surfaces are often assumed to be locked in elevation, i.e., the lowering of the erosion surface by chemical weathering and mass transport is negligibly small compared with stream erosion in the deepening valley. Because of continuous stream erosion in nonkarstic rocks, old surfaces are usually highly dissected, and recognition of the surface in accordant hill tops and valley interfluves can be subject to much interpretation, debate, and argument.

In contrast, karst surfaces are usually well preserved. Because of internal drainage, there is no dissection of the surface by downcutting authigenic surface streams, although allogenic streams are common. As a result, limestone valleys and limestone uplands often have a "peneplain" character although they may be dotted with sinkholes. However, there is a spatially averaged lowering of the surface by chemical denudation and it is not correct to assume that the karst surface is locked in elevation.

Such a valley upland surface is found in Centre County, Pennsylvania, in the drainage divide area between the Spruce Creek, Penns Creek, and Spring Creek drainage basins. Fifty to 100 m of residual soil occur on the interfluve area. With the assumption that all insoluble residue remains in the soil, a simple mass balance argument demands that 425 m of Gatesburg dolomite must have been removed to produce the observed 50 m of residual material (Parizek and White, 1985). This means that what appears as an upland erosion surface is actually 425 m lower than the initial surface. Applying equilibrium denudation rates to the amount of carbonate rock removed provides an age of 14 Myr (mid-Miocene time) for the initiation of the surface. The present-day relief between the valley uplands and the incised surface stream valleys is about 100 m. Taken as a proportion of the 14 Myr that the valley floor has been deepening, the present-day upland surface and the cave fragments that exist immediately under it should have an age of about 3 Myr, a number of consistent with the late Pliocene to early Pleistocene ages proposed for caves in the Appalachian Highlands.

The continuous and rather rapid solutional lowering of karst plains, while maintaining the planar surface, leads to certain difficulties in fitting these surfaces into the context of the regional geomorphology. Little work has been done on this problem, but two examples serve to illustrate agreement and disagreement.

Agreement is found in the Mammoth Cave area (see White and White, 1989, for detailed description of hydrogeology of the area). The present-day catchment for a regional karst drainage

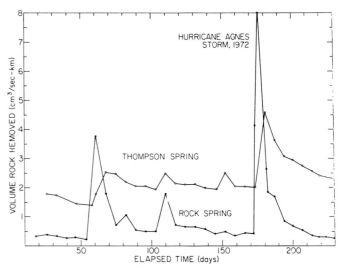

Figure 15. Flux of dissolved carbonate rock removed from Thompson Spring and Rock Spring Basins in Centre County, Pennsylvania. Time zero is January 1, 1972. Fluxes calculated from data of Jacobson (1973).

system is the Pennyroyal Plateau, a doline karst at an elevation of 200 to 220 m. The drainage is entirely subsurface. Active conduits extend beneath the Mammoth Cave Plateau to carry water to a series of springs on Green River, discharging below river level at an elevation of 120 m. The highest abandoned cave levels near Green River are directly beneath the sandstone caprock at an elevation of 215 m. Geomorphic interpretation (Miotke and Palmer, 1972; Palmer, 1989) and paleomagnetic measurements on the clastic sediments in the caves (Schmidt, 1982) agree that the highest cave levels date from late Pliocene or earliest Pleistocene time. Applying a denudation rate of 30 mm/kyr (30 m/Myr) over a nominal 3-Myr period yields 90 m of lowering of the Sinkhole Plain since the time when it was draining through the highest cave levels. This is in excellent agreement (considering the rough character of the numbers) with the spacing between the highest abandoned cave passages and the presently active drainage conduits.

An example of disagreement is the Harrisburg Surface in the Appalachian Highlands. The Harrisburg Surface is identified with karst surfaces many places in the Appalachians, including the Great Valley of Pennsylvania and Maryland, the Big and Little Levels in eastern West Virginia, and the Highland Rim of Tennessee. The denudation rate in the temperate humid Appalachians can again be taken as 30 mm/kyr. However, the Harrisburg Surface is thought to be old, with dates extending from late Eocene to Late Cretaceous (Sevon, 1985). Using ages of 40 to 60 Myr and assuming a constant denudation rate yields 1,200 to 1,800 m of limestone removed while still preserving a recognizable "Harrisburg Surface." These thicknesses of carbonate rock are not available in the Appalachian Plateau locations of the surface, and it seems unlikely in the extreme that the topographic representation of the surface in the Valley and Ridge Province would be preserved while these thicknesses of structurally and lithologically varying rocks were removed.

These examples illustrate a problem that must be faced as karst processes are interfaced with other aspects of geomorphology. Karst processes, in general, often seem to be too fast; both chemical and mass transport kinetics allow caves and surface land features to develop more quickly than is consistent with other geologic evidence.

SUBSURFACE KARST IN THE VADOSE ZONE

Vadose zone hydrology

The problem of geochemical and hydrodynamic modeling of karst is essentially one of mass transport. Limestone is removed preferentially from certain parts of the bedrock mass to etch out blind valleys, closed depressions, gorges, and other features of negative relief. One factor controlling the way in which this is done is the source of water available to do the transporting. Of importance here is allogenic versus authigenic recharge (Fig. 16).

Allogenic recharge is water that enters the karst system from nonkarstic border lands. Surface basins collect runoff from non-

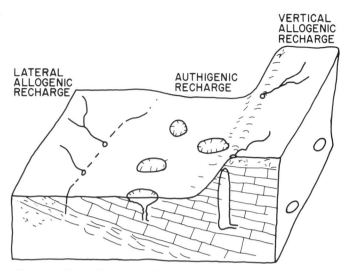

Figure 16. Block diagram showing source of recharge into fluviokarst.

carbonate rocks and then drain laterally into the karst as sinking streams. This recharge source may be called lateral allogenic recharge. In many karst areas of the United States, particularly those of the Cumberland Plateau of Tennessee, Alabama, Kentucky, and West Virginia, and the Plateau area of the Illinois Basin, including the Mammoth Cave area, the carbonate rocks are capped with resistant sandstones. These act as catchments for surface streams that flow off the plateau and enter the karst from the top as a vertical allogenic recharge.

Authigenic recharge is derived from rain that falls directly onto the karst surface. The recharge may enter the subsurface through small fractures and joints and is sometimes seen in caves as drip waters emerging from cracks in the roof where it deposits secondary carbonates—dripstone and flowstone. When rainfall saturates the soils, overland flow into closed depressions enters the subsurface at very localized points, which then rather rapidly enlarge by solution into open pathways. Chemically and conceptually, there is value in making a distinction between diffuse infiltration through tight fractures and intergranular porosity and internal runoff draining vertically through open cavities. This is essentially the same idea as Thrailkill's (1968) vertical seeps and vertical flows.

There has been a tendency, implicit rather than explicit, to regard infiltration through the vadose zone of the karst as an extremely rapid process. This unwarranted assumption no doubt has been influenced by the observation of water streaming down open shafts and large open joints just below the limestone surface. In the past several years (Williams, 1983, 1985), it has been realized that this intuitively obvious notion is usually not correct. If the limestones contain insoluble clastic material, there may be a breakup of the limestone at the base of the soil to form a well-defined C horizon of this rubbly material. In more pure limestones, the soil-bedrock contact is usually very sharp, and limestone rubble is rapidly leached from the soil. However, the

Figure 17. The subcutaneous zone, east of Mammoth Cave National Park, south-central Kentucky karst.

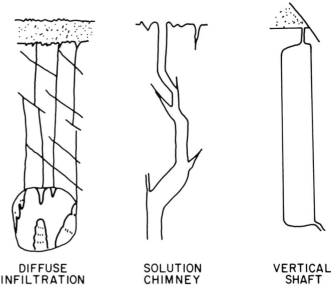

Figure 18. Sketch showing types of vertical drains in the vadose zone. Left: solutionally widened fractures. Middle: solution chimney. Right: vertical shaft.

bedrock surface itself is sculptured into an elaborate network of solutionally widened joints and small channels along bedding planes. These solution openings are usually choked with soil (Fig. 17) and provide a permeable zone for temporary storage of perched groundwater in what Williams refers to as the "subcutaneous zone," or what French workers call the "epikarstic zone."

Throughput of water from its origin as rainfall to its appearance in the open-conduit system of the karst may be delayed for periods of days to weeks because of hold-up in the subcutaneous zone. This plays a role both in the interpretation of the hydrology, particularly the water balance, and also in the geochemical modeling of the karst landforms.

Some understanding has been achieved on the hydrologic function of different flow paths in the subcutaneous zone, including evidence for extensive lateral flow in some instances (Gunn, 1981, 1983). Tracer experiments show that the flow paths are complex, branching, and have greatly varied travel times (Friederich and Smart, 1981). The geochemistry also varies with flow path and with magnitude of discharge along any given flow path (Kogovsek and Habic, 1980; Kogovsek, 1981).

The deep drainage in the vadose zone is through several types of permeability (Fig. 18). If the water has time to become saturated with respect to calcite in the reaction zone at the base of the soil, it moves to the subsurface through fractures and joints with little additional reaction with the wall rock. Joints and fractures carrying saturated waters are at best only slightly enlarged by solution. Undersaturated waters may enlarge the pathway while maintaining the overall geometry of the original joint and bedding plane sets. The aperture of these vertical pathways varies from a centimeter or less, through typical sinkhole drains of 10 to 20 cm, to solution chimneys of a size large enough for human exploration. When the path becomes sufficiently open that the hydrodynamics of vertically moving water films becomes dominant, the vertical shaft appears (Brucker and others, 1972). The vertical distance between the bottom of the subcutaneous zone and the top of the saturated zone (water table) ranges from a few meters in low-relief karst to more than 1,000 m in some high plateaus. The vadose zone drainage often takes the form of vertical shafts from 1 to 10 m in diameter and from 10 m to several hundred meters deep (Fig. 19).

Thus, in addition to the surface landforms of karst areas, there are subsurface contemporaneous landforms that are linked to the caves and conduit systems at depth. The vadose zone cavernous features, such as solution chimneys and vertical shafts, are in-feeders for the underground drainage system. The inputs are linked to the contemporary landscape. Their outlets are through cave systems that may be much older. Surface catchments and subsurface drains together make up the groundwater flow system in the vadose zone.

Initiation of vertical flow paths: Kinetic thresholds

The chemistry of vertically moving waters can be sampled where it moves in thin sheets down the walls of chimneys and shafts, and it can be sampled as drip water from cave ceilings. The contrast is remarkable (Fig. 20). The waters were collected from a variety of sites in Virginia (Burnsville Cove), Kentucky (Mammoth Cave area), West Virginia (Swago Creek Basin, Pocohontas

Figure 19. Vertical shaft in Newberry Cave, Virginia. Photo courtesy of Ronald Simmons.

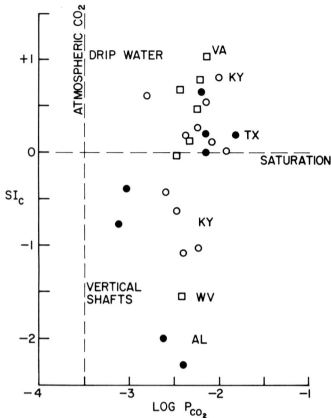

Figure 20 Comparison between chemistry of drip waters from cave ceilings with waters sampled from the walls of vertical shafts. Points at and above the saturation line are drip waters; those below the line are shaft waters. Symbols refer to the states where samples were collected as marked.

County), and Texas (various sites along the Balcones Escarpment), analyzed, and the saturation indices and carbon dioxide partial pressures calculated. Drip waters are saturated to highly supersaturated. Shaft waters range from slightly to highly undersaturated. The carbon dioxide partial pressures span a relatively narrow range of values between 0.003 and 0.01 atmospheres. Yet both sets of waters are derived ultimately from the transient aquifer in the epikarstic zone. The strong contrast in chemistry suggests a threshold in the solution of carbonate rock in the vadose zone.

Examination of solutionally widened joints at the base of the soil is instructive (Fig. 21). As the original joint is followed downward, the widening due to solution abruptly stops but the joint continued downward little modified by solution. The various rate equations proposed to describe the dissolution of calcium carbonate by tradition are cast in terms of the change in concentration with respect to time. However, one can describe equally well the progress of solution in terms of a penetration distance. This is a combination of dissolution kinetics and hydrodynamics, which results in a statement of how far the water will travel before reaching some specified level of saturation. See Dreybrodt (1988) for detailed calculation of the penetration distance for various choices of rate equation.

If the specified level of saturation is set as $SI = -0.3$, which is the value at which dissolution rate begins to decrease rapidly, then the time required for the penetration distance to equal the distance from the base of the soil to underlying cave passages constitutes a threshold for the system. Below the threshold, water, which is nearly saturated with respect to the carbon dioxide partial pressure at the base of the soil (and which may be supersaturated with respect to the carbon dioxide partial pressure of the cave atmosphere) slowly enlarges the joint or fracture following the largely unknown dissolution kinetics that obtain near saturation. As the aperture of the joint slowly increases, the penetration distance also increases, and more highly undersaturated water penetrates farther into the bedrock. When water with a saturation index less than -0.3 penetrates the entire thickness of

Figure 21. A solutionally widened joint in the southern Indiana karst showing the abrupt narrowing with depth.

rock, the particular joint or fracture will begin to dissolve at rates approaching those given by the horizontal regions of the curves in Figure 12. This particular pathway enlarges more rapidly to become a solution chimney or a vertical shaft.

Fractures, solution chimneys, and vertical shafts: Hydrodynamic thresholds

There is a second threshold that is crossed when the solutionally widened fracture becomes large enough to permit transport of soil to the subsurface without choking up. This threshold may be taken as marking the boundary between a solutionally widened fracture and a solution chimney. The development of such soil-transmitting pathways permits soil piping and the development of cover-collapse sinkholes and related land-use hazards in the karst.

A third threshold is crossed with the transition from a solution chimney to a vertical shaft. Water percolates down the sides of fractures, chimneys, and shafts as films only 1 or 2 mm thick (Fig. 22). As the system opens up, these films become thicker and move faster according to the relation:

$$\bar{v} = \frac{g\,\delta^2}{3\,\nu}$$

where δ is the film thickness, g is acceleration due to gravity, and ν is kinematic viscosity, all in some consistent system of units. Measured velocities of flowing-water films in vertical shafts vary from 1 to 10 m/sec (Brucker and others, 1972). At the higher end of the velocity scale the Froude number of the water films becomes greater than unity, and the water goes into supercritical flow. Any projections or irregularities on the wall break up the supercritical flow. Energy release due to the hydraulic jump increases the erosive and corrosive power at the irregularity. As a result, the walls ultimately become streamlined, and the irregular fracture surfaces of the solution chimney are transformed into the nearly cylindrical, vertical-walled structures of the vertical shaft.

Rate of vertical shaft enlargement

From those sparse measurements of the water chemistry at the top and bottom of vertical shafts, it is seen that most shaft waters are highly undersaturated but at carbon dioxide partial pressures typical of cave atmospheres (Fig. 23). There is a small decrease in CO_2 pressure from top to bottom which may indicate some CO_2 loss to the atmosphere as the water film moves down the shaft. Because of the high degree of undersaturation in the most active shafts, the kinetic regime is well into the CO_2-independent horizontal stage of the rate curves in Figure 12. Thus, vertical shaft enlargement can be rather easily calculated by simply assuming that any unit area of limestone wall, bathed in a continuously streaming water film, is dissolved at a rate of approximately 10 to 7 millimoles/cm^2/sec (taken from Fig. 12). Taking into account the density of limestone, this number becomes 0.12 cm/yr of bedrock removed. At this rate, a vertical shaft 10 m in diameter could be dissolved out in roughly 4,000 yr. This would be a maximum rate because it requires a continuous supply of water and does not allow for the insulating effects of clay minerals and other insoluble material in the limestone. Even so, it is apparent that development of major solution openings in the vadose zone would have no trouble keeping up with such surface processes as slope retreat and erosion of protective caprock.

CLOSED DEPRESSIONS

Taxonomy

Closed depressions are the characteristic features of karst landscapes. The discussion here is concerned only with the smaller depressions, those called sinkholes or dolines in the literature. Large closed depressions such as the poljes of the Adriatic karst usually are the result of many different processes operating simultaneously in different parts of the depression, and generic discussion is more difficult. Closed depressions are important in karst hydrology because to a great extent they represent the primary inlet points in the karst drainage system. Each closed depression has an associated catchment area from which precipitation is captured and focused as internal runoff into the drain in the depression.

Sinkholes or dolines have three components: (1) the bowl-like depression dissolved into the underlying bedrock, (2) a mantle of soil or other insoluble material draped over the bedrock basin, and (3) a drain connecting the depression to the conduit drainage system in the subsurface. There are three processes operating in the evolution and development of closed depressions: (1) dissolution of the carbonate bedrock in the basin and drain; (2) transport of soil to the subsurface by in-wash, piping, and collapse; and (3) mechanical breakdown of bedrock either by direct collapse of shallow cavities or by upward stoping from deep cavities. Auxiliary controlling hydrologic and geologic factors include: the input of water, controlled by regional precipitation and size of catchment, and input of clastic material, determined by thickness of residual soils and the presence of clastic overburden of alluvium, colluvium, glacial tills, volcanic ash, or other material. Each of the operating processes is a function of time. Any given sinkhole is seen at an intermediate point in the time sequence of development, and quite different forms appear depending on the relative importance of the competing processes.

Of the three processes, stoping and collapse are not easily described by geochemical and mass-transport models. The process depends on the preexisting cavities in the rock, which themselves depend on antecedent processes and on the lithology, degree of induration, and fracturing. Stoping and collapse is also a threshold process. There is no movement of material until some critical weakness is developed, and then mass movement occurs very rapidly, to be followed by another period of quiescence. The present discussion is limited to the process of dissolution of the bedrock.

Catchment cells and bedrock depressions

As a basis for modeling solution sinkholes, suppose there exists a massive bedrock, uniformly fractured on a regular grid (Fig. 24). The intersections of the fractures will be loci of high vertical permeability and will provide the preferred pathways for infiltration water moving to the subsurface. Therefore, they will be the initiation sites for sinkhole development. By drawing bisectors to the fracture lines, it is possible to subdivide the surface into catchment cells, each defining an area from which runoff would reach the fracture intersection at its center. Assume that the only inlet is at the fracture intersection, and all runoff from within the cell reaches it. Depending on how the runoff water interacts with the underlying carbonate bedrock and the characteristics of the soil cover, if any, it is possible to perform some rough calculations on sinkhole development.

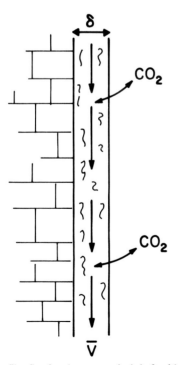

Figure 22. Water film flowing down a vertical shaft with velocity, V, and thickness delta.

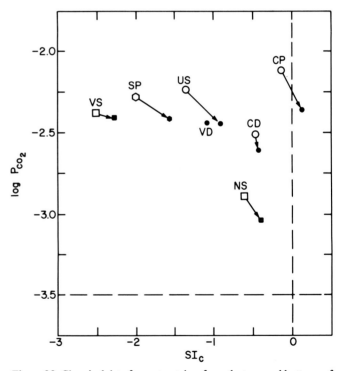

Figure 23. Chemical data for waters taken from the tops and bottoms of some vertical shafts in Kentucky, West Virginia, and Alabama. Data from Brucker and others (1972). Letters indicate name of shaft: CD = Colossal Dome, CP = Cascade Pit, VD = Vaughns Dome, and US = unnamed shaft, all in Mammoth Cave System, Kentucky. NS = Neversink shaft and VS = Valhalla shaft in Jackson County, Alabama. SP = Swago Pit, Pocohantas County, West Virginia.

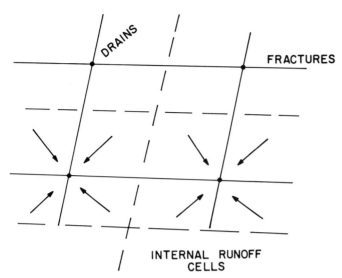

Figure 24. Catchment cells bounding points of high vertical permeability as the initiation points for sinkholes.

Mass-balance arguments permit the calculation of the shape of an idealized sinkhole. Within the boundaries defined by the catchment cell, it is assumed that the rate of attack on the limestone bedrock by internal runoff is a constant defined by the uptake of carbonate into solution, Δc (the internal runoff is taken to be highly undersaturated). The volume of flow available to dissolve limestone increases radially from the boundary of the catchment cell to the drain. The rate of vertical lowering of the limestone mass is given by:

$$z = \frac{\pi \Delta c}{\rho} (R^2 - r^2)(P - E)$$

where ρ is the rock density, R is the radius of the catchment cell, r is the radius of the sinkhole, and $(P - E)$ is precipitation less evapotranspiration. The model is crude both in terms of the assumption of constant chemistry, and in ignoring lithologic variations and the effect of soils in retaining and redistributing internal runoff. It does predict a generally parabolic shape.

CONE AND TOWER KARST

Differential solution along fracture systems

Cone and tower karst has often been equated with "tropical karst" because most of the known regions of cone and tower karst are in tropical environments. Cone and tower karst forms on a variety of geologic substrates (Williams, 1987). The discussion here is concerned only with cone and tower karst on massive limestones. Before considering possible geochemical mechanisms for these striking landforms, certain geologic criteria must be met.

(1) Cone and tower karst typically has a relief ranging from 100 to 1,000 m. The limestones must be at least this thick to provide adequate bedrock for the development of the landscape.

(2) The best examples of cone and tower karst have formed on massive pure limestones. These leave little residual soil and provide little clastic load for the subsurface drainage. Note particularly the contrasting landscape development on pure limestones and chalky limestones in Puerto Rico as described earlier.

(3) Well-developed fracture systems that provide good primary pathways to the subsurface seem to be important. Interior drainage needs to develop before surface drainage can be organized.

Although much cone and tower karst is developed on low-dip young limestones with good primary permeability, these do not appear to be primary criteria. In the south China karst, cones and towers are developed on Devonian and Carboniferous limestones that have been extensively deformed and fractured. Superficially at least, the geometric form of the cones and towers do not reflect the internal structure and lithology, except that vertical fractures appear to play some role in the oversteepening of the towers.

Data are sparse to nonexistent, but it does seem that the differential solution between the tops of the towers and the gorges between the towers that leaves them standing in relief is more one of equilibrium reaction than of differential kinetics.

In temperate climates where silica is not leached from soils, or in geologic settings where the rocks have a large component of insoluble clastic material, the karst surface is mantled with a soil layer that controls and distributes CO_2 production (Fig. 25). In tropical environments, soils tend to be thin, silica and clay minerals are leached, and there is a topography of exposed bedrock alternating with depressions containing soil and organic debris. CO_2 pressures may reach $10^{-1.5}$ or higher. Leaving aside questions of kinetics and whether or not there is a difference in relative rate of reaction between the tower summit and the intermediate gorge, just the equilibrium solubility alone would provide a factor of roughly 5 between limestone solubility on the exposed core of the tower and the solubility at the base of the wet organic-rich mass that accumulates in the gorge.

Taking 100 m as a typical relief in cone and tower karst, the relief can be accounted for by allowing the removal of 20 m of limestone from the tower summits while removing 100 m of limestone from the gorges. This simple-minded calculation assumes uniform precipitation over the karst surface. The generation of relief would be accelerated by slope wash and other factors that would concentrate water into the gorges. Using denudation rates discussed earlier, relief on this order of magnitude could be generated in a few million years. Although there is geomorphic evidence that some regions of cone and tower karst, such as those of south China, are Tertiary or older, these rough mass-transport calculations seem to assure that there are no chemical barriers to landscape development.

CONCLUSIONS

Chemical weathering in the vadose zone is responsible for the characteristic sculpturing of karst landscapes. The rate at

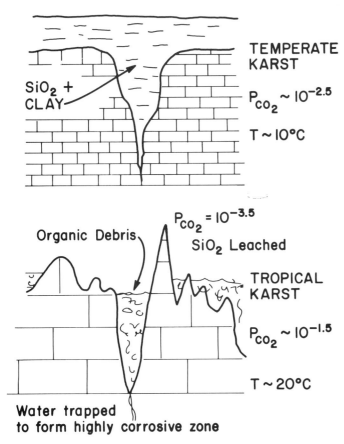

Figure 25. Sketch showing bedrock surface in (A) temperate and (B) tropical karst regions. Typical CO_2 pressures are given.

which this process proceeds is largely determined by mass transport, i.e., by the rate at which water moves through the system. Sculpturing at the soil-bedrock contact is by water, which is highly undersaturated with calcium carbonate, and the rate of reaction is in a chemical kinetics regime that is relatively insensitive to environmental factors.

The overall lowering of carbonate terranes, the denudation rate, can be accounted for reasonably well by assuming that, on a regional scale, most karst runoff comes into chemical equilibrium with the carbonate bedrock. Because equilibrium saturation levels are strongly dependent on carbon dioxide partial pressure, contrasts between arctic, temperate, and tropical denudation rates that are not attributable to differences in available runoff can be accounted for by regional or climatic variations in CO_2 pressure in the soils or at the bedrock contact.

Doline karsts, or sinkhole plains, are preserved because of the interior drainage and are not dissected by authigenic streams. Because of the relatively great lowering rates, the reduction in elevation in karst plains since Tertiary time brings present-day elevations of these features into disagreement with other estimates of the ages of Tertiary erosion surfaces.

Interior features of the vadose zone, such as solution chimneys and vertical shafts, require consideration of chemical kinetics. It is argued that formation of these features is a very rapid process kinetically and that mature vertical shafts can be created in a few thousand years. In contrast, such surface karst features as dolines and cone and tower karst are less influenced by the details of chemical kinetics and can best be interpreted in terms of mainly mass-transfer processes.

REFERENCES CITED

Atkinson, T. C., and Smith, D. I., 1976, The erosion of limestones, *in* Ford, T. D., and Cullingford, C.H.D., eds., The science of speleology: London, Academic Press, p. 151–177.

Back, W., Hanshaw, B. B., Pyle, T. E., Plummer, L. N., and Weidie, A. E., 1979, Geochemical significance of groundwater discharge and carbonate solution to the formation of Caleta Xel Ha, Quintana Roo, Mexico: Water Resources Research, v. 15, p. 1521–1535.

Baumann, J., Buhmann, D., Dreybrodt, W., and Schulz, H. D., 1985, Calcite dissolution kinetics in porous media: Chemical Geology, v. 53, p. 219–228.

Berner, R. A., and Morse, J. W., 1974, Dissolution kinetics of calcium carbonate in sea water; 4, Theory of calcite dissolution: American Journal of Science, v. 274, p. 108–134.

Bögli, A., 1960, Kalklösung und Karrenbildung: Zeitschfift für Geomorphologie, Supplement Band 2, p. 4–21.

——, 1980, Karst hydrology and physical speleology: Berlin, Springer-Verlag, 284 p.

Brucker, R. W., Hess, J. W., and White, W. B., 1972, Role of vertical shafts in the movement of ground water in carbonate aquifers: Ground Water, v. 10, no. 6, p. 5–13.

Buhmann, D., and Dreybrodt, W., 1985a, The kinetics of calcite dissolution and precipitation in geologically relevant situations of karst areas; 1, Open system: Chemical Geology, v. 48, p. 189–211.

——, 1985b, The kinetics of calcite dissolution and precipitation in geologically relevant situations of karst areas; 2, Closed system: Chemical Geology, v. 53, p. 109–124.

Busenberg, E., and Plummer, L. N., 1982, The kinetics of dissolution of dolomite in CO_2–H_2O systems at 1.5 to 65°C and 0 to 1 atm P_{CO_2}: American Journal of Science, v. 282, p. 45–78.

Compton, R. G., and Daly, P. J., 1984, The dissolution kinetics of Iceland spar single crystals: Journal of Colloidal and Interface Science, v. 101, p. 159–166.

Corbel, J., 1957, Les Karsts du Nord-Ouest de l'Europe: Institut des Études Rhodaniennes de l'Université de Lyon Mémoires et Documents, v. 12, 541 p.

——, 1959, Vitesse de l'erosion: Zeitschrift für Geomorphologie, v. 3, p. 1–28.

Cramer, H., 1941, Die Systematik der Karstdolinen: Neues Jahrbuch für Mineralogie, Geologie, und Paläeontologie, v. 85, p. 293–382.

deJong, E., and Schappert, H.J.V., 1972, Calculation of soil respiration and activity from CO_2 profiles in the soil: Soil Science, v. 113, p. 328–333.

Drake, J. J., 1984, Theory and model for global carbonate solution by groundwater, *in* LaFleur, R. G., ed., Groundwater as a geomorphic agent: Boston, Massachusetts, Allen and Unwin, p. 210–226.

Drake, J. J., and Harmon, R. S., 1973, Hydrochemical environments of carbonate terrains: Water Resources Research, v. 9, p. 949–957.

Dreybrodt, W., 1988, Processes in karst systems: Berlin, Springer-Verlag, 288 p.

Edwards, N. T., 1975, Effects of temperature and moisture on carbon dioxide evolution in a mixed deciduous forest floor: Soil Science Society of America Proceedings, v. 39, p. 361–365.

Folk, R. L., Roberts, H. H., and Moore, C. H., 1973, Black phytokarst from Hell, Cayman Islands, British West Indies: Geological Society of America Bulletin, v. 84, p. 2351–2360.

Ford, D. C., 1979, A review of alpine karst in the southern Rocky Mountains of Canada: National Speleological Society Bulletin, v. 41, p. 53–65.

Ford, D. C., and Williams, P. W., 1989, Karst geomorphology and hydrology: London, Unwin Hyman, 601 p.

Friederich, H., and Smart, P. L., 1981, Dye tracer studies of the unsaturated zone recharge of the carboniferous limestone aquifer of the Mendip Hills, England, in Proceedings of the 8th International Congress of Speleology: British Cave Research Association, p. 283–286.

Girou, A., and 5 others, 1982, Dissolution experimentale d'un rhomboedre de calcite soumis a une contrainte: Bulletin de Mineralogie, v. 105, p. 301–306.

Gunn, J., 1981, Hydrological processes in karst depressions: Zeitschrift für Geomorphologie, v. 25, p. 313–331.

—— , 1983, Point-recharge of limestone squifers; A model from New Zealand karst: Journal of Hydrology, v. 61, p. 19–29.

Herak, M., and Stringfield, V. T., 1972, Karst; Important karst regions of the Northern Hemisphere: Amsterdam, Elsevier, 551 p.

Herman, J. S., 1982, The dissolution kinetics of calcite, dolomite, and dolomitic rocks in the CO_2-water system [Ph.D. thesis]: University Park, Pennsylvania State University, 214 p.

Herman, J. S., and White, W. B., 1985, Dissolution kinetics of dolomite; Effects of lithology and fluid flow velocity: Geochimica et Cosmochimica Acta, v. 49, p. 2017–2026.

House, W. A., Howard, J. R., and Skirrow, G., 1984, Kinetics of carbon dioxide transfer across the air/water interface: Faraday Discussions of the Chemical Society, v. 77, p. 33–46.

Jacobson, R. L., 1973, Controls on the quality of some carbonate ground waters; Dissociation constants of calcite and $CaHCO_3^+$ from 0 to 50°C [Ph.D. thesis]: University Park, Pennsylvania State University, 131 p.

Jacobson, R. L., and Langmuir, D., 1974, Controls on the quality variations of some carbonate spring waters: Journal of Hydrology, v. 23, p. 247–265.

Jakucs, L., 1977, Morphogenetics of karst regions: New York, John Wiley and Sons, 284 p.

James, N. P., and Choquette, P. W., 1988, Paleokarst: New York, Springer-Verlag, 416 p.

Jennings, J. N., 1983, Karst landforms: American Scientist, v. 71, p. 578–586.

—— , 1985, Karst geomorphology: Oxford, Basil Blackwell, 293 p.

Jennings, J. N., and Bik, M. J., 1962, Karst morphology in Australian New Guinea: Nature, v. 194, p. 1036–1038.

Keir, R. S., 1980, The dissolution kinetics of biogenic calcium carbonates in seawater: Geochimica et Cosmochimica Acta, v. 44, p. 241–252.

Kemmerly, P. R., 1982, Spatial analysis of a karst depression population; Clues to genesis: Geological Society of America Bulletin, v. 93, p. 1078–1086.

Kern, D. M., 1960, The hydration of carbon dioxide: Journal of Chemical Education, v. 37, p. 14–23.

Kogovsek, J., 1981, Vertical percolation in Planina Cave in the period 1980/81 (in Slovene): Acta Carsologica, v. 10, p. 111–125.

Kogovsek, J., and Habic, P., 1980, The study of vertical water percolations in the case of Planina and Postojna Caves (in Slovene): Acta Carsologica, v. 9, p. 133–148.

Langmuir, D., 1971, The geochemistry of some carbonate ground waters in central Pennsylvania: Geochimica et Cosmochimica Acta, v. 35, p. 1023–1045.

Massard, P., and Desplan, A., 1980, La cinetique de dissolution a 20°C du calcaire oolithique du Dogger: Bulletin de Mineralogie, v. 103, p. 317–323.

Medville, D. M., Hempel, J. C., Plantz, C., and Werner, E., 1979, Solutional landforms on carbonates of the southern Teton Range, Wyoming: National Speleological Society Bulletin, v. 41, p. 70–79.

Milanović, P. T., 1981, Karst hydrogeology: Littleton, Colorado, Water Resources Publications, 434 p.

Miotke, F. D., and Palmer, A. N., 1972, Genetic relationship between caves and landforms in the Mammoth Cave National Park area: Wurzburg, Böhler, 69 p.

Monroe, W. H., 1976, The karst landforms of Puerto Rico: U.S. Geological Survey Professional Paper 899, 67 p.

Mylroie, J. E., 1988, Field guide to the karst geology of San Salvador Island, Bahamas: Mississippi State, Mississippi State University Department of Geology and Geography, 108 p.

Pälmer, A. N., 1989, Geomorphic history of the Mammoth Cave system, in White, W. B., and White, E. L., eds., Karst hydrology; Concepts from the Mammoth Cave area: New York, Van Nostrand Reinhold, p. 317–337.

Parizek, R. R., and White, W. B., 1985, Application of Quaternary and Tertiary geological factors to environmental problems in central Pennsylvania, in Guidebook to 50th Annual Field Conference of Pennsylvania Geologists: Pennsylvania Bureau of Topographic and Geologic Survey, p. 63–119.

Philip, J. R., 1969, Theory of infiltration, in Ven Te Chow, ed., Advances in hydroscience, v. 5; New York, Academic Press, p. 215–296.

Plummer, L. N., and Busenberg, E., 1982, The solubilities of calcite, aragonite, and vaterite in CO_2-H_2O solutions between 0 and 90°C, and an evaluation of the aqueous model for the system $CaCO_3$-CO_2-H_2O: Geochimica et Cosmochimica Acta, v. 46, p. 1011–1040.

Plummer, L. N., and Wigley, T.M.L., 1976, The dissolution of calcite in CO_2-saturated solutions at 25°C and 1 atmosphere total pressure: Geochimica et Cosmochimica Acta, v. 40, p. 191–202.

Plummer, L. N., Wigley, T.M.L., and Parkhurst, D. L., 1978, The kinetics of calcite dissolution in CO_2-water systems at 5° to 60°C and 0.0 to 1.0 atm CO_2: American Journal of Science, v. 278, p. 179–216.

Plummer, L. N., Parkhurst, D. L., and Wigley, T.M.L., 1979, Critical review of the kinetics of calcite dissolution and precipitation in Jenne, E. A., ed., Chemical modeling in aqueous systems: American Chemical Society Symposium Series 93, p. 537–573.

Quinlan, J. F., 1967, Classification of karst types; A review and synthesis emphasizing the North American literature 1941–1966: National Speleological Society Bulletin, v. 29, p. 107–108.

—— , 1978, Types of karst, with emphasis on cover beds in their classification and development [Ph.D. thesis]: Austin, University of Texas, 323 p.

Reddy, M. M., Plummer, L. N., and Busenberg, E., 1981, Crystal growth of calcite from calcium bicarbonate solutions at constant P_{CO_2} and 25°C; A test of a calcite dissolution model: Geochimica et Cosmochimica Acta, v. 45, p. 1281–1289.

Rickard, D., and Sjöberg, E. L., 1983, Mixed kinetic control of calcite dissolution rates: American Journal of Science, v. 283, p. 815–830.

Schmidt, V. A., 1982, Magnetostratigraphy of sediments in Mammoth Cave, Kentucky: Science, v. 217, p. 827–829.

Sevon, W. D., 1985, Pennsylvania's polygenetic landscape, in Guidebook for the 4th Annual Field Trip: Harrisburg, Pennsylvania, Harrisburg Area Geological Society, 55 p.

Sjöberg, E. L., 1976, A fundamental equation for calcite dissolution kinetics: Geochimica et Cosmochimica Acta, v. 40, p. 441–447.

Smart, P., Waltham, T., Yang, M., and Zhang, Y., 1986, Karst geomorphology of western Guizhou, China: Cave Science, v. 13, p. 89–103.

Smith, D. I., and Atkinson, T. C., 1976, Process, landforms, and climate in limestone regions, in Derbyshire, E., ed., Geomorphology and climate: London, John Wiley and Sons, p. 367–409.

Song, L., 1986, Karst geomorphology and subterranean drainage in South Dushan, Guizhou Province, China: Cave Science, v. 13, p. 49–63.

Sweeting, M. M., 1972, Karst landforms: London, Macmillan, 362 p.

Thrailkill, J., 1968, Chemical and hydrologic factors in the excavation of limestone caves: Geological Society of America Bulletin, v. 79, p. 19–46.

Troester, J. W., and White, W. B., 1984, Seasonal fluctuations in the carbon dioxide partial pressure in a cave atmosphere: Water Resources Research, v. 20, p. 153–156.

Troester, J. W., White, E. L., and White, W. B., 1984, A comparison of sinkhole depth frequency distributions in temperate and tropical karst regions, in

Beck, B. F., ed., Sinkholes; Their geology, engineering, and environmental impact: Rotterdam, A. A. Balkema, p. 65–73.

Trudgill, S., 1985, Limestone geomorphology: London, Longman, 196 p.

White, E. L., and White, W. B., 1979, Quantitative morphology of landforms in carbonate rock basins in the Appalachian Highlands: Geological Society of America Bulletin, v. 90, p. 385–396.

White, W. B., 1979, Karst landforms in the Wasatch and Uinta Mountains, Utah: National Speleological Society Bulletin, v. 41, p. 80–88.

—— , 1984, Rate processes; Chemical kinetics and karst landform development, *in* LaFleur, R. G., ed., Groundwater as a geomorphic agent: Boston, Massachusetts, Allen and Unwin, p. 227–248.

—— , 1988, Geomorphology and hydrology of karst terrains: New York, Oxford University Press, 464 p.

White, W. B., and White, E. L., eds., 1989, Karst hydrology; Concepts from the Mammoth Cave area: New York, Van Nostrand Reinhold, 346 p.

Williams, P. W., 1971, Illustrating morphometric analysis of karst with examples from New Guinea: Zeitschrift für Geomorphologie, v. 15, p. 40–61.

—— , 1972a, Morphometric analysis of polygonal karst in New Guinea: Geological Society of America Bulletin, v. 83, p. 761–796.

—— , 1972b, The analysis of spatial characteristics of karst terrains, *in* Chorley, R. J., ed., Spatial analysis in geomorphology: London, Methuen, p. 135–163.

—— , 1983, The role of the subcutaneous zone in karst hydrology: Journal of Hydrology, v. 61, p. 45–67.

—— , 1985, Subcutaneous hydrology and the development of doline and cockpit karst: Zeitschrift für Geomorphologie, v. 29, p. 463–482.

—— , 1987, Geomorphic inheritance and the development of tower karst: Earth Surface Processes and Landforms, v. 12, p. 453–465.

Yuan, D., 1985, New observations in tower karst, *in* 1st International Conference on Geomorphology: Manchester: England, 14 p.

MANUSCRIPT ACCEPTED BY THE SOCIETY NOVEMBER 14, 1989

Printed in U.S.A.

Chapter 8

Groundwater processes in karst terranes

Arthur N. Palmer
Department of Earth Sciences, State University of New York, College at Oneonta, Oneonta, New York 13820-4015

INTRODUCTION

The solutional processes described in the preceding chapter for surface and near-surface environments extend downward varying distances into the karst aquifer below. Where recharge from the surface is most concentrated, solution conduits eventually form that are capable of transmitting turbulent flow through the aquifer to springs at lower levels. Surface karst features and subsurface conduits therefore are intimately related, and their development is simultaneous and interdependent. Topography helps to determine the pattern of groundwater recharge, which in turn controls the morphology of solution conduits. On the other hand, underground solution causes local subsidence of overlying rock and regolith, forming or enlarging surface depressions. A more subtle but equally important agent of surface degradation is the transport of detrital material by turbulent groundwater. This chapter examines these and other groundwater processes in karst, the features they produce, and their effect on the overlying karst surface.

Among geologists who are concerned with subsurface processes, karst researchers have an advantage in being able to view them first-hand under natural conditions by visiting caves, which are solutional openings of traversable size. Access to caves and their associated groundwater systems can be gained through openings such as sinkholes (dolines), sinking streams, or springs. Underground travel, of course, is confined to the larger conduits dissolved by groundwater, so conditions in the surrounding inaccessible openings must still be inferred from indirect methods. However, one's view of karst processes is enhanced considerably by extending hydrologic, chemical, and geomorphic measurements into these underground flow systems.

There is a large body of literature on karst, mostly from European countries. The major comprehensive works are cited in the previous chapter, so only those that deal specifically with underground karst processes are described here.

The earliest karst studies (e.g., Cvijić, 1893; Grund, 1903; Katzer, 1909) treated underground solution in rather general terms, integrating it with the overall geomorphic and hydrologic setting but with little emphasis on specific factors that control cave origin. The pioneering speleological work of Martel (1921) helped to reveal the morphology of caves and their hydrologic function. In a more technical vein, Lehmann (1932) discussed the hydraulics of solution conduits, emphasizing contemporary groundwater conditions rather than geomorphic processes.

The 1930s and early 1940s saw a burst of enthusiasm among American geologists for devising a comprehensive theory of cave origin. Their particular concern was determining which hydrologic zone is the most favorable to groundwater solution. Davis (1930) and Bretz (1942) proposed that caves form by slow-moving water deep beneath the water table, where the bedrock is exposed to solution for the greatest length of time. Swinnerton (1932) and Rhoades and Sinacori (1941) argued that caves are more likely to form in the zone of most rapid groundwater flow, namely at or near the water table. Piper (1932) and Malott (1938) also favored an origin at shallow depth but stressed the importance of invasion by surface streams. Gardner (1935) considered cave development to be concentrated within favorable strata, with progressively lower cave levels forming as strata are exposed by valley entrenchment. The question of cave origin was revived later by Sweeting (1950) and Davies (1958), who cited the low gradient of many caves and their correlation with fluvial terraces as evidence for a shallow-phreatic origin. Woodward (1961) interpreted this correlation as the result of cave development during short-lived episodes of rejuvenation, when groundwater solution would presumably be most vigorous.

In a break with tradition, White and Longyear (1962) stated that all previous theories had some merit, but that each was valid only under a narrow range of conditions. There is no single favorable zone of groundwater solution. Instead, caves develop wherever the movement of groundwater happens to be greatest. Ford (1965, 1971) and Ford and Ewers (1978) clarified this point further by showing that the depth of cave development depends on the availability of fissures penetrable by aggressive groundwater.

In recent years, several hydrochemical models have been devised for the development of groundwater solution conduits. Some of the chemical and hydraulic constraints on cave development have been discussed by Kaye (1957), Weyl (1958), Howard (1964), Curl (1968), and Thrailkill (1968). Measurements of limestone and dolomite solution have been made in the

Palmer, A. N., 1990, Groundwater processes in karst terranes, in Higgins, C. G., and Coates, D. R., eds., Groundwater geomorphology; The role of subsurface water in Earth-surface processes and landforms: Boulder, Colorado, Geological Society of America Special Paper 252.

Figure 1. A tubular conduit in prominently bedded St. Louis Limestone (Mississippian), Mammoth Cave, Kentucky. The stream in this passage represents the local water table; it is at virtually the same elevation as the spring in the entrenched Green River valley. The passage ceiling drops below water level several hundred meters downstream from this point.

laboratory by Roques (1962), Howard and Howard (1967), Ek and Roques (1972), Rauch and White (1977), and Herman and White (1985) and in caves by High (1970) and Coward (1975). Using heat-flow principles, Wigley (1973) has devised a model of solution-conduit growth, and Palmer (1972, 1975) has applied hydraulics to the origin of maze caves. Laboratory data on calcite solution kinetics by Berner and Morse (1974), Plummer and Wigley (1976), and Plummer and others (1978) have been translated into rates of cave origin by White (1977), Palmer (1981, 1984), and Dreybrodt (1981a, 1981b). On the basis of their own experiments, Sjöberg (1978) and Buhmann and Dreybrodt (1984, 1985) have presented additional models for calcite dissolution and precipitation. With salt and plaster models, Ewers (1982) has illustrated the mechanisms by which the solution conduits in a karst aquifer link together into an integrated system. Some of this work is described in detail later in this chapter.

A great deal of field information on karst hydrology is available as maps of groundwater conduits, including accessible caves as well as inaccessible conduits delineated by dye or other tracers. In addition, long-term measurements of water budget and chemistry provide a detailed picture of the flow patterns within karst aquifers. The most extensive of these projects is being carried out by the Underground Laboratory of the Centre National de la Récherche Scientifique, in Moulis, France, where the hydrologic and chemical mass balances have been modeled on the basis of decades of measurements within the same basin (Mangin, 1974–1975; Bakalowicz, 1979).

So far, little of this knowledge has been applied to groundwater management. Lacking familiarity with the quantitative methods of describing the heterogeneous permeability of a soluble aquifer, many hydrologists treat karst groundwater in the same manner as diffuse flow through a uniformly porous medium. When this approach is used to predict water supply or to track contaminants the interpretation can be seriously flawed. It is therefore appropriate to consider some of the environmental aspects of karst along with the geomorphic ones. Predictive techniques in karst are still in their infancy, but general concepts of how the geologic setting controls the pattern of karst groundwater are beginning to emerge.

SUBSURFACE KARST PROCESSES

The nature of karst groundwater systems

A karst aquifer is one in which solutional widening of intergranular pores, fractures, and partings has caused significant modification of the initial flow paths. Solution conduits form integrated systems linking upland recharge areas to discharge points at lower elevations. This relation is clear in accessible caves, and dye traces show this to be true for inaccessible solution conduits as well. Underground solution is hardly ever a random process creating scattered voids, but conforms instead to the most favorable paths of groundwater flow.

Turbulent flow occurs in those openings that have been sufficiently enlarged by solution. Laminar flow in the surrounding smaller openings flows toward the larger ones in much the same way that it does toward entrenched surface streams. Most openings enlarged by the solvent action of flowing water (solution conduits) are not of cave size. Although "cave" and "solution conduit" are often used interchangeably, any use of the former term in this chapter implies knowledge gained from direct observation within traversable openings.

The development of solution conduits requires throughflow of water to carry away dissolved material, which in turn depends on a preexisting interconnected network of nonsolutional openings between recharge points and outlets. These include intergranular pores, fractures, and bedding-plane partings. Their relative importance in the development of conduits depends on the nature of the host rock. Fractures (particularly joints) and bedding-plane partings are of nearly equal importance, and most caves show a combination of fracture and bedding control. Faults, despite their tectonic significance, rarely exert more than local control over cave patterns, although a few caves are oriented entirely along single faults. Caves formed in prominently bedded rocks are mostly sinuous, curvilinear tubes with wide, lenticular cross sections (Fig. 1). Those enlarged along major fractures tend to be linear fissures with high, narrow cross sections (Fig. 2). A multitude of small fractures will produce a pattern intermediate between the two, usually with angular bends superimposed on the sinuous pattern. Solutional enlargement of intergranular pores is significant only in young, poorly indurated carbonates or in reef rocks, where they form irregular voids like the pores in a sponge (Fig. 3).

In most karst aquifers, dissolution by groundwater is highly selective. Despite the abundance of presolutional openings, very few are enlarged significantly. The result is a sharp discontinuity in the scale of underground voids, with large caves surrounded by a network of tiny openings enlarged very little, if at all, by solution, as shown in Figures 1 and 2. This characteristic is less pronounced at and near the land surface but becomes increasingly prominent with distance along the paths of groundwater flow. Most conduit flow, whether open-channel or closed-conduit flow, retains a solutional aggressiveness toward the bedrock that persists all the way to the spring. In contrast, the surrounding fractures and pores contain slow, laminar flow close to saturation with dissolved bedrock (Thrailkill, 1968; Shuster and White, 1971).

Every opening in the bedrock surface receives some infiltrating water, whether or not there is a soil cover. A shallow zone of many interconnecting solutional openings, known as the epikarstic or subcutaneous zone, is common in the upper few meters of bedrock (Williams, 1983). Although many of these openings undergo solutional widening at their upper ends (Fig. 4), relatively few receive enough water to enlarge into solution conduits that penetrate deeply into the aquifer below. Sinkholes develop in the land surface at the inputs to the larger conduits as the result of concentrated dissolution, collapse, and transport of overburden through the conduits by groundwater. Most solutionally aggressive water enters through these relatively few points. Surface

Figure 2. A joint-controlled fissure in the floor of a wider passage in thick-bedded Salem Limestone (Mississippian), Blue Spring Cave, Indiana.

depressions and conduits evolve interdependently. Sinkholes depend on the presence of underlying conduits, although the converse is not always true.

Within most karst aquifers, conduits tend to form a branching system in which tributaries join to form higher-order passages with greater discharge and generally larger cross section. In this way, cave systems can develop that contain many kilometers of interconnected traversable passages.

Almost all groundwater emerges at discrete springs in nearby entrenched river valleys (Fig. 5). Only a few of the largest streams are able to persist at the surface in a mature karst region. These maintain an effluent relation with the surrounding groundwater and so are able to entrench their channels, unlike smaller streams perched at higher elevations, which lose water to the ground and evolve into sinking streams.

Figure 3. A sponge-like cave pattern produced by solutional enlargement of primary pores in massive reef limestone (Capitan Limestone of Permian age), Carlsbad Caverns, New Mexico.

Characteristics of the karst water table

Whatever determines the position of springs exerts the major influence on the level of the water table in a mature karst area. This control is most commonly provided by entrenched rivers (at least in humid regions), sea level, or stratigraphic perching horizons. Solution conduits feeding the springs are so efficient in transmitting water that they possess very little hydraulic gradient. The potentiometric surface within or above the conduits lies only slightly higher than spring level.

During low flow, heads in these openings are lower than in their less efficient neighbors, so water tends to flow toward the conduits from the surrounding narrow fissures and pores. This trend is often reversed during floods, when the larger openings are exposed to sudden surges of water from the surface (Palmer, 1984). The head in the conduits quickly rises to values higher than those of the surroundings, forcing water into the surrounding cracks as a form of bank storage, which is later released slowly as the floodwaters subside. Major groundwater conduits thus behave like subsurface analogs of surface rivers.

The water table in most karst regions is highly irregular and discontinuous, owing to great differences in the size of underground openings and in the spatial and temporal distribution of groundwater discharge. A karst aquifer can be viewed as an elaborate underground plumbing system through which water flows in discrete conduits. Water may stand at different levels in nearby wells, and dry or poorly productive wells may occur in the same area as successful wells. Mining and tunneling operations often intersect isolated pockets of water surrounded by dry rock. Dye tracing (e.g., by Zötl, 1961) shows instances where flow paths cross one another without mixing. From this evidence, some researchers deny the existence of a karst water table. However, some of these characteristics are found in any kind of bed-

Figure 4. Solutional widening of joints in the epikarstic zone, exposed by quarrying of the Salem Limestone in Indiana.

Figure 5. A karst spring at fluvial base level in the Green River valley, Kentucky.

rock. Owing to stratal irregularities and facies changes, it is normal to have slight water-table discontinuities and perched phreatic zones even in unconsolidated sediment. However, the scale of such irregularities is larger in soluble rock than in other materials. The water-table concept is perfectly valid in karst regions, but only if applied regionally or as a generalized concept, rather than at the scale of individual solution conduits or wells.

Whenever there is a change in elevation of the entrenched river, or of any other control of spring levels, the position of the water table tends to adjust accordingly. As the major rivers deepen their channels into soluble rock, there is a progressive diversion of groundwater flow to lower levels or entrenchment of existing conduits. Where diversion takes place, formerly active conduits are abandoned in favor of new routes (Fig. 6). Older passages remain as relics, which at first are periodically reactivated during high flow but eventually are abandoned by all but diffuse seepage. The pattern of relict conduits indicates former groundwater flow directions. Ultimately they are destroyed by surface weathering and erosion.

A rise in fluvial base level will raise the water table, flooding conduits that were formerly air filled. Deeper conduits eventually may become choked with detrital sediment. As alluviation takes place in a valley, karst springs are able to maintain their identity for a considerable length of time because their high-velocity outflow keeps sediment from engulfing them (Powell, 1963). Water rises upward through conical openings in the sediment known as alluviated springs (Fig. 7).

Comparison of vadose and phreatic solution features

While forming, most cave passages have an upstream vadose section leading to a downstream phreatic section. Some early researchers (e.g., Bretz, 1942) envisioned primary development of caves below the water table, with later modification by invading vadose water. This is true in a few areas, but in general the vadose and phreatic sections of a passage develop at the same time as geomorphic contemporaries. The phreatic sections of a passage may later be drained by a drop in the water table, exposing them to modification in the vadose zone, but the genetic continuity between them and their vadose infeeders is still markedly visible in relict conduits (Fig. 6). Although there is a clear distinction between vadose and phreatic processes in a karst area (cf., Palmer, 1984), the two are intimately related. To understand the evolution of a karst groundwater system it is necessary to trace groundwater and its chemical evolution over the entire length of flow.

The terms "vadose" and "phreatic" are out of fashion among hydrologists, who prefer the terms "unsaturated" and "saturated." However, the latter terms are unsuited to karst studies because they confuse the distinction between hydrologic and chemical saturation. The former terms must be carefully defined, however, as their ambiguity is what has turned hydrologists away from them. Vadose and phreatic cave features are discussed at length by Bretz (1942), Ford (1965), Palmer (1972, 1984), and White (1976), so the topic needs only a brief introduction here.

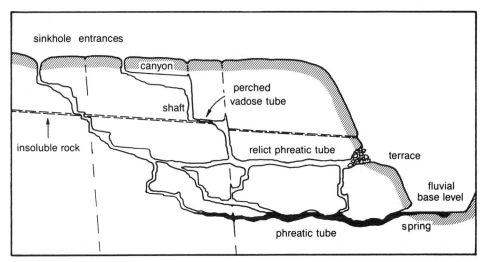

Figure 6. Vertical pattern of a typical cave draining water from uplands into an entrenched surface valley. Approximately 80 percent of all caves consist of conduits such as these arranged in a branchwork pattern. Canyons and shafts are of vadose origin; tubes are generally phreatic, except where perched on insoluble beds in the vadose zone.

In karst studies the most useful distinction between vadose and phreatic water is in the forces that govern their flow: vadose water moves mainly by gravity and by differences in capillary potential; phreatic water moves in response to a combination of gravity and hydrostatic pressure. The geomorphic distinction is obvious even in relict conduits. Vadose solution conduits are formed by gravitational water draining downward along the steepest available openings, so they exhibit continuously descending profiles with no rises in the downstream direction. Where water follows vertical fractures or partings, it forms well-like shafts (Fig. 8). If the openings are inclined, as are most bedding-plane partings, the water forms narrow, roofed canyons with downward-entrenching floors (Fig. 9). Vadose water tends to flow directly down the dip of the controlling structure, except where constriction of the opening or accumulation of sediment causes either local ponding or deviation around the blockage. Water usually does not fill the entire passage, except possibly during floods, so the ceiling is dissolved upward only slightly, if at all. Vadose streams are independent of one another, as there is no hydraulic continuity between them.

In contrast, phreatic water concentrates along paths of greatest hydraulic efficiency; i.e., those paths that can conduct the available flow with the least expenditure of energy. Phreatic conduits are usually tubular (Fig. 1), and their patterns show no consistent relation to the dip, unless by chance that direction is the most efficient. Because they undergo much of their development while water filled, most phreatic passages have irregular profiles with sections that rise in the downstream direction.

Ford and Ewers (1978) distinguish two types of vadose conduits: (1) the drawdown type, which originate as phreatic conduits above the fluvial base level in a low-permeability rock, and which eventually develop vadose characteristics as the conduit enlarges and acquires open-channel flow; and (2) the invasion type, which are formed by recharge into openings that have already been evacuated by a drop in the water table. Inherent in their idea is the premise that the water table lies at or near the land surface during the initial stages of karst development, owing to the low permeability of the original bedrock, and that only with the solutional enlargement of conduits does the water table drop enough to provide a substantial vadose zone. In some caves, however, even the oldest passages show a sharp transition from dip-oriented vadose canyons to phreatic tubes with irregular profiles, suggesting that a deep vadose zone either predated the passages or developed after only a slight amount of solutional enlargement (Palmer, 1972; Mylroie, 1977).

The distinction between vadose and phreatic conduits can be obscured in several ways. Structural traps may cause local ponding of water above the water table, causing perched phreatic conditions; this condition is usually short-lived, however, owing to erosion or solution of the perching threshold. More significantly, seasonal and storm-related fluctuations in groundwater discharge can cause sudden and often large rises in water level within conduits, imposing temporary phreatic solution on zones that are vadose in character during the rest of the year.

The phreatic section of a solution conduit can be missing entirely if the vadose water exits at a perched spring before reaching the water table, but this condition is common only where the soluble rock is underlain by thick insoluble rocks and the contact lies above the local fluvial base level. On the other hand, some caves formed by deep-seated processes exhibit no distinct vadose passages, particularly if they are preserved as relics within an arid environment where there is little recharge from the surface.

Dissolution processes in groundwater

The most essential process in the origin of karst is the solution of bedrock by groundwater. All rocks are soluble to some

Figure 7. An alluviated karst spring in Cobleskill, New York. Water rises from a solution conduit located several tens of meters below the spring level. Part of the valley fill is of glacial origin.

extent, but only evaporites and carbonates dissolve rapidly enough to form karst topography. Rare examples of solutional karst in siliceous rocks are known in areas of extremely long-term weathering, but on a global scale, these are of negligible importance.

Evaporites dissolve by simple dissociation in water, whereas carbonate rocks are able to dissolve substantially only in acidic water. Carbonic acid derived from carbon dioxide in the atmosphere and soil is the most common source of groundwater acidity. Pertinent equilibria for this reaction are described in Chapter 7 (this volume). Figure 10 shows the saturation concentration of dissolved calcite at different initial partial pressures of carbon dioxide, calculated with equilibrium constants recommended by Plummer and Busenburg (1982).

Several other reactions can be locally significant in karst areas and can even be the dominant chemical processes. These include the dissolution of carbonate rocks by: (1) aqueous hydrogen sulfide in reducing environments (e.g., in basinal brines); (2) sulfuric acid from oxidation of hydrogen sulfide, where water from a reducing environment mixes with oxygenated water or becomes aerated at the water table; and (3) carbonic acid from deep basinal or volcanic sources of carbon dioxide. Less important are: (4) sulfuric acid from the oxidation of metallic sulfides; (5) acids produced at depth by the heating of hydrocarbons and the high-temperature hydrolysis of salts; and (6) organic acids.

Sediment transport through karst aquifers

Karst processes involve a three-fold mass transfer in which water, dissolved load, and detrital sediment play an inseparable

Figure 8. A vertical shaft 20 m deep in Mammoth Cave, Kentucky. The view is straight downward. Note vertical flutes formed by dripping vadose water and abandoned drains at various levels.

Figure 9. A canyon passage in Mammoth Cave, showing the typical high, narrow cross section produced by an entrenching vadose stream. The upper-level tube represents an earlier flow route.

role (White, 1978). As the initial openings enlarge by solution, unconsolidated detrital material enters the solutional openings by subsidence and aqueous transport. These processes create or enlarge surface depressions, which in turn control the pattern and regime of groundwater recharge. Where flow velocities are high, mechanical abrasion by clastic load can add significantly to the enlargement of conduits, particularly during floods (Newson, 1971). Groundwater velocities in karst can range up to several meters per second in large solution conduits, even during low-flow conditions. In mixed terranes of soluble and insoluble rock, an enormous volume of sediment is transported from uplands through the entire length of underground conduits to entrenched valleys.

Cave sediments are similar to those of alluvial deposits in surface valleys. Gravel and sand are the dominant materials in stream beds, unless these materials are not present in the drainage basin, whereas finer materials are deposited along the banks and in relict passages subject to periodic flooding by slow-moving water. Figure 1 shows banks of detrital clay, silt, and sand deposited by a cave stream during high flow. These deposits are mixed with collapse blocks that have spalled off the ceiling and walls.

Karst groundwater hydraulics

The size contrast between solution conduits and the unenlarged fractures, partings, and pores in a karst aquifer requires a nontraditional approach to the interpretation of groundwater flow (Boulton and Streltsova, 1977; Cullen and LaFleur, 1984). Quantifying a system that contains two functionally disparate parts is a difficult task even when the exact pattern of solution conduits is known, an ideal that is rarely, if ever, achieved. It may never be possible to predict local head distributions and well yields in a specific karst aquifer with laboratory or digital models constructed from regional geologic maps and well data. On the other hand, conceptual views of karst aquifers can explain known geomorphic and hydrologic conditions and even provide crude predictions of groundwater flow paths.

Despite the hydraulic importance of solution conduits, most of the other openings in a karst aquifer have undergone little or no enlargement. The arrangement of these openings is similar to that in any competent sedimentary rock. Laminar flow prevails in those parts of an aquifer containing nonsolutional voids, and to some extent, Darcy's Law can be applied within them. Darcy's Law and its various integral forms assume that the water behaves as a continuous medium. This is untrue even in granular materials, but it is a valid assumption where the aquifer cross section being considered is much larger than the individual pores and the space separating them. Laminar flow in fractured aquifers can be modeled on a small scale only by accounting for the distribution and width of individual fractures.

The functional relations within a single fracture are sufficient to illustrate many geomorphic concepts, although to understand an entire karst system it is necessary to extrapolate to three dimensions. Within a single fracture the relations are:

$$Q = \frac{w^3 b \ \gamma \ dh}{12\mu \ dL} \qquad (1)$$

where Q = discharge, w = fracture width, b = long dimension of fracture cross section, γ and μ = specific weight and dynamic viscosity of water, and dh/dL = hydraulic gradient. The combined terms $w^2\gamma/12\mu$ are analogous to the hydraulic conductivity in Darcy's Law.

Note the strong influence of fracture width in determining the relation between discharge and hydraulic gradient. Owing to the tendency for fractures to group in certain preferred directions, a strong anisotropy results. Consequently the hydraulic gradient shown by water-table maps in prominently fractured rocks is not a faithful indicator of actual flow directions. Dye tests and con-

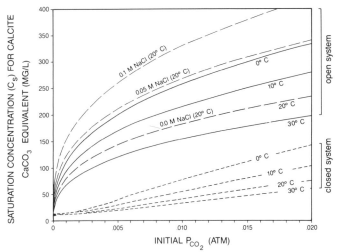

Figure 10. Saturation concentration of dissolved calcite as a function of initial CO$_2$ partial pressure in open and closed systems. The variation of C_s with salinity is also shown in an open system at 20°C.

taminant traces in bedrock aquifers often show groundwater flow at large angles to the hydraulic gradient.

Solutional widening of fractures not only enhances the anisotropy but causes Darcy's Law to break down entirely as turbulent flow develops. Turbulence sets in when the Reynolds Number ($\rho vw/\mu$ in fissures, where v and ρ = velocity and density of water) rises above 1,100. Laminar flow can persist at higher Reynolds Numbers only in the absence of wall irregularities and flow transients. At typical groundwater temperatures, this transition takes place at fracture widths that vary from a minimum of 0.12 cm for vadose flow in vertical openings to 1 or 2 cm in low-gradient phreatic flow.

Turbulence has little effect on solution rate at the pH (approximately 7 to 8.5) of most karst groundwater (Plummer and Wigley, 1976; White, 1984), but the flow dynamics are quite different from those of laminar flow. Turbulent flow must be analyzed at the scale of individual conduits, since no meaningful regional average can be stated for hydraulic geometry and conductivity. The functional relations for turbulent flow in both closed conduits and open channels are:

$$Q = A \sqrt{\frac{8Rgi}{f}} \quad (2)$$

where A = cross-sectional area of flow, R = hydraulic radius (A/wetted perimeter), i = hydraulic gradient, expressed as the positive ratio of head loss over flow distance, and f is a friction factor (approximately 0.03 to 0.1 in natural solution conduits). For further details see Gale (1984) and Lauritzen and others (1985). Solutional widening of a fracture or parting tends to concentrate along a narrow zone, so by the time turbulence develops, the opening has acquired a discrete lenticular or elliptical cross section usually no more than 1 or 2 m in width.

Equation 2 can be applied to an entire conduit system only by subdividing each conduit into segments of uniform geometry (Palmer, 1976). This approach is best suited to caves that have been carefully mapped, but it also allows the construction of idealized aquifer models (Cullen and LaFleur, 1984). However, if hydraulic principles are to be applied to the geomorphic evolution of a karst aquifer, it is first necessary to integrate them with solution kinetics.

Kinetics of limestone dissolution

According to popular view, the growth of a cave proceeds at an unimaginably slow rate. An impression of great antiquity may be valid in some caves, but under optimum conditions the solution of carbonate bedrock can be surprisingly fast. Rates of surface denudation have been calculated from the chemical mass balance in rivers draining karst areas (Corbel, 1959; Williams, 1968; Atkinson and Smith, 1976). These vary from only a few millimeters per thousand years in tundra to several hundred millimeters per thousand years in mountainous or tropical areas with high rainfall. Direct measurements over several years using micrometer probes show that local solution rates can be even greater: as much as 0.4 to 0.8 mm/yr in perennially active cave passages (High, 1970; Coward, 1975). Compatible results have been produced by laboratory experiments in which water is run through artificial fractures and tubes in limestone under controlled carbon-dioxide pressures (Howard and Howard, 1967; Rauch and White, 1977).

From these measurements it might appear that the development of solution conduits and associated surface features is a rapid and inevitable process. Ironically, this rapid solution may be a serious handicap to the origin of solution conduits. Measured rates are so high that the water approaches saturation after only a short distance of travel. Extrapolating the data of Howard and Howard (1967), groundwater in fractures 0.001 to 0.01 cm wide at typical ranges of P_{CO_2} (0.001 to 0.01 atm) and hydraulic gradient (0.0001 to 0.01) would become 90 percent saturated after flowing only about 0.001 to 10 m—hardly an inspiring solutional history. Exposing fractured rock to such conditions would cause rapid denudation but could not produce traversable conduits within a geomorphically feasible time, unless the initial openings were at least 1mm wide (Palmer, 1991). Obviously this scenario is not valid, because karst and caves are widespread through nearly every exposed limestone formation.

Until recently, few measurements of limestone dissolution rate had been extended to saturation. In reality the rate drops sharply as the water approaches equilibrium (Berner and Morse, 1974; Plummer and Wigley, 1976). As a result, water is able to penetrate large distances into limestone while still retaining a large part of its aggressiveness. Although operating at a decreased rate, solutional widening is able to extend throughout the length of all major fractures.

White (1977) applied the Berner and Morse data to the development of solution conduits. He considered the change from low to high solution rate, as the concentration of dissolved load in a conduit decreases with time, to be a kinetic threshold that allows only a few favorable paths to enlarge to cave size. In a typical solution conduit this change takes place roughly at the same time that turbulence sets in, as well as when the velocity first exceeds the critical tractive force necessary to transport sediment.

Palmer (1991) has calculated rate coefficients and reaction orders from the data of Rauch and White (1977), Plummer and Wigley (1976), and Plummer and others (1978) for the solution of calcite and applied them to the standard equation for surface-controlled chemical reactions:

$$\frac{dC}{dt} = \frac{A' k}{V} \left(1 - C/C_s\right)^n \quad (3)$$

where dC/dt = change in solute concentration with time, A' = solid/liquid surface area, V = volume of solvent, and C_s = saturation concentration, k = reaction coefficient, and n = reaction order. For typical karst groundwater in limestone (impure calcite) at 15°C and P_{CO_2} = 0.01 atm, k = 0.015, and n = 1.8. The transition from rapid to slow solution rate takes place at C/C_s = 0.7, beyond which $k \sim 0.25$ and $n \sim 4$. Reaction order (n) decreases with P_{CO_2}, whereas k and the transition C/C_s increase with temperature and P_{CO_2}. One reason why the sudden decrease in solution rate at high C/C_s was not observed until recently is that the change does not take place until 90 percent saturation at the standard temperature of 25°C.

Plummer and others (1978) constructed a rate equation from their data that utilizes the activities of dissolved chemical species (see Chapter 7, this volume). Equation 3, although less deterministic, fits the experimental data more closely and is simpler to use in geomorphic studies. It is also compatible in overall trend to the dissolutional models of Sjöberg (1978) and Buhmann and Dreybrodt (1984, 1985).

To be applicable to the growth of solution conduits, Equation 3 must be combined with the solutional mass balance, which is, in essence, the concept that the mass dissolved from a conduit must equal the mass carried away in solution. Stated in more applicable terms:

$$\frac{dA}{dt} = \frac{Q}{\rho_r} \frac{dC}{dL} \quad (4)$$

where dA/dt = rate of increase in conduit cross-sectional area, dL = incremental distance of flow, and ρ_r = bulk density of bedrock.

Equations 1 through 4 can be combined in finite-difference analyses to simulate the growth of individual conduits or conduit systems (Palmer, 1991). The results in simple branching-conduit systems show the following: (1) Where the water exiting the downstream end of a system lies close to saturation with respect to dissolved calcite, the rate of solutional wall retreat is controlled mainly by discharge. Unless the discharge increases during the initial stages of growth, the enlargement rate stagnates or even diminishes with time. (2) As C/C_s decreases at the conduit terminus, solution kinetics become dominant, and the rate of wall retreat (annual mean) approaches a constant value, theoretically about 0.1 cm/yr. The theoretical maximum is rarely reached because water usually acquires much of its dissolved load before it enters a conduit. (3) Where hydraulic gradients are steep, as during intense flooding, all fractures and partings greater than about 0.01 cm wide can enlarge simultaneously at almost the same rate.

Open versus closed systems

As water passes through soluble rock it encounters a constantly changing chemical environment. The rate at which karst processes respond to these changes depends on the reaction kinetics, which falls into three categories: equilibria within the water (very fast, except for the hydration of CO_2, which is only moderately fast), gas/water reactions governed by Henry's Law (moderately fast), and dissolution of solid species (slow in the case of carbonate bedrock). Because dissolution is the slowest process, it is usually the rate-limiting step, except in highly acidic water.

Water exposed to a high-volume gas phase (open system) approaches equilibrium with the various gases in proportion to their partial pressure. Carbon dioxide is of greatest concern to karst processes, as it determines the saturation concentration of carbonate bedrock in contact with the water (Fig. 10). Most initial dissolution in a developing conduit takes place under fully or quasi-closed conditions with respect to carbon dioxide, even in the vadose zone. Most vadose conduits eventually enlarge enough to allow substantial aeration.

In an open system, as dissolution consumes aqueous CO_2, additional CO_2 is added to maintain equilibrium. However, if the water fills the openings as closed-conduit flow, it is cut off from its supply of gaseous CO_2, becoming a closed system. As further solution consumes dissolved CO_2, equilibrium is approached at a CO_2 content lower than that of the original open system. As a result, the equilibrium concentration of dissolved carbonate is lower than it would have been in the open system (Fig. 11). Intermediate situations are common: water may alternate be-

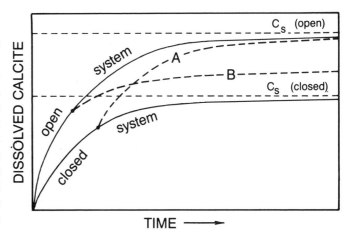

Figure 11. Chemical evolution of karst water in open and closed systems. A = change from closed to open conditions; B = change from open to closed conditions.

tween open systems and closed systems along its path of flow, or small pockets of air can become partly depleted of CO_2 (quasi-open systems).

A change from open to closed conditions does not cause a precipitous decrease in solution rate, because the CO_2 content of the water is not immediately affected. But as the dissolved load increases in the closed system and aqueous CO_2 is used up, the effective saturation concentration of the solute diminishes toward the final equilibrium value. Therefore, the value C/C_s in Equation 3 increases because of a drop in C_s as well as a rise in C. The solution rate diminishes accordingly. Because of the lower solution rate, the remaining aggressiveness is prolonged, so in comparison with an open system the passage growth is merely slower, without a significant difference in the spatial distribution of that growth.

Renewal and reduction of aggressiveness in subsurface karst water

Degassing and uptake of carbon dioxide. Streams that flow across carbonate rocks before sinking underground are often close to saturation with respect to calcite, although they tend to be more aggressive during high discharge. It might seem that solutionally exhausted recharge would be unable to contribute to the growth of underground conduits. However, CO_2 partial pressures are usually far greater in air-filled subsurface openings because of downward diffusion from the soil and transport of CO_2 in solution by infiltrating drip waters (Jacobson and Langmuir, 1970; Thrailkill, 1971). As a result, the calcite-saturated water regains its solutional aggressiveness after passing underground by absorbing CO_2 from the cave atmosphere. In addition, organic matter carried into the cave continues to be oxidized, contributing to the supply of CO_2 (Bray, 1975). Organic material carried downward into vadose pores by infiltrating water accomplishes a similar feat and accounts for the enlargement of pores at depth below the surface (Wood, 1985). Thus, the solutional history of water passing through underground conduits is prolonged, particularly in those that are well aerated.

If vadose water emerges into aerated openings in which the CO_2 partial pressure is in disequilibrium with the aqueous CO_2, there is a moderately fast uptake or degassing of CO_2 toward equilibrium. Infiltrating water is typically adjusted to a P_{CO_2} higher than that in aerated caves. Under these conditions it loses CO_2 to the cave air and its C/C_s increases, diminishing the solutional capacity or driving the water to supersaturation. Typical results include travertine speleothems in caves (Fig. 12) and meniscus cements within smaller pores. On the other hand, capillary water in the vadose zone can fill narrow cracks entirely, creating local closed systems in which the water approaches saturation at low CO_2 content. Even if the initial P_{CO_2} is greater than that of the air in underlying caves, the CO_2 level can easily drop below that of the caves under these conditions. When the water emerges into a cave it acquires renewed aggressiveness, like a diver coming up for air, as it absorbs CO_2 from the cave atmo-

Figure 12. Travertine speleothems in McFail's Cave, New York. Infiltrating vadose water acquires a high CO_2 content from the overlying soil, approaches equilibrium with respect to calcite, and deposits travertine when it degasses in the lower P_{CO_2} of the cave atmosphere. An open system with respect to gaseous CO_2 is required in the overlying vadose zone, implying the presence of fairly transmissive, aerated openings.

sphere. In many caves, trickles of water emerge from narrow fissures that have no discernible solutional widening, yet where they run down the cave wall the bedrock has been etched to depths as much as several centimeters where the water is absorbing CO_2.

Common-ion effects. The common-ion effect can cause the solutional capacity of water to decrease. For example, solution of gypsum or anhydrite will increase the saturation index of calcite and dolomite by increasing the calcium concentration. In water saturated with calcite, solution of dolomite can still proceed if the magnesium concentration is low, but the calcite is driven to supersaturation by the addition of calcium from the dolomite. Precipitation of calcite has been shown to occur through this

Figure 13. Jewel Cave, in the Black Hills of South Dakota, is a vast network maze along prominent joints in the Pahasapa Limestone (Mississippian). It is thought to have formed in part by rising thermal water (Bakalowicz and others, 1987). A 15-cm-thick wall crust of calcite crystals was deposited by nearly static water late in the development of the cave.

time periods this process can produce solutional pores and caves. The best examples of active thermal cave origin can be observed in Hungary (Ozoray, 1961). The decrease in hydrostatic pressure as the water rises tends to allow degassing of CO_2 near the water table, resulting in wall crusts of calcite (Fig. 13). Degassing takes place below the water table if the equilibrium P_{CO_2} in the water exceeds the hydrostatic pressure, and at the water table if it exceeds the P_{CO_2} of the cave atmosphere.

Solution in mixing zones. Although most underground solution conduits owe their origin to the activity of CO_2-rich water from surface recharge, several other processes can affect the solutional capacity of groundwater, particularly in areas where surface-derived CO_2 is limited. There are several ways in which the aggressiveness from surface-derived carbonic acid can be rejuvenated. Many of these produce deep-seated porosity that is not ordinarily considered a karst phenomenon but is important as oil reservoirs, permeability zones, and sites of ore emplacement.

Mixing of waters of different chemical character is perhaps the most common origin for such porosity. For example, water that infiltrates into a porous, poorly cemented limestone approaches saturation within a few centimeters or meters of the surface. However, if it encounters groundwater of different (usually lower) CO_2 content, the nonlinear relation between C_s and P_{CO_2} shown in Figure 10 causes the solutional aggressiveness of the mixture to exceed that of either of the two sources. Even if both of the initial solutions are saturated, the mixture will be undersaturated. This process was first applied to karst solution by Bögli (1964). It is less effective, and often of no consequence at all, if either or both of the solutions is supersaturated (Thrailkill, 1968). Similar effects can occur where waters of different H_2S content mix.

Mixing of waters of different salinity can achieve a similar effect, even if the saline components do not share equilibrium relations with carbonate species, because the equilibrium concentration of calcite increases with ionic strength (Runnels, 1969; Plummer, 1975; Wigley and Plummer, 1976; Back and others, 1984). The increase in calcite C_s with salinity is shown in Figure 10. The nonlinearity of C_s with salinity is not as great as with carbon dioxide or hydrogen sulfide, however.

Palmer and others (1977) found that fresh phreatic water in Bermuda is nearly always undersaturated, as in artificial collector galleries used for water supply, and that undersaturation is proportional to the rate of mixing with infiltrating water. The phreatic water usually has a lower CO_2 content and greater salinity than the infiltrating water. Brackish water in caves is usually saturated with calcite and has a seawater content of 80 to 90 percent, but during periods of high infiltration the seawater content at the water surface drops to about 20 to 40 percent and the mixture is aggressive to depths of about 5 m. This appears to be the zone of maximum cave development (Fig. 14). During periods of low infiltration the water at sea level can become supersaturated, but the high magnesium content of the water retards the precipitation of calcite. Aragonite precipitates at depths below about 10 m.

mechanism in Castleguard Cave, directly beneath the Columbia Ice Field, even though the water is in equilibrium with a CO_2 content below that of the cave air (Atkinson, 1983).

Solution by rising thermal water. The equilibrium concentration of dissolved calcite increases as groundwater cools (Fig. 10). Water rising from depth will show a systematic drop in temperature as it descends the geothermal gradient, and over long

Figure 14. Shark's Hole, Bermuda, a sea-level cave formed by solution in the zone of mixing between infiltrating fresh water and seawater. The shoreline niche shares this origin to some extent but is partly biogenic.

Solution in oxidation zones. Hydrogen sulfide is commonly produced by anaerobic bacteria in the decomposition of organic matter and reduction of gypsum. Oxidation of rising hydrogen sulfide (either gaseous or aqueous) produces sulfuric acid, which has a potent solutional effect on carbonate rocks. Although the effect on surface karst is rather small, this process has produced extensive porosity, including some very large caves. For example, the Kane Caves, near Lovell, Wyoming, were formed by sulfuric acid where H_2S-rich water rising from depth in the Bighorn Basin reached the water table (Egemeier, 1981). The lower of the two caves is still actively enlarging in this way at the level of the Bighorn River. Caves in the Guadalupe Mountains, such as Carlsbad Caverns, are thought to have formed either in the same manner (Davis, 1980) or by the rising of gaseous H_2S (Hill, 1981). Oxidation of metal sulfides, particularly those of iron, also releases sulfuric acid. This process, which is the well-known source of acid mine drainage, takes place slowly enough under natural conditions that its effect is usually limited to scattered porosity of little geomorphic significance.

Evolution of solution conduits in a karst aquifer

On the basis of the field and laboratory data described in previous paragraphs, it is possible to reconstruct the solutional history of the various kinds of carbonate aquifers. When soluble rock is first exposed at the surface by uplift or by breaching of overlying insoluble rocks, little water infiltrates, and yet the water table is rather high because the initial fissures are narrow and the permeability low. The hydrologic setting is similar to that of any competent insoluble rock. Solutionally enlarged pores may be present as the result of deep-seated solution, but in most carbonate aquifers they appear to have little influence on flow patterns.

Solution rates are most rapid immediately beneath the soil cover, where water has a high carbon dioxide content and low degree of saturation. Water penetrating the carbonate rock soon reaches the transition to a much lower solution rate, and the remaining slow solution is spread out over long distances within the fissure network. The upstream ends of the more prominent flow routes enlarge rapidly into funnel-shaped openings, but until the conduits develop turbulent flow the entrances tend to remain choked with soil and have little surface expression, as shown in Figure 4. Openings in the floors of perched surface streams possess the most favorable growth conditions. Although the water in such openings has a lower carbon dioxide content than that of soil water, they possess the greatest recharge rate.

Eventually one or more flow paths enlarge enough that water is able to pass through to a surface discharge point while still retaining much of its aggressiveness. Those few select fissure systems begin to enlarge rapidly along their entire length. After their lengthy gestation period, the conduits enlarge rapidly at a maximum rate of about 0.01 to 0.1 cm/yr. Finite-difference analyses of this solutional growth in idealized fissures using Equations 1 through 4 (Palmer, 1991) suggest that this stage is reached after about 10,000 to 100,000 yr, which agrees well with geomorphic field evidence. This requisite time varies directly with the length of the fissure system and inversely with the discharge and saturation concentration. Many trials under a variety of conditions show that the time is only poorly dependent on the nature of the solution kinetics. This is fortunate, because the kinetics near equilibrium are poorly understood.

These studies show that the conduits average only about 1 cm wide by the time rapid solution is achieved over their entire length. Their cross-sectional areas diminish considerably within the upstream few meters, but downstream from this entrance area there is negligible taper.

The increasing hydraulic conductivity of the karst aquifer causes the level of ponding in conduits to drop almost to the level of the spring outlets. Groundwater in surrounding openings is drawn toward these zones of low head, and as additional flow routes enlarge they tend to form a convergent pattern of conduits. Other less common patterns can develop where groundwater conditions are favorable, as shown later.

Sinkholes begin to form when the flow in conduits becomes great enough to transport the soil that chokes the entrances. Increasing amounts of water are funneled into the few large conduits at the expense of neighboring fissures. The large openings continue to grow rapidly, while surrounding openings stagnate with little flow and negligible enlargement.

Caves are eventually destroyed by degradation of the overlying surface. Their passages are interrupted by collapse and reduced to discontinuous fragments. These may still be accessible from the surface or through underlying passages, so that the original flow pattern can be interpreted.

Cave patterns

The pattern of solution conduits in a karst aquifer depends on several interacting geologic and hydrologic variables. At the most fundamental level, the location and extent of a cave depends on the position of its recharge and discharge areas. The pattern of passage interconnections depends on the nature of groundwater recharge to the karst aquifer. The exact layout and direction of individual passageways is controlled by local stratigraphy and geologic structure, and by the position of the water table. The vertical pattern is further influenced by the geomorphic history of the region.

Influence of groundwater recharge. Solutional cave patterns fall into two major categories: branchworks and mazes (Fig. 15). These patterns correlate with the mode of groundwater recharge to the aquifer (Palmer, 1975, 1984). Although recharge, in turn, is controlled by many aspects of the geologic and topographic setting, it is through this single variable that each of them exerts its influence on the type of cave that will form.

Recharge through an exposed surface of soluble rock produces branchwork caves. These are the most common, accounting for about 80 percent of all mapped caves. The selective processes by which only a few favored solution conduits develop, as described in the preceding section, are most active in this setting, whether or not the bedrock surface is bare or soil covered. Such caves develop simultaneously and interactively with surface karst depressions. Vadose passages descending from these surface inputs develop a branching pattern where the geologic structure forces the water to converge, as in a syncline or at fracture intersections. Otherwise they remain independent until they reach the phreatic zone, and it is common for vadose canyons to extend subparallel to one another along the local dip for distances of several hundred or even thousands of meters without converging. In the phreatic zone, under normal conditions, the low head in the largest passages attracts surrounding phreatic water. As the initial openings grow by solution, a branchwork cave is formed.

Ewers (1982) has simulated the development of solution conduits with laboratory models of highly soluble materials such as gypsum and plaster. A typical experiment consisted of preparing a slabbed face of the soluble material and pressing a transparent water-filled bag against it to form a slightly transmissive planar contact. Water was forced into the aperture at various locations and the solutional modification of the block was viewed through the bag. Solution channels developed from all inputs, growing in the downstream direction with time. When one eventually penetrated to the opposite end of the model, its hydraulic head diminished sharply, causing all others to grow toward it down the new potential gradient. A dendritic pattern was created in this way. Although the time scale and the specific form of conduits cannot be translated to that of actual karst aquifers because of unnaturally steep hydraulic gradients and different solution kinetics, these models dramatically illustrate the linking mechanism of phreatic conduits.

The convergent tendency of phreatic water is often reversed during severe floods. In a karst region, flooding is not confined to surface channels but is equally intense in caves fed by sinking streams. Aggressive water in areas of flood-water ponding is

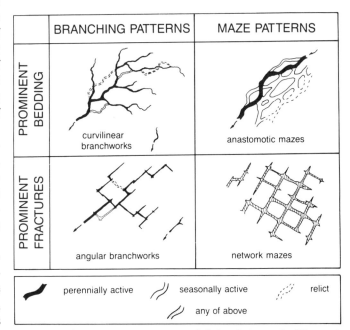

Figure 15. Common cave patterns and their relation to bedding and fractures. Branchwork patterns are generally formed by drainage from karst depressions. Solution by floodwater may superimpose local maze patterns of either type on branchwork caves. Network mazes can also be formed by diffuse recharge. Sponge-like patterns of enlarged pores (Fig. 3), which are usually formed by diffuse flow in mixing zones, rarely make up entire caves.

Figure 16. Slab breakdown in the thin-bedded Paoli Member of the Girkin Formation, Mammoth Cave.

forced into any available openings under steep hydraulic gradients, often enlarging a labyrinth of interconnecting passages. Despite the fact that flood-water solution operates only a small percentage of the time, the solutional capacity of this water is great enough that a traversable maze of passages can be formed in less than 10,000 yr. Mazes of this type are usually associated with allogenic recharge that accumulates on a broad area of insoluble rock and exhibits the flashiness typical of surface streams. Flooding and flow reversal are less severe in caves fed by sinkholes of limited drainage area, because a greater percentage of water is retained in the soil or held as ponds in surface depressions. Where the flooding and the structural setting are not sufficient to produce a maze, the solutional effect of flood water may be limited to ceiling and wall pockets or dead-end fissures.

Flood-water mazes are commonly superimposed on branchwork caves as diversion routes around blockages, where the build-up of head in cave streams is most severe. Ceiling collapse (Fig. 16), sediment accumulation, and insoluble beds are the usual causes of hydraulically inefficient constrictions. The additional head loss caused by a constriction is proportional to the square of the flow velocity through it, as well as to the change in cross-sectional area. Head loss across a constriction therefore increases as the square of the discharge. Since the flow increases many tens or hundreds of times during a typical flood, the head loss can be considerable.

In many such caves, flood-water conduits do not bypass the constriction, but simply form a subsidiary labyrinth of widened fissures (Figs. 17 and 18). These are partly or completely dry during low flow but receive overflow from stream passages at or near the water table during the rising limbs of floods. This water slowly drains back into the stream passages as the floods subside. This flow pattern has been documented in several caves by Palmer (1975). Hydrographs of wells in and adjacent to underground stream courses confirm this flow pattern (James Quinlan, Mammoth Cave, Kentucky, personal communication, 1983). The presence of such underground storage systems, or annexes, has been inferred by Mangin (1974–1975) from the analysis of karst spring hydrographs, and their presence has been further verified by chemical variation in the same springs (Bakalowicz, 1979).

Diffuse infiltration lies at the other end of the recharge spectrum and accounts for 10 to 15 percent of all known caves. Where water infiltrates into soluble rock through a competent cap of permeable but insoluble material such as sandstone, each fracture in the soluble rock receives comparable amounts of water and enlarges at rates similar to those of its neighbors (Fig. 19). A rather uniform network maze is produced with passages that follow every major fracture within the zone of maximum infiltration. The pattern is similar to that of city streets (Fig. 15). Sandstone is by far the most common permeable caprock of this type.

Diffuse infiltration also takes place through a surface of porous carbonate rock, but the water approaches saturation within a very short distance. This water produces caves only at depth where mixing is able to renew the solutional aggressiveness. Either a sponge-like pattern or a network of solutional openings is produced, depending on whether intergranular pores or fractures dominate.

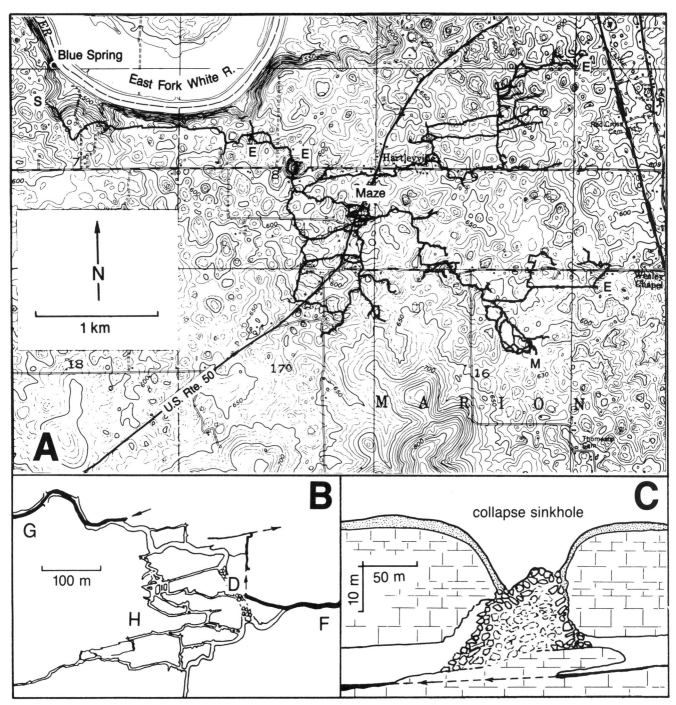

Figure 17. (A) Map of Blue Spring Cave, a branchwork cave fed by sinkholes in the Mitchell Plain of Indiana. A floodwater network maze circumvents a collapse in the main stream passage. (B) Enlarged map of the Maze. (C) Profile through the Maze and the overlying collapse sinkhole. E = entrances; M = upstream end of known part of main stream passage; S = sump in main stream passage; D = collapse zone that caused maze development; F, G = main stream passage; H = maze of flood-water passages around the collapse.

Figure 18. A view in the Maze of Blue Spring Cave, Indiana, showing the irregular solutional development typical of flood-water passages.

Rising groundwater is usually in the form of diffuse flow through fractures. It produces network or sponge-like mazes by renewing or augmenting its initial aggressiveness, through cooling, mixing, or sulfide oxidation. In the few places where rising water enters carbonate aquifers at isolated points, single-passage stream caves are formed.

Stratigraphic and structural controls. The configuration of groundwater divides, as shown by well data and dye tests, can be a faithful predictor of general flow directions in karst aquifers (cf., Quinlan and Ray, 1981). However, specific flow paths rarely follow the steepest component of the hydraulic gradient. Also, the routes taken by vadose flow can be highly devious, so that groundwater and topographic divides rarely coincide. Stratigraphic and structural data are necessary to predict (or, more realistically, to explain) local patterns of subsurface flow. In karst areas, much of this information comes from geologic mapping of caves (see White, 1969, and Palmer, 1984). Although caves do not represent all subsurface flow paths, their distribution has a major effect on the movement of the surrounding groundwater. Moreover, caves form only along the most favorable presolutional flow routes, so they can provide an intimate view of geologic influences on both vadose and phreatic water.

The geologic variable most pertinent to determining subsurface flow patterns in a karst region, and perhaps in all indurated sedimentary rocks, is the relative influence of bedding versus discordant fractures. At one extreme is the rock with thin but prominent beds separated by distinct bedding-plane partings. Such formations usually exhibit considerable lithologic variety. Very thin, shaley interbeds may enhance the tendency for beds to separate from one another, whereas thicker ones provide hydrologic barriers that inhibit flow across the strata. At the other extreme is massive, prominently jointed rock of fairly uniform lithology. Most rocks are of intermediate or composite character.

In massive, jointed rocks, caves have fissure-like passages with sharp, angular bends (Figs. 2 and 15). Vadose water is able to descend across the bedding in large steps. For example, the Coeymans Formation in New York is a massive, gently dipping Devonian biosparite. Vadose water descends through joint-controlled fissures and vertical shafts that almost invariably transect the entire formation in a single drop. Most phreatic groundwater flow in such rocks is moderately discordant to the bedding and follows rather direct paths to nearby entrenched valleys, although mechanical widening of joints along steep slopes by unloading can cause phreatic water to follow devious paths subparallel to the valley wall just before emerging at a spring (Renault, 1970). Where the recharge conditions favor maze caves, they take on a network pattern if fractures are prominent (Fig. 15).

Where bedding is prominent, flow patterns are more responsive to variations in dip and strike. Vadose perching is very common, as most fractures are not continuous enough to allow water to descend across the beds for great distances. Shaley beds provide excellent perching horizons, although a simple parting is sufficient. The preferred direction of gravitational water is, of course, vertically downward, but at any given point the vertical fractures or pores may not be sufficient to transmit all the incoming water. The excess is shunted along less steeply inclined openings, usually down the dip of stratal partings, thin insoluble beds, or more highly fractured beds. Most vadose flow follows a stair-step pattern of short vertical pitches alternating with lengthy downdip passageways (Fig. 6). In well-bedded, lithologically varied rocks such as the Ste. Genevieve Limestone of Kentucky, some vadose streams are known to follow the gentle dip for several kilometers before reaching the water table.

The effect of prominent bedding on phreatic water is equally strong. Rather than being able to flow down the steepest available gradient, water is usually confined along the same parting or favorable bed that it occupied where it first reached the water table. Because of the limited opportunity for discordant flow, plus a low-relief water table caused by the steep hydraulic conductivity, the most efficient flow route is nearly parallel to the strike of the beds, in the direction that leads most directly to a surface outlet. Along this course, phreatic water normally forms gently undulant passageways that loop up and down the dip of the controlling beds at and just below the water table while maintaining an overall strike orientation. Many karst aquifers are sufficiently fractured to allow flow to cut discordantly across the beds, and the tendency for strike-oriented flow, though still present, is greatly reduced.

Faulting can disrupt the tendency for perching or stratal confinement in a well-bedded rock. For example, in Ellison's Cave in northwestern Georgia, vadose streams remain perched for hundreds of meters on shaley Mississippian limestone more than 350 m above a nearby valley. The water encounters a zone of strike-slip faulting and originally plunged 200 m in a single vertical shaft, although because of headward retreat the water now drops in two steps of 40 and 155 m. This is the deepest known natural shaft in the United States.

Maze caves in prominently bedded rocks are mostly of the anastomotic type produced by floodwater (Fig. 15). Diffuse recharge into such rocks usually results in interstratal solution and may dissolve sponge-like cavities that sprawl in two dimensions along the bedding.

The depth of cave development beneath the water table was a topic of lively debate among early karst geomorphologists. Some, including Davis (1930) and Bretz (1942), favored a deep-phreatic origin, whereas others, such as Swinnerton (1932), favored an origin at or near the water table. More recent evidence (White and Longyear, 1962; Ford, 1971) shows that caves can form at a variety of depths, depending on where the greatest flow occurs.

Ford (1971) classified typical stream caves according to their depth of penetration below the original water table (Fig. 20). He also envisioned the depth of solution to be greatest during the earliest stages and becoming shallower with time as fractures and partings increase in number, owing to release of stress through unloading and cave development. Palmer (1969), based on the strong dependence of groundwater discharge on the width of openings in Equation 1, considered depth of solution to be limited by the downward reduction of fracture and parting widths, and bathyphreatic solution to be prevalent only where faulting or folding creates deep openings that are fortuitously greater in width than shallower ones. The two viewpoints, emphasizing fissure frequency and width, respectively, are complementary.

Figure 19. A cave dissolved in the Girkin Formation by water infiltrating through permeable quartzose sandstone of the Big Clifty Formation, James Cave, Kentucky. The bed at the top of the photograph is sandstone that has subsided slightly into the enlarging void.

Influence of fluvial base level. The vertical arrangement of passages in a cave is strongly influenced by the erosional history of the region. When entrenched rivers experience a long period of rather static base level, cave springs along the river banks tend to remain stable throughout that time without shifting elevation. Caves at that base level remain active long enough to grow by solution to a relatively large size (Fig. 6). On the other hand, rapid valley deepening causes frequent diversion of groundwater to lower elevations. Passages are quickly abandoned by their water or acquire entrenched canyons in their floors, so

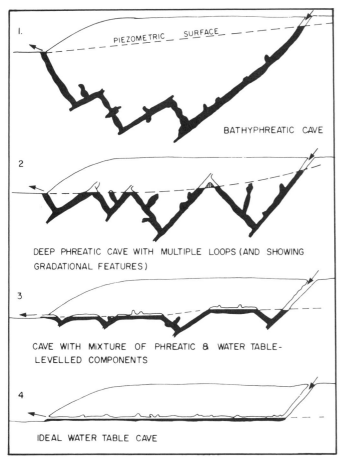

Figure 20. Relation between the vertical relief of phreatic cave passages and the spatial frequency of transmissive partings and fractures, according to Ford (1971). Phreatic development tends to evolve from older bathyphreatic passages toward younger water-table passages as erosional unloading and cave development increase the frequency of penetrable openings in the bedrock. (Reproduced with permission of D. C. Ford.)

growth of the original water-filled sections is arrested. Thus, the major passages in an extensive cave are commonly arranged in fairly persistent levels of development and are surrounded by more profuse narrower passages.

Water in phreatic conduits tends to divert to lower flow paths only if there is a drop in base level. Most of the new conduits form at or near the current fluvial base level, and the old high-level ones are gradually abandoned. Relict passages may continue to receive vadose water, but this water is out of adjustment with the original phreatic passage configuration and only modifies it by entrenchment or by depositing precipitates.

In contrast to the long-term stability of phreatic conduits, all channels in the vadose zone are continuously susceptible to diversion to new routes, except for those very few that extend vertically the entire distance from the surface to the water table. Inclined passages are influent, losing water to less efficient fissures in their floors. As these lower routes enlarge by solution, they pirate progressively greater amounts of water from the original passage, until it is active only during high flow or is abandoned entirely. The initial stages of this process may be very slow, because many vadose passages are perched on relatively insoluble beds. However, a great many abandoned vadose routes may be present, especially in a thin-bedded rock with prominent bedding-plane partings.

Interpretation of relict cave passages

Relict passages can persist for a long time without modification by surface degradation. As a result, a complex cave with passages at many levels can provide a great deal of information about the erosional/depositional history of the region.

Former stands of base level are most often recognized by the presence of fluvial terraces, but older terraces are easily obscured by destructive surface processes. In karst areas a more reliable indicator of a past base level is the elevation of former vadose-phreatic transition points in relict cave passages. The change from steeply inclined downdip canyons to tubes with ungraded profiles is most diagnostic.

The bedrock walls of a solutional cave show a variety of features that indicate the nature of the original flow and cave development. Most common are solutional scallops, which are asymmetrical hollows dissolved in the bedrock surface by turbulent flow (Fig. 21). In the same manner as dunes and ripples, their asymmetry indicates the flow direction during the last solutional episode. Scallop length is inversely proportional to the flow velocity, and Curl (1974) has provided a means of calculating the mean velocity on the basis of dimensional analysis and experiment. As a crude approximation at temperate groundwater temperatures, the velocity in cm/sec is obtained by dividing the constant 350 by the mean scallop length in centimeters.

Vertical flutes are solutional wall grooves that indicate the presence of vadose drips or waterfalls (Fig. 8). These resemble the flutes in the sides of classical columns and are abundant in most vadose shafts and passages intersected by descending vadose water.

The walls of cave passages may contain a myriad of small, interconnected tubes along bedding-plane partings, called anastomoses. These have a braided pattern similar to that of the anastomotic maze shown in Figure 15, except that they are only a few centimeters or tens of centimeters in diameter. They appear to be similar in origin, also, in that anastomoses are most common in passages that have undergone periodic flooding and probably have been dissolved by the spelean equivalent of bank storage in surface streams. A different origin is favored by most karst researchers, i.e., that they represent the initial stage of conduit development, and that the cave passage itself represents the coalescing of the larger tubes, with only remnants of anastomoses left in the walls. The former origin is more likely, however, because the solutional enlargement of many competing flow paths to form a maze pattern is strongly favored by aggressive high-gradient water far from saturation. Anastomoses differ from anastomotic caves not only in size, but also in their origin by

Figure 21. Scallops in the walls of a relict cave passage indicate flow direction and approximate velocity. The direction is indicated by scallop asymmetry, as in a sand dune. Velocity is inversely proportional to scallop length. The latest solvent flow at this point was toward the left at about 120 cm/sec.

in-and-out surges of water rather than by the high-velocity unidirectional flow that forms traversable mazes.

Solutional ceiling and wall pockets are generally concentrated in those parts of a cave most susceptible to flooding. Their origin is similar to that of anastomoses, except that most are simple dead-end indentations dissolved along joints, rather than bedding planes. Where they are large and profuse, they create a highly irregular passage profile and cross section (Fig. 18).

Effects of karst groundwater processes on the overlying surface

Most of the larger surface features in a karst terrane owe their origin either directly or indirectly to the solutional activity of subsurface water. Surface processes triggered by groundwater activity include subsidence of regolith and loose bedrock blocks into subjacent voids by mass wasting, collapse of cavern roofs, and accelerated soil erosion due to sharp local relief. But the process most essential to the origin of surface karst features is often overlooked: the ability of subsurface cave streams to erode and transport detrital material. Without this important ingredient the subsidence and collapse of surficial material into underground voids would be self-limiting.

Development of karst valleys. The first solution conduits to form in a typical karst aquifer are fed by influent surface streams perched above base level. Diversion of the surface streams into the growing conduits progresses in the headward direction until the streams are reduced to only a few ephemeral upstream remnants, or until a contact with insoluble rock is reached. During this evolution, the active parts of the stream valley continue to be eroded downward, while the abandoned channel remains at approximately the original level. A great deal of dissolved and detrital material is carried underground through conduits, forming a closed valley with no surface outlet. The valley floor is pocked by sinkholes formed either directly by the sinking stream or later by less concentrated infiltration. It is common for such a karst valley to encompass millions or even billions of cubic meters within depression contour lines. This entire rock volume, plus a great deal more from higher parts of the catchment area, has been transported through subsurface conduits by groundwater flow.

Subsidence of regolith. Once the initial solution conduits form by diversion of surface streams, less concentrated infiltration elsewhere in the drainage basin begins to develop sinkholes and tributary conduits. The sinking stream phase is not always present, and if not, the origin of conduits is greatly delayed. In areas of thick soil and much biogenic CO_2, dissolution at the soil/bedrock interface can account for up to 80 percent of the dissolution within a karst drainage basin. Solutionally widened fissures can grow to considerable size in the bedrock surface; yet they have no topographic expression because mass wasting tends to fill most of the voids with regolith as rapidly as they grow (Fig. 4). Only where solution is concentrated along many major fractures or the overburden is thin or absent will there be substantial expression at the surface.

However, when solution conduits enlarge enough that underground water can carry detrital material to surface streams at lower elevations, surface depressions begin to develop rapidly as regolith is sapped underground. As the depressions enlarge and acquire greater catchment area, erosion by ephemeral streams and overland flow increases the amount of detrital material entering the conduits, further enlarging the depressions.

A shallow cave may be roofed by unconsolidated material over part of its length. As the lower parts drop into the cave, a regolith arch forms over the conduit. Eventually the arch fails, forming a steep-walled sinkhole.

Where regolith subsides into an underlying dry cave and is not carried away by subsurface water, a detrital cone may partly fill the underlying passage, eventually reaching the ceiling and clogging the inlet openings (Fig. 22). Such a process usually has no surface expression, and the few sinkholes that form in this way are small.

The subsidence of regolith into underlying voids is commonly known among karst researchers as "suffosion." However, the primary geomorphic usage of this term applies to upheavals of water-saturated regolith by subsurface processes, such as those related to periglacial activity. Its use should be accompanied by a description of how the term is being applied.

Bedrock collapse. Nearly all large karst depressions originate at least in part by sudden collapse or slow subsidence of bedrock fragments. The resulting depressions are usually termed "collapse sinkholes," although without subsurface evidence this origin is usually inferred simply from their size and irregularity. The surface expression of a collapse sinkhole can be indistinguishable from that of a purely solutional sinkhole.

Where shallow underground stream caves are roofed by thin regolith, usually a very local phenomenon, the eventual collapse produces a steep-walled sinkhole whose connection with the underground conduit is most commonly obscured by the collapse debris. With time, much of this material is removed from below, while slumping and surface erosion soften the contours of the depression. A new regolith arch may stabilize above the level of the subsurface stream, providing temporary stability. If not, a karst window is formed that allows direct access to the underground conduit.

The great majority of subsurface stream conduits are roofed by bedrock. As the conduits grow, bedrock blocks sag and drop from wide, unsupported ceilings (Fig. 16). Any single episode of breakdown usually involves individual pieces or packets of several thin slabs, but rarely an entire cave roof. As fragments accumulate on the cave floor, they occupy a volume roughly 25 to 40 percent greater than they did as solid bedrock. Therefore, as this process continues, it tends to be self-limiting: collapse reaches an upward limit where the fragments fill the void (Šušteršič, 1984). The last few ceiling slabs to come loose simply sag onto the top of the pile a tiny distance without actually coming loose from the ceiling. Collapse of a cave 5 m high could theoretically propagate upward no more than 12 to 20 m, depending on how tightly the blocks are packed.

Although the above situation may be valid for collapse of an abandoned relict passage, it is modified by the solutional and erosive capacity of karst groundwater in active caves. The collapse material offers a much greater surface area upon which solution can operate. Increased flow velocities around and

Figure 22. Cone of detrital sediment from subsidence of soil through a solutionally enlarged fracture in the Salem Limestone, Blue Spring Cave, Indiana. A small sinkhole overlies this point at the surface.

through the constriction sweep away most detrital sediment that might otherwise retard solution of the rock. Ponding of floodwaters upstream from the blockage can enlarge existing routes and form new diversion routes at a very fast rate, as shown previously. Small fragments are carried downstream away from the collapse zone. As a result of these processes, much of the material that falls from the ceiling is dissolved or carried away, leaving room for additional collapse. Under favorable circumstances the collapse can propagate all the way to the surface. If the overburden is thick it can temporarily mask all trace of the breakdown beneath it; but as detrital material subsides into the interfragmental voids beneath, some of it too is carried away by cave streams.

Bedrock collapse and regolith subsidence are most frequent during floods, when subsurface streams are high and their erosive capacity is at a maximum. Eyewitness reports tell of large collapse sinkholes developing within a few hours. Examination from below, in accessible caves, shows that many of these have undergone prior collapse, but that their surface expression had previously been masked by overburden. Erosion of the collapse blocks and their detrital matrix by cave streams produces an unstable arch capable of sudden collapse on a scale larger than would be expected for primary collapse of a solid cave roof.

A collapse sinkhole normally extends no deeper than the active cave passage that controls its development. The widest active passages, those most susceptible to collapse, usually lie at or just beneath the water table, although some are perched on insoluble rocks within the vadose zone. To a developing sinkhole the level of cave streams is analogous to the fluvial base level. Until the cave streams divert to a lower level, the sinkhole is limited in depth but may grow laterally by continued collapse, solution, and underground erosion. Mazes of diversion passages may develop around the collapse, promoting further breakdown. In the rare cases where these processes continue long enough, a flat-floored depression is eventually created that resembles a polje (see Chapter 7, this volume). Although groundwater tends to

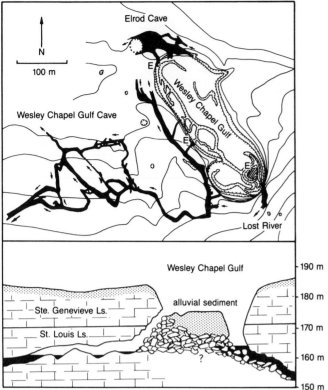

Figure 23. Map and cross section of Wesley Chapel Gulf, a polje-like depression in the Lost River basin of the Mitchell Plain, Indiana. E = cave entrance. (Map by C. A. Malott and R. L. Powell.)

circumvent the collapse and detritus beneath a sinkhole, by the time it evolves far enough to acquire a flat floor, it is common for cave streams to emerge at springs on one side of the depression and sink once more on the opposite side (Fig. 23). Periodic floods cover the depression floor with alluvial sediment, just as on the flood plain of a river.

The best-publicized examples of catastrophic collapse are those in the Tertiary limestones of Florida and other southeastern states, where large cavities are spanned by massive limestone and overlying siltstone. These materials are more competent than regolith but less so than most older and more highly indurated bedrock. Large spans can be produced beneath it, and primary collapse is frequently sudden and destructive. Sinkhole collapse in these limestones, in contrast to those in well-indurated rocks, appears to correlate with lowering of the water table by droughts or pumping, rather than by floods. When the overlying material experiences a reduction of pore pressure or a drop in the water table, it is much more liable to collapse. Most solution conduits in these areas lie below the water table, according to reports by scuba divers, and although there are local zones of high velocity, groundwater movement is generally rather slow and non-erosive.

In porous, poorly indurated limestones such as those of Bermuda, where much of the solution is concentrated in zones of mixing at the water table, the overlying limestone breaks into large blocks that subside into the enlarging voids, producing broad, shallow, soil-floored depressions. Lateral growth of depressions is aided by spalling of bedrock blocks into peripheral caves along irregular tension fractures concentric to the depression walls. The fractures and many of the peripheral caves dip away from the depressions at an average of 50 to 70 degrees. The main island is deeply penetrated by bays up to several kilometers in diameter, which probably owe their origin to karst solution and collapse. Similar features have been noted in the Bahamas and in the Yucatan (Back and others, 1984).

The spring outlet of an active cave is affected by the same processes that form collapse sinkholes. A spring tends to be rather unstable because of blockage by talus, colluvium, and ceiling collapse. Water ponds upstream from a blocked spring during floods, just as it does at any constriction in the passage, and commonly forms diversion passages and distributary mazes. Collapse around the outlet causes the valley wall to retreat, forming an embayment called a spring alcove. An example is shown in the walls of the East Fork of White River immediately southeast of Blue Spring (Fig. 17). Abandoned alcoves are common and represent the location of former spring outlets. Collapse can proceed far enough upstream at a spring that a steep-walled canyon is formed with the spring at its head. These are appropriately known as "steepheads." The larger ones present the unusual spectacle of a mature river valley abruptly ending upstream in a blank wall.

Subterranean meander cutoffs. In a surface river with entrenched meanders the hydraulic gradient through a meander neck can be great enough to form solutional piracy routes (Malott, 1922). The meander loop may become entirely dry except during flood periods when the river flow exceeds the capacity of the subsurface conduits. Concentrated solution and collapse at both the inlet and outlet accelerate breaching of the meander neck. Eventually the diversion route becomes entirely exposed and the cut-off is complete. It is likely that nearly all cut-offs of entrenched meanders in karst areas owe part of their origin to subsurface diversion of this type.

Spatial and temporal variations in karst

It is apparent from the previous sections that the surface depressions so characteristic of a karst landscape owe their existence to the solution conduits into which they drain. A discussion of spatial and temporal variations in karst, therefore, should focus on the conditions that either foster or limit the development of conduits. In doing so, many secondary processes are being ignored that modify and add to the array of karst features (see full description in Chapter 7, this volume).

Lithology is the dominant control over the distribution of karst. Virtually all karst areas are limited to areas of exposure of carbonate rocks and gypsum. Evaporites other than gypsum are rarely exposed at the surface, except in arid basins where solutional features are small and short-lived. Solution conduits are rare in rocks having more than 50 percent insolubles, but this limitation can be overridden in the presence of other favorable conditions. Dolomite dissolves rapidly in highly aggressive water, but near saturation the solution rate diminishes greatly (Rauch and White, 1977; White, 1984). As a result, conduits develop very slowly in dolomite aquifers. Most dolomite karst surfaces are rubbly with poor sinkhole development. However, dolomite interbedded with limestone may actually be favored by dissolution, because water saturated with calcite is still aggressive toward dolomite at most groundwater temperatures.

Exposure of the soluble rock above the fluvial base level is necessary for a karst surface to form. The degree of karst development is proportional to the amount of relief and length of time of exposure. Greater topographic relief does not imply greater hydraulic gradients, except possibly during the initial stages of karst development, when the water table is relatively high. Gradients in the vadose zone are essentially equal to the slope of the conduits, since there is no pressure build-up, and gradients in the phreatic zone are adjusted to the size of the conduits and the discharge. However, the frequency and width of available fractures and partings increases with relief. Finite-difference modeling of solution kinetics shows that the time required for a conduit to develop a rapid solution rate and turbulent flow is inversely proportional to the square or cube of the initial fissure width.

This required time is also directly proportional to the distance of flow between input and outlet. Areas remote from entrenched surface valleys require a great amount of time to develop extensive karst features with well-developed subsurface drainage. Some of the bare limestone surfaces of Yugoslavia are so remote from their outlets that karst is limited to widened surface fissures, with no accessible caves leading to the springs. Water tables in these areas are high, indicating limited solutional permeability.

The degree of undersaturation of the infiltrating water has only a minor effect on the time required for conduits to form. In narrow fissures, water loses nearly all its aggressiveness in the first few meters of flow, so most of the length of a conduit undergoes its initial solution very close to saturation. Although the maximum solution rate in a large conduit is limited by the degree of undersaturation, differences in rate of enlargement are less significant than the great range of times required for various conduits to undergo their initial development near chemical equilibrium.

Caves and large karst depressions are most common where infiltration from the surface is concentrated by topographic convergence (e.g., in valley floors or glacial cirques) or by proximity to a recharge area on insoluble materials. A soil cover on soluble rock inhibits concentrated recharge at first, but with the appearance of sinkholes the soil helps to concentrate overland flow into discrete points of recharge, aiding the growth of selected conduits. This process is especially significant because of the low infiltration capacity of most karst soils.

Climate has a great influence on many of the variables described above. Its effect is greatest on the form and distribution of surface karst features. However, despite differences in solution rate, the specific groundwater processes and morphology of the resulting forms are virtually independent of climate.

Temperature affects saturation values in two ways: carbon dioxide is more soluble in water at lower temperatures, and the saturation concentration of dissolved carbonate is greater. On the other hand, the production of carbon dioxide in the soil generally increases with temperature (Harmon and others, 1973; Miotke, 1974), so that the amount of dissolved carbonate in karst groundwater is greater, not less, at higher temperatures. Furthermore, the rate of solution increases with temperature.

Climate also has a great effect on recharge rates from the surface. Variations in the amount of rainfall and snow are as significant as temperature in controlling solution rates. The net rate of carbonate removal from a drainage basin is directly proportional to the annual precipitation. In alpine karst, for example, the disadvantages of low carbon dioxide production and temperature are offset by the large rainfall and snowmelt, and such areas rival subtropical and tropical regions in having the greatest rates of solutional denudation (Bögli, 1980).

The absence of soil, as in an alpine setting, allows a large percentage of available water to enter the bedrock. Little of the solutional potential is used up at the surface because of rapid runoff into fissures, and a dense array of subsurface conduits forms readily. Where there is a thick soil, water disperses slowly through it in response to differences in capillary potential, and much of its aggressiveness is lost at the soil/bedrock surface. Evapotranspiration further reduces the amount of aggressive water that is free to penetrate into the bedrock to form conduits.

Variation in the depth of snow cover exerts a strong control over the distribution and intensity of karst, especially in mountainous regions where windward slopes may be swept almost free of snow, whereas those on the downwind side accumulate large drifts that persist year round. Meltwater during warm seasons, in combination with rainfall, concentrates much greater amounts of infiltration in these areas than could be provided by uniformly distributed precipitation.

Soluble rocks beneath semi-permanent glaciers tend to stagnate in their karst development, although seasonal meltwaters can be significant in forming and enlarging caves. For example, roughly half of 20-km-long Castleguard Cave, including the major recharge areas, extends beneath the Columbia Ice Field (Ford and others, 1983). Glacial meltwater continues to enlarge the cave today and was probably also responsible for its origin.

Mechanical weathering in arid climates tends to overwhelm the solutional processes, and large surface karst features are rare. Those that do exist are mainly relics of more humid climates or caves of deep-seated origin exposed fortuitously at the surface. However, karst springs are not uncommon in arid regions, and some that drain extensive and generally mountainous catchments are among the largest in the world.

Current status of research in karst groundwater geomorphology

Karst geomorphology is presently at a crossroads. Whereas it has for some time been considered an esoteric subject of little practical value, geologists in many diverse fields are suddenly realizing that is is of considerable importance to their research. Geochemists, economic geologists, environmental scientists, and hydrologists are the most concerned with karst problems and are entering into cooperative ventures with geomorphologists having a background in the field. Some of the topics of greatest interest today are outlined below.

Geochemists are concerned with resolving the problem of reaction kinetics of carbonate solution and deposition, and carbonate sedimentologists are applying these principles to diagenetic problems. The relative importance of mass transport of dissolved species through the solvent water versus the reaction at the rock/water interface is still not certain in natural settings. The influence of microbial activity on karst groundwater processes has barely been touched.

Many ores, particularly those of suspected hydrothermal origin, have been emplaced in carbonate rocks in association with karst solution and brecciation. Paleokarst is a particularly common host for lead-zinc ores. The interaction between deep-seated fluids and carbonate rocks is still poorly understood.

Hydrologists and engineers are coming to grips with a particularly insidious problem in karst, that of deteriorating groundwater quality. Not only does the rapid and somewhat unpredictable flow of karst water transmit bacteria easily and with very little filtering, but the spread of contaminants from toxic wastes and accidental spills poses a serious health threat. In some areas of the country the accumulation of volatiles in underground conduits has reached explosive levels. The problems of predicting flow paths and monitoring contaminants in karst are just now being addressed. Another popular environmental issue is land stability in karst, particularly the prediction, prevention, and amelioration of sinkhole development.

Models of groundwater flow and quality are being developed on the basis of long-term monitoring of karst drainage basins. Despite the complexity of karst groundwater flow, the fact that it varies both in discharge and in chemistry provides internal checks on the validity of groundwater models.

In the realm of pure research, karst studies are providing considerable insight into geomorphic history and changes in climates and sea level. These studies have been aided by advances in cave geochronology, such as the analysis of paleomagnetism of cave sediments, Uranium-series and electron-spin resonance dating of speleothems, and isotopic composition of crystalline deposits.

There is no question that an understanding of karst groundwater processes is essential for every geologist, hydrologist, or engineer who contemplates working in a soluble-rock terrane.

CASE STUDY: KARST OF THE ILLINOIS BASIN IN KENTUCKY AND INDIANA

Mississippian limestones crop out in a broad band around the eastern and southern perimeter of the Illinois Basin in Indiana and Kentucky, forming one of the most extensive examples of karst topography in the United States (Fig. 24). Nearly all the processes and forms described earlier in this and the previous chapter can be illustrated by examples from this area. Geologically it has some limitations, as its relief is rather uniform and low and there is little structural deformation. Many of the structural relations can be recognized only through careful surveying.

There has been more study and interpretation of this karst region than any other in the country. The most detailed work includes that of Malott (1945), Powell (1964, 1970), Howard (1968), LaValle (1968), White and others (1970), Quinlan (1970), Miotke and Papenberg (1972), Palmer and Palmer (1975), Hess (1976), Bassett (1976), George (1976), Wells (1976), Quinlan and Rowe (1977), Woodson (1981), and Quinlan and others (1983). With this extensive background, there is no need for more than a brief summary of the regional setting. The following case study concentrates on selected topics related to groundwater processes rather than attempting to provide a comprehensive geomorphic view of the karst.

Geologic setting

The stratigraphic setting along the southeastern edge of the Illinois Basin is shown in Figure 25, together with the associated topographic features. In this area, the region around Mammoth Cave, Kentucky, the carbonate strata are thickest. A maximum of 200 m of relatively pure Mississippian carbonate rocks occur in this region. These are rather thin in comparison with the massive

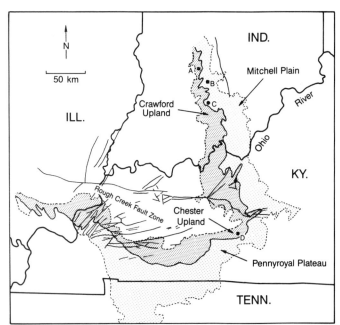

Figure 24. Map of the karst area around the southeastern rim of the Illinois Basin. The Pennyroyal Plateau and Mitchell Plain are low-relief karst plateaus of Mississippian limestones; in the Chester (Crawford) Upland the limestones form dissected ridges capped by predominantly insoluble rocks. Major faults are shown as thin solid lines. A = Popcorn Spring Cave; B = Blue Spring Cave; C = Wesley Chapel Gulf; D = Mammoth Cave.

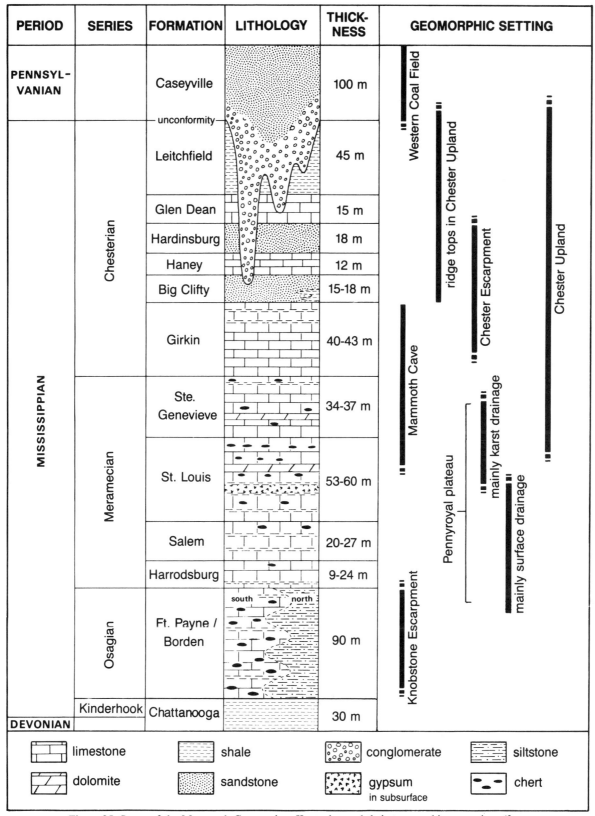

Figure 25. Strata of the Mammoth Cave region, Kentucky, and their topographic expressions (from Palmer, 1981).

Figure 26. The Mitchell Plain, a karst surface developed on the St. Louis Limestone over Blue Spring Cave, Indiana. Outliers of the Crawford Upland rise in the background.

carbonates of most orogenic regions, but the low dip in the Illinois Basin allows them to be exposed in a band of karst topography covering more than 20,000 km^2 in Kentucky and Indiana alone.

The carbonate rocks vary a great deal in lithology and internal structure. To the north and west, the thin, impure beds within the Girkin and upper Ste. Genevieve grade into thick clastic units that partition the limestone into several independent aquifers. In Kentucky, the lowest carbonate formations in this section, the Salem and Harrodsburg, are rather impure and are poor karst formers. In Indiana they are massive, pure, and prominently jointed and contain some of the most extensive caves. Over the entire region these formations grade upward into thin-bedded, shaley, and cherty limestones and dolomites of the upper Salem and the St. Louis Limestone. At depth the St. Louis contains thick beds of gypsum, but interstratal solution by groundwater has removed almost all of it from rocks exposed at or near the surface. The overlying Ste. Genevieve Limestone contains a variety of limestone textures and bedding thicknesses, but with fairly prominent bedding-plane partings. Its upper half contains thin, incompetent beds of shaley and silty limestone interspersed throughout the purer, more competent limestone. In Indiana and western Kentucky, the overlying Chesterian Series consists mainly of detrital rocks of deltaic origin alternating with thin limestones, but in the Mammoth Cave area the carbonates persist uninterrupted through much of the Chesterian as the Girkin Formation.

The best-developed karst in the region is located where relatively pure limestones are thickest and fluvial dissection is deepest. Thus, the deeply dissected Mammoth Cave area shows the most intense karst development, the low-relief plateaus and comparatively thin limestone of western Kentucky the least. Indiana lies between the two in degree of karst development. Although it possesses thinner limestone than the Mammoth Cave area, south-central Indiana has been deeply dissected by several major rivers. Some of this relief is masked by thick glacio-alluvial valley fill of late Pleistocene age.

The carbonate rocks form a broad, low-relief karst plain typified by sinkholes, sinking streams, and caves (Fig. 26). Perennial surface streams are almost entirely absent except for a few major entrenched rivers. This is called the Pennyroyal Plateau in Kentucky and the Mitchell Plain in Indiana. The overlying detrital rocks form a rugged, intensely dissected upland (Chester Upland in Kentucky, Crawford Upland in Indiana), whose updip fringes consist of cavernous limestone ridges capped by insoluble rocks (shown in the background of Fig. 26). The abrupt but irregular Chester Escarpment, which forms the border between the two landscapes, is about 50 to 75 m high.

The stratal dip is radially inward toward the center of the Illinois Basin at an average of about half a degree. As erosion proceeds, the regional landscape pattern gradually retreats in the downdip direction. Therefore, it is possible to obtain a generalized view of the karst evolution simply by scanning outward from the center of the basin. This steady-state view is not entirely valid, however, because the effects of varied rates of erosional dissection are superimposed.

The geologic structure is rather subdued, with only broad folding superimposed on the basinward dip. A major exception is the Rough Creek fault zone, which extends through western Kentucky, with displacements as great as 750 m (Fig. 24). Associated fluorite and lead-zinc deposits suggest movement of water from considerable depth along the faults.

The karst surface

Surface karst is most highly developed in the St. Louis Limestone, where it forms a band of sinkhole topography averaging 5 to 10 km in width, the so-called "sinkhole plain." Surface drainage is almost entirely absent, except for ephemeral streams, most of which drain underground after short distances, and a few major entrenched rivers. The overlying limestones have fewer and generally shallower sinkholes, partly because of colluvial fill from the nearby Chester Escarpment. Sinkholes in the underlying limestones are very sparse because of their greater insoluble content, although local concentrations of narrow, steep sinkholes occur in the massive, pure Salem Limestone in Indiana. Fluvial dissection in some areas has partitioned the lower limestones into isolated remnants, which contain only a few scattered karst features. In parts of Kentucky the limestones below the middle St. Louis are shaley enough to support perched surface streams, which sink wherever they encounter the purer overlying limestones.

In the Crawford and Chester Uplands, surface karst is mainly limited to dry valleys in which streams have breached the insoluble caprock and eventually disappeared underground into the limestone. A few large, steep, and deep collapse sinkholes have breached the insoluble caprock by collapse into underlying solutional voids.

Subsurface processes

To clarify the evolutionary nature of karst groundwater features, the following discussion moves from downdip to updip areas. By doing so, the spatial changes in existing geomorphology can be used as a crude indicator of landscape evolution through time.

Where the limestones lie at great depth within the Illinois Basin, drill cores show that subsurface solution is limited mainly to small pockets a few millimeters or centimeters in diameter. Their irregular geometry and distribution suggest that they are isolated from one another, or at least are not integrated tubes. They may result from deep-seated solution by hydrogen sulfide, an idea supported by the presence of natural gas and petroleum. These openings have little effect on later solution conduits, which instead follow integrated patterns of groundwater flow.

Where the major limestones are first exposed in valley bottoms in the Crawford and Chester Uplands, the conditions are not immediately favorable to conduit development. Widening of fractures and partings by erosional unloading has not progressed far, and the distances between potential recharge and discharge areas are large. In Indiana, however, deep flow has been established in the limestones by solution of interbedded gypsum in the St. Louis Limestone (Ash, 1984). Water emerges from springs in valley bottoms by rising upward through bedrock conduits, which may be fault controlled. Recharge is from the east where the limestone has greater exposure. Spring hydrographs show less variation and more dissolved sulfate in the areas farthest downdip, where the limestone has most recently been exposed, indicating that conduits are still poorly developed. Hydrogen sulfide emerges from the springs when hydraulic gradients are disturbed by unusual events, such as when a nearby gypsum mine at Shoals, Indiana, is flooded or dewatered. Ash (1984) considers the gypsum to be undergoing bacterial reduction at depth, producing hydrogen sulfide and allowing subsurface conduits to form in and adjacent to the gypsum beds. Leaching of the gypsum is so thorough that virtually all of it has been removed by the time the St. Louis is exposed at the surface. The only visible evidence of the former gypsum is local distortion and brecciation of the limestone, but this disruption may aid the development of later karst.

Sulfur springs located along the Rough Creek fault zone in Grayson County, Kentucky, show similar evidence for deep flow (George and others, 1984). Some of the springs are warm, and past hydrothermal activity is indicated by minor lead-zinc mineralization along subsidiary fractures. In western Kentucky, many faults are masked by overburden, and yet their presence can be traced by the alignment of karst features (Kastning, 1984).

The updip edge of the Crawford and Chester Uplands is dissected into numerous ridges of limestone capped by insoluble rock. Despite the paucity of surface karst, cave development is as intense as anywhere in the world. Mammoth Cave in Kentucky and Wyandotte and Marengo Caves in Indiana are among the best known. Water infiltrates through sinkholes and sinking streams on adjacent parts of the sinkhole plain and in karst valleys, the remnants of influent stream valleys, which separate some of the ridges from one another. It also enters along the contact between the clastic caprock and the limestone. Caves are common within the ridges, but only where water from peripheral areas is able to flow through them. Little or no aggressive recharge enters through the thick, insoluble caprock.

Figure 27 is a cross section through the Mammoth Cave area showing the pattern of caves with relation to the topography and geology. Most passages are sinuous, curvilinear, and highly concordant with the strata because of the prominent bedding. Vadose shafts and canyon passages descend steeply from recharge points along the ridge flanks, joining low-gradient tubes that are generally fed by more remote sources. Some of the tubes and canyons are more than 20 m in diameter, although most are much smaller. The largest vertical shafts are about 10 m in diameter and 60 m in vertical extent. The appearance of a strong downdip orientation of passages in Figure 27 is partly illusory, because many of the tubular passages lie at large angles to the cross section, almost parallel to the strike. More than half the passages are vadose canyons, however, and these have a very strong downdip tendency. The insoluble cover greatly delays the destruction of relict passages by surface processes, allowing numerous passage levels to coexist.

As shown by the map of Popcorn Spring Cave (Fig. 28), the passage pattern within the ridges has little relation to the topography but is controlled instead by geologic structure. Surveys by the author show that the main passage is nearly parallel to the local strike.

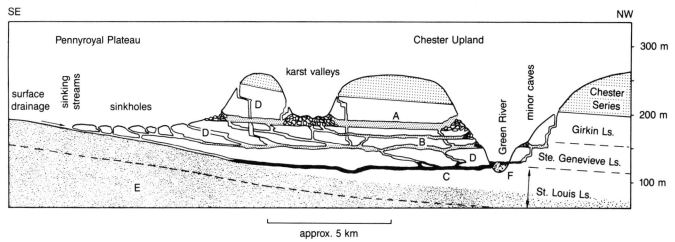

Figure 27. Profile through the Mammoth Cave area, showing the relation of cave passages to the geology and topography. A = large relict Tertiary canyons and tubes; B = relict Quaternary tubes; C = active phreatic tubes; D = active and relict vadose passages; E = impure limestones with poor karst development; F = late Quaternary alluvium.

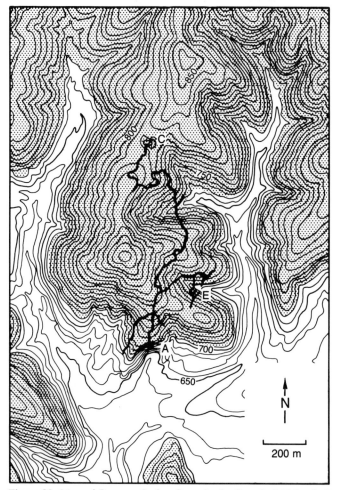

Figure 28. Map of Popcorn Spring Cave, in the Crawford Upland of Indiana, showing its general lack of relation to topographic details. The shaded area represents the sandstone and shale caprock over the cavernous limestone. E = cave entrance; C = collapse sinkhole in the insoluble rock, caused by solution in the underlying limestone; A = spring alcove at Popcorn Spring. Elevations are in feet; contour interval = 10 ft.

Nevertheless, the overall pattern of caves is controlled by the relative location of recharge and discharge areas. For example, the south-flowing headwaters of Indian Creek in the eastern part of the Crawford Upland near Bloomington, Indiana, were pirated underground to the west, in the downdip direction, to the more deeply entrenched Richland Creek (DesMarais, 1973). An extensive series of caves was formed during the various stages of piracy, as well as afterward by water from the resulting karst valley and from ridge flanks.

Caves above base level in the Crawford and Chester Uplands show little evidence for deep-phreatic flow. Identification of the former vadose/phreatic transition in relict passages in Mammoth Cave (Palmer, 1981) shows that most phreatic passages extended no more than a few meters below the water table. The deepest extent was 21 m in rather massive, prominently jointed beds. Evidently, the deep flow through gypsiferous beds in the St. Louis and along major faults is not typical of groundwater flow throughout the region.

The insoluble caprock has not interfered with the branching tendency of these caves, which is shared by most caves that are fed by karst depressions. It controls the distribution of caves by restricting recharge to narrow bands, so the branching pattern is distorted in comparison with a dendritic surface stream. Furthermore, the structural control, inaccessibility of many passages, and superposition of several stages of development all help to mask the branching pattern on cave maps, as in Figure 15. Yet the active passages at any time display this pattern quite distinctly.

As the upland becomes more greatly dissected, the number of recharge sites increases, so the pattern of caves becomes more complex. Newly formed vadose shafts and canyons intersect older passages, creating a complex tangle of passages at all levels. In Mammoth Cave alone, the total surveyed length of interconnected passages is more than 510 km. Caves in remnant outliers of the upland consist almost entirely of vadose canyons and shafts, with a few relict fragments of phreatic tubes. Where the limestone is capped with only a thin bed of sandstone, as in some of the outliers south of Mammoth Cave, the uppermost cave levels are rectilinear mazes formed by diffuse infiltration along joints in the sandstone.

In the Mitchell Plain and Pennyroyal Plateau, where the insoluble caprock has been completely removed, recharge to caves takes place mainly through overlying sinkholes (Fig. 17). Each sinkhole is the feeder for a cave passage, or at least a solution conduit capable of transmitting detrital sediment. Caves exhibit a more open branching pattern than those of the upland ridges because of the uniform distribution of recharge points. Some of the groundwater in the Mitchell Plain follows deep paths through the St. Louis Limestone and rises upward along fracture-controlled openings. This pattern is apparently inherited from earlier solution of gypsum beds. For example, Lost River is a large surface stream that sinks in the western part of the Mitchell Plain (Fig. 29). Its underground course may be seen in several collapse sinkholes. The downstream 3.6 km of the underground route loops downward slightly across the strata and rises 45 m through a vertical bedrock tube into the surface continuation of the river.

Caves in the St. Louis Limestone are not as accessible as those in higher strata. The incompetent, shaley limestone tends to break down easily, and although some large passages are known, most are merely short segments truncated by collapse. Passages are curvilinear like those in the younger limestones of the upland. There are a few exceptionally large caves, including the Hidden River Cave System east of Mammoth Cave, which contains more than 35 km of passages. Its downstream end consists of an extensive distributary network of anastomotic floodwater tubes (Quinlan and Rowe, 1977).

The Salem Limestone of Indiana is so massive and competent that its caves have an unusually high continuity of traversable passages. The largest, Blue Spring Cave near Bedford, has a total passage length of more than 32 km (Fig. 17; Palmer, 1969). Most caves in the Salem have highly angular patterns reflecting the prominent jointing. Where the lower St. Louis beds overlie cavernous parts of the Salem, sinkholes in the St. Louis feed recharge into the caves by rather devious vadose routes, with the result that only the large collapse sinkholes correlate well with the position of mapped cave passages. Where the Salem is exposed at the surface, its sinkholes usually lead directly into underlying caves. At the extreme updip edge of the Mitchell Plain, underground conduits are small and fragmentary.

Almost all caves in the sinkhole plain contain active streams that flood readily. Floodwaters have created diversion routes and mazes in many of them. The best example is that in Blue Spring Cave, where collapse at the junction of two large passages has caused ponded flood waters to dissolve a network maze containing 2 km of fissure passages (Figs. 17 and 18). The fracture pattern followed by the maze is quite different from that in the rest of the cave, apparently formed by stresses related to the breakdown. A large compound sinkhole has developed over the collapse zone, and it will continue to grow as further solution and

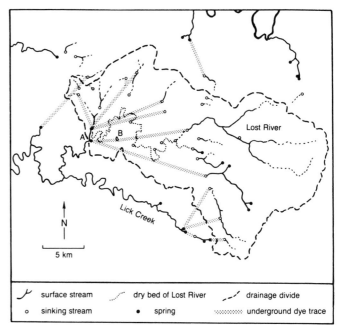

Figure 29. Map of the Lost River karst basin, showing patterns of underground streams as determined by dye traces (from Bassett, 1976). A = main spring of underground Lost River; B = Wesley Chapel Gulf (Fig. 23).

collapse take place. A more advanced stage of sinkhole growth by this process can be seen at Wesley Chapel Gulf, a large, flat-floored depression along the underground course of Lost River (Fig. 23).

It seems reasonable to expect that groundwater systems in the Mitchell Plain and Pennyroyal Plateau are simply inherited from those in the Crawford and Chester Uplands, with their upper levels truncated by erosion and their lower levels expanded by the addition of new recharge areas. By a quirk of geomorphic history, this is not the case. The rather flat surface of the Mitchell Plain and Pennyroyal Plateau was formed during a lengthy period of slow fluvial erosion late in the Tertiary. The surface lay near fluvial base level and supported surface drainage. An overburden of alluvium, colluvium, and insolubles from the limestone accumulated to thicknesses as great as 30 m. Underground drainage was probably very limited. Drainage from this surface formed caves in the Chester Upland, particularly in the Mammoth Cave area, where the base of the caprock lay as much as 60 m above the fluvial base level.

Quaternary glaciation caused major changes in drainage patterns in the region, which included diversion of the pre-glacial Teays River system into the Ohio River (Thornbury, 1965). The drainage area of the Ohio more than doubled, causing its entrenchment rate and that of its tributaries to increase markedly. Pleistocene entrenchment of the Mitchell Plain and Pennyroyal Plateau allowed surface drainage to sink and caves to develop. Caves in the upland acquired lower levels as valleys deepened. Therefore, groundwater systems in the sinkhole plain developed

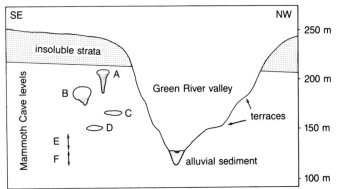

Figure 30. Composite cross section of passage levels in Mammoth Cave, showing passage levels A through F, with their relation to terraces in the Green River valley. The upper terrace is accordant with the local level of the Pennyroyal Plateau and karst valleys in the Chester Upland. A faint terrace at passage level C has been largely obliterated by surface erosion and mass wasting.

independently during the Pleistocene and were not inherited from the upland.

Much of this information comes from the mapping of passage levels in caves, particularly Mammoth Cave, which has the most complete array of Tertiary and Quaternary passages in the region (Miotke and Palmer, 1972; Fig. 30). Late Tertiary passages at and above the Pennyroyal surface are wide canyons and tubes partly filled with alluvial sand and gravel (Fig. 31). These suggest slow fluvial dissection alternating with periods of aggradation. Quaternary passages below the Pennyroyal are smaller and represent rapid dissection. They include several distinct levels of tubes that formed during periods of relatively static base level. Each passage within a given level is located in different beds, which encompass a great variety of rock types, so the levels are not stratigraphically controlled.

The interpretation of karst groundwater systems therefore requires more than a steady-state approach. Although a general view of their evolution can be obtained by scanning the differences between downdip and updip areas in relatively undeformed regions, it is necessary to superimpose the specific history of base-level changes in order to decipher the geomorphic details.

Environmental problems related to karst groundwater

The Mitchell Plain and Pennyroyal Plateau compose a pleasant landscape of farmland, with a few small cities and a growing rural and suburban population. Life on the sinkhole plain has its difficulties, however, particularly with regard to underground drainage. These are described in detail in a symposium volume on environmental problems in karst, sponsored by the National Water Well Association (1986). Some of the more serious problems associated with a karst landscape are outlined below.

1. Water wells show a high spatial variation in yield, even over short distances. This variation is greatest in the prominently jointed Salem Limestone (Palmer, 1969). Many wells have an unacceptably low yield or are entirely dry.

2. Wells with high yield are often contaminated. Effluent from septic tanks, drainage through trash and carcasses in sinkholes, and toxic wastes all pass through solution conduits with virtually no filtration.

Figure 31. Audubon Avenue, in Mammoth Cave, a large relict passage of Tertiary age filled to half its original depth by streamborne sediment.

3. Toxic spills can cause serious and rapid contamination of groundwater over a large area. Leakage of gasoline tanks has killed underground aquatic life, including members of endangered species such as eyeless fish. Fumes from underground conduits have seeped upward into overlying buildings, and water in some wells has been found to contain flammable gases close to the explosive threshold. In 1980 a cyanide spill along Interstate Highway 65 near Cave City, Kentucky, was contained just before it drained into nearby sinkholes.

4. Silting of sinkholes as a result of poor land use causes them to flood during wet periods. In some areas it is necessary to install drain pipes that penetrate to underground solution conduits. In low-relief areas the problem is the opposite: underground floodwater can rise through sinkhole bottoms, inundating the surrounding land.

5. Other land-use problems include instability and subsidence of land, leakage of reservoirs, and cut-and-fill requirements during the construction of highways and foundations.

Such problems can be minimized through the understanding of groundwater processes in karst terranes. Nevertheless, even with considerable regulation and monitoring, the environmental impact of high population density and industrialization is much greater in a karst area than in nearly any other kind of terrain.

CONCLUSIONS

Karst aquifers probably require a broader range of interpretive tools than any other geomorphic system, owing to the complex interaction among hydrologic, chemical, geologic, and topograhpic variables. Nevertheless, a fairly complete and detailed picture of karst groundwater processes is available, thanks to the accessibility provided by underground solution conduits. The applicability of this information to land-use planning and groundwater protection has become apparent only in recent years. It is time for geomorphologists, hydrologists, engineers, and planners to share their knowledge so that karst problems can be approached in the most systematic and effective way.

ACKNOWLEDGMENTS

I wish to thank the following technical reviewers of this chapter and its case study: Derek C. Ford, Department of Geography, McMaster University; and John Thrailkill, Department of Geology, University of Kentucky. Their recommendations have contributed greatly to the text.

REFERENCES CITED

Ash, D. W., 1984, Evidence for deep-seated groundwater movement in Middle Mississippian carbonate lithologies of south-central Indiana [abs.], in Proceedings National Speleological Society Annual Convention, Sheridan, Wyoming: National Speleological Society, p. 28.

Atkinson, T. C., 1983, Growth mechanism of speleothems in Castleguard Cave, Columbia Icefields, Alberta, Canada, in Ford, D. C., ed., Castleguard Cave, Columbia Icefields area, Rocky Mountains of Canada, Alberta, Canada: Arctic and Alpine Research, v. 15, p. 523-536.

Atkinson, T. C., and Smith, D. I., 1976, The erosion of limestones, in Ford, T. D., and Cullingford, C.H.D., eds., The science of speleology: London, Academic Press, p. 151-177.

Back, W., Hanshaw, B., and Van Driel, J. N., 1984, Role of groundwater in shaping the eastern coastline of the Yucatan Peninsula, Mexico, in LaFleur, R. G., cd., Groundwater as a geomorphic agent: Boston, Massachusetts, Allen and Unwin, p. 281-293.

Bakalowicz, M., 1979, Contribution de la géochemie des eaux a la connaissance de l'aquifère karstique et de la karstification [Ph.D. thesis]: Paris, L'Université Pierre et Marie Curie, 269 p.

Bakalowicz, M., Ford, D. C., Miller, T., Palmer, A. N., and Palmer, M. V., 1987, Thermal genesis of dissolution caves in the Black Hills, South Dakota: Geological Society of America Bulletin, v. 99, p. 729-738.

Bassett, J., 1976, Hydrology and geochemistry of the upper Lost River drainage basin, Indiana: National Speleological Society Bulletin, v. 38, p. 79-87.

Berner, R. A., and Morse, J. W., 1974, Dissolution kinetics of calcium carbonate in sea water; 4, Theory of calcite dissolution: American Journal of Science, v. 274, p. 108-134.

Bögli, A., 1964, Mischungskorrosion, ein Beitrag zur Verkarstungsproblem: Erdkunde, v. 18, p. 83-92.

——, 1980, Karst hydrology and physical speleology: Berlin, Springer-Verlag, 284 p.

Boulton, N. S., and Streltsova, T. D., 1977, Flow to a well in an unconfined fractured aquifer, in Dilamarter, R. R., and Csallany, S. C., eds., Hydrologic problems in karst regions: Bowling Green, Western Kentucky University, p. 214-227.

Bray, L. G., 1975, Recent chemical work in the Ogof Ffynnon Ddu System; Further oxidation studies: Transactions of the British Cave Research Association, v. 2, p. 127-132.

Bretz, J H., 1942, Vadose and phreatic features of limestone caverns: Journal of Geology, v. 50, p. 675-811.

Buhmann, D., and Dreybrodt, W., 1984, The kinetics of calcite dissolution and precipitation in geologically relevant situations of karst areas; Part 2, Closed system: West Germany, Universität Bremen, Fachbereich Physik, Report 6, 27 p.

——, 1985, The kinetics of calcite dissolution and precipitation in geologically relevant situations of karst areas; 1, Open system: Chemical Geology, v. 48, p. 189-211.

Corbel, J., 1959, Erosion en terrain calcaire: Annales de Géographie, v. 68, p. 97-116.

Coward, J.M.H., 1975, Paleohydrology and streamflow simulation of three karst basins in southeastern West Virginia [Ph.D. thesis]: Hamilton, Ontario, McMaster University, 394 p.

Cullen, J. J., and LaFleur, R. G., 1984, Theoretical considerations on simulation of karstic aquifers, in LaFleur, R. G., ed., Groundwater as a geomorphic agent: Boston, Massachusetts, Allen and Unwin, p. 248-280.

Curl, R. L., 1968, Solution kinetics of calcite, in Proceedings 4th International Congress of Speleology, Ljubljana, Yugoslavia: International Union of Speleology, p. 61-66.

——, 1974, Deducing flow velocity in cave conduits from scallops: National Speleological Society Bulletin, v. 36, p. 1-5.

Cvijić, J., 1893, Das Karstphänomen: Geographische Abhandlungen, v. 5, no. 3, p. 215-319.

Davies, W. E., 1958, Caverns of West Virginia: West Virginia Geological and Economic Survey, v. 19a, 330 p.

Davis, D. G., 1980, Cave development in the Guadalupe Mountains; A critical

review of recent hypotheses: National Speleological Society Bulletin, v. 42, p. 42–48.
Davis, W. M., 1930, Origin of limestone caverns: Geological Society of America Bulletin, v. 41, p. 475–628.
DesMarais, D., ed., 1973, The Garrison Chapel karst area, *in* National Speleological Society Guidebook to 1973 Annual Convention, Bloomington, Indiana: National Speleological Society, p. 16–34.
Dreybrodt, W., 1981a, Kinetics of the dissolution of calcite and its applications to karstification: Chemical Geology, v. 31, p. 245–269.
—— , 1981b, Mixing corrosion in $CaCO_3–CO_2–H_2O$ systems and its role in the karstification of limestone areas: Chemical Geology, v. 32, p. 221–236.
Egemeier, S. J., 1981, Cavern development by thermal waters: National Speleological Society Bulletin, v. 43, no. 2, p. 31–51.
Ek, C., and Roques, H., 1972, Dissolution expérimentale de calcaires dans une solution de gaz carbonique; Note préliminaire: Transactions of the Cave Research Group of Great Britain, v. 14, no. 2, p. 67–72.
Ewers, R. O., 1982, Cavern development in the dimensions of length and breadth [Ph.D. thesis]: Hamilton, Ontario, McMaster University, 398 p.
Ford, D. C., 1965, The origin of limestone caverns; A model from the central Mendip Hills, England: National Speleological Society Bulletin, v. 22, no. 1, p. 109–132.
—— , 1971, Geologic structure and a new explanation of limestone cavern genesis: Transactions of the Cave Research Group of Great Britain, v. 13, no. 2, p. 81–94.
Ford, D. C., and Ewers, R. O., 1978, The development of limestone cave systems in the dimensions of length and depth: Canadian Journal of Earth Sciences, v. 15, p. 1783–1798.
Ford, D. C., Smart, P. L., and Ewers, R. O., 1983, The physiography and speleogenesis of Castleguard Cave, Columbia Icefields, Alberta, Canada, *in* Ford, D. C., ed., Castleguard Cave, Columbia Icefields area, Rocky Mountains of Canada, Alberta, Canada: Arctic and Alpine Research, v. 15, p. 523–536.
Gale, S. J., 1984, The hydraulics of conduit flow in carbonate aquifers: Journal of Hydrology, v. 70, p. 309–327.
Gardner, J. H., 1935, Origin and development of limestone caverns: Geological Society of America Bulletin, v. 46, p. 1255–1274.
George, A. I., 1976, Karst and cave distribution in north-central Kentucky: National Speleological Society Bulletin, v. 38, p. 93–98.
George, A. I., Quinlan, J. F., Cubbage, J. C., and Ray, J. A., 1984, Tectonic influence in karst development, Grayson, County, Kentucky, *in* National Speleological Society Guidebook to Kentucky Speleofest: National Speleological Society, v. 13, p. 48–65.
Grund, A., 1903, Die Karsthydrographie: Geographische Abhandlungen, v. 7, no. 3, p. 103–200.
Harmon, R., and 9 others, 1973, Geochemistry of karst waters in North America, *in* Proceedings 6th International Speleological Congress, Oloumec, Czechoslovakia: International Union of Speleology, v. 4, p. 103–114.
Herman, J. S., and White, W. B., 1985, Dissolution kinetics of dolomite; Effects of lithology and fluid flow velocity: Geochimica et Cosmochimica Acta, v. 49, p. 2017–2026.
Hess, J. W., 1976, A review of the hydrology of the central Kentucky karst: National Speleological Society Bulletin, v. 38, p. 99–102.
High, C. J., 1970, Aspects of the solutional erosion of limestone with special consideration of lithological factors [Ph.D. thesis]: Bristol, United Kingdom, University of Bristol, 228 p.
Hill, C. A., 1981, Speleogenesis of Carlsbad Cavern and other caves of the Guadalupe Mountains *in* Proceedings 8th International Speleological Congress, Bowling Green, Kentucky: International Union of Speleology, v. 1, p. 143–144.
Howard, A. D., 1964, Processes of limestone cave development: International Journal of Speleology, v. 1, p. 47–60.
—— , 1968, Stratigraphic and structural controls on landform development in the central Kentucky karst: National Speleological Society Bulletin, v. 30, no. 4, p. 95–114.

Howard, A. D., and Howard, B. Y., 1967, Solution of limestone under laminar flow between parallel boundaries: Caves and Karst, v. 9, p. 25–38.
Jacobson, R. L., and Langmuir, D., 1970, The chemical history of some spring waters in carbonate rocks: Ground Water, v. 8, p. 5–9.
Kastning, E. H., 1984, Hydrogeomorphic evolution of karsted plateaus in response to regional tectonism, *in* LaFleur, R. G., ed., Groundwater as a geomorphic agent: Boston, Massachusetts, Allen and Unwin, p. 236–244.
Katzer, F., 1909, Karst und Karsthydrographie: Zur Kunde der Balkanhalbinsel, no. 8, 94 p.
Kaye, C. A., 1957, The effect of solvent motion on limestone solution: Journal of Geology, v. 65, p. 35–46.
Lauritzen, S.-E., and 6 others, 1985, Morphology and hydraulics of an active phreatic conduit: British Cave Research Association Transactions, v. 12, no. 4, p. 139–146.
LaValle, P., 1968, Karst morphology in south-central Kentucky: Geographical Annals, v. 50, p. 94–108.
Lehmann, O., 1932, Die Hydrographie des Karstes: Leipzig, Erdkunde, Band 6, 212 p.
Malott, C. A., 1922, A subterranean cut-off and other karst phenomena along Indian Creek, Lawrence County, Indiana: Indiana Academy of Science Proceedings, v. 31, p. 203–210.
—— , 1938, Invasion theory of cavern development [abs.]: Geological Society of America Proceedings, v. 41, p. 323.
—— , 1945, Significant features of the Indiana karst: Indiana Academy of Science Proceedings, v. 54, p. 8–24.
Mangin, A., 1974–1975, Contribution a l'étude hydrodynamique des aquifères karstiques: Annales de Spéléologie, v. 29, p. 283–601; v. 30, p. 21–124.
Martel, E. A., 1921, Nouveau traité des eaux souterraines: Paris, Librairie Octave Doin, 838 p.
Miotke, F.-D., 1974, Carbon dioxide and the soil atmosphere: Munich, Abhandlungen Karst und Höhlenkunde, Reihe A., Heft 9, 49 p.
Miotke, F.-D., and Palmer, A. N., 1972, Genetic relationship between caves and landforms in the Mammoth Cave National Park area: Würtzburg, Böhler Verlag, 69 p.
Miotke, F.-D., and Papenberg, H., 1972, Geomorphology and hydrology of the sinkhole plain and Glasgow Upland, central Kentucky karst: Caves and Karst, v. 14, p. 25–32.
Mylroie, J. E., 1977, Speleogenesis and karst geomorphology of the Helderberg Plateau, Schoharie County, New York [Ph.D. thesis]: Troy, New York, Rensselaer Polytechnic Institute, 336 p.
National Water Well Association, 1986, Environmental problems in karst terranes and their solutions: Dublin, Ohio, National Water Well Association, 525 p.
Newson, M. D., 1971, The role of abrasion in cavern development: Cave Research Group of Great Britain Transactions, v. 13, p. 101–107.
Ozoray, G., 1961, The mineral filling of the thermal spring caves of Budapest: Symposium Internationale di Speleologia Memoria 2, p. 152–170.
Palmer, A. N., 1969, A hydrologic study of the Indiana karst [Ph.D. thesis]: Bloomington, Indiana University, 181 p.
—— , 1972, Dynamics of a sinking stream system, Onesquethaw Cave, New York: National Speleological Society Bulletin, v. 34, p. 89–110.
—— , 1975, The origin of maze caves: National Speleological Society Bulletin, v. 37, p. 56–76.
—— , 1981, A geological guide to Mammoth Cave National Park: Teaneck, New Jersey, Zephyrus Press, 196 p.
—— , 1984, Geomorphic interpretation of karst features, *in* LaFleur, R. G., ed., Groundwater as a geomorphic agent: Boston, Massachusetts, Allen and Unwin, p. 173–209.
—— , 1991, The origin and morphology of limestone caves: Geological Society of America Bulletin (in press).
Palmer, A. N., Palmer, M. V., and Queen, J. M., 1977, Geology and origin of the caves of Bermuda, *in* Proceedings 7th International Speleological Congress, Sheffield, England: International Union of Speleology, p. 336–339.
Palmer, M. V., 1976, Ground-water flow patterns in limestone solution conduits

[M.A. thesis]: Oneonta, State University of New York, 150 p.

Palmer, M. V., and Palmer, A. N., 1975, Landform development in the Mitchell Plain of southern Indiana: Zeitschrift für Geomorphologie, v. 19, p. 1–39.

Piper, A. M., 1932, Groundwater in north-central Tennessee: U.S. Geological Survey Water Supply Paper 640, 238 p.

Plummer, L. N., 1975, Mixing of seawater with calcium carbonate groundwater; Quantitative studies in the geological sciences, *in* Whitten, E.H.T., ed., Quantitative studies in the geological sciences: Geological Society of America Memoir 142, p. 219–236.

Plummer, L. N., and Busenberg, E., 1982, The solubilities of calcite, aragonite, and vaterite in CO_2–H_2O solutions between 0° and 90°C and an evaluation of the aqueous model for the system $CaCO_3$–CO_2–H_2O: Geochimica et Cosmochimica Acta, v. 46, p. 1011–1040.

Plummer, L. N., and Wigley, T.M.L., 1976, The dissolution of calcite in CO_2-saturated solutions at 25°C and 1 atmosphere total pressure: Geochimica et Cosmochimica Acta, v. 40, p. 191–202.

——, Wigley, T.M.L., and Parkhurst, D. L., 1978, The kinetics of calcite dissolution in CO_2-water systems at 5° to 60°C and 0.0 to 1.0 atm CO_2: American Journal of Science, v. 278, p. 179–216.

Powell, R. L., 1963, Alluviated cave springs in south-central Indiana: Indiana Academy of Science Proceedings, v. 72, p. 182–189.

——, 1964, Origin of the Mitchell Plain in south-central Indiana: Indiana Academy of Science Proceedings, v. 73, p. 177–182.

——, 1970, Base level, lithologic, and climatic controls of karst groundwater zones in south-central Indiana: Indiana Academy of Science Proceedings, v. 79, p. 281–291.

Quinlan, J. F., 1970, Central Kentucky karst, *in* Actes de la Reunion Internationale karstologie en Languedoc-Provence, Mediterranée: Études et Traveaux, v. 7, p. 235–253.

Quinlan, J. F., and Ray, J. A., 1981, Groundwater basins in the Mammoth Cave region, Kentucky: Friends of Karst Occasional Publication 1, map.

Quinlan, J. F., and Rowe, D. R., 1977, Hydrology and water quality in the central Kentucky karst, Phase 1: Bowling Green, Kentucky, Water Resources Research Institute Research Report 101, 93 p.

Quinlan, J. F., Ewers, R. O., Ray, J. A., Powell, R. L., and Krothe, N. C., 1983, Hydrology of the Mammoth Cave region, *in* Geological Society of America Annual Meeting Field Trip Guidebook, Indianapolis: Boulder, Colorado, Geological Society of America, 85 p.

Rauch, H. W., and White, W. B., 1977, Dissolution kinetics of carbonate rocks: Water Resources Research, v. 13, p. 381–394.

Renault, P., 1970, La formation des cavernes: Paris, Presses Universitaires de France, 126 p.

Rhoades, R., and Sinacori, N. M., 1941, Patterns of groundwater flow and solution: Journal of Geology, v. 49, p. 785–794.

Roques, H., 1962, Considerations theoretiques sur la chemie des carbonates: Annales de Spéléologie, 3me Memoires, v. 17, p. 11–41, 241–284, 463–467.

Runnels, D. C., 1969, Diagenesis, chemical sediments, and the mixing of natural waters: Journal of Sedimentary Petrology, v. 39, p. 1188–1201.

Shuster, E. T., and White, W. B., 1971, Seasonal fluctuations in the chemistry of limestone springs; A possible means for characterizing carbonate aquifers: Journal of Hydrology, v. 14, p. 93–128.

Sjöberg, E. L., 1978, Kinetics and mechanism of calcite dissolution in aqueous solutions at low temperatures: Stockholm, Contributions to Geology, v. 32, p. 1–92.

Šušteršič, F., 1984, A simple model of the collapse doline transformation: Acta Carsologica (Ljubljana), v. 12, p. 107–138.

Sweeting, M. M., 1950, Erosion cycles and limestone caverns in the Ingleborough District of Yorkshire: Geographical Journal, v. 124, p. 63–78.

Swinnerton, A. C., 1932, Origin of limestone caverns: Geological Society of America Bulletin, v. 43, p. 663–694.

Thornbury, W. D., 1965, Regional geomorphology of the United States: New York, John Wiley and Sons, 609 p.

Thrailkill, J., 1968, Chemical and hydrologic factors in the excavation of limestone caves: Geological Society of America Bulletin, v. 79, p. 19–46.

——, 1971, Carbonate chemistry of aquifer and stream water in Kentucky: Journal of Hydrology, v. 16, p. 93–104.

Wells, S. G., 1976, Sinkhole plain evolution in the central Kentucky karst: National Speleological Society Bulletin, v. 38, p. 103–106.

Weyl, P. K., 1958, The solution kinetics of calcite: Journal of Geology, v. 66, p. 163–176.

White, W. B., 1969, Conceptual models for carbonate aquifers: Ground Water, v. 7, p. 15–21.

——, 1976, The geology of caves, *in* Geology and biology of Pennsylvania caves: Pennsylvania Geological Survey, General Geology Report 66, 71 p.

——, 1977, Role of solution kinetics in the development of karst aquifers, *in* Tolson, J. S., and Doyle, F. L., eds., Karst hydrogeology: Huntsville, University of Alabama Press, p. 503–517.

——, 1978, Theory of cave origin and karst aquifer development: Yellow Springs, Ohio, Cave Research Foundation Annual Report, p. 36–37.

——, 1984, Rate processes; Chemical kinetics and karst landform development, *in* LaFleur, R. G., ed., Groundwater as a geomorphic agent: London, Allen and Unwin, p. 227–248.

White, W. B., and Longyear, J., 1962, Some limitations on speleo-genetic speculation imposed by the hydraulics of ground water flow in limestone: National Speleological Society, Nittany Grotto Newsletter, v. 10, p. 155–167.

White, W. B., Watson, R. A., Pohl, E. R., and Brucker, R. W., 1970, The central Kentucky karst: Geographical Review, v. 60, p. 88–115.

Wigley, T.M.L., 1973, Speleogenesis; A fundamental approach: Proceedings 6th International Speleological Congress, Olomouc, ČSSR, International Union of Speleology, v. 3, p. 317–324.

Wigley, T.M.L., and Plummer, L. N., 1976, Mixing of carbonate waters: Geochimica et Cosmochimica Acta, v. 40, p. 989–995.

Williams, P. W., 1968, An evaluation of the rate and distribution of limestone solution in the River Fergus basin, Western Ireland, *in* Contributions to the study of karst: Australian National University Department of Geography, p. 1–40.

——, 1983, The role of the subcutaneous zone in karst hydrology: Journal of Hydrology, v. 61, p. 45–67.

Wood, W. W., 1985, Origin of caves and other solutional openings in the unsaturated (vadose) zone of carbonate rocks; A model for CO_2 generation: Geology, v. 13, p. 822–824.

Woodson, F. J., 1981, Lithologic and structural controls on karst landforms of the Mitchell Plain, Indiana, and Pennyroyal Plateau, Kentucky [M.A. thesis]: Terre Haute, Indiana State University, 132 p.

Woodward, H. P., 1961, A stream piracy theory of cave formation: National Speleological Society Bulletin, v. 23, p. 39–58.

Zötl, J., 1961, Die Hydrologie des nordalpinen Karstes: Steirische Beitrag Hydrogeologie für 1960/61, no. 2, p. 54–183.

MANUSCRIPT ACCEPTED BY THE SOCIETY NOVEMBER 14, 1989

Printed in U.S.A.

Chapter 9

*Permafrost and thermokarst; Geomorphic effects of subsurface water on landforms of cold regions**

The editors, *with*
Troy L. Péwé
Department of Geology, Arizona State University, Tempe, Arizona 85287
R.A.M. Schmidt
1402 West 11th Avenue, Anchorage, Alaska 99501
Charles E. Sloan
CH2M Hill, 2550 Denali Street, 8th Floor, Anchorage, Alaska 99503

INTRODUCTION

Permafrost is perennially frozen ground that has a temperature of 0°C or colder continuously for two or more years. In order for permafrost to form and persist, the mean annual air temperature must usually be several degrees below zero. About 20 percent of the land area of the Earth is underlain by permafrost, including about one-half of Canada and nearly 85 percent of Alaska. Its thickness largely depends on the balance between the local geothermal gradient and the mean annual air temperature. It is as much as 1,600 m deep in northern Siberia. The term "permafrost" was first used in 1943 by S. W. Muller (1943), although the existence of perennially frozen ground and some of its effects were known long before.

The permafrost zone

Frozen ground is defined exclusively on the basis of temperature, and does not require the presence of subsurface water; so-called "dry" permafrost has very little water or ice. In some permafrost, part or all of the moisture may be liquid, depending on the chemical composition of the water or the depression of the freezing point by capillary or other forces. For example, highly saline soil water may remain liquid at temperatures far colder than 0°C. Most permafrost, however, is consolidated by ice.

The upper surface of permafrost is called the *permafrost table,* or *frost table.* Seasonally the frost table moves higher and lower in the soil. The surficial layer of ground above the frost table freezes in winter (seasonally frozen ground) and thaws in summer. This is called the *active layer.* The thickness of the active layer depends largely on the soil type, the moisture content, and the vegetation cover. The active layer may be 2 to 3 m thick in well-drained gravels but only 10 to 30 cm thick in wet organic sediments. Thickness also depends on the thermal regime, that is, the quantity of heat affecting the permafrost and the overlying active layer. The thermal regime in turn is determined by climatic, geomorphic, and vegetational factors, such as seasonal and long-term weather cycles, annual snowfall, slope exposure, local drainage, and tree, brush, and tundra cover. It may be especially affected by changes in any one or more of these.

Human structures and activities can be adversely affected by subsidence of the ground caused by man-induced thawing of permafrost, as described in many engineering studies (e.g., Péwé, 1982; Péwé, 1990). The present discussion is concerned with natural processes and landforms associated with freezing and thawing of frozen ground with a high ground-ice content.

Ground ice

The ice content of permafrost varies greatly, from minute amounts in pores to large masses often referred to as *ground ice.* Ground ice tends to be concentrated in the upper levels of permafrost, where it may constitute as much as 60 percent by volume of

---

**Editors' Note: This chapter was to have included a full-length discussion of the role of subsurface water in the geomorphology of cold regions, but for varying reasons a succession of invited authors were unable to provide a manuscript. Because we regard this subject as so important, affecting as it does some 20 percent of the Earth's land surface, and because we believe that *some* discussion is better than none, we have assembled a brief overview of the several important roles that subsurface water plays in landform processes of cold regions, based in part on the text by Charles Sloan for the GSA Centennial Volume, *Hydrogeology* (Higgins and others, 1988), with additions from Troy Péwé (1982, 1990), R.A.M. Schmidt (1966), David M. Hopkins (personal communication, 1989), and the late Robert Black (1976). Péwé, Schmidt, and Sloan have commented on and improved the text. It has further benefited from a careful review by J. Ross Mackay. Readers who wish to pursue the subject beyond this sketchy outline are urged to consult the more recent papers listed at the end of the chapter.*

The editors *with* Péwé, T. L., Schmidt, R.A.M., and Sloan, C. E., 1990, Permafrost and thermokarst; Geomorphic effects of subsurface water on landforms of cold regions, *in* Higgins, C. G., and Coates, D. R., eds., Groundwater geomorphology; The role of subsurface water in Earth-surface processes and landforms: Boulder, Colorado, Geological Society of America Special Paper 252.

Figure 1. Large ice wedge in retransported Wisconsinan loess overlain by Holocene loess. Mamontova Gora, central Yakutia, Siberia. Photograph number 3427 by Troy L. Péwé, July 22, 1973.

the upper 1 to 3 m (French, 1987). It can take many forms, some of the more common being *ice wedges* (Fig. 1), *ice lenses,* and *pingo ice.* The occurrence of groundwater in the permafrost region is discussed in some detail by Sloan and van Everdingen (1988). R.J.E. Brown (1973) has discussed the geomorphic role of ground ice as an initiator of landforms in permafrost regions.

Freezing of subsurface water and thawing of subsurface ice influences the surface processes and landforms of permafrost regions in two distinct ways. First, the formation and growth of ground ice produces aggradational landforms such as ice-wedge polygons, palsas, and pingos (Fig. 2). Second, the thaw of ground ice produces degradational features such as beaded streams, thaw lakes, and alases (or allasy). The resulting hummocky ground surface is called *thermokarst* because of some resemblance to the karst topography of limestone regions.

Some other geomorphic effects of freezing and thawing are only indirectly related to the presence of permafrost. For example, the permafrost table creates an effective barrier to downward percolation of soil water, even though some water from the active layer may percolate down into the permafrost in summertime (Mackay, 1983). The result is periodic saturation and waterlogging of the poorly drained soil of the active layer. This soil then sludges haltingly down gentle slopes to form lobate solifluction sheets, benches, streams, terraces, and other forms. Where the active layer retains much water, the downslope transport may attain rates of several centimeters per year. Some other types of mass wasting in the active layer are summarized by French (1987).

Frost heaving

A variety of forms of patterned ground in cold regions may result from frost cracking, wedging, and heaving, with or without the presence of permafrost. The growth of seasonal ice in pore spaces causes limited volume expansion and therefore elevation of the ground. This is due to a volume increase of about 9 percent when the interstitial water changes from a liquid to a solid. Volume expansions of as much as several hundred percent have been recorded during seasonal freezing of silt and clay when excess water is absorbed by the soil. Some complications of the processes of ice segregation and frost heave are summarized by French (1987).

Fine-grained moist to wet materials promote migration of moisture to the freezing front, forming ice lenses and other clear-ice segregations. The result is an uneven upward displacement of the ground called frost heaving. Such frost action affects human structures in various destructive ways outlined by Péwé (1982, 1990). The frost heaving and stirring that is promoted by freezing and thawing of the active layer may also displace and sort pebbles, boulders, and joint blocks in the soil, with or without the

Figure 2. Collapsed pingo near Tuktoyaktuk, Northwest Territories, Canada. Ice-wedge low-center polygons in foreground; oriented thaw lakes in background. Photograph number 1047 by Troy L. Péwé, August 18, 1954.

presence of permafrost. This contributes to the development of sorted patterned ground, soil churning, and other surficial effects. Some net-like patterns are formed of coarse rubble surrounding nearly circular cores of finer material; others, in more uniform soil, are marked by borders of vegetation. Generally in permafrost areas, the sizes of polygons, circles, and nets tend to reflect the thickness of the active layer and the size of the soil materials being sorted (Black, 1976). A great variety of patterned ground features are described by Washburn (1980).

On slopes, *frost creep,* a slow, rachetlike downslope movement of soil particles promoted by freezing and thawing, may cause patterned ground to be strung out in stripes (Fig. 3) or as transverse ridges and furrows. Another effect of frost creep on the landscape is to form broad convex-upward hillslopes.

AGGRADATIONAL LANDFORMS

Ice-wedge polygons

Ice-wedge polygons are the most extensive cold-region landforms created by freezing of subsurface water. They are initiated by deep thermal contraction cracking of frozen soil, which opens vertical cracks 1 to 10 mm wide and as much as 7 or 8 m deep in roughly polygonal patterns, 5 to 30 m in diameter (Fig. 1). In early spring, water from melting snow and sometimes a little soil fills the open cracks. When the water freezes, it forms a vertical vein of ice that penetrates the permafrost. The following summer, when the permafrost warms and expands, the horizontal compression forces the adjacent sediments upward. Through the perennial cycles of many seasons, repeated cracking and filling widens the ice veins a fraction of a millimeter to several millimeters each year that the ice wedges crack, enlarging them into wedge-shaped masses. The resulting ice wedges are coincident with the borders of the polygons.

The displaced soil may bulge up in the center of the polygons or along the borders. Low-center polygons, with raised borders, are common in poorly drained areas in the zone of continuous permafrost and indicate its presence (Figs. 2 and 4). High-center polygons may develop where drainage is better. Doming of the center by the growth of ice wedges is rare and only occurs north of about latitude 70°. The resulting mounds are 10 to 15 m wide and 1 to 2 m high. During the thaw season, low-center polygons often contain ponds in the central areas, whereas high-center ones hold water in the bordering depressions.

Figure 3. Patterned ground stretched downslope into elongate polygonal nets and stripes on the Seward Peninsula, Alaska. Photograph taken June 18, 1958, copyright by R.A.M. Schmidt.

Figure 4. Raised-edge low-center tundra polygons 2 km south of Barrow, Alaska. Photograph number 1779 by Troy L. Péwé, July 24, 1958.

Figure 5. Thermokarst pit in loess in golf course, Fairbanks, Alaska. Photograph number 1037 by Troy L. Péwé, August 15, 1954.

Mounds

A variety of small to large mounds occur in cold-region soils. Some are closely spaced and regular in size and shape; others are solitary.

Earth hummocks, or *thufur,* are a form of patterned ground. Well-developed ones are normally as large as 0.5 m high and 1 to 2 m in diameter (Washburn, 1980). On level ground they are circular, but may become elongated on slopes, either across the slope as steps, or down the slope as linear ridges. Mackay (1980) has proposed an equilibrium model for the growth of hummocks in arctic Canada, in which upward doming is caused by freeze and thaw of ice lenses at the top and bottom of the active layer. However, different kinds may form in different ways (Beschel, 1966). The general regularity of some suggests that they are initiated in a polygonal frost-cracked network and then modified by frost heaving and other processes. Others seem independent of permafrost but probably do involve frost heaving. Recent studies of hummocks are reviewed by French (1981).

Palsas, or *palsen,* are also moundlike, but tend to be larger and more irregular and solitary than thufur. Small clustered groups typically consist of merely several, not tens, of mounds covering areas of only a few hectares. They range from 1 to 6 m high and 10 to 30 m wide, although elliptical ones on gentle slopes may reach 150 m or even 500 m in length (Washburn, 1980). They represent isolated occurrences of sporadic permafrost in unfrozen bogs, where their elevation results from upward doming by segregated ice. Palsas are characteristic of the subarctic, where they commonly occur in areas of discontinuous permafrost. Some recent studies are reviewed by Worsley (1986).

In contrast with palsas, *pingos* are large perennial ice-cored mounds or hills that are more or less circular in form, generally 10 to a maximum of almost 60 m high, and 50 to 500 m in diameter (Fig. 2). Pingos are necessarily associated with permafrost, and like low-center ice-wedge polygons, are key indicators of polar or subpolar environments. They are notably abundant in the Tuktoyaktuk Peninsula and Richards Island region of the western arctic coast of Canada (Mackay, 1985).

So-called "closed-system" pingos are associated with thaw basins where permafrost is aggrading. Water is confined between the downward-freezing surface, the permafrost beneath, and the frozen margin of the basin. As permafrost aggrades in the saturated sediments, pore water may be expelled beneath the permafrost by the volume expansion of water upon freezing. The water pressure domes up the site with the thinnest permafrost to start the growth of the ice core (Mackey, 1985). "Open-system" pingos occur in zones of discontinuous or sporadic permafrost where a confined artesian system beneath the permafrost bows it up and injects water to form the core. All pingos eventually collapse into doughnut-shaped hillocks known as pingo scars after their ice cores become exposed and thaw (Fig. 2). Flemal (1976) has identified Wisconsinan pingo scars in northern Illinois and summarized occurrences reported elsewhere.

DEGRADATIONAL LANDFORMS

Ground subsidence

The most widespread effect of the thawing of ice-rich permafrost in cold regions is subsidence of the ground surface. Melting of large ground-ice masses produces dramatic differential settlement and can result from natural factors such as coastal

Figure 6. Beaded stream with small thaw lakes in foreground and larger aligned or parallel thaw lakes in the distance, east of Teshepuk Lake, southeast of Point Barrow, Alaska. Photograph taken August 8, 1965, copyright by R.A.M. Schmidt.

erosion, stream erosion, climatic change, or wildfires, but can also be caused by human disturbances that affect the subsurface thermal regime. The geomorphic effects of such thaw-induced subsidence are generally included under the umbrella term "thermokarst." The engineering consequences have been described by Péwé (1982, 1990) and others. Although human-aided thawing is locally significant, however, natural effects are much more widespread. Certainly the magnitude and extent of thaw lakes in the permafrost zone far outweigh the puny effects of human activities.

Thermokarst

Thermokarst is the name given to karstlike topographic features produced by the melting of ground ice or ice-rich permafrost and the subsequent settling or caving of the ground. Thus, it is a thermal effect rather than a chemical one. It is characterized by an uneven ground surface with depressions and other signs of subsidence. In many areas thermokarst is the expression of fairly recent thaw and collapse processes promoted by climate change or destruction of the vegetation cover. Such removal of the insulating vegetation blanket by fire or by clearing and cultivation allows the ground to absorb more solar radiation, thereby upsetting the thermal equilibrium. Well-developed thermokarst features require differential distribution of large ground-ice masses near the surface, so they do not form where the ice occurs only as interstitial cement and there is no release of excess water above saturation when thawed.

On gentle slopes, shallow groundwater circulation may thaw ice masses and hydraulically enlarge the resulting cavities. In rare instances, surface water may be diverted underground where it also contributes to thawing and subterranean erosion. On level surfaces this water forms ponds and lakes in the resulting depressions, and the borders of the depressions cave and retreat under the further thawing effects of waves. Such depressions range in size from small pits or pools along *beaded streams* to large *thaw lakes* and *alases*.

Thermokarst pits. Thermokarst pits are steep-walled, sinkhole-like depressions that are initiated by the melting of ground ice above the water table. They may then be enlarged by the inflow of captured surface water. Some described by Péwé in the Fairbanks area are generally 1.5 to 6 m deep and 1 to 10 m wide, as shown in Figure 5. Headward erosion by running water may extend a pit along a road rut or plowed furrow to form a linear depression. In Alaska, most pits become apparent within 18 to 10 years after clearing of vegetation. Where the water table is shallower, thawing ice masses create pools or larger cave-in lakes rather than dry pits.

Beaded streams. Thaw following polygonal ice wedges along drainage lines produces beaded streams (Fig. 6). These are characteristic of minor stream drainages in areas underlain by ice wedges in frozen peat and silt. They consist of a series of small pools connected by short watercourses. The pools result from thawing wedge ice at polygon intersections. They typically have steep banks, range in depth from 0.5 to 3m, and in diameter from a few to as much as 30 m. The straight connecting drainage segments are generally aligned along the thawing ice wedges, so that they tend to form short, straight sections separated by angular bends.

Thermokarst mounds. Thaw along polygonal ice wedges in some cases forms pools along streamways but elsewhere may form a pattern of polygonal ground in which the centers of the polygons are left in high relief as mounds or little hillocks. Péwé (1982) has described polygonal or circular mounds in the Fairbanks area that have been formed in this manner. Some of these are shown in Figure 7. They are closely packed, 0.3 to 2.5 m high and 3 to 15 m across, and commonly appear in cleared areas where ice-wedge thawing is promoted by vegetation removal:

Figure 7. Thermokarst mounds in an abandoned cleared field at the U.S. Department of Agriculture Experimental Farm at the University of Alaska, Fairbanks. Mounds are 3 to 10 m in diameter. Photograph number PK 10,356 by Robert F. Black and Troy L. Péwé, September 10, 1948.

Trenches form where melting polygonal ground-ice causes differential settlement of the ground surface. Local, unconnected depressions first appear; surface water collects in the depressions and speeds thawing and, after entering cavities left by the melting ice, enlarges them. If the underlying ice is distributed in a polygonal pattern, the center of each polygon will eventually stand out in relief. The fact that the silt in the mounds is not deformed is evidence that the mounds do not originate by frost heaving (Péwé, 1982, p. 36).

The Siberian *bulganyakh* are the same, and appear on sunny, cleared, south- and southwest-facing slopes of the *allasy*, which are large thermokarst depressions that originated as thaw lakes. A resemblance of these cold-region mounds to the "mima" or "pimple" mounds of temperate parts of western North America has prompted some investigators to suggest that the latter are relict periglacial features, but this view has been contested by proponents of other origins.

Thaw lakes. One of the most conspicuous kinds of thermokarst where the ground contains much more ice than mineral matter is the thaw, or thermokarst lake (Figs. 2 and 6). Thawing enlarges initial pits or ponds into large basins and shallow lakes that may be many kilometers across and tens of meters deep. Oriented thaw lakes are elliptical and have a parallel alignment. They are common in some areas, such as the Alaskan Arctic Coastal Plain and the adjacent western arctic coast of Canada, where their orientation is normal to the direction of the prevailing winds. The reason for this alignment is debatable, but one explanation is that wind-directed waves erode the downwind shore, armoring it with a coarse lag concentrate. Thus protected, the downwind shore remains fixed while the lateral shores continue to recede by thaw and collapse of the underlying ice. This cannot explain all oriented lakes because they occur in many areas without a lag concentrate.

Alases. Alases (Siberian *allasy*) are very large thermokarst depressions with steep sides and flat, grassy floors that form where the ground-ice content of the upper part of the permafrost is very high. They are commonly round to oval and may contain one or more shallow lakes. They are particularly well developed in the taiga of the central Yakutia lowland in Siberia (Fig. 8). Some thaw basins are kilometers across and may owe their enlargement in part to seepage erosion of thawing soil at the base of their slopes (see Chapter 14, this volume).

CONCLUSIONS

Much of the characteristic cold-region landscape results from the freezing and thawing of subsurface water. The resulting landforms range in scale from the convex solifluction slopes of entire hillsides to the tiny intricacies of the most recent patterned ground. Some of these features are associated with permanently frozen ground; others are not. Many are so distinctive that the experienced observer can infer their origins and something of the

Figure 8. Part of an alas bordered by thermokarst mounds, all developed in loess, near Maya, 30 km southwest of Yakutsk, Siberia. Photograph number 3454 by Troy L. Péwé, July 25, 1973.

soil texture and ice content. As yet, however, it is possible to make unequivocal identification of only a few types of relict cold-region features in now-temperate regions. This remains a worthy goal in attempts to assess former (and future?) climatic regimes in the most populated parts of the world.

REFERENCES CITED

Beschel, R. L., 1966, Hummocks and their vegetation in the high arctic, *in* Proceedings of the Permafrost International Conference, 1963: Washington, D.C., National Academy of Sciences-National Research Council Publication 1287, p. 13–20.

Black, R. F., 1976, Features indicative of permafrost: Annual Review of Earth and Planetary Sciences, v. 4, p. 75–94.

Brown, R.J.E., 1973, Ground ice as an initiator of landforms in permafrost regions, *in* Fahey, B. D., and Thompson, R. D., eds., Research in polar and alpine geomorphology; 3rd Guelph Symposium in Geomorphology: Norwich, United Kingdom, GeoBooks, p. 25–42.

Flemal, R. C., 1976, Pingos and pingo scars; Their characteristics, distribution, and utility in reconstructing former permafrost environments: Quaternary Research, v. 6, p. 37–53.

French, H. M., 1981, Periglacial geomorphology and permafrost: Progress in Physical Geography, v. 5, p. 267–273.

—— , 1987, Periglacial geomorphology in North America; Current research and future trends: Progress in Physical Geography, v. 11, p. 533–551.

Higgins, C. G., and 12 others, 1988, Landform development, *in* Black, W., Rosenshein, J. S., and Seaber, P. R., eds., Hydrogeology: Boulder, Colorado, Geological Society of America, The Geology of North America, v. O-2, p. 383–400.

Mackay, J. R., 1980, The origin of hummocks, western arctic coast: Canadian Journal of Earth Sciences, v. 17, p. 996–1006.

—— , 1983, Downward water movement into frozen ground, western arctic coast, Canada: Canadian Journal of Earth Sciences, v. 20, p. 120–134.

—— , 1985, Pingo ice of the western arctic coast, Canada: Canadian Journal of Earth Sciences, v. 22, p. 1452–1464.

Muller, S. W., 1943, Permafrost or permanently frozen ground and related engineering problems: U.S. Army Corps of Engineers, Strategic Engineering Study, Special Report 62, 231 p.

Péwé, T. L., 1982, Geologic hazards of the Fairbanks area, Alaska: Alaska Division of Geological and Geophysical Surveys Special Report 15, 109 p.

—— , 1990, Permafrost, *in* Kiersch, G. A., ed., The heritage of engineering geology: Boulder, Colorado, Geological Society of America, Centennial Special Volume 3 (in press).

Schmidt, R.A.M., 1966, Geology color slide set II: Permafrost and frozen ground patterns [text to accompany Kodachrome slides], 7 p.

Sloan, C. E., and van Everdingen, R. O., 1988, Permafrost region, *in* Back, W., Rosenshein, J. S., and Seaber, P. R., eds., Hydrogeology: Boulder, Colorado, Geological Society of America, The Geology of North America, v. O-2, p. 263–270.

Washburn, A. L., 1980, Geocryology; A survey of periglacial processes and environments: New York, John Wiley and Sons, 416 p.

Worsley, P., 1986, Periglacial environments: Progress in Physical Geography, v. 10, p. 265–274.

MANUSCRIPT ACCEPTED BY THE SOCIETY NOVEMBER 14, 1989

Geological Society of America
Special Paper 252
1990

Chapter 10

Land subsidence and earth-fissure formation caused by groundwater withdrawal in Arizona; A review

Troy L. Péwé
Department of Geology, Arizona State University, Tempe, Arizona 85287

INTRODUCTION

Land subsidence is a geologic phenomenon caused by a variety of natural and man-made processes. Although subsidence is generally not very spectacular or catastrophic, it annually causes millions of dollars worth of damage in the United States. Natural processes causing land subsidence include the dissolving of limestone, evaporites, and other soluble materials, and in some instances, seismic activity. Abrupt collapse and gradual subsidence of the ground because of the solution of underlying limestone are widespread in karst areas of the world and are discussed in Chapters 6 and 7 (this volume). The effects of solution of halite, gypsum, and other water-soluble evaporites are less well known but are seen in broad areas.

Man-induced subsidence may be the result of the withdrawal of oil, gas, and groundwater, as well as drainage of marsh lands and especially the removal of rock during underground mining (notably, shallow subsurface coal mining) (Holzer, 1984a). Subsidence caused by fluid withdrawal and subsequent compaction of aquifer sediments has resulted in serious damage to surface and underground facilities such as canals, aqueducts, sewer systems, and buried pipelines, and especially has resulted in extensive damage in coastal cities by flooding from the sea.

Excessive groundwater pumping in the San Joaquin Valley, California, for example, has caused the land to subside locally as much as 9 m (Bull, 1973; Poland and others, 1975) (Fig. 1). The valley contains the world's largest subsidence area; about 13,500 km^2 have been affected. This large area of subsidence is the result of overpumping of groundwater from a confined aquifer and probably represents the greatest single man-made alteration of the configuration of the Earth's surface, an aspect of groundwater geomorphology. Extensive groundwater withdrawal for irrigation purposes began in the 1920s and increased until the mid-1950s. At that time the annual subsidence in the western part of the valley was as much as 0.55 m/yr (Coates, 1981). Since that time, subsidence has slowed with the reduction of groundwater pumping and importation of water from elsewhere in California.

The Santa Clara Valley at the southern end of San Francisco Bay was the first area in the United States where land subsidence caused solely by excessive groundwater removal was recognized (Forrester, 1971). As much as 1 m of subsidence was recognized by 1933, and by 1967, subsidence amounted to 4 m (Poland, 1969). The city of San Jose and its suburbs are in the middle of the subsidence bowl. Millions of dollars have been spent realigning canals and ditches, and repairing roads and buildings. Millions more have been spent to reinforce the levees along San Francisco Bay to prevent flooding in the subsidence areas (Coates, 1981). Subsidence due to groundwater withdrawal is now known in many parts of the U.S., including the Galveston area of Texas, where by 1973 there was maximum subsidence of 2.7 m. Coastal flooding into the subsided areas was reported to cause more than $30 million damage annually in the 1970s, but that amount has since been reduced.

Land subsidence as a result of groundwater withdrawal is also reported from Utah (Cordova and Mower, 1976; Christenson, 1986), Nevada, and Idaho (Holzer, 1986). Subsidence in Las Vegas Valley, Nevada, has been recognized and studied for more than 40 years, and was carefully summarized by Bell (1981). Despite periodic remedial measures, it is a continuing problem that poses a serious potential hazard. By 1980, the center of Las Vegas had subsided about 1.5 m (Viets and others, 1979, p. III-38; Bell, 1981, p. 1). Steel casings of wells are protruding more than 1 m as the land sinks; and earth fissures 1 m wide and up to 800 m long have formed.

Land-subsidence problems resulting from groundwater withdrawal from unconsolidated sediments are not limited to the United States. They are well known in Japan (Yamamato, 1977), Mexico City, and Venice, for example. The most serious problems appear to be in Japan where several areas have subsided. In Tokyo, 2,000,000 people live in an area that has dropped below high-tide level, and in Osaka, more than 600,000 people live in similar conditions (Forrester, 1971).

The most spectacular modification of the land surface due to excessive groundwater withdrawal from unconsolidated sedi-

---

Péwé, T. L., 1990, Land subsidence and earth-fissure formation caused by groundwater withdrawal in Arizona; A review, *in* Higgins, C. G., and Coates, D. R., eds., Groundwater geomorphology; The role of subsurface water in Earth-surface processes and landforms: Boulder, Colorado, Geological Society of America Special Paper 252.

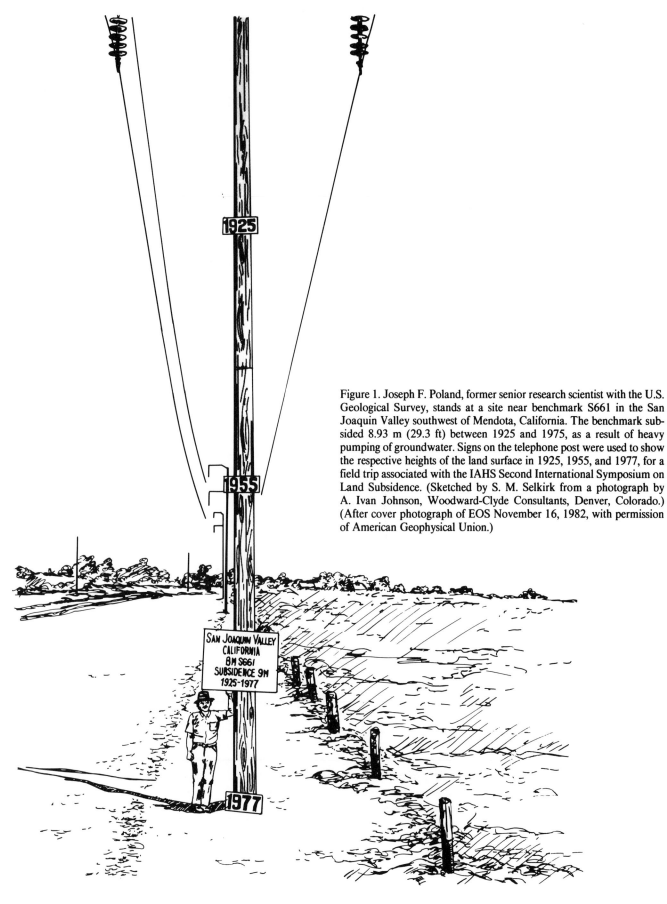

Figure 1. Joseph F. Poland, former senior research scientist with the U.S. Geological Survey, stands at a site near benchmark S661 in the San Joaquin Valley southwest of Mendota, California. The benchmark subsided 8.93 m (29.3 ft) between 1925 and 1975, as a result of heavy pumping of groundwater. Signs on the telephone post were used to show the respective heights of the land surface in 1925, 1955, and 1977, for a field trip associated with the IAHS Second International Symposium on Land Subsidence. (Sketched by S. M. Selkirk from a photograph by A. Ivan Johnson, Woodward-Clyde Consultants, Denver, Colorado.) (After cover photograph of EOS November 16, 1982, with permission of American Geophysical Union.)

ments is the formation of earth fissures that may be more than 1 km long and enlarged by erosion to be 1 to 2 m wide. Fissures are known in six U.S. states; most are in Arizona (Holzer, 1986).

CAUSES OF SUBSIDENCE DUE TO GROUNDWATER WITHDRAWAL

The withdrawal of groundwater causes the subsequent lowering of the water table and reduces the hydraulic pressure, the buoyancy of the water. With the water gone, the intergranular stress increases, resulting in closer packing of grains in the underlying sediments. The deposits thus become more compressed, and the land surface is lowered. Because the pore space is reduced in the alluvium, the underlying storage capacity is also reduced.

Fine-grained sediments (silts and clays) compact more than coarse-grained sediments (sand and gravel) under compressional stress. This compression is nonelastic and nonrecoverable. Because of the lower permeability of fine-grained deposits, pore water is expelled during the consolidation process at a slower rate than from the coarser-grained deposits. Thus, subsidence caused by compaction of silt and clay beds is slow, and may lag some time behind the stress increases produced by water-level decline. In addition to the obvious controlling parameter of water-level decline, the time lag, rate, and total amount of subsidence are also dependent on the thickness, permeability, lithology, and compressibility of the individual alluvial units. Most compaction of sediments in deep alluvial basins that are dewatered can be accounted for within the upper 100 to 400 m of sediments (Poland, 1969; Schumann and Poland, 1969; R. H. Raymond, U.S. Bureau of Reclamation, oral communication, 1980).

LAND SUBSIDENCE AND EARTH-FISSURE FORMATION IN SOUTH-CENTRAL ARIZONA

Land subsidence

Known land-subsidence areas in Arizona contain more than 7,800 km^2 (Strange, 1983; Schumann and Genualdi, 1986a) and lie in the Basin and Range Physiographic Province of the south-central part of the state. It is here that most of the state's population and agricultural and urban development are located. About 2.4475 × 10^5 hm^3 (196 million acre-feet) of groundwater has been withdrawn in this area since 1900 (U.S. Geological Survey, 1985). For example, in 1981, about 6.78 × 10^3 hm^3 (5.4 million acre-feet) of groundwater was pumped from the permeable unconsolidated alluvial deposits (U.S. Geological Survey, 1984). These withdrawals of groundwater exceed the amount of natural recharge by 2.762 × 10^3 hm^3/yr (2.2 million acre-feet per year) (Briggs, 1976). This has caused groundwater-level declines in some basins of more than 170 m (Thomsen and Baldys, 1985) and widespread subsidence of the land surface, locally up to more than 5 m. Decline of the level of the water table was recognized early. About 1 m/yr decline was recorded in 1930–1931 near Eloy, and 2.5 m/yr in 1936–1937 (Smith, 1940, Table II and

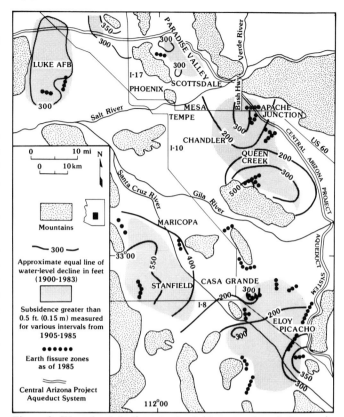

Figure 2. Water-level declines, land subsidence, and earth fissures in south-central Arizona. (Sources: Schumann, 1974; Winikka and Wold, 1976; Laney and others, 1978; Arizona Department of Transportation, 1981; Patino and others, 1985; C. Winikka, oral communication, 1985; Thomsen and Baldys, 1985; Schumann and Genauldi, 1986b. Location of CAPAS from U.S. Bureau of Reclamation, 1980.)

Plate III). Subsidence in south-central Arizona first began sometime between 1934 and 1948 and was probably first detected near Eloy in the Picacho Basin in 1934 (Robinson and Peterson, 1962) (Fig. 2). Repeated leveling since 1952 indicates as much as 3.8 m of subsidence near Eloy by 1977 (Laney and others, 1978) and 5 m by 1985 (C. C. Winikka, Assistant State Engineer, Arizona Department of Transportation, oral communication, June 1985; Anonymous, 1987). Subsidence is continuing, and an area of 625 km^2 is affected (Fig. 3). Maximum average subsidence rate near Eloy has been about 15 cm/yr (Peirce, 1979). The Picacho Basin has the best-documented record by repeated leveling in Arizona (Holzer, 1984b).

In the lower Salt River Valley, there are three areas at the present time where the water table has dropped more than 100 m and land subsidence has been recorded: (1) Queen Creek–Apache Junction area, (2) southern Paradise Valley, and (3) western Salt River Valley (Luke Air Force Base) (Fig. 2). Just west of Apache Junction (Fig. 2), the maximum measured subsidence between 1948 and 1981 of a benchmark near U.S. Highway 60 and the Bush Highway is 1.6 m. During the same interval,

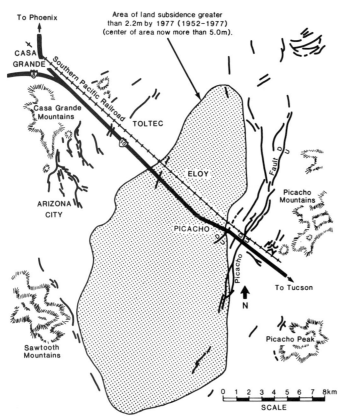

Figure 3. Earth fissures, surface faults, and area of major land subsidence in Picacho Basin, Arizona. (Modified from Schumann, 1974; Laney and others, 1978; Jachens and Holzer, 1982; Holzer, 1984b.)

probably 1.9 m of subsidence has occurred at two benchmarks near the intersection of Williams Field and Powers Road, 6.4 km east of Chandler (Patino and others, 1985). The Central Arizona Project aqueduct near Apache Junction required special design and construction to accommodate land subsidence in the area. More than 0.7 m of land subsidence has been measured since 1971 along the aqueduct route south of Apache Junction (Schumann and others, 1983).

Subsidence of the land in the southern part of Paradise Valley was brought to local and state attention in 1980, when a 130-m-long fissure formed in a new housing area in northeast Phoenix (Péwé, 1984). A detailed geological and gravity study was made of the area in cooperation with the city of Phoenix to evaluate the extent and rate of land subsidence, origin of the earth fissure, and future subsidence and earth fissuring (Péwé, 1980; Larson, 1982; Péwé and Larson, 1982; Larson and Péwé, 1983, 1986). Paradise Valley (Fig. 2) is an elongate alluvial basin lying between the Phoenix Mountains on the southwest and the McDowell Mountains on the east. The alluvium is about 300 m thick along the edges of the basin, but estimated to be greater than 2,700 m near the center (Lausten, 1974).

Water levels remained nearly constant in northeast Phoenix prior to about 1950, generally within 75 m of the surface. Increased pumping in relatively unproductive aquifers has caused rapid water-level decline, particularly in two areas where groundwater had dropped more than 40 m from its original level. These "cones of depression" are centered halfway between Greenway Road and Bell Road at 44th Street, and near 56th Street and Thunderbird Road. Withdrawals of groundwater are many times the natural recharge rate, and this overdraft has resulted in depletion of thin aquifers peripheral to the mountains, and loss of supply to shallow wells. More wells will certainly become dry as pumping in the area continues. Up to 25×10^6 m^3 of groundwater is withdrawn annually.

Since the mid-1950s, water levels have declined, resulting in current water depths of more than 100 m. Subsidence apparently began about a decade later in the vicinity of 52nd Street and Thunderbird Road after water levels declined from 38 to 45 m. Since 1970, the subsidence bowl has increased in size at an average rate of 5 km^2/yr, with early expansion predominantly in a westerly direction, and more recent expansion toward the north and east (Fig. 4).

As of March 1982, the maximum subsidence measured was 1 m at 56th Street and Thunderbird Road (Fig. 4), near the center of the southern cone of water-level depression. At the assumed center of the subsidence area (or subsidence "bowl") 0.8 km to the southwest, there is indirect evidence from topographic and land-survey data for as much as 1.6 m of subsidence. Harmon (1982) noted that the subsidence rate has increased to the south, particularly at 56th Street and Cactus Road, and Tatum Boulevard and Cholla Street, where the ground is subsiding at 13 cm/yr. This occurrence may represent a southward shift in the center of the subsidence bowl.

The growth of the subsidence bowl suggests that it will expand farther, particularly toward the north and east; subsidence has been measured to the east in Scottsdale. The extent of land subsidence to the south into the town of Paradise Valley, however, is not known. There is insufficient data on compaction and material properties of the subsurface to fully evaluate the potential of future land subsidence in northeast Phoenix; however, given known thicknesses of alluvium and present subsidence rates near the center of the subsidence bowl, more than 3 m of land subsidence is possible if this area is completely dewatered.

Reconnaissance information from the city of Scottsdale (Patrick Neal, field engineer, city of Scottsdale, personal communication, January 11, 1984) strongly suggests subsidence up to 20 cm in north-central Scottsdale west of the McDowell Mountains on the east side of Paradise Valley.

In the western Salt River Valley basin, immediately east of the White Tank Mountains, about 1.2 m of subsidence has been noted between 1948 and 1981 (Strange, 1983). Also, 18 cm of subsidence is recorded in the Harquahala Plains; 3 to 8 cm in the basin lying east of Tonopah from 1967 to 1981; and 8 cm in the Gila Bend area (Strange, 1983). As much as 3.6 m of subsidence is measured in the Maricopa-Stanfield area (Laney and others, 1978), and 5 m of subsidence in the Eloy-Picacho area (Fig. 2) in the lower Santa Cruz Valley (Anonymous, 1987).

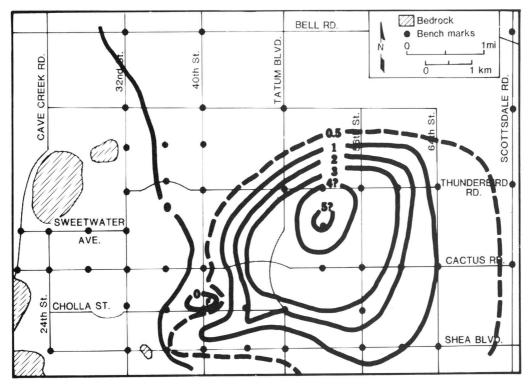

Figure 4. Isolines of land subsidence in feet in northeastern Phoenix, Arizona (1962 to 1982). F indicates location of fissure. Dots indicate location of city benchmarks (from Péwé and Larson, 1982).

In the Wilcox–Bowie–San Simon area of Cochise County, as much as 1.79 m of subsidence is recorded (Holzer, 1980, 1984b). In the Tucson Basin (Platt, 1963), 13 cm of subsidence is known, and in the Arva Valley, 32 cm was recorded between 1948 and 1980 (Strange, 1983).

Earth fissures

Preliminary statement. The most spectacular result of groundwater withdrawal and subsequent aquifer compaction in unconsolidated sediments in Arizona is the deformation of the land surface by aseismic ground failure or rupture. The most common type of failure is tension failure in the unconsolidated sediments that produces long, narrow earth cracks or fissures. There is no surface vertical displacement, but horizontal displacement occurs perpendicular to the plane of failure. A second but much less common failure is a surface fault with a vertical offset and scarp generally of less than 0.5 m. These scarps and earth fissures, especially when the latter extend hundreds of meters and have been eroded by running surface water into gullies 2 to 4 m wide, are indeed striking microgeomorphologic features induced by the action of groundwater.

Description and distribution. The long, narrow, steep-sided, eroded tension earth fissures occur in unconsolidated sediments, typically near the margins of mountains or outlying bedrock outcrops where groundwater levels have dropped from 60 m to more than 135 m. Initial appearance of fissures is noted by: (1) small linear or en echelon hairline cracks, (2) irregularly spaced depressions, or (3) large linear open holes (Fig. 5). Field evidence indicates most fissures are less than 1 km in length; however, one fissure zone near Picacho is about 16 km long (Laney and others, 1978). Fissures are generally perpendicular to drainage, intercepting water that erodes the initial hairline-width cracks into gullies, as much as 4 m wide and 5 m and more deep (Fig. 6). Johnson (1980) measured a depth of 25 m in south-central Arizona.

The first fissure reported in Arizona was observed near Picacho in 1927 (Leonard, 1929). During the last 60 years, accelerated groundwater withdrawal has caused intense fissuring of several alluvial basins in south-central Arizona (Feth, 1951; Heindl and Feth, 1955; Pashley, 1961; Robinson and Peterson, 1962; Anderson, 1973; Schumann, 1974; Laney and others, 1978; Peirce, 1979; Jachens and Holzer, 1982; Strange, 1983; Holzer, 1984b, 1986; and others). Until recently, fissures have been essentially confined to outlying agricultural areas, but in January 1980, a 130-m-long fissure opened in Paradise Valley (northeast Phoenix) at a residential construction site. The Paradise Valley fissure is the first known occurrence in a densely populated, nonagricultural part of the state, and the first in Phoenix (Péwé, 1980). As urban areas continue to enlarge, past and future fissures will occur in what was originally agricultural land.

The number of fissures has multiplied enormously since the 1950s, and now hundreds of fissures are known in the alluvial

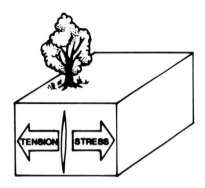
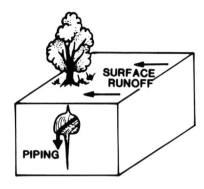

1. Lateral stresses induce tension cracking

2. Surface runoff and infiltration enlarge crack through subsurface piping

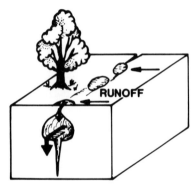
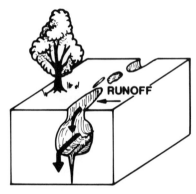

3. As piping continues, fissure begins to appear at surface as series of potholes and small cracks

4. As infiltration and erosion continue, fissure enlarges and completely opens to surface as tunnel roof collapses

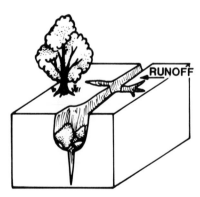
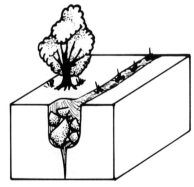

5. The entire fissure is opened to the surface and enlargement continues as fissure walls are widened, extensive slumping and side-stream gullying occur

6. Fissure becomes filled with slump and runoff debris and is marked by vegetation lineament and slight surface depression, it may become reactivated upon renewal of tensile stress

Figure 5. Generalized stages of earth-fissure development (modified from Bell, 1981).

basins of southern Maricopa, western Pinal, western Pima, and northwestern Cochise Counties (Schumann and Genualdi, 1986b). Today, most are concentrated in Pinal and Maricopa Counties (Fig. 2).

In just about every area of known fissuring, they are continuing to increase in number and length. An excellent example that illustrates such activity occurs on the east edge of the Casa Grande Mountains, about 10 km southeast of Casa Grande, Arizona (Fig. 2). Jachens and Holtzer (1982) demonstrate that there were no known fissures in the area in 1949 and only one fissure in 1954. As of 1980, the number of fissures had increased to about 50, and some of them occupy areas now taken from cultivation (Fig. 7).

Another well-documented instance of increasing numbers of earth fissures occurs on the west side of Signal Peak of the San Tan Mountains, 11 km northeast of Casa Grande. A prominent fissure, now more than 1.2 km long at the base of the western side of Signal Peak, was not present in 1971, but was recognized in 1975. It lengthened several meters in the summer of 1976 (Raymond and others, 1978) and by July 1985 had extended a couple hundred additional meters farther south. This fissure is a steep, sharp-walled gully 1 to 2 m wide and 1 to 30 m deep cut into the alluvium.

Recently a 1-km-long, curving, steep-walled fissure formed immediately to the west of the Signal Peak fissure (Fig. 8). Although known to exist in 1975–1976, it was then very short (Laney and others, 1978). Subsequently, it has extended about 1 km and now reaches into an area of new houses. As time passes, these uncontrolled fissures will get larger, wider, and deeper, influenced by surface drainage, and especially lowering of the water table and continued subsidence of land.

On the south side of Apache Junction (Fig. 2), in an area of existing earth fissures, a large new fissure formed rapidly in the summer of 1984 due to the concentration of surface-water runoff 100 m downslope from the Central Arizona Project Aqueduct. In this area, surface runoff blocked by the aqueduct levee is passed over the aqueduct through 2-m-diameter pipes to spread out downslope from the aqueduct. At one locality, 100 m west of Ironwood Street, the water was discharged into an abandoned stock pond (a depression dug to collect water for cattle) 4 m deep and 25 to 30 m across. The pond rapidly filled on July 18, 1984, and around midnight the water escaped into a previously unknown subsurface earth fissure underlying and crossing the south end of the pond site. The water roared downward with the sound of a subterranean waterfall, as quoted by a person living at the site (Troy Vickory, oral communication, August 25, 1984). The pond water developed a whirlpool and drained in a matter of minutes, creating a fissure 3 m wide, 4 m deep, and 200 m long. The fissure intersected the concrete patio of a house, severed a surface access road, and buried electrical lines.

Approximately 2 weeks later, another rainstorm occurred, and water was again channeled and discharged onto the same property. This time the fissure was rapidly enlarged another 200 m to the southeast (Fig. 9). The U.S. Bureau of Reclamation has

Figure 6. Large curving earth fissure caused by land subsidence resulting from groundwater withdrawal. South side of Chandler Heights on Hunt Highway. View south to the Gold Mine Mountains (see Fig. 11). (Photograph No. 3357 by Troy L. Péwé, July 25, 1972.)

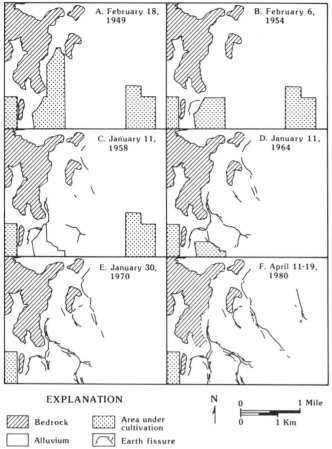

Figure 7. Chronology of fissure formation and area under cultivation in study area based on interpretation of aerial photographs. Map of April 11–19, 1980, is based on ground traverses. Casa Grande Mountains, 10 km southeast of Casa Grande, Arizona (see Fig. 2). (From Jachens and Holzer, 1982.)

Figure 8. Ground subsidence earth fissure in slightly caliche-cemented alluvium 8.5 km northeast of Casa Grande, Arizona. Fissure formed about 1974–1975 and in 1985 was 1 km long. It is actively lengthening into an area of newly constructed houses. (Photograph No. 4868 by Troy L. Péwé, July 27, 1985.)

driven sheet piling 15 m deep across the trace of the fissure to prevent erosion from enlarging the fissure, which might damage the aqueduct (Péwé and others, 1987).

An up-to-date map on a scale of 1:500,000 showing the location of earth fissures, land subsidence, and long-term water-level decline in south-central and southwestern Arizona is now available (Schumann and Genualdi, 1986b), and maps at a scale of 1:250,000 are in preparation. A very useful colored map of southern Arizona at a scale of 1:1,000,000 showing land subsidence, earth fissures, and water-level changes has recently been published by the Arizona Bureau of Geology and Mineral Technology (Schumann and Genualdi, 1986a).

Origin. In the 1960s the writer was working in the field in what is now eastern Mesa and was surprised to see, early in the morning, a landowner grading a former agricultural field to cover an earth fissure in the unconsolidated sediments. The owner stated that the fissure formed the night before during an intense rainstorm and he was now filling and covering it prior to selling the land that day. Such was the writer's introduction to the origin of earth fissures in Arizona. The relation of rainstorms to earth fissures has been repeatedly mentioned in the literature and by laymen since 1927. Because by far most of the earth fissures associated with groundwater withdrawal in the U.S., and perhaps in the world, occur in Arizona, a review of the evolution of ideas of their origin may be in order.

According to Leonard (1929), an earth fissure formed in the evening of September 11, 1927, 4.8 km southeast of Picacho (Figs. 2 and 3). It crossed the highway and railway. A severe rain and windstorm had occurred the afternoon and night before. Leonard (1929) believed the crack was due to seismic vibrations and was the first to recognize that there was tensional strain in the unconsolidated sediments.

The seismic idea for origin was suggested because Leonard received word of an earthquake within a radius of 250 km of Tucson. No record of this earthquake has ever been found (DuBois and others, 1982). Leonard considered but discarded the theory of land subsidence. The 1927 earth fissure appears exactly at the same place as subsequent earth fissures in the area. Additional fissures appeared in the area in 1949, 1951, and 1952. Violent rainstorms preceded each appearance, but seismic shocks were absent (Fletcher and others, 1954). Feth (1951) believed that the 1949 fissure was essentially, if not identically, in the same place as the 1927 one.

The fissure reported by Leonard (1929) is not believed by all to be related in origin to subsequent fissures; Holzer (1984b) believes the fissure to be the result of natural forces. However, the writer, as well as Robinson and Peterson (1962), believe it perhaps was a conventional earth fissure caused by land subsidence because of its size, shape, and locality, and that some land subsidence, even though small, had occurred in the Picacho area in the period from 1905 to 1934.

In the 1950s, Fletcher and others (1954) made the suggestion that the fissures were due to piping in unconsolidated sediments. This was rejected by Heindl and Feth (1955). It was in 1951 (Feth, 1951) that we have the first publication of the idea that fissuring possibly was the result of strain in the unconsolidated sediments caused by ground subsidence, the result of groundwater withdrawal. Feth (1951, p. 26) attributes the idea of the origin by tension due to subsidence to S. F. Turner of the U.S. Geological Survey in 1949; Turner is also credited with the idea that the breaks may be due to differential compaction of the sediments over the edge of a buried pediment marking a fault scarp. Heindl and Feth (1955) note that heavy summer storms may be related to fissure formation, but also that many such storms occur with no fissure formation.

In the late 1950s, Pashley (1961) observed fissures in the Casa Grande area (Figs. 2 and 3) and noted that their formation was associated with violent rain and windstorms. He appears to be the first to report that the surface water entering the fissures does not flow laterally to much extent, but moves downward through outlets in the bottom of the cracks, possibly recharging the groundwater.

The last report of the pioneer period of earth-fissure study in Arizona was by Robinson and Peterson (1962). Their report was the first to demonstrate clearly with data and diagrams that fissure formation was caused by land subsidence due to groundwater withdrawal. Generally overlooked in this report is the statement that in 1961 Peterson made a gravimetric survey in the Picacho area and related earth fissures to differential settlement over buried bedrock highs (edge of concealed [buried] pediment). This is the first report of what later was to become a common study.

Recent work utilizing geophysical and geodetic surveying techniques supports the hypothesis that many fissures are caused by differential subsidence and compaction over buried bedrock hills, ridges, or fault scarps (Schumann and Poland, 1969; Sauck, 1971, 1975; Jennings, 1977; Christie, 1978; Pankratz and others, 1978; Jachens and Holzer, 1979, 1982; Péwé and Larson 1982; Raymond, 1985; Holzer, 1986). Some fissure locations occur for other reasons, such as where there are variations in type and thickness of alluvium, and variations in water-level decline (Larson, 1982; Raymond, 1986). For example, Schumann and Tosline (1983) show that during investigations of earth fissures in

Figure 9. Earth fissure that opened on Junker's Ranch in Apache Junction, Arizona, on July 18, 1984, as a result of water filling a stock pond diverted from the CAP Aqueduct construction. Fissure opened at midnight and immediately drained the filled stock pond. (Photograph No. 4815 by Troy L. Péwé, August 25, 1984.)

relation to the route of the Central Arizona Project aqueduct, geophysical investigations and test drilling indicated that some earth fissures west of the Picacho Mountains (Fig. 2) occur above irregularities on the buried bedrock surface and at abrupt changes in texture or thickness of the overlying alluvial sediments.

An informative example of a fissure that formed over a buried bedrock hill is well documented in northeastern Phoenix by geological and gravity studies (Larson and Péwé, 1986). Well-drilling records and gravity data provide the basis for a depth-to-bedrock map. The map shows the relation of past and potential land subsidence and earth fissuring to the buried bedrock topography.

The underground bedrock slopes gently toward the northeast from the Phoenix Mountains. The inner part of this area is buried less than 165 m, and extends at least 4 km into the Paradise Valley Basin (Fig. 2), with a series of hills and ridges with relief of 30 to 90 m. The buried bedrock features follow the same NE-SW direction as the foliation and topographic expression in the adjacent Phoenix Mountains. One can visualize the buried bedrock topography as that which would exist if the present Papago Park near Tempe (Fig. 2) (4 km southeast of the Phoenix

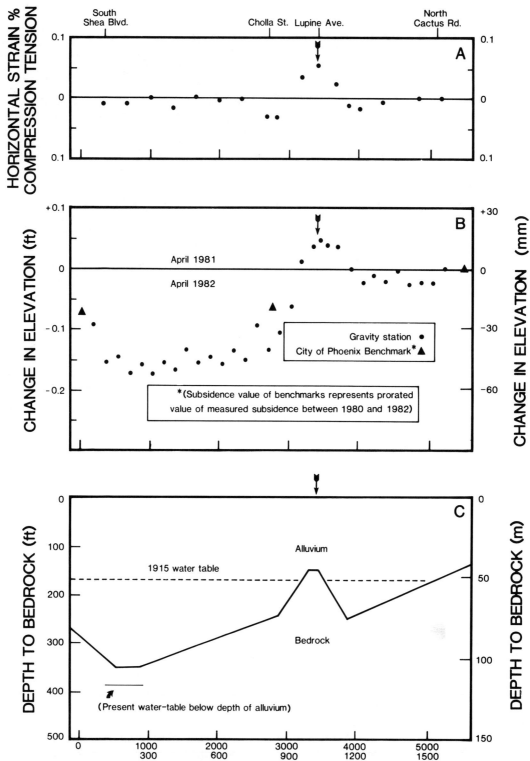

Figure 10. Computed horizontal surface strain (1980) at time of fissuring, land subsidence (April 1981 to April 1982), and interpreted depth to bedrock, 40th Street, northeastern Phoenix. Arrows indicate location of fissure. a. Horizontal surface strain; b. land subsidence; c. depth to bedrock. Based on two-dimensional gravity modeling and depth-to-bedrock map. Vertical exaggeration = 6.6 ×. Numbers at bottom of figure indicates horizontal distances in feet and meters (from Larson and Péwé, 1986).

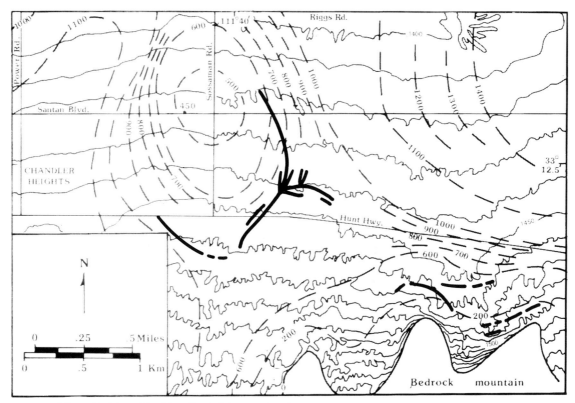

Figure 11. Map of the Chandler Heights area, Maricopa and Pinal Counties, showing depth to bedrock in feet beneath alluvium (heavy dashed lines), surface topography in feet (light solid lines), and location of earth fissures (heavy solid lines). Depth to buried bedrock surface estimated from gravity and well data (map slightly modified from Jennings, 1977).

Mountains) were to be buried beneath 100 to 165 m of silt, sand, and gravel.

Bordering the inner surface is an outer, more deeply buried, low-relief topography, sloping gently northeastward at a depth of 165 to 300 m. A major NW-SE Basin and Range fault in Paradise Valley separates this gently sloping surface from thick deposits of consolidated sediments.

The subsurface geologic conditions control patterns of water-level decline and land subsidence. Maximum subsidence and water-level decline have occurred on the deeper outer surface, whereas minimal subsidence has occurred and little or no water has been obtained from wells drilled on the shallow buried inner surface. Subsidence generally increases wherever the thickness of alluvium increases.

Gravity data indicate that a small bedrock hill underlies the fissure at a depth of about 50 m, with at least 30 m of relief (Fig. 10C). Differential compaction induced by dewatering of sediments across this buried knoll was sufficient to cause ground failure. Continued differential subsidence has been measured (April 1981 to April 1982) along 40th Street between Shea Boulevard and Cactus Road, with as much as 40 mm of subsidence south of the fissure (Fig. 10B). The striking similarity between the subsidence curve and an interpreted depth-to-bedrock profile along 40th Street supports the argument that fissuring is associated with the crests of buried hills. Finite-element analysis indicates surficial tensile stress of 3,400 psf (1.4 kg/cm^2) and strain of 0.19 percent when fissuring occurred (Ragan, 1986).

Measured differential subsidence and calculated horizontal strain strongly suggest a reopening of the entire fissure. Continued displacement is indicated by small cracks that have lengthened and become more numerous in the newly constructed paved road and concrete wall across the original fissure trace. On the basis of detailed gravity traverses, a future westward extension of the fissure is probable, with less than 200 m of eastward extension possible. Several fissures subparallel to the original could form in the vicinity of 40th Street and Lupine Avenue.

The history of fissured basins in southern Arizona bears ample evidence that the initial fissure is later followed by complex patterns of multiple fissuring. In northeast Phoenix, future fissuring may be localized in three geologic settings: (1) buried bedrock topographic highs, (2) at the hinge line of subsiding areas controlled by bedrock depth, and (3) buried fault scarps. Gravity data suggest that there are several buried hills between 30th and 42nd Streets (Fig. 4), with a high probability of fissuring, particularly near those hills directly north of the fissure. Another area of potential fissuring is near the hinge line of subsidence between Shea Boulevard and the Phoenix Mountains east of 34th Street. Differential subsidence and fissuring are also possible across an

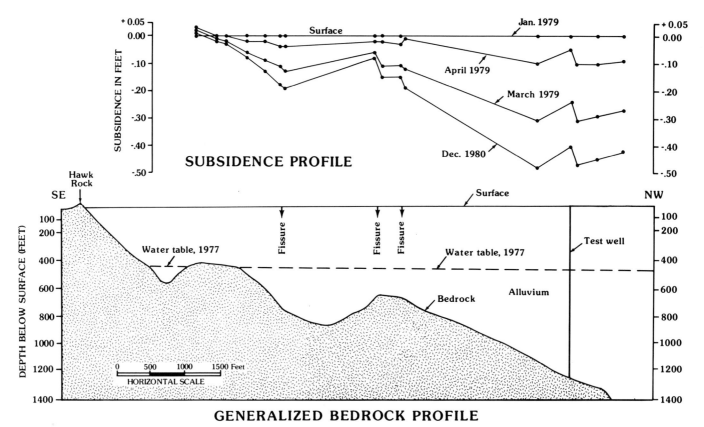

Figure 12. Relation of subsurface geology, water table, earth fissures, and land subsidence near Apache Junction Arizona (modified from Raymond, 1985).

inferred buried basin-and-range fault scarp in the eastern part of the study area; however, because most water-level decline and land subsidence has occurred on the upthrown rather than downthrown fault block, fissuring seems less likely in this area at the present time.

About 15 km southeast of Chandler, on the southern side of the Queen Creek area (Fig. 2), occurs one of the deepest and widest earth fissures. It is 5 m wide and is open as a small crack to a depth of at least 16 m. It is a multiple curved fissure, and several adjacent parallel fissures are present (Péwé and others, 1987). The fissure formed in 1961 and intercepts surface drainage (Robinson and Peterson, 1962). The water table in the area has declined 160 m, permitting the development of "hinge-type origin" fissures near the mountains on the south side of Hunt Highway (Fig. 11) in the dewatered area, and large fissures on the side of a buried bedrock knob detected by gravity studies (Jennings, 1977) on the north side of the highway.

Perhaps the most detailed study of land subsidence and earth-fissure formation was a multi-year investigation undertaken by the U.S. Bureau of Reclamation and U.S. Geological Survey starting in 1976 near Apache Junction (Fig. 2) in connection with the alignment of the Central Arizona Project Aqueduct.

It was clearly demonstrated that the water table has declined about 62 m since the turn of the century (Fig. 2) and has declined about 1.6 m from 1976 to 1986 (Raymond, 1985). The fissures are directly related to the convex bedrock irregularities (Fig. 12; Pankratz and others, 1978; Hassemer and Dansereau, 1980). Level data indicate that differential subsidence occurred on either side of the buried ridge (Fig. 12), and horizontal surveys showed that land-surface extension occurred over the ridge and normal to the fissure (Raymond and others, 1979).

Surface faults. It has been demonstrated by Holzer (1977, 1984b, 1986; Holzer and others, 1979; Holzer and Thatcher, 1978) that locally the differential compaction of sediment caused by groundwater withdrawal may result in small surface displacements (faults) in addition to earth fissures. Such faults have scarps less than a 0.5 m high and may be as much as 17 km long. Holzer (1978, 1980) shows that this surface faulting takes place along preexisting faults.

The scarps increase in size and length very slowly and are not related to seismic activity. The best known such fault in Arizona is the Picacho fault in south-central Arizona (Figs. 2, 3, and 13). It is about 17 km long and has a scarp about 0.5 m high. It crosses Interstate 10 and the Southern Pacific Railroad. Constant maintenance is required for both the highway and the railway.

Figure 13. Oblique aerial view looking northeast of the Picacho fault and associated earth fissures crossing Interstate 10 and the Southern Pacific Railroad, 10 km east of Eloy, Arizona. Lineament to the north can be seen in the middle ground. See Figures 2 and 3. (Photograph No. PK 27517 by Troy L. Péwé, May 16, 1985.)

Land subsidence has been recognized since the early 1950s, and more than 5 m was recorded in the Picacho Basin 60 km southeast of Phoenix by 1985. The subsidence areas are mainly agricultural land, but urban areas are now expanding into the subsiding regions. Parts of northeastern Phoenix have subsided as much as 1.6 m since the mid-1960s.

The most spectacular and costly result of groundwater withdrawal in the alluvial basins of Arizona is the deformation of the land surface by aseismic ground failure or rupture. The most common type produces long, narrow earth fissures by tension failure. There is no vertical displacement, but horizontal displacement occurs perpendicular to the plane of failure. A second but much less common failure is a surface fault with a vertical offset less than 0.5 m. The earth fissures extend to hundreds of meters and are eroded by surface water that runs into gullies 2 to 4 m wide.

It has been demonstrated since the 1950s and early 1960s that earth fissures are caused by differential compaction of unconsolidated sediments resulting from withdrawal of underground water and subsequent decline of the water table. It has been further demonstrated since the 1970s, and especially in the 1980s, that the differential compaction occurs perhaps most commonly over buried upward-convex irregularities in the bedrock topography (bedrock hills or ridges) where the depth to bedrock is generally less than 300 or 400 m.

The tension cracks are present prior to reaching the surface (Fig. 5), and running water at or under the surface enlarges the opening until part or all of the surface cover over the fissure collapses. This is promoted by a considerable amount of running water. Many fissures are thus "unroofed" during violent rainstorms.

A few of the ground-failure features due to subsidence are surface faults with minor scarps. Movement is along preexisting faults.

SUMMARY

More than 7,800 km^2 of land in alluvial basins in south-central Arizona has subsided due to withdrawal of groundwater and subsequent lowering of the water table. Withdrawal of groundwater greatly exceeds recharge, and the water table has declined in some basins more than 170 m, mainly since the late 1940s and 1950s.

ACKNOWLEDGMENTS

The writer deeply appreciates the careful and thoughtful review of this chapter ty T. L. Holzer, U.S. Geological Survey; R. H. Raymond, U.S. Bureau of Reclamation; D. Ragan, Arizona State University; and C. G. Higgins, University of California. Holzer and Higgins were especially helpful with the organization of the chapter.

REFERENCES CITED

Anderson, S. L., 1973, Investigation of the Mesa earth crack, Arizona, attributed to differential subsidence due to groundwater withdrawal [M.S. thesis]: Tempe, Arizona State University, 111 p.

Anonymous, 1987, Subsidence areas and earth fissure zones: Arizona Bureau of Geology and Mineral Technology Field Notes, v. 17, no. 1, p. 6–9.

Arizona Department of Transportation, Highways Division, Photogrammetry and Mapping Services, 1981, 1981 NGS Level Line, Apache Junction to I-10 at Miller road: ADOT Preliminary Adjustment, unpublished report, 192 p.

Bell, J. W., 1981, Subsidence in Las Vegas Valley: Reno, University of Nevada, Mackay School of Mines, Nevada Bureau of Mines and Geology Bulletin 95, 84 p.

Briggs, P. C., 1976, Arizona groundwater resources, supplies, and problems, in Proceedings 20th Arizona Watershed Symposium, Phoenix, September

1976: report 8, p. 11–29.

Bull, W. B., 1973, Geologic factors affecting compaction of deposits in land-subsidence area: Geological Society of America Bulletin, v. 84, p. 3783–3802.

Christensen, G. E., 1986, Utah's geologic hazards: Utah Geological and Mineral Survey Survey Notes 20, no. 1, p. 3–8.

Christie, F. J., 1978, Analysis of gravity data from the Picacho Basin, Pinal County, Arizona [M.S. thesis]: Tucson, University of Arizona, 105 p.

Coates, D. R., 1981, Environmental geology: New York, John Wiley and Sons, 701 p.

Cordova, R. M., and Mower, R. W., 1976, Fracturing and subsidence of the land surface caused by the withdrawal of groundwater in the Milford area, Utah: U.S. Geological Survey Journal of Research, v. 4, no. 5, p. 505–510.

DuBois, S. M., Smith, A. W., Nye, N. K., and Nowak, T. A., Jr., 1982, Arizona earthquakes, 1776–1980: Arizona Bureau of Geology and Mineral Technology Bulletin 193, 456 p.

Feth, J. H., 1951, Structural reconnaissance of the Red Rock Quadrangle, Arizona: U.S. Geological Survey Open-File Report, 32 p.

Fletcher, J. E., Harris, K., Peterson, H. R., and Chandler, V. N., 1954, Symposium of land erosion and "piping": EOS Transactions of the American Geophysical Union, v. 35, p. 258–262.

Forrester, F., 1971, Land subsidence: California Geology, v. 24, p. 148–149.

Harmon, D. B., 1982, Subsidence in northeast Phoenix; A new problem for engineers: Arizona Bureau of Geology and Mineral Technology Field Notes, v. 12, no. 3, p. 10–11.

Hassemer, J. H., and Dansereau, D. C., 1980, Gravity survey in parts of Maricopa and Pinal Counties, Arizona: U.S. geological Survey Open-File Report 80-1255, 9 p.

Heindl, L. A., and Feth, J. H., 1955, Piping and earthcracks; A discussion: EOS Transactions of the American Geophysical Union, v. 36, no. 2, p. 342–345.

Holzer, T. L., 1977, Ground failure in areas of subsidence due to groundwater decline in the United States: International Association of Hydrological Sciences Publication 121, p. 423–433.

—— , 1978, Results and interpretation of exploratory drilling near the Picacho Fault, south-central Arizona: U.S. Geological Survey Open-File Report 78-1016, 17 p.

—— , 1980, Earth fissures and land subsidence, Bowie and Wilcox areas, Arizona: U.S. Geological Survey Miscellaneous Field Studies Map MF-1156, scale 1:24,000.

—— , 1984a, Man-induced land subsidence: Boulder, Colorado, Geological Society of America Reviews in Engineering Geology, v. 6, 221 p.

—— , 1984b, Ground failure induced by ground water withdrawal from unconsolidated sediment, in Holzer, T. L., ed., Man-induced land subsidence: Boulder, Colorado, Geological Society of America Reviews in Engineering Geology, v. 6, p. 67–105.

—— , 1986, Ground failure caused by underground water withdrawal from unconsolidated sediments—United States, in Johnson, I. A., and others, ed., Land subsidence: International Association of Hydrological Sciences Publication 151, p. 747–756.

Holzer, T. L., and Thatcher, W., 1978, Modeling deformation due to subsidence faulting, in Proceedings Evaluations and Prediction of Subsidence: American Association of Civil Engineers, p. 349–357.

Holzer, T. L., Davis, S. N., and Lofgren, B. E., 1979, Faulting caused by groundwater extraction in south-central Arizona: Journal of Geophysical Research, v. 84, no. 2, p. 603–612.

Jachens, R. C., and Holzer, T. L., 1979, Geophysical investigations of ground failure related to groundwater withdrawal, Picacho Basin, Arizona: Groundwater, v. 17, no. 6, p. 574–484.

—— , 1982, Differential compaction mechanisms for earth fissures near Casa Grande, Arizona: Geological Society of America Bulletin, v. 93, p. 998–1012.

Jennings, M. D., 1977, Geophysical investigations near subsidence fissures in northern Pinal and southern Maricopa Counties, Arizona [M.S. thesis]: Tempe, Arizona State University, 102 p.

Johnson, N. M., 1980, The relation between ephemeral stream regime and earth fissuring in south-central Arizona [M.S. thesis]: Tucson, University of Arizona, 158 p.

Laney, R. L., Raymond, R. H., and Winikka, C. C., 1978, Maps showing water level declines, land subsidence, and earth fissures in south-central Arizona: U.S. Geological Survey Water Resource Investigations 78-83, scale 1:250,000.

Larson, M. K., 1982, Origin of land subsidence and earth fissures, northeast Phoenix, Arizona [M.S. thesis]: Tempe, Arizona State University, 151 p.

Larson, M. K., and Péwé, T. L., 1983, Earth fissures and land subsidence in northeast Phoenix: Arizona Bureau of Geology and Mineral Technology Field Notes, v. 13, no. 2, p. 8–11.

—— , 1986, Origin of land subsidence and earth fissuring northeast Phoenix, Arizona: Association of Engineering Geologists Bulletin, v. 23, no. 2, p. 139–165.

Lausten, C. D., 1974, Gravity methods applied to geology and hydrogeology of Paradise Valley, Maricopa County, Arizona [M.S. thesis]: Tempe, Arizona State University, 137 p.

Leonard, R. J., 1929, An earth fissure in southern Arizona: Journal of Geology, v. 37, no. 8, p. 765–774.

Pankratz, L. W., Hassamer, J. H., and Ackermann, H. D., 1978, Geophysical studies related to earth fissures in central Arizona [abs.]: Geophysics, v. 44, no. 3, p. 367.

Pashley, E. F., Jr., 1961, Subsidence cracks in alluvium near Casa Grande, Arizona: Arizona Geological Society, v. 4, p. 95–101.

Patino, F. H., Robertson, J., and Péwé, T. L., 1985, Geological hazards, in Environmental geology of the Chandler Quadrangle, Maricopa County, Arizona: Report to the City of Chandler, Arizona, scale 1:24,000.

Peirce, H. W., 1979, Subsidence-fissures and faults in Arizona: Arizona Bureau of Geology and Mineral Technology Field Notes, v. 9, no. 2, p. 1–2, 6–7.

Péwé, T. L., 1980, Earth crack, 40th Street and Lupine, Phoenix, Arizona: unpublished report for City of Phoenix, 13 p.

—— , 1984, Fissures in Arizona: Earth Science, summer, p. 19–21.

Péwé, T. L., and Larson, M. K., 1982, Origin of land subsidence and earth fissures in northeast Phoenix: City of Phoenix, 98 p. Appendices A-G.

Péwé, T. L., Raymond, R. H., and Schumann, H. H., 1987, Land subsidence and earth fissure formation in eastern Phoenix metropolitan area, Arizona, in Davis, G. H., and VandenDolder, E., eds., Geologic diversity of Arizona and its margins; Excursions to choice areas: Arizona Bureau of Geology and Mineral Technology Special Paper 6, p. 199–204.

Platt, W. S., 1963, Land-surface subsidence in the Tucson area [M.S. thesis]: Tucson, University of Arizona, 38 p.

Poland, J. F., 1969, Status of present knowledge and needs for additional research in compaction of aquifer systems, in Proceedings of Tokyo Symposium on Land Subsidence: International Association of Scientific Hydrology-UNESCO, p. 11–21.

Poland, J. F., Lofgren, B. E., Ireland, R. L., and Pugh, R. G., 1975, Land subsidence in the San Joaquin Valley, California, as of 1972: U.S. Geological Survey Professional Paper 437-H, 78 p.

Ragan, D. M., 1986, Model of the Phoenix earth crack, in Larson, M. K., and Péwé, T. L., Origin of land subsidence and earth fissuring, northeast Phoenix, Arizona: Association of Engineering Geologists Bulletin, v. 23, no. 2, p. 163–165.

Raymond, R. H., 1985, Earth fissure investigations for Reach 2A, Salt-Gila Aqueduct, Central Arizona Project, Arizona: U.S. Bureau of Reclamation Arizona Projects Office Internal Report, 24 p.

—— , 1986, Seven geologic settings for earth fissures in south-central Arizona [abs.]: Journal of the Arizona-Nevada Academy of Science Proceedings Supplement, v. 21, p. 51.

Raymond, R. H., Winikka, C. E., and Lanney, R. L., 1978, Earth fissures and land subsidence, eastern Maricopa and northern Pinal Counties, Arizona, in Burt, D. M., and Péwé, T. L., eds., Guidebook to the geology of central Arizona: Arizona Bureau of Geology and Mineral Technology Special Paper 2, p. 107–114.

Raymond, R. H., Laney, R. L., Pankrantz, L. W., Riley, F. S., and Carpenter,

M. C., 1979, Relationship of earth fissures in alluvial basins in south-central Arizona to iregularities in the underlying formations: Geological Society of America Abstracts with Programs, v. 11, no. 7, p. 501.

Robinson, G. M., and Peterson, D. E., 1962, Notes on earth fissures in southern Arizona: U.S. Geological Survey Circular 466, 7 p.

Sauck, W. A., 1971, Land subsidence and earth cracks: Tempe, Arizona State University Research and Special Programs, p. 29.

—— , 1975, Geophysical studies near subsidence fissures in central Arizona [abs.]: EOS Transactions of the American Geophysical Union, v. 56, no. 12, p. 984–985.

Schumann, H. H., 1974, Land subsidence and earth fissures in alluvial deposits in the Phoenix area, Arizona: U.S. Geological Survey Miscellaneous Investigation Series Map I-845, scale 1:250,000.

Schumann, H. H., and Genauldi, R., 1986a, Land subsidence, earth fissures, and water-level change in southern Arizona: Arizona Bureau of Geology and Mineral Technology Map 23, scale 1:1,000,000.

—— , 1986b, Land subsidence, earth fissures, and water-level change in southern Arizona: Arizona Bureau of Geology and Mineral Technology Open-File Report 86-14, scale 1:500,000.

Schumann, H. H., and Poland, J. F., 1969, Land subsidence, earth fissures, and groundwater withdrawal in south-central Arizona, U.S.A., in Land subsidence: Tokyo International Association of Science Hydrology Publication 88, v. 1, p. 295–302.

Schumann, H. H., and Tosline, D. J., 1983, Occurrence and prediction of earth-fissures hazards by groundwater depletion in south-central Pinal County, Arizona [abs.], in 2nd Arizona Symposium on Subsidence: Arizona Consulting Engineers Association in Phoenix.

Schumann, H. H., Litton, C. R., and Wallace, B. L., 1983, Monitoring water-level change, aquifer compaction, and subsidence along the Salt-Gila Aqueduct in south-central Arizona [abs.], in 2nd Arizona Symposium on Subsidence: Arizona Consulting Engineers Association in Phoenix.

Smith, C.E.P., 1940, The groundwater supply of the Eloy District in Pinal County, Arizona: Tucson, University of Arizona Agriculture Experiment Station Technical Bulletin 87, 42 p.

Strange, E., 1983, Subsidence monitoring for State of Arizona: National Oceanic and Atmospheric Administration and National Geodetic Survey, 74 p.

Thomsen, B. W., and Baldys, S., III, 1985, Groundwater conditions in and near the Gila River Indian Reservation, south-central Arizona: U.S. Geological Survey Water Resources Investigations Report 85-4073, scale 1:250,000.

U.S. Bureau of Reclamation, 1980, Central Arizona Project, Aqueduct System; Location Map: Phoenix, U.S. Bureau of Reclamation, scale 1:250,000.

U.S. Geological Survey, 1984, Annual summary of groundwater conditions in Arizona, Spring 1982, to Spring 1983: U.S. Geological Survey Open-File Report 84-428, scale 1:250,000.

—— , 1985, Annual summary of groundwater conditions in Arizona, Spring 1983 to Spring 1984: U.S. Water Resources Investigations 85-410, scale 1:250,000.

Viets, V. F., Vaughan, C. K., and Harding, R. L., 1979, Environmental and economic effects of subsidence: Berkeley, University of California Lawrence–Berkeley Laboratory, part 3, 88 p.

Winikka, C. C., and Wold, D. D., 1976, Land subsidence in central Arizona, in 2nd International Symposium on Land Subsidence: International Association of Hydrological Sciences Publication 181, p. 95–103.

Yamamoto, S., 1977, Recent trend of land subsidence in Japan: Science Reports of the Tokyo Kyoiku Dalgaku, section c, v. 13, no. 125-126, p. 1–7.

MANUSCRIPT ACCEPTED BY THE SOCIETY NOVEMBER 14, 1989

Printed in U.S.A.

Chapter 11

Spring sapping and valley network development

Victor R. Baker
Department of Geosciences, University of Arizona, Tucson, Arizona 85721
with case studies by
R. Craig Kochel
Department of Geology, Bucknell University, Lewisburg, Pennsylvania 17837
Victor R. Baker
Department of Geosciences, University of Arizona, Tucson, Arizona 85721
Julie E. Laity
Department of Geography, California State University, Northridge, California 91330
Alan D. Howard
Department of Environmental Sciences, University of Virginia, Charlottesville, Virginia 22903

INTRODUCTION

Victor R. Baker

Spring sapping occurs where groundwater outflow undermines slopes and, where appropriately concentrated, contributes to the development of valleys. Higgins (1984) distinguishes between "spring sapping" caused by concentrated groundwater discharge and "seepage erosion" caused by diffuse discharge at rock boundaries of various kinds. The action of groundwater in sapping may be concentrated at valley heads, leading to headward growth. Both enhanced weathering and direct erosion by the concentrated fluid flow lead to slope undermining and collapse at sites of groundwater outflow (Laity and Malin, 1985). When the exact process of subsurface flow and its erosive action can be specified, then a precise terminology may be used, as discussed by Dunne (this volume). However, the analysis of valley forms usually requires subjective interpretation of relict features for which the exact genetic process is unclear even though subsurface flow is involved. It is hoped that further research will permit refinements of the terminology as proposed by Dunne (this volume).

It is important to recognize that sapping can contribute both to channel and valley development. The relation to channel networks has been reviewed by Dunne (1980). In a summary of field-experimental and theoretical analyses, Dunne (1980) outlined a model of headward channel growth and branching by sapping that contrasts with channel network development by overland flow. Models for the latter predict network development by piracy and cross-grading (Horton, 1945), or by headward growth through abstraction (Abrahams, 1984). Jones (1987) reviews the current literature on the initiation of natural drainage networks by both surface and subsurface flow processes.

Unfortunately, geomorphic evidence for channel network evolutionary processes is rarely discernible in the complex hillslope-channel relations of valleys. Valley processes involve a considerable component of nonfluvial degradation. Although fluvial incision may drive the hillslope systems of valleys, the picture can be confused by the action of greatly enhanced past processes that result in a relict valley morphology. Sapping has been underappreciated as a geomorphic process (Higgins, 1984), probably because lowered water tables or desiccating climatic conditions have reduced its influence in many Holocene valleys. In other valleys the results of spring sapping processes are obscured by modification of valley forms by nonsapping morphogenetic processes.

Morphology of sapping valleys

Excellent examples of sapping valleys have been described: (1) on the Colorado Plateau, where massive sandstone units are eroded by perched water emerging from bedding-plane boundaries; and (2) in the Hawaiian Islands, where basalt flows flanking shield volcanoes are dissected by both runoff and sapping valleys. Two case studies following this introduction describe these contrasting situations. However, several distinct morphologic attributes seem to characterize sapping valleys in numerous associations. The Hawaiian example is particularly instructive in this regard because volcanic ash mantles on the older shield volcanoes (such as Mauna Kea, Kohala, and Haleakala) resulted in the development of long, parallel V-shaped valleys by overland flow processes (Fig. 1). With continued deep incision, underlying permeable lava flows were encountered, and the valley morphology changed to U-shaped in cross section. These steep-walled

Baker, V. R., 1990, Spring sapping and valley network development, with case studies by Kochel, R. C., Baker, V. R., Laity, J. E., and Howard, A. D., *in* Higgins, C. G., and Coates, D. R., eds., Groundwater geomorphology; The role of subsurface water in Earth-surface processes and landforms: Boulder, Colorado, Geological Society of America Special Paper 252.

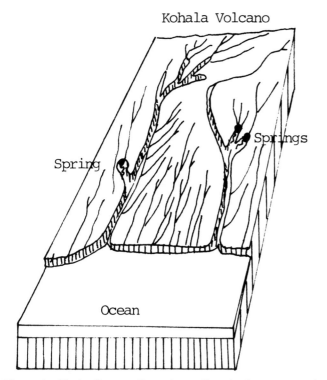

Figure 1. Block diagram illustrating valley development on the northeast-facing slopes of Kohala Volcano, at the northern end of the island of Hawaii. Elongate, narrow ravines with a V-shaped cross section compose the high density of parallel valleys dissecting the upland surfaces. The wide, deeply incised troughs with U-shaped cross sections are interpreted as sapping valleys, as described in the case study following this chapter.

valleys are the result of enhanced weathering at the water table (Wentworth, 1928) and development of cirque-like valley heads (Hinds, 1925). Perennial flow is maintained by large springs. Hanging valleys developed as the deep U-shaped valleys receded headward and captured the more shallowly incised V-shaped valleys.

The Hawaiian sapping valleys, in a fashion similar to the Colorado Plateau valleys, show elongate basin shape, low network drainage density, low degree of interfluve dissection, widely spaced and short tributaries to main trunk valleys, theater-like valley heads, local structural control, local examples of long and narrow interfluves between adjacent valley segments, steep-sided valley walls meeting valley floors at a sharp angle, irregular variation in valley width as a function of valley length, relatively high drainage densities in upstream portions of basins, and local examples of hanging valleys.

Sapping processes

The basal undercutting processes that constitute sapping vary with numerous factors, depending on the scale of concern (Fig. 2). At the largest scales (megascale), sapping is facilitated by rock type, structure, and climate. For example, scarp recession by sapping occurs most readily where a massive, resistant lithology overlies a relatively weak, incompetent one. Undermining of the resistant layer occurs by weathering and groundwater flow along the contact between the two lithologies. The climate must promote sufficient recharge for the groundwater system to be active. Details of rock structure directly influence the nature of ground-

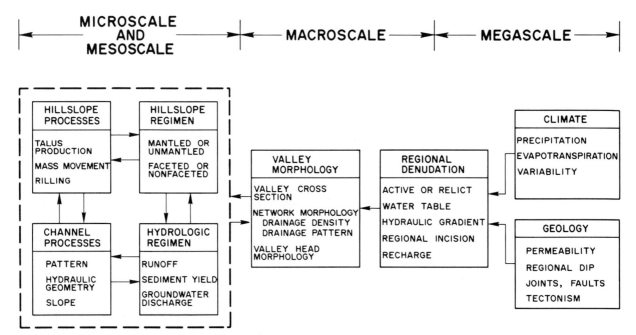

Figure 2. Factors important in valley morphogenesis by spring sapping operating at various scales of geomorphic concern. Feedback relations are indicated by arrows pointing in opposing directions. Local processes involving hillslopes and adjacent channels constitute an especially complex interrelated system (dashed lines).

Figure 3. Tafoni development in Hermansburg Sandstone, Fink River Gorge, central Australia.

water flow, as described in the case studies that follow this introduction. Especially important is the concentration of sapping processes at local zones, rather than at diffuse locations along escarpments.

At regional scales, the water table conditions, base-level effects, and nature of the developing valleys all influence the process. Feedback occurs because developing valley forms influence process and vice versa.

The details of valley development by sapping occur at the mesoscale, such as on individual valley floors and hillslopes, and at the microscale, such as on an individual seepage face. For example, the microclimate of the seepage area exerts an influence on the weathering process of basal slope zone. This, in turn, affects the development of the entire slope containing the valley wall.

A common manifestation of concentrated weathering by crystal growth is alveolar weathering, which results in tafoni (Fig.

Figure 4. Cavernous development at the base of a sandstone cliff (Kombolgie Formation) at Nourlangie Rock, Arhemland, Northern Territory, Australia.

Figure 5. Alcove development in Navajo Sandstone, Betatakin Ruin, Navajo National Monument, northeastern Arizona.

3). Tafoni development occurs in sandstone, tuff, granite, and other massive rocks (Mustoe, 1982, 1983; Twidale, 1982). The crystal growth of various salts contributes both to accelerated erosion of hollows and case hardening of exposed portions of rock (Conca and Rossman, 1982). Calcite deposition and spalling are most prominent in sapping processes for Colorado Plateau sandstones (Laity, 1983). Because salts can be drawn up by capillary action from deep zones of saturation, alveolar weathering can persist during dry periods when groundwater outflow ceases. Thus, spring heads that were formerly active in pluvial periods commonly continue to enlarge slowly by salt-related processes, even though the spring sapping morphology is relict.

In cold climates, ice-related processes can exert a major control on basal sapping. Freeze and thaw will loosen material at the seepage face for subsequent removal by water flow. Laity and Malin (1985) describe sapping of moist seepage faces aided by the weight of accumulated ice in winter.

Cavernous zones at slope bases (Fig. 4) can be eroded when climate and rock type permit sufficient strength in overlying slope elements. In the Colorado Plateau sandstones, immense alcoves may develop (Fig. 5), enlarging until the undercut cliff faces collapse by slab failure (Bradley, 1963). In contrast, the Hawaiian

Figure 6. Ingrown bedrock meanders developed in Navajo Sandstone in lower Harris Wash, Escalante River basin, south-central Utah. Undercutting on the outside bends, enhanced by spring sapping, results in unusually large cavernous alcoves.

valleys, incised in basalt under a tropical climate, have very active upper slopes subject to soil avalanches, slumps, and other mass movement (Wentworth, 1943). Enhanced weathering at the water table (Wentworth, 1928) leads to maintenance of the steep slopes that favor mass movement and high-gradient fluvial action (Macdonald and others, 1983).

The interaction of fluvial processes with groundwater sapping can lead to additional complexities in the genetic explanation of the resulting landforms. An excellent example occurs in the deeply incised fluvial canyons of the Escalante River basin in south-central Utah. In the downstream reaches of Harris Wash, an Escalante tributary, ingrown meanders have developed in the massive Navajo Sandstone (Fig. 6). The outer bends of those bedrock meanders are most deeply incised and serve as loci for concentrated groundwater flow through the sandstone. Sapping erosion is concentrated at these sites, enlarging the outer bends of the meanders and enhancing the cavernous undercutting of the valley walls (Fig. 7). During major floods, the sapping-enlarged meander bends become depositional sites as water flows more efficiently in a straighter route that bypass the outer bends (Patton and Boison, 1986).

Small valley networks on Mars

Perhaps the most exotic valley systems that have been attributed to sapping are the ancient networks that dissect heavily cratered uplands on the planet Mars (Milton, 1973; Pieri, 1980; Baker, 1982). Nirgal Vallis (Fig. 8) illustrates a Martian valley typical of many that formed early in the history of the planet when liquid water was an important agent of denudation (Baker, 1985). The network shows strong structural control, theater-headed valleys, and numerous other attributes of sapping mor-

phogenesis, as noted above. The paleoenvironmental implications of past sapping processes on Mars center on the need for an active hydrologic cycle to maintain spring discharge at the valley heads (Mars Channel Working Group, 1983). The necessary energy could have been supplied by precipitation during a paleoclimatic phase of Mars very different than that prevailing today (Pollack, 1979). Alternatively, hydrothermal systems related to impacting or volcanism may be invoked to explain some networks (Brakenridge and others, 1985; Gulick and Baker, 1989).

Ages for the valley networks on Mars have been interpreted from terrain associations and crater counting. More than 99 percent of the valleys occur on the densely cratered upland terrains of Mars (Pieri, 1976; Carr and Clow, 1981) and seem to have formed coincident with the emplacement of large impact basins (Schultz and others, 1982). These associations and direct crater-counting techniques (Pieri, 1980; Baker and Partridge, 1986) yield an age of 3.8 to 3.9 billion years. Valley formation seems to have been favored in a relatively brief time period of Martian history, perhaps lasting about 10^8 yr (Baker and Partridge, 1986). Because of numerous uncertainties in deriving absolute ages from crater counts, these figures are considered to be approximate.

Models of sapping valley development

The conceptual model of Dunne (1980) provides a useful picture of channel network evolution in which spring sapping is a dominant process. The initial condition in Dunne's model is a water table with a regional slope toward a hydraulic sink provided by a depressed region. Water emerging along a spring line

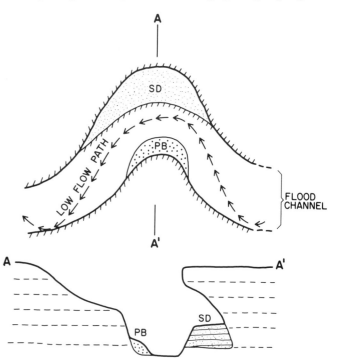

Figure 7. Sketch map (top) and cross section showing meander development as in Figure 6. Flood slackwater deposits (SD) have accumulated in the sapping-enlarged outer meander bend, and a normal point bar (PB) is located on the inside of the bend.

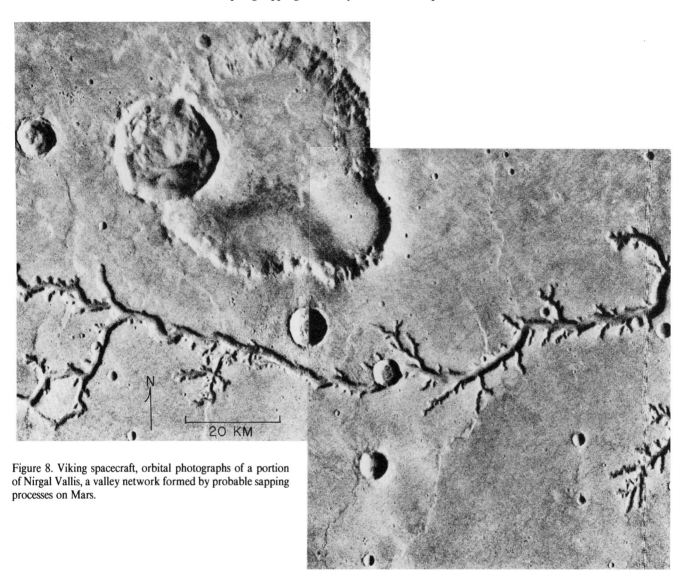

Figure 8. Viking spacecraft, orbital photographs of a portion of Nirgal Vallis, a valley network formed by probable sapping processes on Mars.

would then foster chemical weathering, thereby increasing the porosity of the seepage zone, reducing the local rock tensile strength, and rendering the weathering zone more susceptible to erosional undercutting of adjacent slopes. Local zones of heterogeneity in the rock will result in some zones achieving the critical conditions necessary for such undermining before other zones achieve them. Joints, faults, and folds serve this function. These critical zones then experience enhanced undermining. Once initiated, this process becomes self-enhancing because the groundwater flow lines converge on the spring head. The increased flow accelerated chemical weathering, which leads to further piping at the same site.

The farther a spring head retreats, the greater the flow convergence that it generates, thereby increasing the rate of headward erosion. Headward sapping proceeds faster than valley widening because the valley head is the site of greatest flow convergence (Dunne, 1980). However, headward growth may intersect other zones that are highly susceptible to sapping. A particularly favorable zone will result in a tributary, which also experiences headward growth, and which may generate tributaries of its own. Thus, sapping that develops in a zone of jointing or faulting will develop a pattern aligned with those structures. It will, however, be organized by the hydraulic controls on the groundwater flow.

This process of sapping, headward retreat, and branching eventually will form a network of valleys. The developing network works to counteract the self-enhancing effect of flow concentration mentioned above. As spring heads migrate to the neighborhood of one another, their demands for the available groundwater compete with each other. Eventually an equilibrium will be achieved at some optimum drainage density.

A variety of small-scale analog models have been invoked to understand the processes involved in drainage system development by sapping. Because of material, time, and scale differences, these studies must be considered to be suggestive of process operation rather than definitive in identifying the precise genesis of complex large-scale valleys. Thus, studies of gully-head recession

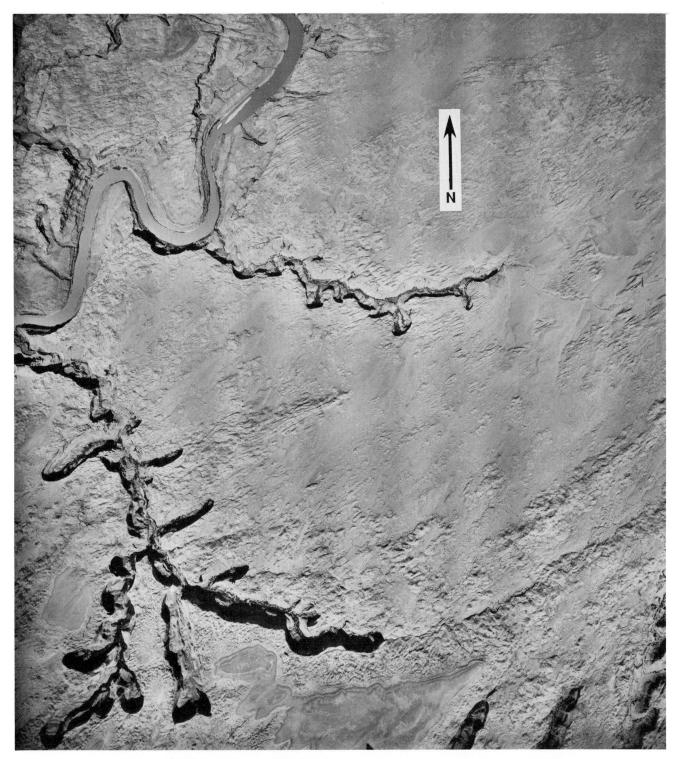

Figure 9. Iceberg and Slick Rock Canyons. The theater-headed terminations of both canyons are indicative of sapping processes. Iceberg Canyon is larger (approximately 6 km in length) and is more symmetrically branched. It is developed within a gently dipping syncline where groundwater is focused toward the canyon headwalls. Slick Rock Canyon is asymmetrically branched with short tributaries entering from the south. The contributing runoff channels on the plateau are exceedingly short and poorly incised, suggesting low overland-flow volume, with little power for headward retreat by waterfall erosion. Substantial groundwater seepage is indicated by the presence of a heavy vegetation cover at the theater heads of the tributaries.

in the Appalachian Piedmont can be used as analog models (Ireland and others, 1939; Woodruff and Gergel, 1969) despite the probable action of vadose water or perched water-table conditions. Similarly, tidal channels (Woodruff, 1971) and beach-face channels (Higgins, 1982, 1984) provide models of network development as water emerges from sand during falling tides. Controlled experiments on sapping have only recently been attempted, but these show promise in the modeling of headwardly growing networks (Kochel and others, 1985; Howard and McLane, 1988).

Results from a large experimental sapping facility (R. C. Kochel and A. D. Howard, written communication, 1984) succeeded in reproducing the following attributes of sapping networks: structural control, U-shaped cross section, theater-like valley heads, stubby tributaries induced by groundwater flow capture, elongate basin shapes, low degree of interfluve dissection, and headward growth and bifurcation as hypothesized by Dunne (1980). More recent results are discussed in the subsequent case study by Alan D. Howard.

The various simulations of valley development by sapping have not yet adequately modeled the complex interplay of weathering, slope, and channel processes responsible for valleys. Nor have these tools addressed questions of climatic change, variable lithologies, and relict phenomena. Nevertheless, it is clear that spring sapping has profound implications for morphogenesis on both Earth and Mars. Paleohydrologic, paleoclimatic, and denudation studies of valleys on the two planets must give special consideration to the operation of this fundamental geomorphic process.

CASE STUDY: THEATER-HEADED VALLEYS OF THE COLORADO PLATEAU

Julie E. Laity

The geomorphic significance of sapping in the maintenance of sandstone cliffs in the American Southwest has long been recognized by individuals working in this region (Ahnert, 1960; Bryan, 1928, 1954; Schumm and Chorley, 1964; Watson and Wright, 1963). However, the role of groundwater in the headward retreat of canyons has received much less attention. The Glen Canyon region of the Colorado Plateau is characterized by numerous deeply entrenched, theater-headed valleys draining to the Colorado River and its principal tributaries (Fig. 9). Many such valleys occur in the Navajo Sandstone, a highly transmissive aquifer underlain by essentially impermeable rocks. Development of these valleys has been interpreted in terms of eolian action (Stokes, 1964), waterfall erosion (Hunt, 1953), or groundwater sapping processes (Gregory, 1917; Stetson, 1936). The latter have received considerable recent attention (Laity, 1980, 1983; Laity and Malin, 1985; Howard and Kochel, 1988).

Environment

The average altitude of the Colorado Plateau near Glen Canyon is approximately 1,600 m. Annual rainfall is as low as 13

Figure 10. Theater-headed tributaries to Long Canyon as viewed from the plateau surface. These canyons are 1 km long and 0.5 km wide.

cm, with precipitation increasing to 38 cm/yr at higher elevations in the Henry Mountains. Summer precipitation is sporadic and convectional in nature, and may result in flash floods. By contrast, winter precipitation is chiefly frontal and is more evenly distributed areally. Long, low-intensity rainfall allows for groundwater recharge.

Sedimentary units of the Glen Canyon region are mainly flat-lying or are gently warped over large areas. Theater-headed valley development appears to be related principally to the nature, occurrence, and attitude of the Navajo Sandstone (Triassic/Jurassic?), considered to represent an extensive dune deposit of an ancient interior desert. The Navajo Sandstone is an excellent aquifer, yielding water to springs and seeps throughout the region. It is underlain by the Kayenta Formation of mudstone, siltstone, and sandstone that acts as an aquiclude for downward-moving groundwater. The recharge potential of the Navajo Sandstone is high because of its widespread exposure at low dip angles, relatively uniform permeability of the rock, and pervasive fracturing. Small, discontinuous bodies of perched groundwater occur extensively throughout the formation above thin, relatively impermeable units of limestone, siltstone, or shale. Alcove development associated with perched water levels is important in modifying canyon walls. However, the evolution of canyon networks is dependent primarily on sustained seepage at valley headwalls fed by regional aquifers.

Within the Navajo Sandstone, two morphologically distinct valley types are observed. Theater-headed canyons are large in scale, ranging up to 150 m in depth, 11 km in length, and 0.75 km in width. The valleys maintain a constant width from source to outlet, have steep-walled headcuts that are cuspate in form (Fig. 10), relatively straight longitudinal profiles, and show strong structural controls on network pattern. A second population of valleys develops where surface runoff dominates. This group is characterized by narrow valleys with tapered terminations of

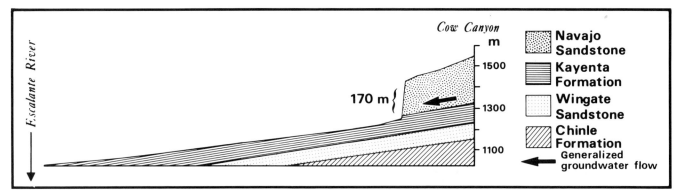

Figure 11. Longitudinal profile of Cow Canyon, a westward-flowing tributary to the Escalante River. Groundwater seepage emerges at the base of the headscarp, above a permeability boundary formed by the Kayenta Formation. The canyon grows headward along this lithologic contact and maintains an essentially straight, lower-profile segment.

first-order tributaries, a relatively smooth, concave-up profile, and a more dendritic network pattern.

Evidence for groundwater sapping

In some arid regions, theater-headed valley forms are relicts of sapping processes in previously more humid climatic conditions. Abundant seeps and collapse features in the Navajo Sandstone indicate that groundwater erosion remains an ongoing process of valley formation on the Colorado Plateau under present semiarid conditions, although discharge and rates of cliff retreat are probably reduced relative to previously wetter periods. Other evidence for groundwater sapping in this region includes the high transmissivity of the aquifer, observed alcove formation and/or undermining of cliff faces at sites of observed seepage, the form of the longitudinal profile (Fig. 11), a spatial distribution of seepage outflow along canyon walls that supports theoretical groundwater outflow models, a strong degree of structural control of canyon growth, and drainage networks that differ considerably in pattern and evolutionary development from those formed by surface-water erosion. The significance of some of these factors is discussed below.

Transmissivity of the aquifer. The Navajo Sandstone is highly transmissive owing to its permeability and its geometric configuration and continuity (Jobin, 1956, 1962). The sandstone yields water in part from intergranular openings and in part from fractures. Immediately above the contact with the Kayenta Formation, however, the Navajo Sandstone become mineralogically more complex, with increased amounts of carbonate cementation and pore-filling clays. The result is that maximum groundwater outflow occurs in a zone 1 or 2 m above the actual contact (Laity, 1983).

Preliminary investigations of groundwater emergence at the seepage face suggest that flow may be through macroscopic tunnels 200 to 600 μm in diameter that appear to be fairly straight, rather than through intergranular pores (Fig. 12). The observed increase in size and decrease in tortuosity of tunnel flow relative to intergranular flow indicate a means by which the volume of converging groundwater at the seepage face can be accommodated.

Erosion at the seepage face. Groundwater emerging at seeps slowly weakens the material that provides the basal support for cliffs. Whereas groundwater erosion of weakly consolidated materials found at beaches or in laboratory sapping chambers results in the entrainment of individual grains, erosion at sites of groundwater emergence in sandstones of the Colorado Plateau involves multiple processes and commonly the sloughing of spalls. The action of seepage pressure on individual grains in part is reduced by a vegetal mat of ferns, mosses, and algae that covers the surface. The algae in particular acts to bind surface grains together at active seeps.

Loss of wall rock occurs primarily through spalls, which are commonly several centimeters in thickness. Weakening and breakup of the sandstone arise from physical and chemical processes. These include the growth of algae in near-surface sandstone pores and within microsheeting joints (centimeters in thickness) that develop parallel to the wall; the deposition of calcite from groundwater as the evaporative surface is approached, acting to wedge grains apart and thereby decreasing the strength of the sandstone; and the deposition of efflorescent salts 100 to 200 μm in thickness on the wall rock surface. Loss of material from the seepage face proceeds by sloughing as the layer of calcite-separated grains thickens; by weakening and failure of microsheeting joints through wetting and drying or biological processes; and by the formation of icicles in winter, which fall by the combined action of thawing and the weight of ice, removing sand grains by a plucking action.

As the weathering surface grows in thickness, the weakened material sloughs off in flakes. The removal of this material may be enhanced in winter by freeze-thaw cycles. Icicles form on seepage faces on a diurnal cycle, and when they fall as a result of the combined effect of thawing and the weight of ice, additional material is removed by plucking action.

The geomorphic consequence of sloughing, grain release,

Figure 12. Scanning electron micrograph showing microscopic tunnels that accommodate the volume of converging groundwater at a seepage face.

and plucking is the enlargement of cavities and alcoves in cliff faces. These undermine basal support and cause eventual slab failure, facilitated by pressure-release joints that develop parallel to the canyon walls. The blocks shatter on impact, are reduced by weathering processes, and the material is moved down valley during floods and by the wind.

Seepage discharge. Groundwater discharge is greatest at headwall seeps (20 to 40 percent of total flow). There is a decreasing rate of inflow in a downstream direction from secondary sources, including sidewall springs and lateral channel inflow. These observations are compatible with a general model of drainage extension by sapping processes wherein groundwater flow is focused at the head of valleys and the relative outflow of groundwater decreases downstream (Dunne, 1980; Dunne, this volume).

Structural controls. Networks of theater-headed canyons suggest strong control by structural factors, as demonstrated by patterns that are highly asymmetric; these networks show unusual constancy of tributary junction angles into the mainstream, and they exhibit pervasive parallelism of tributary orientation over large geographic areas (Fig. 13). This results because groundwater flow is sensitive to changes in hydraulic gradient resulting from regional folding, and to deep fractures in the Navajo Sandstone.

Theater-headed canyons are found principally on surfaces with low dip angles (1 to 4°) that favor a large proportion of infiltrated precipitation. As the bed dip or topographic slope steepens, surface runoff increases, and there is a resultant change in canyon morphology toward narrower valleys with tapered heads. Likewise, where less permeable rocks are situated near the perimeter of the basin, the greater amounts of runoff generated tend to degrade the headcut.

Structural analyses comparing primary direction of canyon growth to joint orientations and bed dips for six drainage basins show a tendency of canyons to grow in an updip direction and parallel to the strike of exposed joints (Laity and Malin, 1985). The fractures act at depth to increase the transmissivity of the bedrock, and it is likely that laterally flowing groundwater exploits the major joints, which act as conduits (Campbell, 1973). Theater-headed valleys are never observed to grow in a downdip direction, regardless of surface topography.

The actual network pattern of theater-headed valleys

strongly reflects the structural controls of dip direction and regional jointing. Valleys that lie transverse to the slope of a monocline develop an asymmetrically branched pattern, having either right- or left-handed tributaries (Fig. 13). Valleys that grow directly up a monocline are stubbily branched or unbranched. The development of a more elongate and symmetrically branched network may occur within a gently plunging syncline (e.g., Iceberg Canyon, Fig. 9). Finally, where groundwater flows away from valley heads and overland flow processes predominate, the pattern that develops is more dendritic, with tapered terminations (Dunne, this volume, Fig. 10).

The drainage basins of theater-headed valleys are markedly elongate, having a length-to-width ratio that averages 3.8:1. The networks themselves are simple in form. Of 30 networks examined, more than 90 percent were Strahler order 1 or 2. Unbranched valleys are most common in elongate basins whose axes lie parallel to the dip of the bed, and in systems in early stages of extension. Under these conditions, maximum groundwater outflow is focused at the headwalls. Branching is favored by groundwater contribution along the entire length of a valley. This occurs where canyons develop in synclines or extend transverse to the dip of the beds.

Case study discussion

Groundwater outflow and sapping are proposed as the most significant processes in the formation of theater-headed valleys in the Navajo Sandstone. These valleys develop where large exposures of gently dipping and highly fractured sandstone allow high rainfall infiltration rates. Groundwater encounters a permeability boundary (the Kayenta Formation) and moves laterally down the hydraulic gradient through intergranular pores and fractures. Where groundwater is intercepted by valley heads, the flow converges and emerges in concentrated zones of seepage, where small-scale erosional processes slowly reduce the support of steep cliffs and contribute to their collapse. Valleys grow headward by successive slab failure, aided by processes that break down the debris produced by slope collapse and transport away the disintegrated materials.

In the initial stages of valley development, growth of the main canyon proceeds more rapidly than does that of tributary canyons, owing to a larger subsurface drainage area. In time, a network of canyons is formed by headward retreat. Given a constant climate, headward growth rates probably decline as the system enlarges and the drainage area of the spring heads lessen. In an advanced stage of valley development, lateral retreat by sidewall seepage approaches that of headward retreat, and valleys widen (Dunne, this volume, Fig. 12). They may continue to grow until adjacent tributaries merge, leaving only isolated buttes and remnants of the original surface. A new drainage network is then initiated on the exposed Kayenta Formation, which bears no relation to the previous system.

The process of network development by sapping in sandstone aquifers of the Colorado Plateau is limited by the many constraints on concentrated groundwater seepage and erosion. These include the qualities of the aquifer (areal extent, geometrical configuration and continuity, uniformity of permeability, structural and stratigraphic relations, zones of recharge); jointing, which affects local canyon-wall morphology, enhances permeability at the surface, and causes local concentrations of discharge at a seepage face; material properties that facilitate rock breakup and removal; surface hydrology (surface runoff in large quantities degrades the cuspate morphology of headcuts); and the presence of a free face at which groundwater emerges. As a result of a complex interplay of conditions, extensive and widespread valley growth by groundwater sapping is limited to the Navajo Sandstone and is not prevalent in other aquifers of the Colorado Plateau.

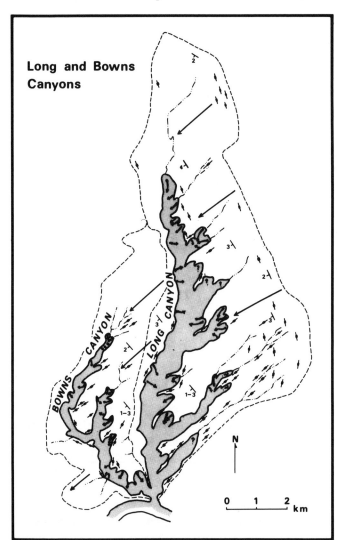

Figure 13. Structural relations in the development of Bowns and Long Canyons. The networks are asymmetrically branched and strongly aligned with regional jointing. Growth is up the gently dipping (1 to 4°) slope of the Waterpocket Fold. Presumed groundwater flow directions are indicated by long arrows; sites of groundwater seepage, mapped in the field and from aerial photographs, are indicated by thicker arrows within canyons. Shaded areas represent canyon floors.

CASE STUDY: GROUNDWATER SAPPING AND THE GEOMORPHIC DEVELOPMENT OF LARGE HAWAIIAN VALLEYS

R. Craig Kochel and Victor R. Baker

Introduction

Studies of the Viking images of Mars have resulted in identification of at least four major classes of channels (Fig. 14): (1) large outflow channels formed by cataclysmic flooding (e.g., Baker and Kochel, 1979); (2) dry valley networks with dendritic patterns formed on ancient hilly and cratered terrain; (3) longitudinal valley networks with short, stubby tributaries with amphitheater heads; and (4) slope valley networks with short amphitheater heads along major Martian escarpments such as in the Valles Marineris. The origins of the valley networks are less certain than the outflow channels, and have led to such different hypotheses as ancient activity of Martian rainfall-runoff regimes and suggestions that they may be the product of groundwater sapping processes. Attention was first focused on groundwater sapping as a possible process of Martian valley formation by Sharp (1973) and Milton (1973) in studies of fretted and chaotic terrain. Recently, we have begun to study the morphology of valley networks in a variety of terrestrial settings where a strong argument can be made for the comparison of valleys formed dominantly by runoff processes (runoff-dominated) with neighboring valleys that have field geomorphic evidence of significant influence by groundwater sapping during their development.

Although the dominance of sapping versus runoff processes has not been quantified in the terrestrial analog studies, we carefully selected valleys where there is evidence of significant spring sapping for comparison to valleys where groundwater contributes relatively little to channel discharge. Field observations of spring sapping serve as the ground truth for subsequent comparisons. The purpose of the analog studies is to determine whether process dominance—runoff versus sapping—is detectable and if it can be quantified from analysis of aerial imagery. If successful, these kinds of studies may be useful in helping to understand the origins of various valley networks on Mars. In addition, we hope to establish a set of morphologic criteria useful in discriminating sapping valleys from runoff valleys in experimental flume settings that will serve as a model for interpreting the significance of the field and imagery studies (Kochel and others, 1985; Kochel and Piper, 1986; Kochel and others, 1988).

MARTIAN VALLEY TYPES

	SIZE	MORPHOLOGY	DISTRIBUTION
OUTFLOW CHANNELS	10's km long 10's km wide	anastomosis, grooves, streamlined hills, massive flood features, distinct chaos sources, lack tributaries	mostly equatorial regions
*** VALLEY NETWORKS ***			
DRY VALLEYS	10's km long < few km wide	dendritic pattern w/ numerous tributaries, fresh & weathered	ancient hilly & cratered terrain
LONGI-TUDINAL VALLEYS	100's km long few km wide	flat-floored main channel, short & stubby tributaries, amphitheater heads	mostly equatorial regions
SLOPE VALLEYS	10's km long few km wide	flat-floored & V-shaped cross-sections, amphitheater heads, stubby tribs. w/ high junction <'s	equatorial along Valles Marineris chasma

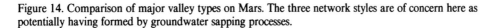

Figure 14. Comparison of major valley types on Mars. The three network styles are of concern here as potentially having formed by groundwater sapping processes.

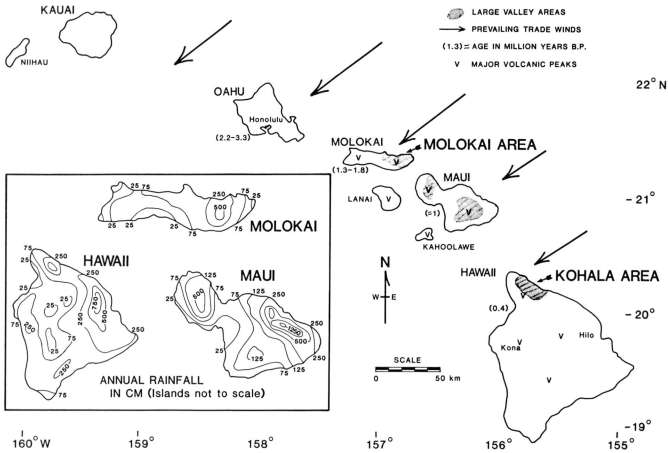

Figure 15. Index map of the study area showing the major volcanic cones, major cities, dominant trade winds, mean annual rainfall, and approximate age of the islands.

The focus of this case study is on the origin of exceedingly large valleys in Hawaii. The southern islands of Hawaii, Maui, and Molokai contain numerous extremely deep, flat-floored valleys with amphitheater heads (Fig. 15, shaded areas) characteristic of the valleys formed by sapping in our flume experiments, and similar to slope valleys along Valles Marineris on Mars, described by Kochel and others (1985). Large, deeply incised valleys display U-shaped cross sections and blunt, theater-like terminations. We hypothesize that these valleys were strongly influenced by groundwater sapping because of their morphologic similarity to other sapping-dominated valleys (Higgins, 1982; Laity and Malin, 1985) and because of their resemblance to valleys formed by sapping processes in experimental networks (Kochel and others, 1985; Kochel and Piper, 1986). Evidence supporting this hypothesis comes from a combination of studies of imagery, topographic maps, field observations, and laboratory experiments. We will concentrate here on valleys in the Kohala area of the island of Hawaii and only provide occasional references to valleys on the other islands.

Geologic, climatologic, and hydrologic conditions

The Hawaiian Islands comprise a chain of volcanic islands (Fig. 15) with ages increasing progressively to the northwest, from Hawaii with its active volcanoes, Kilauea and Mauna Kea (Fig. 16). In general, the dissection of the Hawaiian volcanoes also increases with age to the northwest, but the details of dissection are considerably influenced by climate, parent-material factors, and changes in process. Nevertheless, a remarkable opportunity to study valley development with time is afforded by the phenomenon of northwestward movement of the Pacific plate, carrying a succession of volcanoes away from the stationary mantle plume located at the southern tip of Hawaii (Macdonald and others, 1983).

Rainfall is heaviest on the northeastern slopes of the volcanoes because of the prevailing trade winds (Fig. 15). Although this results in generally higher drainage densities on windward than on leeward slopes of islands such as Hawaii (Fig. 16), there are important exceptions. Mauna Loa, for example, lacks dissection on its northeastern flanks, despite equivalent rainfall to highly dissected parts of Mauna Kea. This results because the basaltic lava flows of the volcanoes are so permeable that drainage will not develop until a less permeable ash mantle is emplaced or until weathering reduces infiltration. Examples of both phenomena occur in Hawaii. Kilauea Volcano, youngest of the Hawaiian shields, displays essentially no dissection except where the 1790 Keanakakoi ash was emplaced (Malin and others,

1983). The older Mauna Loa and Mauna Kea shields display V-shaped ravines only where their flanks were mantled by Pahala ash. Dissection is more pronounced on Mauna Kea, which is older than Mauna Loa. Kohala Volcano, dated at 700,000 B.P., is the oldest shield on the island of Hawaii. Deep weathering of its basalt has reduced infiltration sufficiently to promote high-density drainage on its northeastern slopes (Fig. 17).

On shield volcanoes, parallel ravine systems incise to expose deeper layers where groundwater activity becomes more important. The deepest incision produces trough-shaped, theater-headed valleys (Fig. 18). The contribution of groundwater to the formation of such valleys was suggested by Stearns (1966) and Macdonald and others (1983). Because the layered basalt flows are most permeable parallel to dip, there is efficient groundwater movement toward the sea. The regional water table on the islands is near sea level, with a slight bulge in the central parts of the islands, and thus, has a gentle seaward slope in all directions. In addition, Stearns (1966) showed that central volcanic dike swarms caused significant volumes of groundwater to be perched at high elevations in the central areas of the major volcanoes (Fig. 19). This perched dike water is thought to be the recharge zone and the cause of large-scale groundwater sapping in the big valleys on Hawaii.

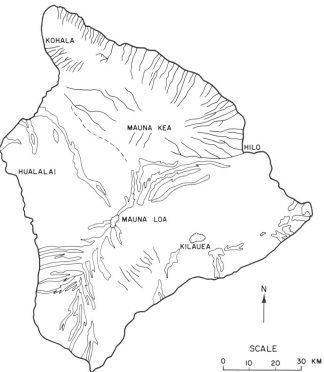

Figure 16. Sketch map of the island of Hawaii showing the five major shield volcanoes, drainage pattern, and recent lava flows (stippled pattern).

Figure 17. Drainage map of Kohala Volcano at the northern end of the island of Hawaii. This map emphasizes V-shaped ravine development and runoff channels developed on broad alluvial valley floors. The former have the highest drainage densities, generally 4 to 5 km/km^2 in the north-central part of the volcano. On the drier northwest and southwest flanks, drainage densities are only 0.3 to 2 km/km^2.

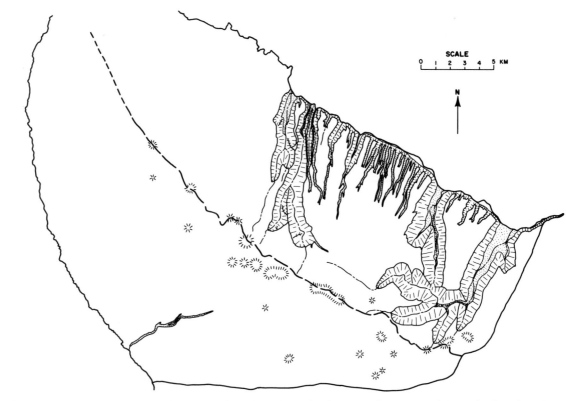

Figure 18. Drainage map of Kohala Volcano showing areas of deep entrenchment of valleys through the cover of volcanic ash. The largest valleys are heavily influenced by spring sapping processes.

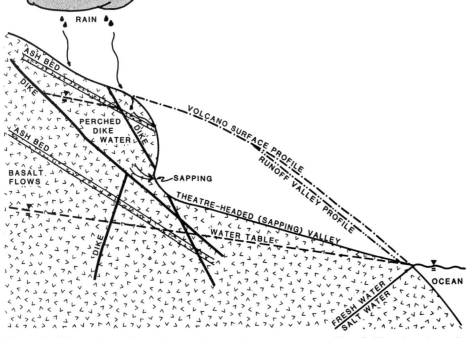

Figure 19. Schematic showing the large-scale groundwater patterns on Hawaii. The view shown is thought to represent the situation for single volcanoes such as the Kohala region of Hawaii. Note the dikes and high-level water perched hundreds of meters above the regional groundwater table. The dashed line shows the profile of the volcano before dissection began. Dissection is interpreted to have started as incision from surface runoff. Once runoff channels cut down far enough to tap into the perched dike water, groundwater sapping is thought to have caused these valleys to enlarge rapidly. (Modified from Stearns, 1966, and Macdonald and others, 1983.)

Figure 20. (a) Oblique aerial photograph of runoff valleys on the west (leeward) slope of Kohala. With few exceptions, these valleys are not greatly incised and exhibit relatively dense parallel drainage networks. (b) Ground view of typical leeward side channel near Kawaihae.

Field relations in the Kohala region

Leeward slopes of Kohala (and Mauna Kea) are drained by a complex of shallowly incised valleys exhibiting parallel drainage patterns due to the steep slope (6 to 8°) of the volcanoes (Fig. 20). Leeward valleys are ephemeral, owing to the semiarid climate, and contain no significant spring discharge. These ephemeral streams occasionally convey large discharges evidenced by the presence of well-imbricated boulders on their bedrock channel floors, but no flows were observed during two field study periods in 1985 and 1987. As a result, these runoff valleys are characterized by relatively small, V-shaped cross sections and terminate upstream by tapering into the rolling topography of the Kohala hill country (Fig. 20).

Valleys on the windward slopes of Kohala exhibit two distinct morphologies, which we feel reflect differences in the importance of spring sapping processes (Fig. 21). The large valleys such as Waipio (Fig. 22) have extremely large cross sections—an order of magnitude larger than neighboring valleys—and exhibit great relief. The heads of these valleys terminate abruptly at large cirque-like, or theater-like terminations (Figs. 23a and 24d). Waterfalls may be present at valley heads and especially along valley walls. Plunge pools are generally poorly developed at the base of these valley-head falls (Fig. 23a) and are absent at the base of falls entering the valley along the walls. Slight grooves along valley walls have been interpreted as resulting from waterfall erosion (Macdonald and others, 1983) (Fig. 23b). Figure 23c shows a major side-valley waterfall that exhibits a step-like descent into the main valley. When abandoned, this feature probably will result in a groove. Valley cross sections are typically U-shaped and exhibit sharp floor/slope junctions (Fig. 23d). Such valleys typically widen toward their heads, in contrast to runoff valleys that normally become narrower. These large sapping valleys are cut to sea level at their mouths along the spectacular wave-cut Kohala Coast (Fig. 24b).

These sapping-dominated valleys (Fig. 21, valleys A through H) typically have steep, relatively straight walls with local relief in excess of 0.5 km. Colluvium is notably absent along the base of the valley walls, indicating that it is efficiently removed by chemical weathering and downstream fluvial processes as rapidly as it accumulates on valley floors. This leaves wall/floor junction angles sharp (Fig. 23d) in spite of active mass wasting of valley heads suggested by the abundance of rockfall and debris-avalanche scars. Valley walls are slightly convex along their upper half and often concave along their lower portions. The importance and active nature of mass wasting along steep valley walls in Hawaii has been discussed elsewhere (Scott and Street, 1976; Jones and others, 1984) and will not be elaborated on here except to note that many recent debris-avalanche scars were observed in field reconnaissance in all large valleys. Valley extension by mass wasting may be facilitated by basal undercutting by spring sapping. Large springs occur along the base of valley walls, in particular at the base of the valley-head amphitheaters (Fig. 25). These springs provide permanent base flow discharge to the big valleys and appear to cause valley walls to be oversteepened near their base because of erosion accelerated by groundwater sapping. Significant chemical weathering of the basalts by the groundwater also appears to be occurring, which may contribute to the efficiency of sapping erosion (Kochel and Piper, 1986).

The majority of large springs observed in the field inspection occur in the vicinity of sites where major dike swarms have been mapped by Stearns (1946; see maps in Kochel and Piper, 1986). A prominent dike is visible in Figure 23a near the head of a tributary to Waimanu Valley. We envision that these large valleys developed by progressive incision of runoff-dominated, V-shaped ravines until one or more intersected high-level, perched dike waters (Fig. 19). From that point in this suggested model, the enlargement of the valleys became much more rapid because of the augmentation of flow and mass wasting related to the

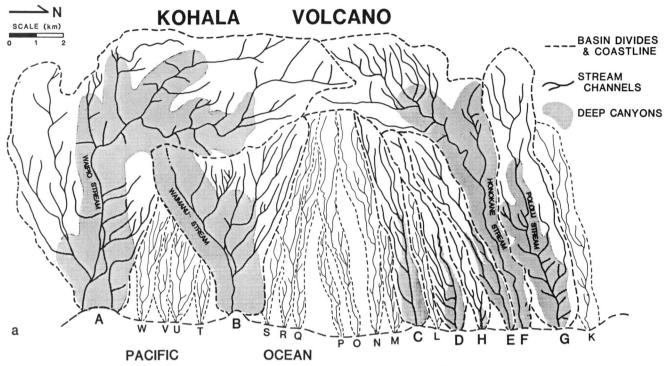

Figure 21. Map of valleys and drainage channels on the east slope of Kohala. A through K are sapping-dominated channels.

groundwater sapping processes. Valleys influenced by sapping tend to enlarge more rapidly than their neighboring runoff-generated ravines. Continued valley-head expansion and subsurface piracy of groundwater appears to have resulted in further retardation of the erosion of neighboring runoff valleys due to subsurface piracy. Similar patterns of piracy and valley enlargement have been observed in our flume experiments as valley heads progressively tapped into high-level aquifers in weakly cemented sand (Kochel and Piper, 1986; Kochel and others, 1988). The deepest, largest valleys probably grow fastest, since they tap the largest supplies of groundwater. The combination of local spring sapping and enhanced chemical disintegration of basalt at the water table, which intersects the slope toes of the valley wall/floor junction, would lead to the exceptionally steep valley walls noted by Wentworth (1928). The growth of these valleys isolates uplands drained by V-shaped ravines, producing hanging valleys. Divides may eventually be reduced to knife-like ridges, as in the older pali of northeastern Oahu (Macdonald and others, 1983).

Interspersed between the large spring-fed valleys on windward slopes of Kohala are numerous smaller valleys with V-shaped cross sections (Fig. 21, valleys K through W; Figs. 24a and c). These valleys contain relatively few springs, which appear to be localized in their downstream reaches when present. Although they usually contain significant discharge due to the frequent tropical rains, their flow is highly variable subject to prevailing winds and may periodically be dry.

These runoff-dominated valleys along the eastern Kohala coast exhibit starkly contrasting valley morphologies in the field compared to the sapping-dominated valleys. Runoff valleys are not as deeply incised and have not succeeded in cutting to sea level in spite of their identical climate, similar lithology, and age compared to the neighboring sapping valleys. Runoff valleys enter the sea via spectacular waterfalls over 100 m high (Fig. 24a). Valley cross sections are V-shaped, and valley floors contain more colluvium than sapping valleys. Runoff valley heads are tapered and merge gently with the undissected slopes of the volcano (Fig. 24c).

Figure 22. Oblique aerial photographs of Waipio Valley illustrating typical morphology of sapping-dominated valleys.

Figure 23. Morphology of sapping-dominated valleys on Kohala. (a) Two views of head falls and minor plunge-pool erosion in tributary to Waimanu Valley. (b) Walls grooves near mouth of Waimanu. (c) Stepped descent of tributary falls into Waimanu Valley. (d) View of contact between valley floor and wall in Honokane Valley on Kohala. Note sharp angle that characterizes valley-wall junctions in valleys influenced by sapping.

Figure 24. Comparison of sapping- and runoff-dominated valley morphology. (a) Mouths of runoff-dominated valleys between Waimanu and Honokane Valleys. These valleys terminate seaward as hanging valleys and enter the ocean via spectacular waterfalls. (b) Mouths of sapping-dominated Waimanu and Wapio valleys. These valleys are cut to sea level. (c) Tapered heads of runoff valleys shown in "a." (d) Amphitheater heads of some of the sapping valleys shown in "b."

These morphologic differences between the large valleys influenced by sapping and the smaller runoff valleys, in an area where climate, lithology, and age are relatively constant, argue for contrasting morphogenetic processes between the two types of valleys. Absolute proof of the sapping hypothesis requires a demonstration of groundwater contribution to channel flow and an extensive field hydrology monitoring program. The absence of gaging stations plus logistical difficulties in measuring valley head discharge before and after it flows over the waterfalls result in a paucity of quantitative data. We feel, however, that the presence of numerous large springs in the sapping-dominated valleys is sufficient for providing the grounds for comparison of valley morphology from aerial imagery. Furthermore, the results of the morphometric analyses yield favorable comparisons to valleys formed in controlled laboratory experiments in models with high-level aquifers contrasted to runoff networks formed in similar sediments (Kochel and others, 1988).

Basin morphology

Fifty-three basins were selected for comparison of basin morphometry between sapping- and runoff-dominated valleys on Kohala and Molokai using the following parameters (measured on 1:24,000-scale topographic maps): cross-axial relief, drainage density, basin shape, first-order stream frequency, ratio of canyon area to basin area, and junction angles. Morphometric indices may be useful in classifying valleys into runoff- versus sapping-dominated, since these types appear to be morphologically distinct.

Morphometric data were analyzed using multivariate principal components analysis (PCA) and by discriminant analysis. Due to space limitations, we will report only on the results obtained from the PCA analysis on Kohala. See Kochel and Piper (1986) for more discussion of the significance and constraints of the PCA analysis. We found similar trends from our analysis of

Figure 25. Hiilawe Stream, a major tributary to Waipio Valley in August 1985. Formerly the highest free-falling waterfall on Hawaii (approximately 600 m), all of the flow has recently been diverted for irrigation. Significant discharge (estimated at about 0.5 m^3/sec) emanates from springs along the base of the headwall (above), resulting in the flow visible in the foreground of the photo on the right.

basins on Molokai and when both data sets were combined. Table 1 summarizes the PCA statistics, and Figure 26 summarizes the plots of various parameters and principal components for windward Kohala valleys.

Table 2 summarizes the major morphologic and morphometric differences between sapping and runoff valleys on Hawaii, based on studies of Molokai and Kohala. PCA analysis shows that there are strong differences in the morphometry of runoff and sapping valleys on the windward slopes of Kohala. Figure 26 illustrates how well PCA separates runoff-dominated valleys from sapping-dominated valleys using basin morphometric parameters.

Most of the variance in the morphometric data is explained by the first principal component (PC1). PC1 appears to be dominantly a vector based on junction angle reflected by the high loading for this variable (Table 1). PC2 appears to be a vector controlled by the ratio of canyon area to basin area, basin shape, and junction angle (Table 1). Interpretation of PC2 suggests that the canyon area/basin area ratio is high for elongate basins with low junction angle.

These statistical analyses simply show that there are distinct differences between the two valley types on windward slopes of Kohala. They say nothing about the cause of these differences, i.e., whether it is due to a dominance of sapping versus runoff processes. If the qualitative interpretations from the field observations of valley morphology and presence of springs are correct in distinguishing sapping-dominated from runoff-dominated valleys,

TABLE 1. PRINCIPAL COMPONENTS STATISTICAL SUMMARY

Principle Component	Variance Summary	
	Total Variance Explained (%)	Cumulative Total Variance Explained (%)
1	97	97
2	2	99
3	1	100

Variable*	Variable Loadings Summary by Vector		
	Eigenvectors		
	1	2	3
K	0.16	0.37	0.88
TDD	0.07	0.1	0.17
MJA	0.97	-0.22	-0.09
FI	0.05	0.10	0.07
R2	0.00	-0.02	-0.02
BCR	0.16	0.89	-0.42

*K = basin shape computed as a lemniscate; TDD = drainage density; MJA = mean junction angle; FI = first-order channel frequency; R2 = cross-axial relief at mid-basin; BCR = canyon area/basin area.

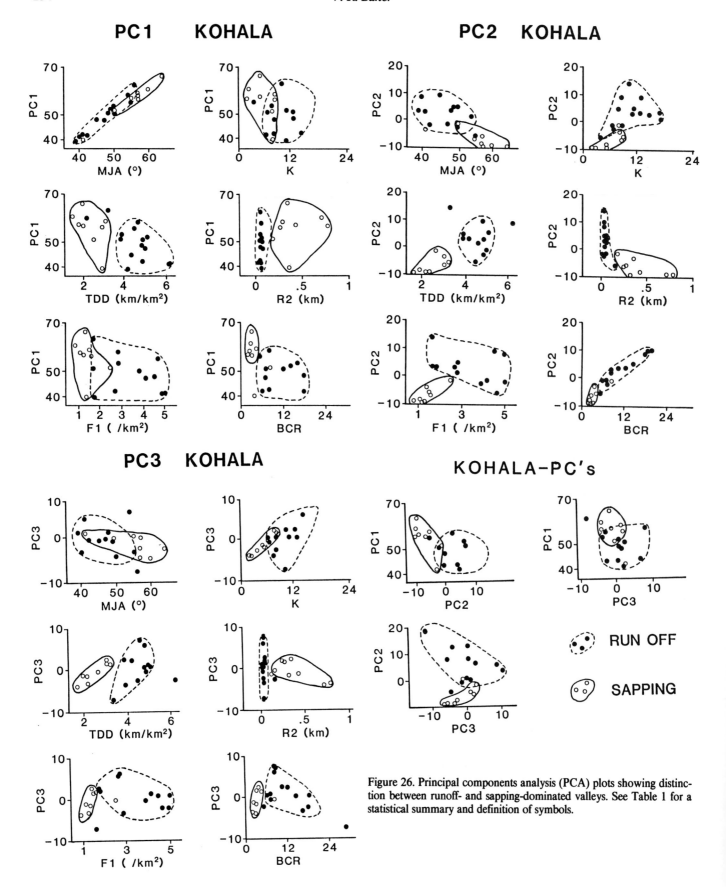

Figure 26. Principal components analysis (PCA) plots showing distinction between runoff- and sapping-dominated valleys. See Table 1 for a statistical summary and definition of symbols.

then support exists for the process differences appearing at the scale of morphometric features visible in aerial imagery with potential application to Martian valleys. Support for these interpretations is provided by similar differences observed in our laboratory studies, where process dominance was controlled.

Valley evolution

Comparison of sapping-dominated valleys on the younger Kohala Volcano with those of older volcanoes on Molokai suggest an evolution of sapping valley development through time (Fig. 27). Initially, Hawaiian valleys originate from the runoff process as very elongated valleys with high-drainage-density, parallel drainage networks (Fig. 27, stage 1). Valleys on the slopes of Mauna Kea, Mauna Loa, and the leeward slope of Kohala and Molokai exemplify this stage. Many of the runoff-dominated valleys on the windward slope of Kohala (e.g., Fig. 21, valleys K–W) also exhibit similar morphology.

As valley incision proceeds, runoff valleys gradually tap into perched groundwater (Fig. 27, stages 2 and 3). This stage marks the initiation of groundwater sapping influence in valley development and is characterized by the headward growth of a deeply incised inner valley with sapping morphology and head springs. Honokane Valley on Kohala (Fig. 21, valley F) and Halawa Valley on Molokai (Fig. 28, valley A) are examples of this stage. The transition from runoff dominance to sapping dominance in these valleys is marked by the downstream decrease in drainage density, paucity of tributaries, and tendency toward more dendritic to rectangular drainage patterns.

The mature phase of valleys influenced by sapping is stage 4, where valley enlargement has proceeded to the surface drainage divides. Between stages 3 and 4, subsurface (groundwater) piracy appears to be important at the basin heads, resulting in significant head widening. Groundwater piracy is probably instrumental in producing the light bulb–shaped basins common in many of these mature sapping basins on Hawaii (i.e., Waipio and Waimanu Valleys on Kohala; Fig. 21) and most of the Molokai sapping

TABLE 2. COMPARISON OF VALLEY GEOMORPHOLOGY

Parameter	Runoff-Dominated	Sapping-Dominated
Basin shape	Very elongate	Light-bulb shaped
Head termination	Tapered, gradual	Amphitheater, abrupt
Channel trend	Uniform	Variable
Pattern	Parallel	Dendritic
Junction angle	Low (40-50)	Higher (55-65)
Downstream tribs.	Frequent	Rare
Relief	Low	High
Drainage density	High	Low
Drainage symmetry	Symmetrical	Asymmetrical (low down-dip)
Canyon area/ basin area	Very low	High

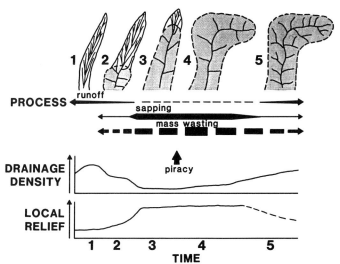

Figure 27. Schematic showing suggested evolution of sapping-dominated valleys in Hawaii.

valleys (Fig. 28, valleys C, D, F, G). Groundwater piracy beheads subsurface and surface drainage to neighboring valleys. In Figure 21, Waipio and Polulu Valley divides have merged, capturing much of the flow to intervening valleys. These captures resulted in dramatic retardation of the development of the intervening valleys on Kohala (Fig. 21). The recent pirating of Waimanu Valley headwaters by Waipio Valley has markedly affected the morphology of upper Waimanu Valley. Waimanu Valley floor, in its valley-head area, is filling with colluvium because there is no groundwater flow to remove this material. This appears in sharp contrast to the active northern tributary of Waimanu (Fig. 21) and other actively growing sapping valleys on Kohala and Molokai, which have broad cross sections and much less colluvium.

Several of the older valleys on Molokai have proceeded to stages 4 and 5, where the sapping form has extended to the basin divide (Fig. 28). Once this stage is reached, the channels receive increasing influence by surface runoff processes again, shown by gradual increase in drainage density and dendritic network pattern (Fig. 27). Figure 29 shows the broader morphologic character of sapping valleys on Molokai. Halawa Valley walls are more dissected and uneven than the younger sapping valleys on Kohala. Valley width is greater for Halawa, and significant talus and debris fan deposits occur along the valley margins. Ultimately, piracy will continue until adjacent valleys merge in a complex, degraded drainage system like the valleys seen on Oahu and the older Hawaiian islands.

Case study discussion and summary

Field observations suggest that the anomalously large valleys dominating windward slopes on Hawaii are strongly influenced by groundwater sapping processes. Spring discharge is exceedingly high in valley heads compared to the seeps measured in sapping valley heads in the Colorado Plateau (Laity and Malin, 1985). Morphologic characteristics indicating the dominance of sapping in the formation of these valleys include their deep dissection, U-shaped cross section, theater-like valley heads, and large springs at the base of valley headwalls. Additional evidence favoring sapping versus runoff dominance for the large valleys includes:

1. Paucity of downstream tributaries due to subsurface piracy in headward regions of their drainage (i.e. Honokane Valley, Fig. 21).

2. Lack of tributaries on downdip sides of valleys (i.e., Waimanu Valley, Fig. 21).

3. Lack of plunge pools at tributary waterfalls.

4. Lack of headward recession (reentrants) along valley walls at tributaries, which would be expected if the valleys retreated dominantly by plunge-pool erosion and fall recession.

5. Coincidence of dikes in heads of sapping valleys.

Additional support for the difference in dominant valley-forming processes is provided by quantitative analysis of basin morphometry. Sapping-dominated valley networks appear to be distinctly different from runoff-dominated networks. Hawaiian sapping valleys are much less elongate and have blunt theater-like heads, lower drainage density, more variable drainage orientations (are more dendritic), higher junction angles, greater cross-axial relief, and higher canyon area/basin area ratios than runoff valleys. It is important to recognize that some of these relations may vary greatly with similar studies in other physiographic and climatic regions. For example, the junction-angle relation will be strongly influenced by regional slope, regardless of channel-forming processes. These morphometric relations may prove useful in developing criteria for interpreting the relative influence of subsurface and surface processes in the origin of valley networks on both Mars and Earth.

We are continuing to investigate similar relations in diverse geologic terrains on Earth where runoff-dominated valleys can be compared with those strongly influenced by sapping. Once a broad database becomes available in combination with flume simulations of these settings, this may lead to criteria useful for inferring process genesis based on morphometric observations detectable in aerial imagery of Mars. Ground-truth field observations in terrestrial analog areas, like the present case study (this chapter), also promise to permit inferences about gross geologic structure in Viking images for areas of Mars characterized by sapping-dominated valley networks.

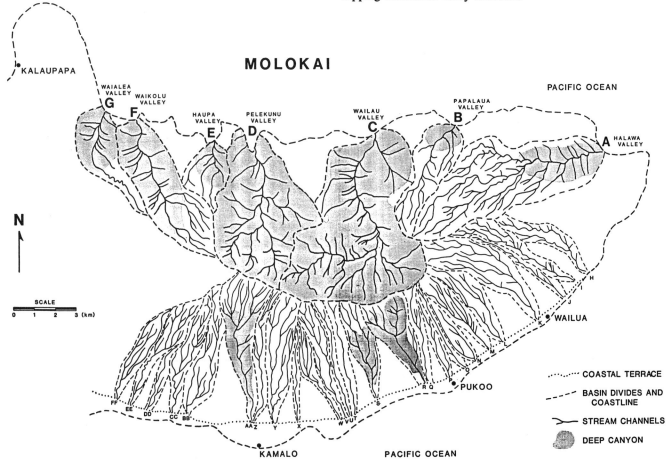

Figure 28. Drainage channels mapped on eastern Molokai.

Figure 29. Photo of Halawa Valley on Molokai taken from the mouth looking westward. Note the broad valley and extensive valley floor deposits compared to younger valleys on Kohala.

CASE STUDY: MODEL STUDIES OF GROUNDWATER SAPPING

Alan D. Howard

Introduction

This case study focuses on experimental development of three-dimensional valley networks by seepage erosion and sapping processes, and on numerical simulation modeling of network development.

Experiments

Initial experiments of sapping erosion of cohesionless sediments were conducted in a narrow, essentially two-dimensional chamber (Howard and McLane, 1988). Three distinct zones occur at a sapping face (Fig. 30). Toward the headward end, the backcutting of the sapping face causes intermittent undermining of the dry and damp sand at intervals of several minutes. In this "undermining zone" the dry sand is at the angle of repose, while below this is damp sand in the capillary fringe, where gradients can become vertical or overhanging. Most of the hydraulic erosion is concentrated in the uppermost zone of seepage outflow, where there are steep surface gradients and strong upward seepage. This "sapping zone" is about 2.5 to 25 cm in length, depending on flow conditions and the type of sand, with gradients of 13 to 17 degrees. The surface of this zone is smooth and wet, although flow depths are barely sufficient to cover the grains. Individual surface grains move downstream, but the predominant mode of movement is by intermittent bulk-flow failures a few millimeters thick occurring at intervals of tens of seconds to a few minutes, depending on grain size and flow conditions. Downstream from the sapping zone is the "fluvial zone," in which the water flow is more than several grain diameters thick, and grains move individually. Channel gradients range from 5 to 7° at the outflow to 10 to 13° at the transition to the sapping zone. In contrast with the sapping zone, low bedforms, generally oblique bars, are common. Whereas the height of the zone of undermining decreases during the course of each experiment and the length of the sapping zone remains relatively constant, the fluvial zone expands as the sapping face retreats (Fig. 31). Most of the experiments were conducted by maintaining a fixed groundwater head at the upstream end of the chamber (Fig. 31). Three experiments were conducted in an angular, uniform, coarse sand of crushed quartzite (#3 Q-Rock, Fig. 31), and two in a fine sand of crushed quartzite (#1 Fine Sand).

Processes acting in the two-dimensional experiments were analyzed by computer simulation modeling of the evolution of the experiments based on a theoretical model of sediment entrainment by the combined forces of gravity, surface flow, and outward seepage (Kochel and others, 1985; Howard and McLane, 1988). The simulations show that the rate of backwasting of the sapping face is determined by the rate of sediment transport through the fluvial zone, which is a function of the sediment characteristics, morphology of the sapping face, and imposed hydraulic gradient through the sand. The flow pattern, in turn, depends on the morphology of the sapping face, so that the temporal evolution of the sapping face results from a complicated interaction of groundwater flow, surface flow, sediment transport, and morphology of the sapping face. The sapping zone, characterized by intermittent shallow bulk-flow failures, has gradients near the threshold of failure as determined by the balance of seepage and gravity moments, and the angle of internal friction of the sediment (surface flow tractive stresses are unimportant in this zone).

The three-dimensional experiments have been conducted in an aluminum tank 1.52 m square and 0.61 m high (see Howard, 1988, Fig. 3). Flow originates from an upstream reservoir whose head can be controlled. A screen separates the sand from the reservoir. After flowing through the sand, the flow, and any sediment in transport, exit into a level trough and then out of the tank. A grid of piezometers on the bottom of the tank allows a rough mapping of the water-table configuration.

Most of the experiments have utilized essentially homogeneous, isotropic sand mixtures. In most experiments the sand surface was constructed with a slight slope (the upland surface)

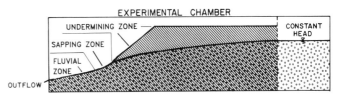

Figure 30. Experimental two-dimensional groundwater sapping chamber. Sand areas are ruled, and water areas are shown by dots. The vertical dashed line is the permeable separator between the water reservoir and the sand-filled area.

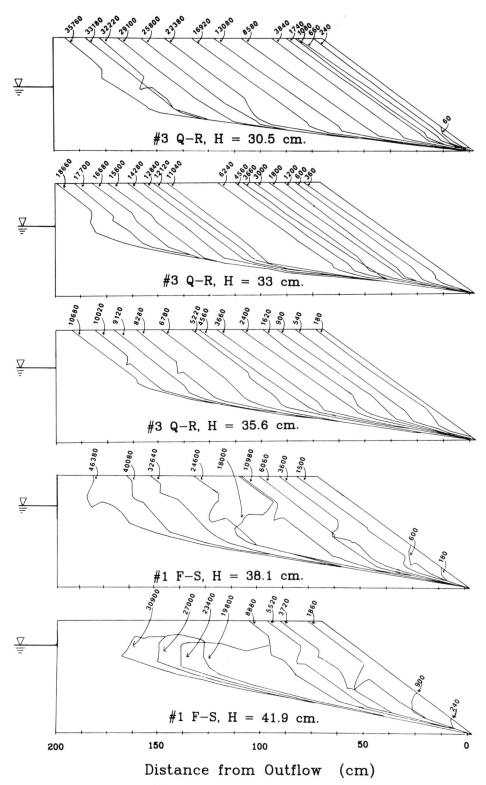

Figure 31. Evolution of the channels and sapping face developed by sapping in experimental chamber for 5 experiments. Elapsed time is given in seconds along the upper margin of each diagram. The line and triangular symbol to the left show the fixed hydraulic head relative to the outflow that was maintained during the experiment. Uniform, angular crushed quartzite was used in the experiments; #3 Q-Rock was a coarse sand and #1 Fine Sand a medium sand.

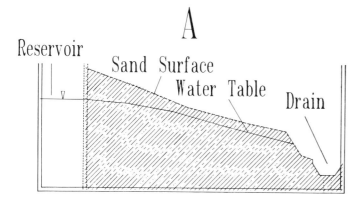
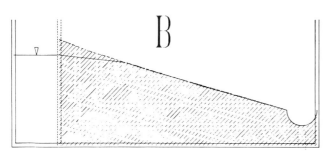

Figure 32. Cross sections through three-dimensional groundwater sapping chamber with (A) and without (B) a downstream scarp molded in the sediment. The water reservoir is on the left end and the drain for water and sediment is on the right end of the tank. In "A," seepage emerges at the scarp face (as in Figs. 33–35), whereas with a high reservoir level or the absence of a scarp (as in "B") seepage occurs along most of the upland surface, causing development of a dense rill network (as in Fig. 36).

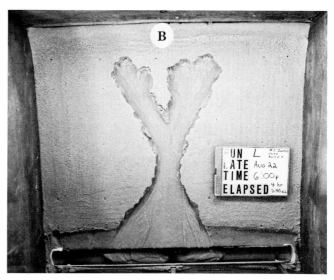

Figure 33. Photographs showing the temporal evolution of drainage networks developed in coarse cohesionless sand in the three-dimensional sapping chamber with discharge slightly above the erosion threshold. In this and Figures 34, 35, and 36, the groundwater enters the sand from the reservoir beyond the top of the picture through the permeable back wall. The water and sediment drain into the gutter at the bottom of the photograph. Note the development of stubby tributaries.

from the reservoir to the vicinity of the collecting trough. Also, for most experiments a scarp was placed near the trough (Fig. 32). The experiments are conducted by raising the head in the reservoir until erosion commences. Several distinct channels are initially formed. However, competition between adjacent channels is very strong, so that after a few inches of headcutting along the main channels, only 1 to 3 channels remain active. The channels that by chance have stronger flow, due to slightly higher permeability of the sand feeding the channel, erode more rapidly. As suggested by Dunne (1980), these channels, by virtue of their greater deepness and length, cause convergence of flow lines and resultant "capture" of groundwater from adjacent, less advanced channels. As the channels extend headward and flow convergence increases the discharge, erosion rates increase. During most experiments this was offset by decreasing the head in the reservoir so as to maintain a roughly constant erosion rate. Competition becomes less strong after the 1 to 3 main ("trunk") channels are established. The drainage density of 1 to 3 major channels in the box is controlled primarily by the geometry of the groundwater flow through the sand mass, particularly the depth of flow relative to the length of flow, as well as the depth of the channel; these factors, plus the magnitude of the critical flow needed to initiate erosion, determine the lateral influence of a given channel on groundwater flow.

In experiments in which flow rates were only slightly above the threshold for sediment entrainment, short tributaries often developed from the trunk channels (Fig. 33). Sometimes both channels remained active after the bifurcation, but often one channel stopped eroding after a time due to competitive disadvantage. As a result, tributary channels are generally short and stubby. Three factors act concurrently to create tributaries. First,

Figure 34. Late-stage development of valley heads in coarse cohesionless sand with discharge well above the erosion threshold. Most intervalley divides have been consumed by valley widening and scarp retreat.

as the trunk channels erode headward, the flow convergence becomes stronger, and there is a tendency for widening of the valley head. This factor alone would probably not produce bifurcation. Small differences in permeability at the valley headwall can create two zones of headwall extension surrounding a zone of slow erosion where permeability is lower. Groundwater competition can then continue this process, with the two channels growing along diverging paths. Another factor that is important in the sand box is the occurrence of headwall slumps. As with the two-dimensional box, the valley headwalls are eroded episodically by mass wasting. When undermining creates a local slump, the infilling of material raises the channel bed and pushes the zone of groundwater emergence downstream, temporarily reducing the rate of erosion. Adjacent portions of the headwall that have not recently experienced slumps, thus, may gain a competitive advantage, causing a bifurcation.

Even with low flow rates and slow rates of erosion the channels become wide relative to their depth in cohesionless sand. The reason for this is the lateral meandering and small-scale avulsion around bars that occurs in the shallow, wide alluvial channels below the headscarp, undermining the valley walls.

At higher flow rates and with correspondingly more rapid erosion the competition between adjacent channels becomes less pronounced. More trunk valleys are active, and valley heads are wider. Lateral erosion often destroys the inter-valley divides a short distance below the headwall (Fig. 34). Tributaries are rare.

Portland cement in concentrations of 1:125 to 1:175 was added to the sand in several experiments to study the effects of cohesion on channel form. The cement provides moderate cohesion to the well-sorted coarse sand matrix without a strong effect on permeability. Experiments with cemented sand in which the groundwater flow regime was similar to the experiments with cohesionless sand produced drainage patterns similar to those with low flow rates in cohesionless sand, in that 1 to 5 trunk channels developed. However, the valley form differed in being narrower and deeper (Fig. 35). Erosion rates were much lower for the same discharge, so that lower channel gradients were required. Also, near their head the channels were steeper than an alluvial channel would be for the same regime of water and sediment, so that the channel floor lacked an alluvial cover. That is, the headward portions were "bedrock" channels in the sense described by Howard (1980). These bedrock channel sections commonly were irregular in gradient with occasional potholes formed in zones of below-average cohesion. The overall rate of erosion was determined by the rate of detachment at the headwall and along the upper bedrock channel rather than by the sediment transport capacity, as in the case of cohesionless sand. Much less lateral erosion of valley walls occurred along the lower portions of the trunk channels than was the case for cohesionless sand. This was due both to the cohesion of the sand and to the lower channel gradient in the lower parts of the trunk valleys.

Because of the cohesion and the somewhat reduced permeability with the admixed Portland cement, hydraulic gradients about 20 to 50 percent higher than those for cohesionless sand were required to initiate erosion. As a result, in some of these experiments, groundwater flowed from the reservoir through the sand, emerged as surface flow on the middle portions of the upland surface, and then once again sank into the sand near the scarp close to the exit trough. The flow then reemerged near the base of the scarp. As a result, a rill system formed on the upland surface. These rills were initially nearly parallel, but as erosion progressed, they became integrated into a crude dendritic pattern. The initial rills died out as they approached the lower scarp due to loss of water through the channel bottom, and the sediment was redeposited as marginal and terminal levees. Head-scarp re-

Figure 35. Drainage networks developed in slightly cohesive coarse sand. Note narrow, steep-walled valleys with slab failures where walls are undermined by seepage.

Figure 36. Late-stage-development drainage networks in slightly cohesive, coarse sand when the upland seepage is strong enough to pass over the basal scarp. The network develops jointly by rill erosion on the upland and headward erosion of gullies from the scarp, partly from water contributed from the rills, and partly from seepage emerging at the scarp face.

cession along trunk channels that eroded headward from the scarp occurred relatively independently of the rill development on the upland surface.

The development of upland channels was further investigated by varying the sand surface morphology and the hydraulic gradients. When the hydraulic gradient was increased, or when the slope of the upland surface was increased (generally requiring a decrease in the height of the basal scarp), the drainage over the upland surface did not sink into the sand, but passed over the basal scarp. As a result, rapid rill incision occurred on the steep scarp surface, and the rill network grew by a combination of downward cutting of the upland portions of the rills and headward retreat of the gullies cutting the lower scarp (Fig. 36). As erosion continued, lower portions of the rills on the upland surface steepened somewhat and the gully heads became more gentle, so that the rill profiles were more uniformly graded. However, local scarps and scour holes occurred along most rills, due to inhomogeneities in the cement binding. When the scarp was eliminated, and the upland surface sloped smoothly down to the trough, rill development was fairly uniform over the upland surface within the zone of emergent seepage. Rill incision was most rapid near the lower end of the upland surface where discharges were greatest, so that the rills developed generally concave profiles. Lateral channel migration and divide erosion caused local captures that produced a crudely dendritic network.

The differences in rill and gully development as the slope of the upland surface and the height of the basal scarp are varied are similar to the variations in drainage pattern developed in runoff rill networks on slopes of different steepnesses described by Phillips and Schumm (1987). Where the upland surface is nearly flat and a relatively large lower scarp occurs, runoff (produced either by emergent groundwater or by excess precipitation) erodes primarily as channels cutting headward as a "wave of dissection." Where the scarp is lower or absent and the upland surface is steep, rills develop and subsequently deepen over most of the upland surface; little erosion occurs through headward scarp retreat from the lower end of the slope. When channel development is solely by headward scarp retreat, only 1 to 3 trunk channels eventually survive, whereas a much denser network of 20 to 30 rills forms where groundwater emerges near the top of a sloping plain.

In conclusion, the experiments described above afford insights into groundwater sapping processes at three levels of generality:

1. The flow tank experiments have direct analogies in similar materials and at similar physical scales in a few natural environments, notably on beaches and fluvial terraces in sandy materials where falling tides or diminishing river stages cause outward seepage (Higgins, 1982; Howard and McLane, 1988).

2. The experiments serve to elucidate the physical processes involved in sapping of cohesionless sediment and to validate theoretical models such as the simulation models discussed below. When the theory is adequate, processes and landforms can be extrapolated to larger groundwater-flow systems or to different sediments, such as might occur in headwater hollows in natural drainage basins.

3. The experiments with cohesive sands afford some insight into the origin of groundwater sapping features in cohesive or consolidated sediments. In particular, erosion of scarps by groundwater sapping is likely to produce narrow, deep gullies or valleys with abrupt headwalls and few tributaries, because both the hydraulic erosion of the experiments and the weathering processes at natural seeps occur at rates proportional to the flow rate of groundwater. Secondly, emergence of groundwater along a sloping upland surface is likely to erode numerous shallow, nearly parallel channels due to the high scour potential of the resulting overland flow, as compared to the lower drainage density of low-gradient channels emerging at the base of scarps. Such upland seepage may contribute to rill or gully development on lakebeds exposed by drawdown and at seeps on hillsides in natural drainage basins. Nevertheless, one must be careful in extrapolating from such experiments to natural environments. The processes and materials producing the small-scale channels differ from those acting in natural networks several orders of magnitude larger. Scarp retreat in the experimental networks occurs by hydraulic erosion together with mass wasting at the scarp face. Sapping of the headwalls in indurated rocks exposed in large valley networks involves physical and chemical weathering, as occurs in the Colorado Plateau and Hawaiian valleys. Thus, extrapolation must be justified in terms of equivalency of process rates and spatial patterns.

Simulation modeling

One way of gaining insight into the interactions of sapping processes, materials, structure, topography, and flow patterns that form valley networks is through theoretical modeling. Theoretical approaches have the advantage over experiments of being able to numerically scale process and material properties. In addition, temporal and spatial boundary conditions can be varied. On the other hand, theoretical modeling is limited to circumstances where the physical processes are understood, and it can be compromised by numerical errors or inadequacies of model structure.

Our initial simulation models of valley development by groundwater sapping utilized simple assumptions about sapping processes, materials, and flow conditions.

The flow model. The numerical experiments used a regional finite-difference numerical flow model for an unconfined aquifer of finite thickness, and permeability that is vertically uniform but areally variable. The gradients of the piezometric surface are assumed to be small enough to permit the DuPuit assumption. Erosion rates are assumed to be slow enough that a steady-state flow model is appropriate. Under these assumptions the governing flow equation is:

$$\frac{\partial}{\partial x}\left(K h \frac{\partial h}{\partial x}\right) + \frac{\partial}{\partial y}\left(K h \frac{\partial h}{\partial y}\right) = R \quad (1)$$

where h is the hydraulic head, x and y are the horizontal axes, K is the hydraulic conductivity, and R is the recharge (positive) or withdrawal (negative) rate. The groundwater flow pattern is determined using Gauss-Seidel iteration with successive over relaxation (Wang and Anderson, 1982, Chapter 3).

For the simulations discussed here a 50 × 50 point square grid defines the active flow area. Groundwater divides form the left, right, and top margins, with a fixed-head boundary on the bottom (the outflow). Uniform areal recharge was assumed. A nominal hydraulic conductivity, K_n, was assumed, and the actual conductivity of each cell, K_a, was defined as:

$$K_a = K_n e^{CN} \quad (2)$$

where N is a random normal deviate (zero mean and unity standard deviation) and C is a scaling factor. Once assigned, the conductivity remained constant. The randomization of conductivity was included to simulate natural areal variability.

The erosion model. For simplicity, the simulated area was assumed to have uniform erodibility, and valley walls and heads were assumed to erode horizontally (zero downstream gradient). Erosion was assumed to occur only along the valley walls and heads, where groundwater emerges to the surface. This simulation corresponds to very low-gradient valleys eroding headward into a level upland. The rate of valley wall erosion was assumed to be proportional to the excess of lateral inflow of groundwater per unit width of valley wall, q, above a critical flow rate, q_c:

$$E = E_c (q - q_c), \quad (3)$$

where E_c is a scaling factor.

Initially the eroding scarp face was assumed to be along the lower, fixed-head boundary of the flow domain. As valleys eroded headward, the head and sidewalls were likewise treated as fixed-head locations. Further details of the erosion and flow simulation are presented in Howard (1988).

Trial simulations. The evolution of the valley network and corresponding flow network was simulated by an iteration, as described by Howard (1988). Figure 37 shows a simulation for an assumed value of q_c of zero. Under these circumstances all active nodes undergo some erosion. However, even for this case, there is a competitive effect such that erosion becomes concentrated along a few wide valleys. The flow net at the close of the simulation is shown in Figure 37B. The final result is similar to the pattern that occurs for high discharge rates (high q relative to q_c) for cohesionless sand (Fig. 34).

An additional simulation was run for a value of q_c that was about 0.8 times the maximum value of q at the start of the simulation. Under these circumstances, only a small fraction of the active nodes have sufficient discharge to undergo erosion. Competition is very strong, with only a few narrow valleys forming (Fig. 38). There is a weak tendency to form tributary valleys. The competition is due to groundwater capture, as discussed above. Many short valley heads and sidewalls experience a decrease in discharge through time as more advanced portions of the drainage network receive their former drainage. The resultant network has narrower valleys than any of the networks in cohesionless sand (e.g., Figs. 33 and 34). This is due to two factors. First, it is difficult to run the experiments at flow conditions very close to the critical value because of the length of time required to conduct the experiment and the necessity for very diligent control by the experimenter. Secondly, lateral planation by the fluvial channels tends to widen the valleys, as discussed previously. Planation effects are not included in the simulation model. The valley networks produced by scarp retreat using cohesive sand are more similar to the simulation with high q_c (Fig. 35).

Future research

The simulation model is being generalized to allow sloping water tables and more realistic modeling of hydraulic conductivity, plus a more generalized rate law governing erosion. The model is also being validated by comparison with the experiments. The model will be used to investigate the effects of changing material parameters, differing assumptions pertaining to the rate laws governing erosion, and different boundary conditions (topography, sources and sinks of water, thickness and layering of

Spring sapping and valley network development

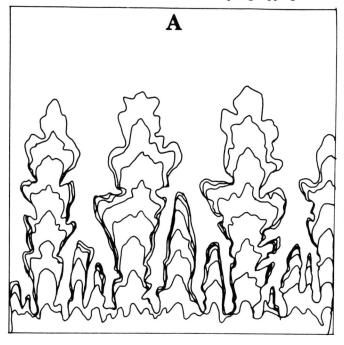

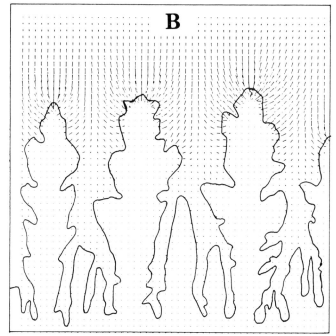

Figure 37. Plan view of a simulated valley network developed by groundwater sapping with a critical discharge, q_c, of zero: (A) the lines show the position of the edge of the valley at intervals of 25 iterations, with erosion progressing from the bottom edge of the matrix; (B) vectors showing the flow field at the end of the simulation (the vectors are assumed to be of zero length within the valleys).

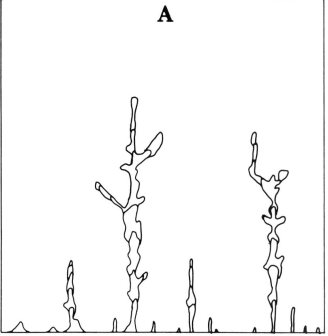

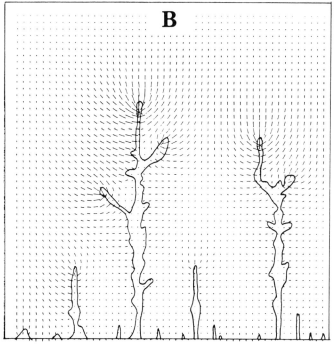

Figure 38. Plan view of a simulated valley network developed by groundwater sapping with a high value of the critical discharge, q_c.

sedimentary layers, and structural features such as faults, craters, etc.). Of particular interest are the effects of varying model parameters on the morphology of simulated valleys. Such comparisons can then provide an improved basis for inferring the hydrologic and geologic setting of terrestrial and Martian valleys of sapping origin.

CHAPTER ACKNOWLEDGMENTS

Research was supported by the National Aeronautics and Space Administration Planetary Geology and Geophysics Program (various grants to the individual authors). Jay Piper contributed greatly to the collection of morphometric data of the Hawaiian valleys. Andy Mason and John Partridge drafted the figures for the Hawaiian case study.

REFERENCES CITED

Abrahams, A., 1984, Channel networks; A geomorphological perspective: Water Resources Research, v. 20, p. 161–188.

Ahnert, F., 1960, The influence of Pleistocene climates upon the morphology of cuesta scarps on the Colorado Plateau: Association of American Geographers Annals, v. 50, p. 139–156.

Baker, V. R., 1982, The channels of Mars: Austin, University of Texas Press, 198 p.

—— , 1985, Models of fluvial activity on Mars, in Woldenberg, M., ed., Models in geomorphology: London, Allen and Unwin, p. 287–312.

Baker, V. R., and Kochel, R. C., 1979, Martian channel morphology; Maja and Kasei Vallis: Journal of Geophysical Research, v. 84, p. 7961–7983.

Baker, V. R., and Partridge, J. B., 1986, Small Martian valleys; Pristine and degraded morphology: Journal of Geophysical Research, v. 91, p. 3561–3572.

Bradley, W. C., 1963, Large-scale exfoliation in massive sandstones of the Colorado Plateau: Geological Society of America Bulletin, v. 75, p. 519–528.

Brakenridge, G. R., Newson, H. E., and Baker, V. R., 1985, Ancient hot springs on Mars; Origins and paleoenvironmental significance of small Martian valleys: Geology, v. 13, p. 859–862.

Bryan, K., 1928, Niches and other cavities in sandstone at Chaco Canyon, New Mexico: Zeitschrift fur Geomorphologie, v. 3, p. 128–140.

—— , 1954, The geology of Chaco Canyon, New Mexico, in relation to the life and remains of the historic peoples of Pueblo Bonito: Smithsonian Miscellaneous Collection, v. 122, no. 7, 65 p.

Campbell, I. A., 1973, Controls of canyon and meander form by jointing: Area, v. 5, no. 4, p. 291–296.

Carr, M. H., and Clow, G. D., 1981, Martian channels and valleys; Their characteristics, distributions, and age: Icarus, v. 48, p. 91–117.

Conca, J. L., and Rossman, G. R., 1982, Case hardening of sandstone: Geology, v. 10, p. 520–533.

Dunne, T., 1980, Formation and controls of channel networks: Progress in Physical Geography, v. 4, p. 211–239.

Gregory, H. E., 1917, Geology of the Navajo country: U.S. Geological Survey Professional Paper 93, 161 p.

Gulick, V. C., and Baker, V. R., 1989, Fluvial valleys and Martian palaeoclimates: Nature, v. 341, p. 514–516.

Higgins, C. G., 1982, Drainage systems developed by sapping on Earth and Mars: Geology, v. 10, p. 147–152.

—— , 1984, Piping and sapping; Development of landforms by groundwater outflow, in LaFleur, R. G., ed., Groundwater as a geomorphic agent: London, Allen and Unwin, p. 18–58.

Hinds, N.E.A., 1925, Amphitheater valley heads: Journal of Geology, v. 33, p. 816–818.

Horton, R. E., 1945, Erosional development of streams and their drainage basins: Geological Society of America Bulletin, v. 56, p. 275–370.

Howard, A. D., 1980, Thresholds in river regime, in Coates, D. R., and Vitek, J. D., eds., Thresholds of geomorphology: Boston, Massachusetts, Allen and Unwin, p. 227–258.

—— , 1988, Groundwater sapping experiments and modelling at the University of Virginia, in Howard, A. D., Kochel, R. C., and Holt, H., eds., Sapping features of the Colorado Plateau; Proceedings and Field Guide for the NASA Groundwater Sapping Conference: National Aeronautic and Space Administration Special Publication SP-491, p. 71–83.

Howard, A. D., and Kochel, R. C., 1988, Introduction to cuesta landforms and sapping processes on the Colorado Plateau, in Howard, A. D., Kochel, R. C., and Holt, H. E., eds., Sapping features of the Colorado Plateau; Proceedings and Field Guide for the NASA Groundwater Sapping Conference: National Aeronautic and Space Administration Special Publication SP-491, p. 6–56.

Howard, A. D., and McLane, C. F. III, 1988, Erosion of cohesionless sediment by groundwater sapping: Water Resources Research, v. 24, p. 1659–1674.

Hunt, C. B., 1953, Geology and geography of the Henry Mountains region, Utah: U.S. Geological Survey Professional Paper 228, 234 p.

Ireland, H. A., Sharpe, C.F.S., and Eargle, D. H., 1939, Principles of gully erosion in the piedmont of South Carolina: U.S. Department of Agriculture Technical Bulletin 633.

Jobin, D. A., 1956, Regional transmissivity of the exposed sediments of the Colorado Plateau as related to uranium deposits: U.S. Geological Survey Professional Paper 300, p. 207–211.

—— , 1962, Relation of the transmissive character of the sedimentary rocks of the Colorado Plateau to the distribution of uranium deposits: U.S. Geological Survey Bulletin 1125, 151 p.

Jones, J.A.A., 1987, The initiation of natural drainage networks: Progress in Physical Geography, v. 11, p. 207–245.

Jones, R. J., and others, 1984, Olokele rock avalanche, Island of Kauai, Hawaii: Geology, v. 12, p. 209–211.

Kochel, R. C., and Piper, J. F., 1986, Morphology of large valleys on Hawaii; Evidence for groundwater sapping and comparisons with Martian valleys: Journal of Geophysical Research, v. 91, p. E175–E192.

Kochel, R. C., Howard, A. D., and McLane, C., 1985, Channel networks developed by groundwater sapping in fine-grained sediments; Analogs to some Martian valleys, in Woldenberg, M., ed., Models in geomorphology: London, Allen and Unwin, p. 313–341.

Kochel, R. C., Simmons, D. W., and Piper, J. F., 1988, Groundwater sapping experiments in weakly consolidated layered sediments; A qualitative summary, in Howard, A. D., Kochel, R. C., and Holt, H., eds., Sapping features of the Colorado Plateau; Proceedings and Field Guide for the NASA Groundwater Sapping Conference: National Aeronautic and Space Administration Special Publication SP-491, p. 161–183.

Laity, J. E., 1980, Groundwater sapping on the Colorado Plateau, in Reports of Planetary Geology Program 1980: National Aeronautics and Space Administration Technical Memorandum 82385, p. 358–360.

—— , 1983, Diagenetic controls on groundwater sapping and valley formation, Colorado Plateau, revealed by optical and electron microscopy: Physical Geography, v. 4, p. 103–125.

Laity, J., and Malin, M. C., 1985, Sapping processes and the development of theater-headed valley networks in the Colorado Plateau: Geological Society of America Bulletin, v. 96, p. 203–217.

Macdonald, G. A., Abbott, A. T., Peterson, F. L., 1983, Volcanoes in the sea: Honolulu, University of Hawaii Press, 517 p.

Malin, M. C., Dzurisin, D., and Sharp, R. P., 1983, Stripping and Keanakakoi tephra on Kilauea Volcano, Hawaii: Geological Society of America Bulletin, v. 94, p. 1148–1158.

Mars Channel Working Group, 1983, Channels and valleys on Mars: Geological Society of America Bulletin, v. 94, p. 1035–1054.

Milton, D. J., 1973, Water and processes of degradation in the Martian landscape: Journal of Geophysical Research, v. 78, p. 4037–4047.

Mustoe, G. E., 1982, The origin of honeycomb weathering: Geological Society of

America Bulletin, v. 93, p. 108–115.
—— , 1983, Cavernous weathering in the Capitol Reef Desert, Utah: Earth Surface Processes and Landforms, v. 8, p. 517–526.
Patton, P. C., and Boison, P. J., 1986, Processes and rates of Holocene terrace formation in Harris Wash, Escalante River Basin, south-central Utah: Geological Society of America Bulletin, v. 97, p. 369–378.
Phillips, L. F., and Schumm, S. A., 1987, Effect of regional slope on drainage networks: Geology, v. 15, p. 813–816.
Pieri, D. C., 1976, Martain channels; Distribution of small channels in the Martian surface: Icarus, v. 27, p. 25–50.
—— , 1980, Martian valleys; Morphology, distribution, age, and origin: Science, v. 210, p. 895–897.
Pollack, J. B., 1979, Climatic change on the terrestrial planets: Icarus, v. 37, p. 479–553.
Schultz, P. M., Schultz, R. A., and Rogers, J., 1982, The structure and evolution of ancient impact basins on Mars: Journal of Geophysical Research, v. 87, p. 9803–9820.
Schumm, S. A., and Chorley, R. J., 1964, The fall of Threatening Rock: American Journal of Science, v. 262, p. 1041–1054.
Scott, G.A.J., and Street, J. M., 1976, The role of chemical weathering in the formation of Hawaiian amphitheater-headed valleys: Zeitschrift fur Geomorphologie, v. 20, p. 171–189.
Sharp, R. P., 1973, Mars; Fretted and chaotic terrains: Journal of Geophysical Research, v. 78, p. 4063–4072.

Stearns, H. T., 1946, Geology of the Hawaiian Islands: Honolulu, Hawaii, Division of Hydrology Bulletin 8, 106 p.
—— , 1966, Geology of the State of Hawaii: Palo Alto, California, Pacific Books, 266 p.
Stetson, H. C., 1936, Geology and paleontology of the Georges Bank canyons; Part I, Geology, Geological Society of America Bulletin, v. 47, p. 339–366.
Stokes, W. L., 1964, Incised, wind-aligned stream patterns of the Colorado Plateau: American Journal of Science, v. 262, p. 808–816.
Twidale, C. R., 1982, Granite landforms: Amsterdam, Elsevier, 372 p.
Wang, H. F., and Anderson, M. P., 1982, Introduction to groundwater modeling: San Francisco, California, W. H. Freeman, 237 p.
Watson, R. A., and Wright, H. E., 1963, Landslides on the east flank of the Chuska Mountains, northwestern New Mexico: American Journal of Science, v. 261, p. 525–548.
Wentworth, C. K., 1928, Principles of stream erosion in Hawaii: Journal of Geology, v. 36, p. 385–410.
—— , 1943, Soil avalanches on Oahu, Hawaii: Geological Society of America Bulletin, v. 54, p. 53–64.
Woodruff, J. F., 1971, Debris avalanches as an erosional agent in the Appalachian Mountains: Journal of Geography, v. 70, p. 399–406.
Woodruff, J. F., and Gergel, T. J., 1969, The origin and headward extension of first-order channels: Journal of Geography, v. 68, p. 100–105.

MANUSCRIPT ACCEPTED BY THE SOCIETY NOVEMBER 14, 1989

Printed in U.S.A.

Chapter 12

Groundwater processes in the submarine environment

James M. Robb
U.S. Geological Survey, Woods Hole, Massachusetts 02543

INTRODUCTION

The geomorphic significance of groundwater does not evaporate at the seashore. Submarine groundwaters do not remain immobile and inactive in their saturated environment. In the following pages, I summarize how submarine groundwater (interstitial water, either fresh or saline) can contribute to the modification of the sea-bottom landscape, and cite some examples. Relatively few studies of submarine groundwater geomorphology have been published, at least in English, so I include some speculation.

This discussion emphasizes processes of continental margin regions (many studies cited are from the U.S. East Coast), but it includes subduction zones, mid-ocean ridges, and even abyssal plains, where groundwaters have a role in shaping features of the sea floor. Groundwater processes of coastal areas are dealt with in other chapters of this book. For our purposes here, I do not try to distinguish between connate and phreatic groundwater, if such a distinction can be made successfully within the saturated, commonly depositional, deep-ocean bottom.

Although submarine groundwater has been largely ignored in recent years as a geomorphic agent, it has a long history relative to modern studies of marine geology. In the early days, shortly after acoustic echo sounding methods had begun to reveal details of subsea features, Stetson (1936) and Stetson and Smith (1938) suggested that high sediment pore pressures and discharge of groundwater could cause submarine erosion and landslides on the continental slope. Douglas Johnson (1938–39; 1939) enrolled continental-margin groundwaters in one of the active geomorphologic controversies of the 1930s and 40s. He proposed that groundwater could pass under the continental shelves in confined aquifers from recharge areas on land to discharge at the continental slope, and that submarine canyons were eroded by artesian spring sapping. His hypothesis was rejected (Bucher, 1940; Shepard and Emery, 1941; Rich, 1941; Kuenen, 1947), largely because discharging groundwater was seen to have limited erosional capability underwater and could occur in relatively few places. By comparison, turbidity currents (Daly, 1936) satisfied the requirements for a vigorous process to erode submarine canyons worldwide. Subsequently the complexities of submarine canyon origin have become more apparent, and an understanding of canyon erosion by multiple causes has evolved (Shepard, 1981). Recent studies indicate that mass-wasting processes are important. Canyons appear to erode headward, upslope, and reach the slope break at an advanced stage of their development (Twichell and Roberts, 1982; Farre and others, 1983).

Johnson's concept of artesian spring sapping as the primary cause of submarine canyons remains untenable, but most of his ideas about submarine groundwater should not be dismissed. More modern studies of subsea sediments, structure, and processes show that groundwater does participate as one of many complementary agents of sea-bottom landscape development. Manheim (1967) cited the discovery of fresh groundwater offshore Florida during early JOIDES drilling (Bunce and others, 1965) to suggest that groundwater may have discharged on the continental slope during low sea levels of the Pleistocene. In a study of continental slope structure using seismic profiles, Rona (1969) pointed out that high groundwater pressures in subshelf aquifers might cause massive submarine slumping on the U.S. East Coast slope. Ryan and others (1970) suggested that submarine slope failures along the Mediterranean coast may have been caused by sediment weakening at "soft layers" where freshwater aquifers are found at depths of 35 to 1,800 m below sea level.

More recently, long-range sidescan sonar systems have given us acoustic images of sea-floor features that are evocative of subaerial groundwater sapping (Robb and others, 1982; Belderson, 1983; Twichell and Parson, 1986). Submersible dives reveal that chemical erosion due to discharge of brines may be a key to the development of the steep carbonate escarpments around Florida (Paull and Neumann, 1987), and may create amphitheatric canyons in those escarpments (Paull and others, 1987). In other environments, sediment dewatering and discharge of pore fluids caused by tectonic compression contribute to faulting and surficial erosion in subduction-zone trench systems (Arthur and others, 1980), and cause diapiric folding and mud volcanoes (Westbrook and Smith, 1983; Barber and others, 1986). Spectacular constructs of hydrothermal discharge are found during explorations of spreading centers on mid-ocean ridges (Haymon and MacDonald, 1985).

Robb, J. M., 1990, Groundwater processes in the submarine environment, *in* Higgins, C. G., and Coates, D. R., eds., Groundwater geomorphology; The role of subsurface water in Earth-surface processes and landforms: Boulder, Colorado, Geological Society of America Special Paper 252.

GEOMORPHIC PROCESSES OF SUBMARINE GROUNDWATER

The sources of energy for geomorphic work of submarine interstitial water are gravity, geothermal heat, and tectonic compression. There are a number of mechanisms that induce differential pore pressures and the potential for groundwater flow (Fig. 1):

1. Lowering sea level reduces the external hydrostatic load. It can thereby induce excess pore pressures (greater than hydrostatic pressure) in in-place sediments of low hydraulic conductivity (Morgenstern, 1967). Pore water in sediments left above sea level will potentially drain (either by subaerial discharge or by subsurface drainage), and pore water below sea level will potentially discharge.

2. Rapid deposition of fine-grained sediment that does not drain rapidly can cause excess pressures by entrapment of pore waters (Booth, 1979).

3. Subaerial recharge of artesian (confined) aquifers that extend below submarine regions may result in submarine discharge (Johnson, 1939). Coastal plain sediments of passive continental margins are the classical model, but other conditions could serve the same function. In arctic regions, offshore permafrost relict from Pleistocene lower sea levels (Washburn, 1980) could serve as a seal.

4. Externally applied compression, as in an accretionary wedge along a subduction zone, can cause high pore-water pressures (Arthur and others, 1980; Bray and Karig, 1985).

5. Density contrasts between pore fluids, leading to convection, can be caused by constituents in solution (salinity), compositional differences (hydrocarbon-water), or temperature differences (geothermal or volcanic heat).

Excess pressure of interstitial water is a critical condition in many cases of submarine slope failure (Morgenstern, 1967; Nardin and others, 1979; Booth, 1979). In a column of saturated sediment, excess pore pressure has a buoyant effect to support part of the overburden load. It thereby reduces the effective stress (the gravitational force of grain-to-grain contact within the sediment) and thus reduces the sediment shear strength (the frictional force that resists slope failure). When the buoyant force becomes significant relative to the weight of the overlying material, the shear strength becomes inadequate with respect to the shear stress, and failure occurs (Morgenstern, 1967; Booth, 1979). As a marine erosional mechanism, excess pore pressure interplays with other factors such as oversteepening, earthquakes, tides, or cyclic loading by storm waves (Clukey and others, 1985). On continental slopes, mass wasting is recognized as the most significant erosional process (Nardin and others, 1979; Farre, 1985; Prior and Coleman, 1986).

Where water actively discharges from the sea bottom, mechanical and chemical erosion can take place, including grain-by-grain removal of surface materials, seepage erosion, possibly piping or tunnel erosion, and sapping (undercutting) of slopes or cliffs (Higgins, 1984), as well as dissolution of carbonate grains and cements, and solute transport. Discharge is most likely to take place where particularly permeable strata or channel deposits crop out or where fractures or jointing provide avenues for subbottom water movement.

However, the processes of downhill flow and fluvial transportation that are intimately associated with groundwater discharge on a subaerial slope to form channels and valleys do not normally occur with the same intimate association in a submarine environment. The potential energy of water that is discharged on a subaerial slope is absent where water is discharged into water (or the potential energy is minimal, dependent on small density differences). Without some transport mechanism and immediate removal of material, the erosional effect of submarine groundwater discharge can only be local.

In comparison with the subaerial environment, channelized fluvial transport is accomplished underwater by sediment gravity flows. As distinguished by Middleton and Hampton (1976) on the basis of how grains are supported above the bed, the general term "sediment gravity flow" includes turbidity currents, fluidized sediment flows, grain flows, and debris flows. Submarine flows clearly have the capacity to create extensive, elaborate channel systems on the sea bottom. Channels in deep-water places have been observed and mapped using high-resolution echo-sounding profiles and seismic profiles for years. Recent long-range sidescan-sonar images show large regions with extensive dendritic networks, braided channels, meanders, levees, and crevasse splays, among other features found in fluvial systems (McGregor and others, 1982; Robb and others, 1987; EEZ-SCAN 84 Scientific Staff, 1988). However, although we know of these channels and can map their configuration, the submarine transport processes that these features represent remain enigmas (Nittrouer and others, 1988).

Large, or catastrophic, high-velocity sediment gravity flows are thought to originate with slope failures, slides, or slumps. Most of the water integral to a submarine sediment gravity flow comes from pore water with minor additional entrained water (Morgenstern, 1967; Hampton, 1972). Morgenstern (1967) believed that excess pore pressure was one factor contributing to the inception of submarine flow and transport following slope failure. He pointed out that large sediment masses would transform more readily than small masses into flows because preexisting pore pressure would not dissipate. Hampton (1972) thought the unruly jostling of a sliding or slumping mass within the water column to be the entraining mechanism leading to its transformation into a sediment gravity flow. The transformation process is not satisfactorily understood (Morgenstern, 1967; Parker, 1982; Kirwan and others, 1986).

Thus, there may be some long-term geomorphic effect of excess pore pressure to increase the frequency of occurrence of sediment gravity flows and the submarine channel forms that they, in turn, create, but such a relation is indistinct. It would appear at this time that the primary mechanical effect of subma-

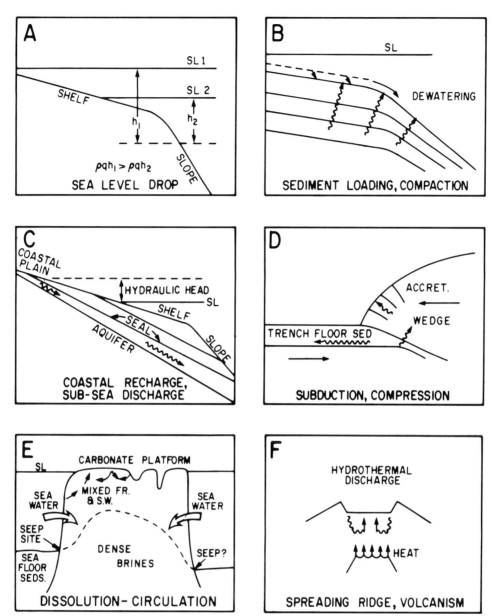

Figure 1. Diagrams of mechanisms that can induce pressure fields in sub-sea-floor pore waters. SL = sea level. A. Relative sea-level drop: significant effect occurs where rate of sea-level change is much greater than the rate of dewatering allowed by ambient hydraulic conductivity. B. Rapid sediment influx, compaction, and dewatering. C. Subaerial recharge of sealed subsea aquifers, submarine discharge. D. Compression: subducting sea floor and accretionary wedge. Dewatering within wedge; if seal is present pore water can be forced seaward. E. Fluid density differences. This example is of the Florida-Bahamas platform, which supports two systems (see text): seawater versus brine (deep) and geothermally heated (shallow) (after Paull and Neumann, 1987). F. Mid-ocean ridge crest: rifting, volcanism, fracturing, hydrothermal flow; simplified diagram of complex and variable flow configurations.

rine interstitial-water processes is to influence the location and form of submarine topography by enhancement of mass wasting, local sapping, or dissolution.

In regions of discharge where fresh groundwater mixes with seawater, rapid dissolution of carbonates can occur (Back and others, 1984). Whereas seawater dissolves carbonates only slowly in the upper ocean, between a surficial supersaturated layer and above the lysocline (Kennett, 1982, p. 464), the mixture of seawater and fresh water is more corrosive than either of the component waters and has been cited by Hanshaw and Back (1980) and Back and others (1984) as an active sculpting mechanism of the coast of Yucatan (cf. Chapter 13, this volume).

Karst-like features have been found in evaporites on the floor of the Mediterranean Sea (Belderson and others, 1978;

Kastens and Spiess, 1984). Kastens and Spiess (1984) suggested that subsurface dissolution (i.e., by "groundwater") took place by means of a circulation mechanism as proposed by Anderson and Kirkland (1980) for a terrestrial application. They propose that dissolution of the evaporite creates a heavy brine, which trickles down through fractures, to be replaced by undersaturated water. They point out that the density contrast between brine and seawater would be adequate to drive the process.

A regional brine-driven circulation system within the Florida-Bahamas platform was proposed by Paull and others (1984) and Paull and Neumann (1987), who observed brine discharge leading to acidic conditions and dissolution at the base of the West Florida escarpment in the Gulf of Mexico. The base of the escarpment is in water deeper than 3 km. A geothermally driven groundwater circulation system was suggested to occur within shallower depths of the Florida peninsula by Kohout (1967).

EROSION AT CONTINENTAL MARGINS

Groundwater and sea-level changes

Eustatic sea-level change creates the potential for massive flux of groundwater within the continental blocks (Hay and Leslie, 1985). Sea-level regressions, in particular, set up conditions for pore overpressures, submarine discharge, and fresh-water throughflow at continental margins. Onset of glaciation is thought to be the condition under which the most rapid eustatic sea-level falls might occur (Pitman, 1978). Although the evidence does not appear to be well established and the rates are controversial, interpretation of the geologic record has suggested to some researchers that regressions are generally faster than transgressions (Vail and others, 1977; Sloss, 1984). Local rate of relative sea-level change and hydraulic conductivities are the important considerations with respect to groundwater activity.

Geomorphically significant groundwater fluxes have been linked to sea-level regressions in some places. Oldale (1985) inferred that during the late Pleistocene in the Gulf of Maine, a combination of glacial rebound and eustatic sea-level drop caused a local rate possibly as rapid as 4.8 m/100 yr. D'Amore (1983) attributed the development of a valley network in marine clays overlying glacial drift near Wells, Maine, to sapping from drainage of groundwater in response to rapid sea-level fall. Mullins and others (1986) mapped buried ancient slumping along the upper West Florida escarpment, which they suggest may have been caused by groundwater pressure during a Miocene fall of sea level. Although they did not remark on groundwater erosional processes, Miller and others (1985), who made a detailed stratigraphic study of an Eocene to Miocene interval on the North Atlantic margins of the U.S. and Ireland, emphasized that the rate of sea-level fall appears to be more important than lowered sea level per se in canyon cutting and erosion of unconformities on continental slopes. Subshelf groundwater offshore New Jersey is still equilibrating to present sea levels, following the late Pleistocene low (Meisler and others, 1984).

Lower sea levels on continental shelves

Most evidence on continental shelves of former groundwater erosion that occurred during Pleistocene lowstands will have been masked by other vigorous processes, particularly those of the recently transgressive shoreline. On the U.S. East Coast margin, the few stillstand shorelines that survive have been modified or covered by a mobile sand sheet (Dillon and Oldale, 1978; Knebel, 1981). Erosional features of the Quaternary regressions are imperfectly known to us, shown in seismic profiles as subshelf reflecting surfaces, and cut in a few places by filled channels (Swift and others, 1980). Outcrops of semiconsolidated materials at the shelf break and in the canyon heads are densely burrowed and modified by submarine animals (Warme and others, 1978; Twichell and others, 1985). Most evidence of a groundwater erosional process, like the miniature valleys on beaches described by Higgins (1982), is ephemeral. Only large topographic features are likely to survive on the continental shelf or upper slope.

One could speculate that discharging groundwater had a role in forming the gully-like headward extensions (10 to 30 km long) of many submarine canyons shoreward of the shelf break along the northeastern U.S. (Fig. 2). During lower sea levels, the now submerged canyon heads were partially exposed. Pleistocene shorelines that were a few kilometers shoreward of the shelf break are documented by the relict Franklin and Nichols shores, now in water depths of 140 to 160 m. (Dillon and Oldale, 1978). Drainage of groundwater from the shelf during sea-level regression and low stands would have been confined in places within subshelf aquifers. Extensive deposits of stiff clay that could serve as confining beds commonly crop out in canyon heads (Stanley and Freeland, 1978; Knebel and Spiker, 1977; Knebel, 1979; Valentine and others, 1984). The topographic lows where canyons of the continental slope break the edge of the shelf would concentrate local groundwater flow, as outlined by Dunne (1980; this volume), leading to sapping and mass wasting. Submarine as well as subaerial groundwater sapping could occur.

As an analog for the continental shelf canyon-head valleys, composite origins involving both groundwater and fluvial processes were cited for valleys cut into coastal terraces in New Zealand by Schumm and Phillips (1986). During the Pleistocene on the New England continental shelf, groundwater processes would complement Pleistocene surficial drainage from the continent, or from glaciers on Georges Bank. On the present-day continental shelf, there are processes of tidal currents, internal waves, and biological erosion that work within the submerged valleys, so the present form of the submarine canyon heads has many components.

Mobility of groundwater under continental shelves

The concept of groundwater transport under the continental shelf from subaerial recharge areas to points of submarine discharge on the continental slopes, beyond the shelf break, was central to Johnson's (1939) hypothesis of submarine canyon

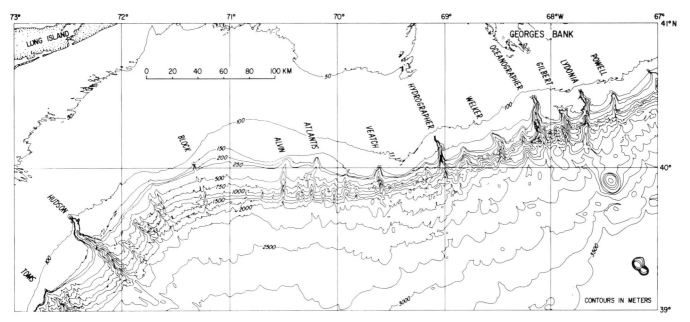

Figure 2. Continental margin off northeastern U.S. showing submarine canyon heads in shelf edge. Contour interval 50 to 250 m, then 250 m.

origin. Since his time, studies of the continental margin off the eastern United States bear out many of his ideas. The continuity of strata that stretch from the subaerial Coastal Plain below the continental shelf to be truncated at the continental slope has been documented by seismic profiling and drilling (Hathaway and others, 1979; Grow and others, 1979; Robb and others, 1981; Poag, 1985). Submarine springs of fresh or brackish water have been found on the continental shelf of Florida at depths of 500 m as well as in shallow nearshore waters (Manheim, 1967; Kohout and others, 1977b, 1988). Manheim and Horn (1968) inferred from salinities of subshelf pore-fluid samples that the upper 1,000 m of submerged coastal plain strata are greatly influenced by meteoric waters; the clastic sections of South Carolina, Georgia, northern Florida, and New Jersey show the most prominent movement of fresh water from recharge areas to deep strata near the coast. Manheim and Horn (1968) also suggested that clayey strata of the coastal plain act as semipermeable membranes that promote osmotic flushing of salty strata by fresher waters and increase the pore pressure of the fresher water intervals. Kohout and others (1977a, 1988) found fresh water in strata buried 450 m below Nantucket Island. Pleistocene or recent recharge from above is unlikely in that place because impermeable strata overlie the zone of fresh water, and the fresh water is too deep to have been emplaced from the mainland as part of a Ghyben-Herzberg lens at present sea level. Kohout and others (1977a, 1988) suggested that rising Holocene sea level had displaced Pleistocene-recharged groundwater of the continental shelf.

Fresh water has been found in Miocene sediments 200 m below the continental shelf as far as 100 km offshore New Jersey (Hathaway and others, 1979; Kohout and others, 1988). Meisler and others (1984) point out that the formation of a broad transition zone between fresh water and salt water, as is found off New Jersey, requires large-scale long-term fluctuations of sea level during the late Tertiary and Quaternary. That transition zone is now moving slowly landward, toward equilibrium with present sea level. Meisler and others (1984) studied digital finite-element models of the New Jersey subshelf hydrology, which were based on structure from seismic profiles and stratigraphy, and permeability figures from offshore stratigraphic test wells. The modeling implied that groundwater could discharge on the continental slope at lower sea levels. The modeling also suggested that without extensive (unmodeled) secondary porosity, throughflow of fresh water from onshore areas of recharge might require million-year periods of low sea level (Leahy and Meisler, 1982; Meisler and others, 1984; H. Meisler, oral communication, 1984). Pleistocene fluctuations, which were brief compared to late Miocene or late Oligocene sea-level lows, may not have been long enough. No present-day strata-bound lenses of fresh water extending as far seaward as the continental slope have been reported.

Features of the lower continental slope

On the lower continental slope (1,400 to 2,100 m) off New Jersey, sidescan-sonar surveys revealed some valleys with steep-walled amphitheatric basins, which Robb (1984) suggested were sculpted by groundwater sapping. Much of the continental slope off the central East Coast of the U.S. is covered by Pleistocene sediment between 20 and 400 m thick, but on the lower slope between Lindenkohl and Toms Canyons, an Upper Cretaceous to Miocene section crops out (Hampson and Robb, 1985) (Fig. 3). The chalky Eocene rocks and Miocene silty claystones are incised by a set of valleys that have steep-walled amphitheatric basins, which are terraced along stratigraphic levels and have flat floors (Fig. 4). The basins in adjacent valleys along the slope appear to be located in stratigraphically similar (more permeable?) parts of the slightly seaward-dipping section. Observations from the submersible *Alvin* showed near-vertical, joint-controlled walls in the valley basins. Undercut strata are common. Large tafone-like recesses that expose vertical tubes of probable Miocene cerianthid worms were found in a cliff at the head of one of the basins (Fig. 5).

The preservation of the trace fossils in tafone-like cavities, and also of fragile sandstone dikes that project cleanly from cliff faces, demonstrates a long term of erosion (greater than 1 m of cliff retreat) by a slow, particle-by-particle process and the absence of catastrophic mass wasting or current erosion. Although the removal of debris from the valleys is difficult to attribute solely to groundwater action (Farre and Ryan, 1985), there is no evidence of repetitive turbidity currents in the unchanneled area upslope of the valley heads, which lie near 1,400 m water depth (Robb, 1985).

Robb (1984) speculated that pits and rillenkarren-like features on the faces of some Eocene chalk outcrops within the valleys may have been caused by dissolution, where fresh water (or mixed fresh and salt water) rose along the outcrop after discharge at a bedding plane (Fig. 6). Fissures along vertical joints were also observed from *Alvin*. Sets of reticulate trenches resembling kluftkarren—dissolution along joints—were observed on sidescan-sonar images of the continental slope, but were not directly observed from the submersible. If these features were created by submarine discharge of groundwater, they are relict, because there is inadequate head for present-day discharge at these depths, and the midshelf fresh water–salt water transition zone is moving slowly shoreward (Meisler and others, 1984). Dissolution of the outcrops by long-term exposure to oceanic water is another possibility, suggested by Prior and Doyle (1985), but a sapping process does appear to have had an effect within the steep-walled amphitheatric valley basins.

Dissolution of carbonate escarpments

Groundwater discharge at great depths below sea level appears to be important to erosion of the great escarpments around Florida, the Bahamas, and probably Yucatan. The 3,000-m-high escarpments in water depths as much as 4,500 m truncate nearly horizontally bedded carbonate rocks. Back-reef facies crop out at the escarpments, showing that kilometers of material have been removed from the cliff faces (Paull and Dillon, 1980; Dillon and others, 1985; Freeman-Lynde, 1983; Paull and others, 1984; Freeman-Lynde and Ryan, 1985). In most places, seismic profiles across the base of the escarpments show erosional benches or onlapping pelagic deposits rather than base-of-slope colluvial accumulations or slump or slide deposits, but oceanic currents and chemical activity do not seem adequate to either erode or dissolve the entire lost volume of carbonate. Whereas oceanic waters below the carbonate compensation depth are highly corrosive to

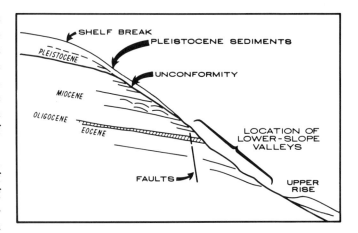

Figure 3. Diagram of continental shelf/slope stratigraphy off New Jersey where lower slope valleys possibly due to groundwater erosion are found (modified after Robb, 1984).

carbonate (Kennett, 1982, p. 465; Freeman-Lynde and Ryan, 1985), the greater parts of the escarpments lie above that level.

There appear to be two mechanisms that circulate pore waters through the Florida-Bahamas platform. Kohout (1967) postulated a massive circulation of groundwater and seawater through the cavernous "Florida boulder zone," 300 to 900 m below the peninsula, driven by geothermal heat. He suggested a deep-ocean influx with outflow through offshore springs, using geochemical data and a reversal of the geothermal gradient as evidence of seawater circulation. This circulation is shallow compared to the more recent observations of Paull and others (1984), who found active discharge of brine at water depths greater than 3,200 m during *Alvin* dives to the base of the West Florida escarpment in the Gulf of Mexico.

Paull and others (1984) and Paull and Neumann (1987) suggest that large volumes of highly saline, sulfide-rich pore fluid created by dissolution of subpeninsular evaporites exit the carbonate platform as it compacts, to spill out over impermeable abyssal sediments where they lap onto the base of the escarpment (Fig. 1E). Paull and others (1984) observed color patterns in the sediments at the base of the escarpment that looked as though they were deposited by fluids flowing downslope, and pointed out that the saline fluids are heavier than seawater.

Populations of organisms like those that surround hydrothermal vents on mid-ocean ridges were found at seep sites. Paull and Neumann (1987) reported that the outcrops they observed within seep zones at the base of the escarpment in the Gulf of Mexico were unlike other exposed strata along the escarpment in that they were not Fe-Mn oxide coated, but rather had extensively pitted, fluted, and corroded surfaces. Paull and Neumann (1987) infer that acidic conditions are created where water, enriched in hydrogen sulfide dissolved from evaporites within the Florida platform, exits into oxygenated seawater. They point out that not only can significant chemical erosion take place, but also that the environment is favorable to hydrocarbon generation from chemosynthetic organic carbon, and that these seep sites could be models for some deposits of sediment-hosted sulfide minerals.

Recent long-range sidescan-sonar images (Twichell and Parson, 1986) (Fig. 7) show that parts of the West Florida escarpment are cut by numbers of canyons with amphitheater-like heads, while other areas show canyons with a more gully-like appearance, and still other places along the escarpment are relatively smooth. The theater-headed canyons and the groupings of canyons showing different erosional styles suggest that groundwater circulation and differences in subpeninsular permeabilities play a role. Paull and others (1987), using the submersible *Alvin,* found sinkhole-like depressions (0.5 km in diameter and >40 m deep) in Florida Canyon and vicinity, in water depths of 2,300 to 3,200 m (David Twichell, personal communication, 1987).

Ancient karstic erosion in Tertiary strata below the continental shelves encircling Florida (Jordan, 1954; Rona and Clay, 1966; Gomberg, 1977; Popenoe, 1985; Hebert, 1985) has been attributed to dissolution by meteoric waters during low sea levels. Malloy and Hurley (1970) suggested submarine dissolution by groundwater outflow as the cause of karstic features on the Pourtales terrace south of Florida. Doyle and others (1985) emphasized the similarity in appearance between two bands of solution features found on the West Florida margin. A band of buried and filled karst features at water depths shallower than 75 to 100 m includes solution valleys extending from Tampa Bay and Charlotte Harbor estuaries. These they attributed to subaerial erosion during low sea levels. The deeper features on the continental slope in water 500 to 800 m deep, were attributed to submarine dissolution by percolating groundwater or to erosion by the Loop current of the Gulf of Mexico.

Pockmarks

A phenomenon that has been attributed to the escape of groundwater or gas from submarine sediments is the pockmark. Fields of generally circular or elliptical depressions, from 15 to 60 m in diameter and 2 to 9 m deep, have been found in fine-grained sediments in flat or gently sloping terrain. On the Nova Scotian shelf, King and MacLean (1970) found between 45 and 200 pockmarks per square kilometer, generally restricted to an area covered by a surficial clay unit. Hovland (1983) found some pockmarks off Norway distributed in strings. Harrington (1985)

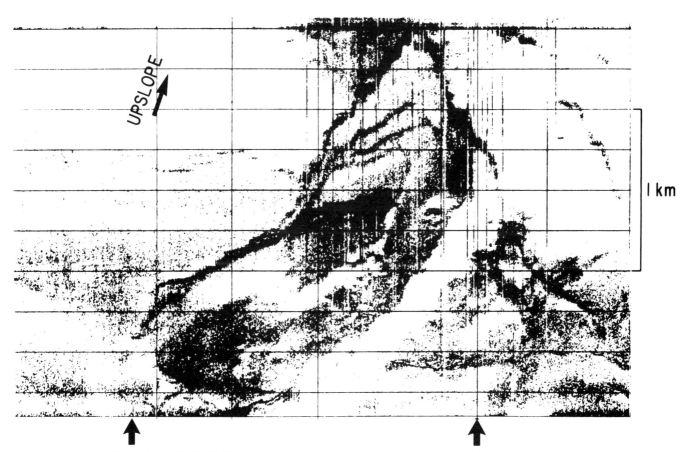

Figure 4. SeaMARC I sidescan sonar image of valley on the lower continental slope off New Jersey. Strong returns on this image are dark, shadows are light; arrows at bottom show look direction. Water depth about 1,700 m at top of image, 2,000 m at bottom. Note basin morphology, cliffed walls, removal of material along stratigraphic levels (from Robb, 1984).

Figure 5. Trace fossils of Miocene cerianthid worms exposed within tafone in a cliff at head of a lower-slope valley. The vertical tubes are 10 to 15 cm in diameter. Water depth about 1,400 m. Photograph from DSRV *Alvin* (from Robb, 1984).

inferred that pockmark form in a North Sea area may follow a progression from groups of smaller pits to smoother, wider depressions and eventual infilling.

Discharge of subsurface gas, as well as subsurface water, can cause pockmarks, but the difference has not been identified from the pockmark form; instead, local geology must be considered. Whiticar and Werner (1981) attributed pockmarks in the western Baltic Sea to fresh-water outflow from glacial-sediment aquifers. King and MacLean (1970) attributed the Scotian shelf pockmarks to escape of groundwater or gas from underlying coastal plain sediments. Harrington (1985) attributed some North Sea pockmarks to groundwater. Hovland (1983) attributed pockmarks associated with elongated depressions on the western slope of the Norwegian Trench to erosion collapse from gas discharge, modified by bottom currents. Hovland and Sommerville (1985) associated a very large depression (700 m × 450 m × 17 m deep) in the North Sea with subsurface seismic-reflection anomalies probably caused by thermogenic gas. Scanlon and Knebel (1989) describe pockmarks in the muddy bottom of Penobscot Bay, Maine. Groundwater seepage is suggested as a cause because many pockmarks penetrate the muddy layer to an underlying coarse stratum known to be an aquifer on land. On the other hand, escape of biogenic gas from within the muddy layer is considered a possibility because seismic profiles show acoustic wipeouts (commonly taken as an indication of interstitial gas) in some places.

SUBDUCTION AND DEWATERING

In subduction zones, fluids forced from sediment that is being compressed into an accretionary prism at the trench wall have important geomorphic roles ranging from basic participation in the underthrusting process to formation of surficial geomorphic features. Highly pressured pore water acts as a force-transmitting fluid to reduce frictional stress on the sediment column (Hubbert and Rubey, 1959; Von Huene and Lee, 1982; Westbrook and Smith, 1983). Multi-channel seismic-reflection profiles, carefully processed to bring out the details of trench-décollement structure in the northeast Pacific, show that thin sequences of unconsolidated oceanic sediments are subducted with little deformation (R. von Huene, personal communication, 1986; Davis and von Huene, 1987). Near-lithostatic pore pressures were measured during Deep Sea Drilling Project (DSDP) drilling on the Barbados Ridge (Moore and others, 1982). The commonly low slopes of the sea floor at the front of an accretionary wedge are an effect of the low shear stresses; high shear stress at the underlying décollement would cause greater slope (Davis and others, 1983).

Dewatering fluids escape along veins or fractures within the accretionary wedge and probably influence the faulting pattern. Mineralization and faulting are attributed to interstitial fluids in both ancient and modern deposits of accretionary prisms (Arthur and others, 1980; Lash, 1985; Ritger, 1985).

Figure 6. Are these solution pits and submarine rillenkarren? Outcrop of Eocene chalk in wall of a valley basin shown in Figure 4. New Jersey lower continental slope. Water depth about 1,950 m. The lineations are vertical. Note the pits along a bedding plane in the lower right of this photograph from DSRV *Alvin* (from Robb, 1984).

The dewatering fluids also enhance mass wasting of the trench wall by slumping, sliding, and debris flow (Moore and others, 1985; Bray and Karig, 1985; Silver and others, 1986). This occurs even on the low slopes at the tongue of an accretionary wedge, and thereby contributes to trench fill (Arthur and others, 1980; Baltuck and others, 1985; Stevens and Moore, 1985).

Dewatering phenomena apparently are not restricted to any one part of an arc-trench complex. Bray and Karig (1985), who studied a number of accretionary prisms, suggested that zones of intense overpressuring would occur near the décollement boundary (just landward of the trench axis), but they also point out that locally permeable zones of fracturing or of coarse granular sediments could determine dewatering routes. Arthur and others (1980), who describe dewatering veins in DSDP core samples from the Japan convergent margin, suggest that tectonic dewatering could induce fracturing and faulting below the inner slope (i.e., in relatively shallow water depths on the trench slope). They find that such fracturing and faulting below an impermeable mud veneer could cause overpressured zones at depths of 200 to 500 m, with local reduction in shear strength and downslope sediment movement.

Superhydrostatic dewatering fluids also contribute to diapir formation (White and others, 1982; Silver and others, 1986) and build mud volcanoes (Stride and others, 1982; Westbrook and Smith, 1983; Silver and others, 1986). Silver and others, (1986) found surficial diapirs and mud volcanoes on all parts of the Flores thrust zone in Indonesia, which they attribute to high fluid pressures. They inferred that circular mud volcanoes on the inner (landward) part of the accretionary wedge had formed in their present setting. Wedge material tends to be squeezed during deformation, so mud volcanoes formed initially at the toe should be elongated parallel to the wedge front as they accrete, and are difficult to distinguish from folds (by using remotely sensed data). Transmission of high pore-fluid pressures horizontally through undeformed oceanic sediments as far as 8 km ahead of a deformation front is illustrated by a mud volcano 60 m high and 2 km across in the Barbados Ridge complex described by Westbrook and Smith (1983).

One could speculate that locally compressional components of transform faulting and dewatering of sediments along the Blanco fracture zone, in the northeast Pacific, contributed to development of an extensive system of submarine canyons, as described by Embley (1985). The canyons incise an isolated band of sediments about 10×70 km in size that lies along the Cascadia Channel where it parallels the fracture zone. Embley attributed the canyon formation to mass wasting triggered by earthquakes and to turbidity current flow from the north flank of the narrow Blanco Ridge, but conceded dissatisfaction with these mechanisms as a complete explanation.

There are phenomena in other deep-ocean physiographic provinces that are attributed to solely gravitational dewatering of sediments. Faulting is observed on seismic-reflection profiles of the flat-lying sediments of the Sohm and Madeira abyssal plains in the Atlantic Ocean (Duin and others, 1984). The faults appear to be associated with basement highs, and Buckley and Grant (1985) suggested that they represent conduits of pore fluids that escape from the compacting sediments where impermeable clay layers terminate against the basement highs. The configuration in plan of the faults that were observed on profiles is uncertain. Buckley and Grant (1985) speculated that the faults ring topographic highs of the basement rocks, associated with particular sediment thicknesses, and have an expression at the sediment surface, which could demonstrate their dewatering origin. They also speculated that dewatering that enhances erosion could be responsible for moats around seamounts. However, in situ measurements of pore-water pressures in the Madeira abyssal plain imply that downward movement of pore-water is occurring over an observed fault and a basement high (Schultheiss and McPhail, 1986). It seems that questions of groundwater circulation in the abyssal sediments remain open, but the investigations show groundwater mobility in deep-ocean sediments that is significant to schemes for deep-water disposal of radioactive waste, as well as to geomorphic modification.

SUBMARINE HYDROTHERMAL FIELDS

A constructional role of submarine groundwater is found at mid-ocean spreading centers, where the heat of submarine volcanism drives a hydrothermal circulation system. Actively discharging hot springs at mid-ocean ridges deposit massive sulfide minerals (Haymon and MacDonald, 1985; Rona and others, 1983). Water chemistry shows that seawater is convected through a ridge-crest network of fractures, dissolves minerals by contact with hot subsurface rocks, and transports them to the surface (Edmond, 1981; Rona and others, 1983, p. 771). Models of circulatory systems have been proposed that vary in such aspects as the width of the ridge-crest influx area, depth of seawater penetration, degree of involvement with underlying magma sources, or number and configuration of separate circulatory cells (Crane and Ballard, 1980; Sleep and others, 1983, p. 56 ff; Lowell and Rona, 1985). The volume of hydrothermal circulation is important to oceanic and global thermal and chemical budgets (Williams and others, 1974; Edmond, 1979; Turekian, 1983). The discoveries of the hydrothermal sites verify hypotheses that submarine volcanism is responsible for massive sulfide mineral deposits (Skinner, 1983).

The first major discoveries of oceanic hydrothermal deposits were made in the Red Sea from 1963 to 1967 (Degens and Ross, 1969), where several basins containing pools of brines as hot as 55°C were discovered. Metalliferous deposits from basin floors contain iron-montmorillonite, amorphous goethite, and sulfides of iron, zinc, copper, and lead, distributed as a blanket that appears to have precipitated from solution on cooling of a subterranean brine that had discharged into seawater (Bischoff, 1969).

In the eastern Pacific, local temperature anomalies at ridge

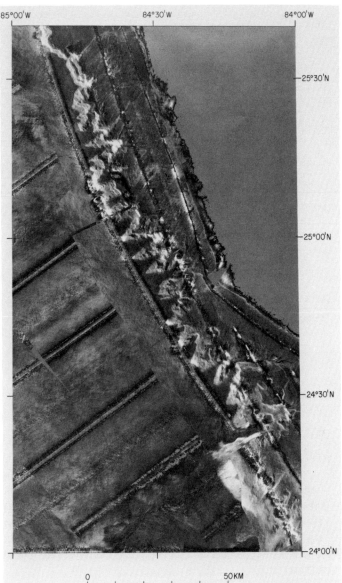

Figure 7. GLORIA sidescan sonar image of part of the West Florida escarpment, eastern Gulf of Mexico. The image is composed of swaths of sidescan sonographs 5 to 25 km wide, geometrically corrected and built into a mosaic, one upon another, on the geographic grid. Strong returns are bright; shadows are dark. Ship tracks lie within the double dark stripes, which show the near range of the surface-towed sensor. Steeper parts of the escarpment are displayed as the jagged bright line trending NNW. Shallow water (200 to 2,000 m) to the east, deep water (about 3,500 m) to west. No image in areas shallower than 200 m. Note that many of the canyons have amphitheatric heads. Paull and others (1984) found brine seeps at the base of the escarpment near 26°N, about 65 km north of this image area. Paull and others (1987) found karst-like features and evidence of spring sapping in the Florida Canyon (near 24° 10′N, 84°05′W). (Photograph courtesy D. Twichell, U.S. Geological Survey.)

crustal areas were investigated using detailed bathymetric mapping, bottom photography, and manned submersibles. Active hydrothermal discharge and sulfide-mineral accumulation was found, as well as places recently active, but now quiescent (Corliss and others, 1979; Francheteau and others, 1979; Williams and others, 1979; Chase and others, 1985; Koski and others, 1985; McConachy and others, 1986). Hydrothermal fields have also been found on the Mid-Atlantic Ridge (Temple and others, 1979; Rona and others, 1984; Rona and others, 1986) and in back-arc basins (Hegerty, 1980; Both and others, 1986). Francheteau and Ballard (1983) suggest that hydrothermal springs along actively spreading ridge crests are to be found about equidistant between transform faults, associated with topographically elevated sections of the axial valley, controlled by heat from shallow sources of magma.

Hydrothermal vents are commonly associated with areas of volcanism, lava flows, and lakes. They are commonly aligned parallel to subparallel with the rift axis—along scarps, faults, or fissures—and they take a variety of forms, including conical or rounded mound-like edifices covered by characteristic fauna, and chimneys venting fluids. Some sites are without surficial deposits, identified only by shimmering water (higher than ambient temperature). Some vents have orifices 10 cm or so in diameter; other vents are diffuse and covered by benthic vent organisms. Active chimneys can discharge black, particulate-laden fluids at temperatures near 350°C (black smokers) or clear fluids usually of lower temperature (Edmund and others, 1979). Ballard and others (1981) found fossil sulfide vents, and low-temperature vents identified only by animal life, along a 24-km segment of the Galapagos spreading center near 21°N. Hekinian and others (1983) mapped a 20-km crustal segment of the East Pacific Rise (13°N) and found more than 80 hydrothermal deposits, 24 of which were active, spaced an average of 100 to 200 m apart. Submersible observations show that chimneys grow quickly and degrade rapidly (Ballard and Francheteau, 1983; McConachy and others, 1986). Hekinian and others (1983) reported the growth rate of one chimney to be 40 cm in 5 days.

Actively discharging vents support communities of extraordinary chemosynthetic benthic organisms, whose metabolism is based on sulfur- or methane-converting bacteria (Corliss and others, 1979; Grassle, 1983; Jannasch, 1983). The megafauna discovered at ridge-crest vents (characteristically large clams and tubeworms) have also been found in subduction zones (Turner and Lutz, 1984; Swinbanks, 1985; Kulm and others, 1986). They have been found associated with discharge of cold, briney groundwater at the base of the West Florida escarpment in the Gulf of Mexico (Paull and others, 1984), with petroleum or methane seeps on the continental slope in the central Gulf of Mexico (Brooks and others, 1985), and with a postulated former seep on the New Jersey continental slope (B. Hecker, personal communication, 1985). The reproductive and disseminative mechanisms of the vent organisms are not well understood, but their global distribution implies that their nourishing fluids are discharged at many, more closely spaced sites than those few that have been discovered.

SUMMARY

Differential pressure gradients can occur in undersea sediment accumulations in spite of the ambient pressures of the ocean. The tectonic environment of a region influences the mechanisms of groundwater activity. For example, gravitational outflow of groundwater is likely to be found along passive margins, compared to dewatering caused by compression along underthrust margins and to hydrothermal circulation in volcanic zones.

Where discharge occurs underwater, the erosional processes take place that are associated with groundwater under subaerial conditions at a seepage face. An important difference is that the immediate removal of eroded material by fluvial transport down a hillslope does not occur underwater as it does on land, but the discharging groundwater will directly remove material from a cliff face by seepage erosion and sapping, and it will enhance rates of other erosional mechanisms such as bottom currents or bioerosion. High gradients of groundwater pore pressures have long been known to reduce slope stability, leading to various forms of mass wasting such as slumps and slides. High pore pressure may also have some effect on increasing the frequency of debris flows and turbidity flows stemming from mass-wasting events, but this is a poorly understood area.

Regressive sea levels create conditions for groundwater drainage at continental margins. Depending on the rate of sea-level change, increased hydraulic head during regression or enhanced artesian conditions during long-term lower sea levels can lead to groundwater overpressure and erosion, thus contributing to mass wasting and submarine canyon formation on the shelves and slopes. The mixture of fresh water and seawater found where subaerially recharged groundwater flows into oceanic waters is a corrosive solution that will rapidly erode a carbonate terrain. Discharge of sulfide-rich brines in water depths as great as 3,500 m or more contributes to the erosion of deep-sea escarpments such as those around Florida, the Bahamas, and Yucatan.

Transmission of load-bearing stresses by interstitial waters is an important aspect of the subduction process, allowing sequences of unconsolidated sediments to be underthrust undisturbed. Dewatering of the sediments in an accretionary prism contributes to mass wasting on trench slopes. Compressional dewatering also causes diapirism and construction of mud volcanoes, especially along the deformation front.

Gravitational dewatering of pore water, constrained by impermeable layers to escape around basement highs, is speculated to cause faulting in otherwise undisturbed flat-lying sediments of abyssal plains, but groundwater influx, rather than efflux, has been measured at some sites. The mobility of submarine groundwater in regions thought to be quiescent is significant to schemes for disposal of radioactive waste.

Hydrothermal circulation and mineral deposition creates elaborate vent structures at ridge crests and probably at other submarine volcanic sites. The hydrothermal fluids support unique communities of chemosynthetic organisms.

The studies cited in this chapter demonstrate that groundwater processes are significant in the submarine environment and should be included in the suite of submarine geologic processes during geomorphologic investigations. The study of submarine landscapes and geomorphic processes has developed as our ability to know the forms of submarine features has advanced. As undersea exploration proceeds over greater areas and to greater resolution of detail, the extent of groundwater-related processes will be better documented. In the future, an understanding of submarine groundwaters and the geomorphic processes of the submarine environment, which is foreign to those of us who grew up as air-breathing land creatures, may help our efforts to explain how the surfaces of planets having heavy atmospheres have evolved.

ACKNOWLEDGMENTS

Charles Higgins has been a marvelously enthusiastic, encouraging, and patient editor during the preparation of this chapter. I thank Brian Tucholke and Robert Belderson for their reviews.

REFERENCES CITED

Anderson, R. Y., and Kirkland, D. W., 1980, Dissolution of salt deposits by brine density flow: Geology, v. 8, p. 66–69.

Arthur, M. A., Carson, B., and Von Huene, R., 1980, Initial tectonic deformation of hemipelagic sediment at the leading edge of the Japan convergent margin, in Initial reports of the Deep-Sea Drilling Project: Washington, D.C., U.S. Government Printing Office, v. 56-57, part 1, p. 569-613.

Back, W., Hanshaw, B. B., and Van Driel, J. N., 1984, Role of groundwater in shaping the eastern coastline of the Yucatan peninsula, Mexico, in LaFleur, R. G., ed., Groundwater as a geomorphic agent: Winchester, Massachusetts, Allen and Unwin, p. 281-293.

Ballard, R. D., and Francheteau, J., 1983, Geological processes of the mid-ocean ridge and their relationship to sulfide deposition, in Rona, P., Bostrom, K., Laubier, L., and Smith, K. L., Hydrothermal processes at seafloor spreading centers; NATO Conference Series, Series 4: Marine Sciences, v. 12, p. 17–25.

Ballard, R. D., Francheteau, J., Juteau, T., Rangin, C., and Normark, W., 1981, East Pacific Rise at 21 degrees N; The volcanic, tectonic, and hydrothermal processes of the central axis: Earth and Planetary Science Letters, v. 55, p. 1–10.

Baltuck, M., McDougall, K., and Taylor, E., 1985, Mass movement along the inner wall of the Middle America Trench, Costa Rica, in Initial reports of the Deep Sea Drilling Project: Washington, D.C., U.S. Government Printing Office, v. 84, p. 551-570.

Barber, A. J., Tjokrosapoetro, S., and Charlton, T. R., 1986, Mud volcanoes, shale diapirs, wrench faults, and mélanges in accretionary complexes, eastern Indonesia: American Association of Petroleum Geologists Bulletin, v. 70, p. 1729-1741.

Belderson, R. H., 1983, Comment on 'Drainage systems developed by sapping on Earth and Mars': Geology, v. 11, p. 55.

Belderson, R. H., Kenyon, N. H., and Stride, A. H., 1978, Local submarine salt-karst formation on the Hellenic outer ridge, eastern Mediterranean: Geology, v. 6, p. 716-720.

Bischoff, J. L., 1969, Red Sea geothermal brine deposits; Their mineralogy, chemistry and genesis, in Degens, E. T., and Ross, D. A., eds., Hot brines and heavy metal deposits in the Red Sea: New York, Springer-Verlag, p. 368–401.

Booth, J. S., 1979, Recent history of mass-wasting on the upper continental slope, northern Gulf of Mexico, as interpreted from the consolidation states of the sediment, in Doyle, L. J., and Pilkey, O. H., eds., Geology of continental slopes: Society of Economic Paleontologists and Mineralogists Special Publication 27, p. 153-164.

Both, R., and 8 others, 1986, Hydrothermal chimneys and associated fauna in the Manus back-arc basin, Papua New Guinea: EOS Transactions of the America Geophysical Union, v. 67, p. 489–490.

Bray, C. J., and Karig, D. E., 1985, Porosity of sediments in accretionary prisms and some implications for dewatering processes: Journal of Geophysical Research, v. 90, p. 768–778.

Brooks, J. M., Kennicutt, M. C., II, Bidigare, R. R., and Fay, R. A., 1985, Hydrates, oil seepage, and chemosynthetic ecosystems on the Gulf of Mexico slope: EOS Transactions of the American Geophysical Union, v. 66, p. 106.

Bucher, W. H., 1940, Submarine valleys and related problems of the North Atlantic: Geological Society of America Bulletin, v. 55, p. 489-511.

Buckley, D. E., and Grant, A. C., 1985, Faultlike features in abyssal plain sediments; Possible dewatering structures: Journal of Geophysical Research, v. 90, no. C5, p. 9173-9180.

Bunce, E. T., and 6 others, 1965, Ocean drilling on the continental margin: Science, v. 150, p. 709-716.

Chase, R. L., and others, 1985, The Canadian American Seamount Expedition 1985, Hydrothermal vents on an axis seamount of the Juan de Fuca ridge: Nature, v. 313, p. 212–214.

Clukey, E. C., Kulhawy, F. H., Liu, P.L.-F., and Tate, G. B., 1985, The impact of wave loads and pore-water pressure generation on initiation of sediment transport: Geo-Marine Letters, v. 5, p. 117-183.

Corliss, J. B., and 10 others, 1979, Submarine thermal springs on the Galapagos Rift: Science, v. 203, p. 1073-1082.

Crane, K., and Ballard, R. D., 1980, The Galapagos Rift at 86 degrees W; 4, Structure and morphology of hydrothermal fields and their relationship to the volcanic and tectonic processes of the rift valley: Journal of Geophysical Research, v. 85, p. 1443-1454.

D'Amore, D. W., 1983, Development of a drainage network completely by groundwater sapping: Geological Society of America Abstracts with Programs, v. 15, p. 176.

Daly, R. A., 1936, Origin of submarine "canyons": American Journal of Science, v. 31, p. 401-420.

Davis, D. M., and Von Huene, R., 1987, Influences on sediment strength and fault friction from structures at the Aleutian Trench: Geology, v. 15, p. 517-522.

Davis, D., Suppe, J., and Dahlen, F. A., 1983, Mechanics of fold-and-thrust belts and accretionary wedges: Journal of Geophysical Research, v. 88, no. B2, p. 1153-1172.

Degens, E. T., and Ross, D. A., eds., 1969, Hot brines and recent heavy metal deposits in the Red Sea: New York, Springer-Verlag, 600 p.

Dillon, W. P., and Oldale, R. N., 1978, Late Quaternary sealevel curve; Reinterpretation based on glaciotectonic evidence: Geology, v. 6, p. 56–60.

Dillon, W. P., Paull, C. K., and Gilbert, L. E., 1985, History of the Atlantic continental margin off Florida; The Blake Plateau Basin, in Poag, C. W., ed., Geologic evolution of the United States Atlantic margin: New York, Van Nostrand Reinhold, p. 189–215.

Doyle, L. J., Brooks, G., and Hebert, J., 1985, Submarine erosion and karstification on the West Florida continental margin; disparate environments and similar features: Geological Society of America Abstracts with Programs, v. 17, p. 565.

Duin, E.J.T., Mesdag, C. S., and Kok, P.T.J., 1984, Faulting in Madeira Abyssal Plain sediments: Marine Geology, v. 56, p. 299–308.

Dunne, T., 1980, Formation and controls of channel networks: Progress in Physical geography, v. 4, p. 211–239.

Edmond, J., 1981, Hydrothermal activity at mid-ocean ridge axes: Nature, v. 290, p. 87–88.

Edmond, J., and 7 others, 1979, Ridge crest hydrothermal activity and the balances of the major and minor elements in the ocean: The Galapagos data: Earth and Planetary Science Letters, v. 46, p. 1–18.

EEZ-SCAN 84 Scientific Staff, 1988, Physiography of the western United States exclusive economic zone: Geology, v. 16, p. 131–134.

Embley, R. W., 1985, A locally formed deep ocean canyon system along the Blanco transform, northeast Pacific: Geo-Marine Letters, v. 5, p. 99–104.

Farre, J. A., 1985, The importance of mass wasting processes on the continental slope [Ph.D. thesis]: New York, Columbia University, 239 p.

Farre, J. A., and Ryan, W.B.F., 1985, Comment on 'Spring sapping on the lower continental slope offshore New Jersey': Geology, v. 13, p. 91–92.

Farre, J. A., McGregor, B. A., Ryan, W.B.F., and Robb, J. M., 1983, Breaching the shelfbreak; Passage from youthful to mature phase in submarine canyon evolution: Society of Economic Paleontologists and Mineralogists Special Publication 33, p. 25–39.

Francheteau, J., and Ballard, R. D., 1983, The East Pacific Rise near 21°N, 13°N, and 20°S; Inferences for along strike variability of axial processes of the Mid-Ocean Ridge: Earth and Planetary Science Letters, v. 64, p. 93–116.

Francheteau, J., and 14 others, 1979, Massive deep-sea sulfide ore deposits discovered on the East Pacific Rise: Nature, v. 277, p. 523–528.

Freeman-Lynde, R. P., 1983, Cretaceous and Tertiary samples dredged from the Florida escarpment, eastern Gulf of Mexico: Gulf Coast Association of Geological Societies Transactions, v. 33, p. 91–99.

Freeman-Lynde, R. P., and Ryan, W.B.F., 1985, Erosional modification of Bahama escarpment: Geological Society of America Bulletin, v. 96, p. 481–494.

Gomberg, D. N., 1977, Neogene karst in the Florida Straits, in Tolson, J. S., and Doyle, F. L., eds., Proceedings of the 12th International Congress on Karst Hydrogeology: University of Alabama at Huntsville Press, p. 213–226.

Grassle, J. F., 1983, Introduction to the biology of hydrothermal vents, in Rona, P., Bostrom, K, Laubier, L., and Smith, K. L., Hydrothermal processes at seafloor spreading centers; NATO Conference Series, Series 4: Marine Sciences, v. 12, p. 665–676.

Grow, J. A., Mattick, R. E., and Schlee, J. S., 1979, Multichannel seismic depth sections and interval velocities over outer continental shelf and upper continental slope between Cape Hatteras and Cape Cod: American Association of Petroleum Geologists Memoir 29, p. 65–83.

Hampson, J. C., Jr., and Robb, J. M., 1984, Geologic map of the continental slope between Lindenkohl and South Toms Canyons, offshore New Jersey: U.S. Geological Survey Miscellaneous Investigation Map I-1608, scale 1:50,000.

Hampton, M. A., 1972, The role of subaqueous debris flows in generating turbidity currents: Journal of Sedimentary Petrology, v. 42, p. 775–793.

Hanshaw, B. B., and Back, W., 1980, Chemical mass-wasting of the northern Yucatan Peninsula by groundwater dissolution: Geology, v. 8, p. 222–224.

Harrington, P. K., 1985, Formation of pockmarks by pore-water escape: Geo-Marine Letters, v. 5, p. 193–197.

Hathaway, J. C., and 8 others, 1979, U.S. Geological Survey core drilling on the Atlantic shelf: Science, v. 206, p. 515–527.

Hay, W. W., and Leslie, M., 1985, Pore space in sediments on continental blocks and sea level change: Geological Society of America Abstracts with Programs, v. 17, p. 605.

Haymon, R. M., and MacDonald, K. C., 1985, The geology of deep-sea hot springs: American Scientist, v. 73, p. 441–449.

Hebert, J. A., 1985, A Miocene karst drainage system; Seismic stratigraphy of the continental shelf west of Florida: Geological Society of America Abstracts with Programs, v. 17, p. 606.

Hegerty, K. A., Jarrard, R. D., and Anderson, R. N., 1980, Detailed multichannel analysis of the structure of a mounds-type hydrothermal area in the Mariana trough [abs.]: EOS Transactions of the American Geophysical Union, v. 61, p. 364.

Hekinian, R., Renard, V., and Cheminee, J. L., 1983, Hydrothermal deposits on the East Pacific Rise near 13° N; Geological setting and distribution of active sulfide chimneys, in Rona, P., Bostrom, K., Laubier, L., and Smith, K. L., Hydrothermal processes at seafloor spreading centers; NATO Conference Series, Series 4: Marine Sciences, v. 12, p. 571–594.

Higgins, C. G., 1982, Drainage systems developed by sapping on Earth and Mars: Geology, v. 10, p. 147–152.

—— , 1984, Piping and sapping; Development of landforms by groundwater outflow, in LaFleur, R. G., ed., Groundwater as a geomorphic agent: Winchester, Massachusetts, Allen and Unwin, p. 18–58.

Hovland, M., 1983, Elongated depressions associated with pockmarks in the western slope of the Norwegian Trench: Marine Geology, v. 51, p. 35–46.

Hovland, M., and Sommerville, J. H., 1985, Characteristics of two natural gas seepages in the North Sea: Marine and Petroleum Geology, v. 2, p. 319–326.

Hubbert, M. K., and Rubey, W. W., 1959, Role of fluid pressure in mechanics of overthrust faulting; 1, Mechanics of fluid-filled porous solids and its application to overthrust faulting: Geological Society of America Bulletin, v. 70, p. 115–166.

Jannasch, H. W., 1983, Microbial processes at deep sea hydrothermal vents, in Rona, P., Bostrom, K., Laubier, L., and Smith, K. L., Hydrothermal processes at seafloor spreading centers; NATO Conference Series, Series 4: Marine Sciences, v. 12, p. 677–711.

Johnson, D. W., 1938–39, Origin of submarine canyons: Journal of Geomorphology, v. 1, p. 111–129, 230–243, 324–340; v. 2, p. 42–60, 133–158, 213–236.

—— , 1939, The origin of submarine canyons; A critical review of hypotheses: New York, Columbia University Press, 126 p.

Jordan, G. F., 1954, Large sink holes in straits of Florida: American Association of Petroleum Geologists Bulletin, v. 38, p. 1810–1817.

Kastens, K. A., and Spiess, F. N., 1984, Dissolution and collapse features on the eastern Mediterreanean Ridge: Marine Geology, v. 56, p. 181–194.

Kennett, J. P., 1982, Marine geology: Englewood Cliffs, New Jersey, Prentice-Hall, Inc., 813 p.

King, L. H., and MacLean, B., 1970, Pockmarks on the Scotian Shelf: Geological Society of America Bulletin, v. 81, p. 3141–3148.

Kirwan, A. D., Doyle, L. J., Bowles, W. D., and Brooks, G. R., 1986, Time-dependent hydrodynamic models of turbidity currents analyzed with data from the Grand Banks and Orleansville events: Journal of Sedimentary Petrology, v. 56, p. 379–386.

Knebel, H. J., 1979, Anomalous topography on the continental shelf around Hudson Canyon: Marine Geology, v. 33, p. M67–M75.

—— , 1981, Processes controlling the characteristics of the surficial sand sheet, U.S. Atlantic outer continental shelf: Marine Geology, v. 42, p. 349–368.

Knebel, H. J., and Spiker, E., 1977, Thickness and age of surficial sand sheet, Baltimore Canyon Trough area: American Association of Petroleum Geologists Bulletin, v. 61, p. 861–871.

Kohout, F. A., 1967, Ground-water flow and the geothermal regime of the Floridian Plateau: Transactions of the Gulf Coast Association of Geological Societies, v. 17, p. 339–354.

Kohout, F. A., and 8 others, 1977a, Fresh ground water stored in aquifers under the continental shelf; Implications from a deep test, Nantucket Island, Massachusetts: Water Resources Bulletin, v. 13, p. 373–386.

Kohout, F. A., Leve, G. W., Smith, F. T., and Manheim, F. T., 1977b, Red Snapper Sink and ground water flow offshore northeastern Florida [abs.], in Tolson, J. S., and Doyle, F. L., eds., Proceedings of the 12th International Congress on Karst Hydrogeology: University of Alabama at Huntsville Press, p. 193–194.

Kohout, F. A., Meisler, H., Meyer, F. W., Johnston, R. H., Leve, G. W., and

Wait, R. L., 1988, Hydrogeology of the Atlantic continental margin, *in* Sheridan, R. E., and Grow, J. A., eds., The geology of North America; The Atlantic continental margin, U.S.: Boulder, Colorado, Geological Society of America, v. I-2, p. 463–480.

Koski, R. A., Lonsdale, P. F., Shanks, W. C., Berndt, M. E., and Howe, S. S., 1985, Mineralogy and geochemistry of a sediment-hosted hydrothermal sulfide deposit from the southern trough of Guaymas Basin, Gulf of California: Journal of Geophysical Research, v. 90, p. 6695–6707.

Kuenen, P. H., 1947, Two problems of marine geology; Atolls and canyons: Koninklijke Nederlandsche Akademie van Wetenschappen, afd. Natuurkunde, 2nd section, v. 43, p. 1–69.

Kulm, L. D., and 13 others, 1986, Oregon subduction zone; Venting, fauna, and carbonates: Science, v. 231, p. 561–566.

Lash, G. G., 1985, Accretion-related deformation of an ancient (early Paleozoic) trench-fill deposit, central Appalachian orogen: Geological Society of America Bulletin, v. 96, p. 1167–1178.

Leahy, P., and Meisler, H., 1982, An analysis of fresh and saline groundwater in the New Jersey coastal plain and continental shelf [abs.]: EOS Transactions of the American Geophysical Union, v. 63, p. 322.

Lowell, R. P., and Rona, P. A., 1985, Hydrothermal models for generation of massive sulfide deposits: Journal of Geophysical Research, v. 90, no. B10, p. 8769–8783.

Malloy, R. J., and Hurley, R. J., 1970, Geomorphology and geologic structure; Straits of Florida: Geological Society of America Bulletin, v. 81, p. 1947–1972.

Manheim, F. T., 1967, Evidence for submarine discharge of water on the Atlantic continental slope of the southern United States, and suggestions for further search: Transactions of the New York Academy of Sciences, series 2, v. 29, p. 839–853.

Manheim, F. T., and Horn, M. K., 1968, Composition of deeper subsurface waters along the Atlantic continental margin: Southeastern Geology, v. 9, p. 215–236.

McConachy, T. F., Ballard, R. D., Mottl, M. J., and Von Herzen, R. P., 1986, Geologic form and setting of a hydrothermal vent field at 10 degrees 56 min N, East Pacific Rise; A detailed study using *Angus* and *Alvin*: Geology, v. 14, p. 295–298.

McGregor, B. A., Stubblefield, W. L., Ryan, W.B.F., and Twichell, D. C., 1982, Wilmington submarine canyon; A marine fluvial-like system: Geology, v. 10, p. 27–30.

Meisler, H., Leahy, P. P., and Knobel, L. L., 1984, Effect of eustatic sea-level changes on saltwater-freshwater in the northern Atlantic Coastal Plain: U.S. Geological Survey Water Supply Paper 2225, 28 p.

Middleton, G. V., and Hampton, M. A., 1976, Subaqueous sediment transport and deposition by sediment gravity flows, *in* Stanley, D. J., and Swift, D.J.P., eds., Marine sediment transport and environmental management: New York, Wiley and Sons, p. 197–218.

Miller, K. G., Mountain, G. S., and Tucholke, B. E,. 1985, Oligocene glacioeustacy and erosion on the margins of the North Atlantic: Geology, v. 13, p. 10–13.

Moore, J. C., and 17 others, 1982, Offscraping and underthrusting of sediment at the deformation front of the Barbados Ridge; Deep Sea Drilling Project Leg 78A: Geological Society of America Bulletin, v. 93, p. 1065–1077.

Moore, J. C., Cowan, D. S., and Karig, D. E., 1985, Structural styles and deformation fabrics of accretionary complexes; Penrose Conference report: Geology, v. 13, p. 77–79.

Morgenstern, N. R., 1967, Submarine slumping and the initiation of turbidity currents, *in* Richards, A. F., ed., Marine geotechnique: Urbana, University of Illinois Press, p. 189–220.

Mullins, H. T., Gardulski, A. F., and Hine, A. C., 1986, Catastrophic collapse of the West Florida carbonate platform margin: Geology, v. 14, p. 167–170.

Nardin, T. R., Hein, F. J., Gorsline, D. S., and Edwards, B. D., 1979, A review of mass movement processes, sediment and acoustic characteristics, and contrasts in slope and base-of-slope systems versus canyon-fan-basin floor systems, *in* Doyle, L. J., and Pilkey, O. H., eds., Geology of continental slopes: Society of Economic Paleontologists and Mineralogists Special Publication 27, p. 61–73.

Nittrouer, C. A., and others, 1988, Sedimentation on continental margins; An integrated program for innovative studies during the 1990s: EOS Transactions of the American Geophysical Union, v. 69, no. 5, p. 58, 59, 67, 68.

Oldale, R. N., 1985, Rapid postglacial shoreline changes in the western Gulf of Maine and the paleo-Indian environments: American Antiquity, v. 50, p. 145–150.

Parker, G., 1982, Conditions for the ignition of catastrophically erosive turbidity currents: Marine Geology, v. 46, p. 307–327.

Paull, C. K., and Dillon, W. P., 1980, Erosional origin of the Blake escarpment; An alternative hypothesis: Geology, v. 8, p. 538–542.

Paull, C. K., and Neumann, A. C., 1987, Continental margin brine seeps; Their geological consequences: Geology, v. 15, p. 545–548.

Paull, C. K., and 9 others, 1984, Biological communities at the Florida escarpment resemble hydrothermal vent taxa: Science, v. 226, p. 965–967.

Paull, C. K., Spiess, F. N., Curray, J. R., and Twichell, D., 1987, Enigmatic submarine box canyons; Florida escarpment [abs.]: EOS Transactions of the American Geophysical Union, v. 68, p. 1316.

Pitman, W. C., III., 1978, Relationship between eustacy and stratigraphic sequences of passive margins: Geological Society of America Bulletin, v. 89, p. 1389–1403.

Poag, C. W., 1985, Depositional history and stratigraphic reference section for central Baltimore Canyon Trough, *in* Poag, C. W., ed., Geologic evolution of the United States Atlantic margin: New York, Van Nostrand Reinhold, p. 217–264.

Popenoe, P., 1985, Seismic-stratigraphic studies of sinkholes and the Boulder Zone under the northeastern Florida continental shelf: Geological Society of America Abstracts with Programs, v. 17, p. 129.

Prior, D. B., and Coleman, J. M., 1986, Submarine slope instability, *in* Brunsden, B., and Prior, D. B., eds., Slope instability: New York, John Wiley, p. 419–455.

Prior, D. B., and Doyle, E. H., 1985, Intra-slope canyon morphology and its modification by rockfall processes, U.S. Atlantic continental margin: Marine Geology, v. 67, p. 177–196.

Rich, J. L., 1941, Review of "The origin of submarine canyons; A critical review of hypotheses," by Douglas Johnson: Journal of Geology, v. 49, p. 107–109.

Ritger, S. D., 1985, Origin of vein structures in the slope deposits of modern accretionary prisms: Geology, v. 13, p. 437–439.

Robb, J. M., 1984, Spring sapping on the lower continental slope, offshore New Jersey: Geology, v. 12, p. 278–282.

——— , 1985, Reply to comment on 'Spring sapping on the lower continental slope offshore New Jersey': Geology, v. 13, p. 92.

Robb, J. M., Hampson, J. C., Jr., and Twichell, D. C., 1981, Geomorphology and sediment stability of a segment of the U.S. continental slope off New Jersey: Science, v. 211, p. 935–937.

Robb, J. M., O'Leary, D. W., Booth, J. S., and Kohout, F. A., 1982, Submarine spring sapping as a geomorphic agent on the east coast continental slope: Geological Society of America Abstracts with Programs, v. 14, p. 600.

Robb, J. M., Schlee, J. S., and Williams, S., 1987, Gloria sidescan-sonar mosaic of the continental margin between Cape Hatteras and Hudson Canyon; Distinct compartments dominated by different processes: Geological Society of America Abstracts with Programs, v. 17, p. 822.

Rona, P. A., 1969, Middle Atlantic continental slope of the United States; Deposition and erosion: American Association of Petroleum Geologists Bulletin, v. 53, p. 1453–1465.

Rona, P. A., and Clay, C. S., 1966, Continuous seismic profiles of the continental terrace off southeastern Florida: Geological Society of America Bulletin, v. 77, p. 31–44.

Rona, P. A., Bostrom, K., Laubier, L., and Smith, K. L., Jr., 1983, Hydrothermal processes at seafloor spreading centers: New York, Plenum Press, 796 p.

Rona, P. A., and 8 others, 1984, Hydrothermal activity at the trans-Atlantic geotraverse hydrothermal field, Mid-Atlantic ridge crest at 26 °N: Journal of Geophysical Research, v. 89, p. 11365–11378.

Rona, P. A., Klinkhammer, G., Nelson, T. A., Trefry, J. H., and Elderfield, H., 1986, Black smokers, massive sulfides, and vent biota at the Mid-Atlantic Ridge: Nature, v. 321, p. 33–37.

Ryan, W.B.F., Stanley, D. J., Hersey, J. B., Fahlquist, D., and Allen, T. D., 1970, The tectonics and geology of the Mediterranean sea, in Maxwell, A. E., ed., The sea, vol. 4, part 2: New York, Interscience Publishers, John Wiley and Sons, p. 387–492.

Scanlon, K. M., and Knebel, H. K., 1989, Pockmarks in the floor of Penobscot Bay, Maine: GeoMarine Letters, v. 9, p. 53–58.

Schultheiss, P. J., and McPhail, S. D., 1986, Direct indication of pore-water advection from pore-pressure measurements in Madeira abyssal plain sediments: Nature, v. 320, p. 348–350.

Schumm, S. A., and Phillips, L., 1986, Composite channels of the Canterbury Plain, New Zealand; A Martian analog?: Geology, v. 14, p. 326–329.

Shepard, F. P., 1981, Submarine canyons; Multiple causes and long-time persistence: American Association of Petroleum Geologists Bulletin, v. 65, p. 1062–1077.

Shepard, F. P., and Emery, K. O., 1941, Submarine topography off the California coast; Canyons and tectonic interpretation. Geological Society of America Special Paper 31, 165 p.

Silver, E. A., Breen, N. A., Prasetyo, H., and Hussong, D. M., 1986, Multibeam study of the Flores backarc thrust belt, Indonesia: Journal of Geophysical Research, v. 91, p. 3489–3500.

Skinner, B. J., 1983, Submarine volcanic exhalations that form mineral deposits; An old idea now proven correct, in Rona, P., Bostrom, K., Laubier, L., and Smith, K. L., Hydrothermal processes at seafloor spreading centers; NATO Conference Series, Series 4: Marine Sciences, v. 12, p. 557–570.

Sleep, N. H., Morton, J. L., Burns, L. E., and Wolery, T. J., 1983, Geophysical constraints on the volume of hydrothermal flow at ridge axes, in Rona, P., Bostrom, K., Laubier, L., and Smith, K. L., Hydrothermal processes at seafloor spreading centers; NATO Conference Series, Series 4: Marine Sciences, v. 12, p. 53–69.

Sloss, L. L., 1984, Comparative anatomy of cratonic unconformities, in Schlee, J., ed., Interregional unconformities and hydrocarbon accumulation: American Association of Petroleum Geologists Memoir 36, p. 1–6.

Stanley, D. J., and Freeland, G. L., 1978, The erosion-deposition boundary in the head of Hudson Submarine Canyon defined on the basis of submarine observations: Marine Geology, v. 26, p. M37–M46.

Stetson, H. C., 1936, Geology and paleontology of the Georges Bank canyons; Part 1, Geology: Geological Society of America Bulletin, v. 47, p. 339–366.

Stetson, H. C., and Smith, J. F., 1938, Behavior of suspension currents and mud slides on the continental slope: American Journal of Science, v. 35, p. 1–13.

Stevens, S. H., and Moore, G. F., 1985, Deformation and sedimentary processes in trench slope basins of the western Sunda Arc, Indonesia: Marine Geology, v. 69, p. 93–112.

Stride, A. H., Belderson, R. H., and Kenyon, N. H., 1982, Structural grain, mud volcanoes, and other features on the Barbados Ridge complex revealed by GLORIA long-range sidescan sonar: Marine Geology, v. 49, p. 187–196.

Swift, D.J.P., Moir, R., and Freeland, G. L., 1980, Quaternary rivers on the New Jersey shelf; Relation of seafloor to buried valleys: Geology, v. 8, p. 276–280.

Swinbanks, D., 1985, Deep-sea clams; New find near Japan's coast: Nature, v. 316, p. 475.

Temple, D. G., Scott, R. B., and Rona, P. A., 1979, Geology of a submarine hydrothermal field, 26 degrees N latitude: Journal of Geophysical Research, v. 84, p. 7453–7466.

Turekian, K. K., 1983, Geochemical mass balances and cycles of the elements, in Rona, P., Bostrom, K., Laubier, L., and Smith, K. L., Hydrothermal processes at seafloor spreading centers; NATO Conference Series, Series 4: Marine Sciences, v. 12, p. 361–367.

Turner, R. D., and Lutz, R. A., 1984, Growth and distribution of molluscs at deep-sea vents and seeps: Oceanus, v. 27, p. 54–62.

Twichell, D. C., and Parson, L. M., 1986, Submarine canyons of the Florida escarpment, eastern Gulf of Mexico [abs.]: Society of Economic Paleontologists and Mineralogists Annual Midyear Meeting Abstracts, v. 3, p. 112.

Twichell, D. C., and Roberts, D. G., 1982, Morphology, distribution, and development of submarine canyons on the United States Atlantic continental slope between Hudson and Baltimore Canyons: Geology, v. 10, p. 408–412.

Twichell, D. C., Grimes, C. B., Jones, R. S., and Able, K. W., 1985, The role of erosion by fish in shaping topography around Hudson Submarine Canyon: Journal of Sedimentary Petrology, v. 55, p. 712–719.

Vail, P. R., and 7 others, 1977, Seismic stratigraphy and global changes of sea level, in Payton, C. E., ed., Seismic stratigraphy; Applications to hydrocarbon exploration: American Association of Petroleum Geologists Memoir 26, p. 49–212.

Valentine, P. C., Uzmann, J. R., and Cooper, R. A., 1984, Submarine topography, surficial geology, and fauna of Oceanographer Canyon, northern part: U.S. Geological Survey Miscellaneous Field Investigation MF-1531, 5 sheets.

Von Huene, R., and Lee, H., 1982, The possible significance of pore fluid pressures in subduction zones, in Watkins, J. S., and Drake, C. L., eds., Studies in continental margin geology: American Association of Petroleum Geologists Memoir 34, p. 781–791.

Warme, J. E., Slater, R. A., and Cooper, R. A., 1978, Bioerosion in submarine canyons, in Stanley, D. J., and Kelling, G., eds., Sedimentation in canyons, fans, and trenches: Stroudsburg, Pennsylvania, Dowden, Hutchinson and Ross, p. 65–70.

Washburn, A. L., 1980, Geocryology: New York, Halsted Press, 406 p.

Westbrook, G. K., and Smith, M. J., 1983, Long décollements and mud volcanoes; Evidence from the Barbados Ridge complex for the role of high pore-fluid pressure in the development of an accretionary complex: Geology, v. 11, p. 279–283.

White, R. S., and Louden, K. E., 1982, The Makran continental margin; Structure of a thickly sedimented convergent plate boundary, in Watkins, J. S., and Drake, C. L., eds., Studies in continental margin geology: American Association of Petroleum Geologists Memoir 34, p. 499–518.

Whiticar, M. J., and Werner, F., 1981, Pockmarks; Submarine vents of natural gas or freshwater seeps?: Geo-Marine Letters, v. 1, p. 193–199.

Williams, D. L., Von Herzen, R. P., Sclater, J. G., and Anderson, R. N., 1974, The Galapagos spreading centre, lithospheric cooling, and hydrothermal circulation: Geophysical Journal of the Royal Astronomical Society, v. 38, p. 587–608.

Williams, D. L., and 5 others, 1979, The hydrothermal mounds of the Galapagos Rift; Observations with DSRV *Alvin* and detailed heat flow studies: Journal of Geophysical Research, v. 84, p. 7467–7484.

MANUSCRIPT ACCEPTED BY THE SOCIETY NOVEMBER 14, 1989

Printed in U.S.A.

Chapter 13

Erosion of seacliffs by groundwater

Robert M. Norris
Department of Geological Sciences, University of California, Santa Barbara, California 93106
with a case study by
William Back
U.S. Geological Survey, 431 National Center, Reston, Virginia 22092

INTRODUCTION

Groundwater acts in several discrete ways to promote erosion of seacliffs. Of these ways, the most widespread is by solution, and the most sensitive materials are carbonate rocks and carbonate cements and, in a few special cases, gypsum. Whenever the groundwater is undersaturated with respect to carbonate, solution will occur. In limestones and dolomites, joint cracks and bedding planes are widened and lengthened, and caves and various other openings and passageways are produced. Solution effects in carbonate-cemented sandstones, conglomerates, and siltstones are typically much less conspicuous and slower, but wherever undersaturated groundwater emerges from a cliff cut in such sedimentary rocks, erosion will occur.

A second way in which groundwater promotes seacliff erosion is by spring sapping, generally regarded as a chiefly mechanical process in which the flow of water from cliff faces undermines portions of the cliff, causing intermittent landsliding. Although mechanical loosening of rock particles may be dominant in true spring sapping, solution of cements and susceptible minerals also plays an important role, and the effects of the two processes are not always clearly separable.

A third way in which groundwater promotes seacliff erosion is perhaps most clearly illustrated by the process called piping (discussed in greater detail in Chapters 3 and 4, this volume). In soils and other unconsolidated or weak surficial deposits, animal burrows, dessication cracks, and channels left by decayed tree roots may provide avenues for water from rain, irrigation, or septic tanks. In soft soils, as well as in weakly consolidated sedimentary rocks, these natural drains may enlarge, lengthen, or coalesce quite rapidly by purely mechanical processes. As such passages become more numerous and larger, they contribute importantly to the erosion of the cliff face. Mechanical effects of this sort are not fundamentally different from such processes as rainwash and sapping, but are distinguished from them here only because the water involved has had some residence time below the surface, albeit a rather short time in most cases. Piping might well be considered a special sort of sapping.

Similarly, emerging groundwater may flow down the face of a seacliff, causing gullying and erosion exactly like that caused by rainwash. Clearly, surface flows and groundwater flows are closely related and have similar consequences.

Groundwater erosion contributes to cliff erosion in several other ways as well, although these are somewhat indirect and are generally of limited importance when all causes of seacliff retreat are considered together.

In many places, groundwater emerging from a seacliff face promotes localized growth of vegetation (Fig. 1). Both the increasing weight of the mass of vegetation and the penetration of growing roots into the rocks causes periodic slab failures, especially in cliffs composed of weak or soft sedimentary rocks.

Weathering caused by sea salt has proved to be a significant weathering process, especially on arid coasts where sea spray is blown onto coastal rocks. In arid and semiarid regions, mineralized groundwaters, undergoing partial evaporation at the seacliff face, have similar consequences, as crystalline efflorescences form around the sites of groundwater discharge. The writer is not aware that this process has been mentioned in the literature, but it should be considered in future investigations.

Rock type is crucially important when considering the role of groundwater in seacliff erosion. It is very clear from the literature that carbonate rocks provide the best and most spectacular examples. Permeable clastic rocks, particularly soft sandstones and siltstones, are much less important so far as obvious effect is concerned, because other erosional processes are apt to outpace groundwater erosion. Shales, unless fractured or interbedded with more permeable beds, are of still less importance as sites for groundwater erosion. In any case, other cliff-forming processes are apt to attack shales so effectively that any effects attributable to groundwater erosion would be completely overwhelmed and thoroughly masked.

Most hard, crystalline igneous and sedimentary rocks—apart from limestones and gypsum—are resistant to solution and spring sapping, but basaltic lavas are commonly highly permea-

Norris, R. M., 1990, Erosion of seacliffs by groundwater, with a case study by Back, W., *in* Higgins, C. G., and Coates, D. R., eds., Groundwater geomorphology; The role of subsurface water in Earth-surface processes and landforms: Boulder, Colorado, Geological Society of America Special Paper 252.

Figure 1. Seacliff west of Santa Barbara, California. Cliff is about 30 m high and cut in Tertiary and Quaternary sedimentary rocks. A prominent line of grass marks the emergence of perched water along a shale parting.

ble, and in areas with abundant groundwater, may provide fine examples of alcove development resulting from spring sapping. Stearns (1936) described such features from the Snake River Canyon of Idaho, noting that some alcoves are 100 m high, 600 m wide, and nearly 3 km long. At Thousand Springs, he reports flows of about 150 million l/sec. While no comparable flows are known to the writer from existing seacliffs, some very large flows are known in Hawaii and are described more fully later in this chapter.

SEACLIFFS IN CARBONATE ROCKS

The effects of groundwater erosion of coastal cliffs are best illustrated by the behavior of carbonate rocks. In many tropical areas where such rocks are widespread, the origin of low cliffs or notching at sea level has been the subject of intense debate. The majority of such features have been attributed to the effects of solution by most workers, many of whom have pointed to the close relation of these features to sea level and to the prevalence of roughened, hackly surfaces in such nips or notches (Higgins, 1980). Despite the fact that study after study has shown that tropical seawater is saturated or supersaturated with respect to calcium carbonate, many investigators, faced with the compelling evidence for solution, have found it necessary to devise ingenious schemes to reconcile that fact with the obvious evidence of solution. Some have suggested nocturnal temperature changes that would permit the increased uptake of carbon dioxide (Fairbridge, 1948). Revelle and Emery (1957) suggested that an increase in carbon dioxide might result from cessation of marine photosynthesis during the hours of darkness. Cloud (1965) cautioned that more and better evidence was needed before solution could be accepted as more effective than the work of intertidal organisms. Folk and others (1973) described the development of phytokarst in humid tropical climates, in which blue-green algae (cyanobacteria) attack carbonate rocks and gypsum, and locally contribute to seacliff erosion.

It now appears that the geochemists have rescued us from such complexity. Plummer (1975), for example, demonstrated that with respect to calcium carbonate the contact of saturated or even supersaturated groundwater with saturated or supersaturated seawater results in a mixture that is undersaturated. Moreover, because of the density difference between fresh and salt water, the zone of mixing and diffusion may remain relatively stable for long periods, allowing the effects of solution to persist in a narrow zone, governed by the typical Ghyben-Herzberg relation (Fig. 2). Back and others, in a number of papers dealing with the Yucatan Peninsula of Mexico, document this process, as does Back's case study at the end of this chapter (Hanshaw and Back, 1980a, b; Back and others, 1979; Back and Hanshaw, 1983).

Solution effects in carbonate rocks are particularly prominent where there is an outflow of fresh groundwater from a Ghyben-Herzberg lens and where the marine environment is sheltered from vigorous wave activity. As Back notes in his case study, the most active solution effects take place underground where the encroaching seawater first comes in contact with the emerging fresh water; solution effects in open waters are insignificant. In such places, a thin lens of brackish water may develop and persist at the sea surface. This lens floats on seawater because of its lower density, and in very sheltered areas may diffuse or mix quite slowly with the underlying normally saline seawater. Even in small, low, coral islands or atolls, discharge of fresh or brackish water must be more or less continuous unless there are substantial withdrawals of groundwater from wells.

If solution by normally saline marine waters were the sole or chief cause of nips and notches, one would expect such features to be prominent and numerous on limestone coasts that are bathed by temperate or cold waters. This does not appear to be the case, and the observations of Higgins (1980) in Greece are instructive. He noted that, although much of the Greek shoreline is cut in calcareous rocks, nips and notches are infrequent. After examining many kilometers of the coast, he found only three places where nips were present. The common characteristic of the three areas is the discharge of fresh groundwater close to sea level, although all three areas are reasonably sheltered from vigorous wave activity as well. Measurements of the salinity of the surface waters in these three areas show a consistent pattern of dilution. Although not reporting nips, notches, or other prominent features, Guilcher (1958) described small corrosion hollows and shallow submarine lapies in some limestone coasts in cold and temperate seas. He also noted that similar features have been observed along lake shores in the Laurentian Shield of Canada.

On the other hand, Davies (1980) described extreme development of notching in some tropical limestone coasts where exaggerated visors result from solution effects. His best examples are found on coasts with relatively weak wave action and small tidal range. Davies refers to the process as "water level weathering," a term invented by Wentworth (1938), and notes that conditions that favor its development include: the presence of permeable, coarsely bedded rocks; high evaporation; low tidal range; and conditions that discourage abrasion and rapid quarrying. Wentworth (1938) described a number of examples of this type of weathering from the tropical Pacific, and Guilcher (1958) reported that well-developed limestone visors are especially prominent only in the warmer seas, perhaps because the rock is usually a porous coralline limestone, citing examples from Hawaii, the Red Sea coast, Western Australia, Madagascar, Bermuda, Fiji, and elsewhere. In some instances, this sort of prominent notching completely surrounds small, isolated sea stacks, resulting in mushroom-shaped remnants. Obviously, a small mushroom rock of this sort cannot hold enough groundwater to sustain a Ghyben-Herzberg lens, so it seems highly probable that close examination of the settings in which these mushroom-shaped rocks occur will reveal that most or all of them form in relatively sheltered waters in which a lens of brackish water on the sea surface is sustained by a nearby discharge from a Ghyben-Herzberg lens or possibly by fresh surface waters.

Although most investigators today would agree that shore nips and notches in calcareous coasts are caused by solution, some workers, recognizing that normal seawater, particularly in the tropics, is saturated or supersaturated with calcium carbonate, have suggested marine abrasion or organic activity as the cause of these features.

Marine abrasion is unlikely for several good reasons. First, most shores where well-developed notches occur are areas of relatively quiet water with very modest wave activity. Second, except in exposed areas subject to storm battering, calcareous rocks rarely produce much pebbly or bouldery rubble, which is needed for abrasion to be effective (Guilcher, 1958). Further, these nips and notches typically show roughened or hackly surfaces with small, delicate detail that could not survive in an environment of abrasion (Higgins, 1980). To some extent, these remarks demolish a straw man, for few knowledgeable investigators today would entertain abrasion as a generally significant process in the development of nips and notches in calcareous coasts. The comments are included here only for the sake of completeness.

Biochemical and biomechanical processes have also been suggested as the cause of nips and notches (Neumann, 1968). There can be no question that some organisms are capable of attacking rocks, particularly soft calcareous ones, but the association of capable organisms and notches is not nearly consistent enough to establish cause and effect. Emery (1962) reminded us that the mere presence of a concentration of organisms in a notch is not proof that the organisms produced the notch, even though they might have been capable of doing so; the organisms may simply prefer such settings and congregate there, regardless of how the notch formed.

Higgins (1980) reviewed in some detail the entire question of groundwater influence on carbonate coasts and persuasively argued that the discharge of groundwater plays an important, if relatively unappreciated, role in the formation of nips and notches. In that paper, however, he does not consider the enhanced solvent effect of the brackish water zone, resulting from diffusion and mixing along the interface between seawater and fresh groundwater.

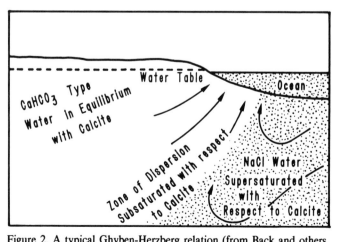

Figure 2. A typical Ghyben-Herzberg relation (from Back and others, 1984).

SEACLIFFS IN VOLCANIC ROCKS

Groundwater erosion of seacliffs is not confined to those formed from calcareous rocks, although such coasts do provide the clearest and most prominent examples. Along volcanic coasts, there are many places where prominent springs emerge from seacliffs, although, as a rule, their contribution to erosion is apt to be much slower than other cliff-forming processes and therefore inconspicuous. Stearns and Macdonald (1946) described several places on the island of Hawaii where lines of springs emerge from the face of a cliff about 300 m high. These springs occur along two thin tuff layers in basaltic rocks (Fig. 3). Moreover, these authors describe a number of large, reentrant, theater-headed valleys, extending inland from the northeast-facing coast of this island. They note that flows of groundwater from the headwalls of these valleys can be as much as 45 million l/day. Although they did not suggest that groundwater played a role in forming the features, it seems probable to the writer that the steep-walled, flat-floored valleys are caused in large measure by spring sapping. Some of these valleys extend inland as much as 8 km, and have nearly vertical headwalls as high as 750 m. Although one can scarcely label these deep canyons as seacliffs today, it is probable that they once were, when repeated Quaternary rises in sea level flooded the valley floors and initiated deposition of the alluvium that now fills them (Stearns and Macdonald, 1946).

The island of Hawaii thus appears to provide a variety of examples of groundwater erosion of volcanic seacliffs. At one end of the continuum are the steep seacliffs with modest outflows of groundwater, illustrated in Figure 3, where marine abrasion and landsliding dominate the erosion. At the other extreme are such features as Waipio Canyon on the northeastern coast where between 85 and 115 million l/day of groundwater is discharged by springs flowing from the headwalls of the canyon. That such large flows occur is no surprise, considering the high permeability of the lava flows and the fact that Kohala Mountain above receives an annual rainfall of more than 5,000 mm in most years. Monthly rainfall amounts range from 300 to 900 mm (Stearns and Macdonald, 1946). Waipio Canyon and other theater-headed gorges on this part of the Hawaiian coast are considered by the writer to be features that result where spring sapping proceeds faster than marine abrasion, leading to the development of deep reentrant valleys with copious discharges of groundwater from their headwalls. Numerous intermediate examples occur on other islands in the Hawaiian group, especially along the northeast coasts facing the trade winds. There is no doubt that formation of these valleys involves some normal stream erosion, but surface streams in Hawaii are apt to be small in relation to their valleys and very flashy with respect to flow, whereas groundwater discharge is more constant.

SEACLIFFS IN CLASTIC SEDIMENTARY ROCKS

My observations of seacliff retreat along part of the southern California coast suggest that even in this semiarid area, nonmarine agencies, collectively, may account for almost half the total retreat (Norris, 1985). Southern California is generally not a very storm-wracked area; parts of the coast, particularly in the Santa Barbara area where most of my observations have been made, are sheltered to considerable degree by the east-west orientation of the mainland coast and the presence of a chain of high mountainous islands offshore. Direct wave attack is proportionally more important on the more exposed west-facing shores to the north and south.

Figure 3. Sketch of springs emerging from seacliffs between Waimanu and Honopue Canyons, island of Hawaii. Springs are localized by two thin tuff layers. Cliff is about 300 m high. From Stearns and Macdonald, 1946.

Nevertheless, even on the relatively sheltered Santa Barbara coast, marine erosion remains dominant as shown by the persistent steepness of the seacliff. Were the nonmarine agencies dominant, material would accumulate at the cliff base, and the slope would gradually flatten.

The chief nonmarine agencies involved in seacliff retreat have been discussed elsewhere (Norris, 1968, 1985), and include: (1) direct rain wash; (2) various forms of weathering, including that caused by sea salt; (3) the activities of organisms, ranging from the growth of plants and cyanobacteria to the burrowing of owls and gophers; (4) piping; and (5) groundwater sapping.

Not only do we have few analyses of the relative importance of marine and nonmarine causes of seacliff retreat, except in the most general way, but we have even less information on the relative importance of the various nonmarine processes. Furthermore, erosion by piping and the activities of organisms are not always clearly separable from groundwater erosion; there is a considerable overlap of process.

These various causative agents are mentioned because the activities of groundwater in promoting erosion of sedimentary rocks need to be understood in context; for the southern California area, at least, groundwater is very seldom the dominant fac-

Because most southern California coastal cliffs are weakly cemented or thin-bedded shales and siltstones of Tertiary or Quaternary age, emerging groundwater readily weakens the rocks by removing interstitial cement, dislodging grains by water pressure, and encouraging the growth of vegetation near the discharge point, which adds weight to the cliff face and also wedges rocks apart as root growth progresses. Observations in the Santa Barbara area are believed representative of the southern California setting generally, and the following examples are taken from there.

Where seacliff retreat is not so rapid as to prevent it, vegetation becomes established on the cliff face wherever groundwater emerges (Fig. 1). For small flows, vegetation may effectively use the entire flow, but if there be any surplus, it will flow down the cliff face below the vegetation and contribute to erosion. In those few areas where flows are as much as 5 to 10 l/min, spring sapping will begin and produce alcove features rather like those known in karst or volcanic areas. If continued, such sapping can produce large features, provided other erosional processes are less active. Sites where active spring sapping occurs are marked by relatively lush patches of vegetation, which are most obvious during the dry summer and fall months. It is probable that many of the short, steep-sided canyons that cut the seacliffs of southern California are due at least in part to intermittent spring sapping, although most of these canyons would be largely the product of normal stream erosion.

Near Santa Barbara, a kind of alcove was eroded beneath a porch on the top of the seacliff as a result of a broken underground drain, which was allowed to go unrepaired for some months (Fig. 5). Although it might be argued that this example is not formed by groundwater in a strict sense, it nonetheless shows how rapid erosion can be where water flows from a cliff face.

Figure 4. Small cliff-face spring emerging from above a thin shale parting, seacliff at More Mesa, west of Santa Barbara, California. Spring is about 3 m above the beach.

tor. Nevertheless, despite the semiarid climate of southern California, examples of groundwater emerging from seacliffs are not difficult to find (Fig. 4). In many cases, flows today may exceed flows before the intensive urbanization of the southern California coastal strip occurred. This is a summer-dry region, and garden and lawn watering may be the equivalent of 1,800 to 2,000 mm/yr of rain. Perhaps even more significant, a number of coastal communities are not sewered, and considerable wastewater is disposed of in septic tank systems; 1,800 to 3,000 l/day per residence is not exceptional. Much of this wastewater becomes shallow groundwater, and where there are permeable zones that communicate with nearby seacliffs, wastewater may be more or less continuously discharged at the cliff face. So far as is known to the writer, there have been no studies quantifying this source of coastal groundwater, particularly with respect to the amount of naturally occurring groundwater.

CASE STUDY: GEOMORPHIC CONSEQUENCES OF THE GROUNDWATER MIXING ZONES ALONG THE EASTERN YUCATAN COAST, MEXICO

William Back

A dynamic mixing zone occurs in the carbonate aquifer along the Yucatan coast because of heavy rainfall (1,000 to 1,500 mm/yr) that rapidly infiltrates the extremely permeable limestone and flows toward the Caribbean, where it discharges as layers of fresh and brackish water above the salt-water interface in coastal areas. From a geochemical study of the Xel Ha caleta, Back and others (1979) concluded that mixing of fresh and saline water and consequent dissolution of limestone are not occurring in open water of the caleta as originally thought, but are occurring in an underground mixing zone before the brackish water discharges (Fig. 6). Their modeling showed that dissolution of limestone does not occur in open bodies of water. The high P_{CO_2} causes outgassing of carbon dioxide to be a reaction more rapid than dissolution of calcite, and the water therefore quickly becomes supersaturated with respect to calcite when it comes in contact with lower P_{CO_2} in the atmosphere.

Figure 5. Alcove-like opening formed beneath a concrete porch at Isla Vista, west of Santa Barbara, California. Excavation took place in about 6 months as a result of a broken stormwater drain. Cliff is about 10 m high.

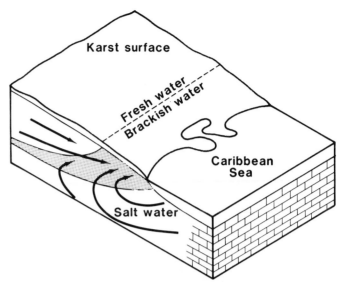

Figure 6. Diagram showing fresh groundwater flow toward coast where it mixes with encroaching seawater to form a mixing zone from which the blended brackish water discharges.

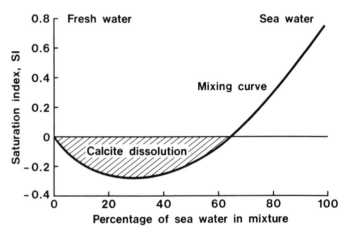

Figure 7. Graph showing typical change of saturation index (SI) for calcite due to simple mixing of seawater with fresh water that is in equilibrium with calcite. SI = log equilibrium constant for calcite/activity of Ca^{++} and $CO_3^=$ in water. Positive values indicate supersaturation. Negative values indicate subsaturation.

The dispersive mixing zone formed by seawater encroachment and groundwater discharge is an extremely active geochemical environment. For example, it is known from thermodynamic principles that when two waters with different dissolved-solids content are mixed, the resulting solution may be sub- or supersaturated with respect to one or more minerals with which they had been in prior equilibrium. Supersaturation with respect to carbonate minerals results from the nonlinearity of mineral solubility with regard to differences in temperature, salinity, partial pressure of carbon dioxide, and ionic activity (Fig. 7).

A cave system is an ideal site for testing the hypothesis of mixing-zone dissolution because caves provide underground access for sampling a range of salinities that occur in the mixing zone. It is possible to obtain samples of water in caves before its contact with the atmosphere, and therefore before carbon dioxide outgasses. It is easy to observe this mixing zone: a layer of fresh brackish water floats on top of more saline water within caves and causes a change in refraction of light between the layers.

The opportunity for observation of dissolution was provided in the sea-level cave of Xcaret about 30 km south of Cancun. Additional testing of the ideas presented here was provided by study of caves in the Balearic Islands, Spain (Herman and others, 1989). The calculations and interpretation based on chemical analyses of water with a range of salinity, obtained from Xcaret Cave, demonstrate that the water is saturated with respect to calcite; the differential dissolution is readily observable within the

Figure 8. Aerial photograph showing two of a series of crescent-shaped beaches that were formed largely by groundwater dissolution and extend along the coast for more than 100 km.

cave (Back and others, 1986). Above the water level, the walls of the cave are quite smooth, with no effect of differential dissolution. Within the mixing zone, however, differential dissolution is quite dramatic and produces extremely high porosity, which causes much of the limestone to have the general appearance of Swiss cheese.

Based on these theoretical considerations and field observations, we hypothesize that evolutionary development of the coastline is as follows (Back and others, 1984). After fracturing of the limestone, groundwater that was flowing from the inland part of the peninsula was channeled along the dominant fractures in its effort to discharge along the coast. This flowing water mixed with seawater that had encroached into the aquifer. The mixing formed a zone of dispersion from which the blended water discharged. The major flow, and therefore the major dissolution, occurred along the fractures with the widest openings; this led to establishing major discharge where dominant fractures intersected the coast. These discharge points are locations of incipient coves. The area of discharge and groundwater flow pattern became a self-perpetuating mechanism that expanded the area of dissolution to form a branching network of subsurface solution channels, which coalesced to form cave systems whose orientation was controlled by fracture patterns. As the dissolution continued, it decreased support for the 1- to 2-m-thick slab of limestone, the bottom of which was the cave roof and the top of which was land surface. Eventually, there remained no support for the roof of the cave, and it collapsed. This process is occurring at the present time. The roof blocks submerge into the zone of dispersion where they are dissolved and eroded as the caleta is enlarged. This enlargement opens the lagoon to wave action that permits physical erosion and provides a habitat for marine organisms, which cause further erosion by biological activity. These processes continue until the lagoon is an open body of water separated from adjacent lagoons by a headland that is gradually eroded by wave action until the coastline develops a serrated configuration resulting from the coalescence of the many crescent-shaped beaches (Fig. 8).

Geologic significance of subsaturation can be emphasized by remembering that every marine limestone now containing fresh water has been subjected to dissolution and diagenesis caused by the mixing-zone phenomenon at least once. Most have probably undergone the effects of this process repeatedly. For example, at any stand of sea level, the zone of dispersion will occupy a certain position within the limestone aquifer. As sea level drops, the zone of dispersion will follow the lowering sea level, thereby subjecting additional carbonate rocks to these processes. Subsequent rise of sea level will permit the zone of dispersion to migrate back up through the aquifer and cause the limestone to undergo diagenesis and differential dissolution.

The position and extent of the mixing zone are controlled not only by sea-level changes, but also by the position and elevation of the water table. Therefore, on a shorter time scale, the position of the salt water–fresh water interface will migrate in response to changes of the water level due to climatic variations, either seasonally or long term, and withdrawal by pumping.

In addition to the mixing zone being a contributing factor to the formation of caves, lagoons, crescent-shaped beaches, nips, and notches, as discussed previously, it may be a significant factor in the sea cliff retreat along some carbonate coasts. For example, Mona Island off the west end of Puerto Rico is characterized by steep sea cliffs with huge blocks of limestone lying on the beach area. A thin lens of fresh water floats on seawater that extends beneath the island (Jordan, 1973). I suggest that the collapse of these blocks is caused by a process similar to that of the collapse of cave roofs in the Yucatan. Dissolution of the limestone and dolomite within the mixing zone, which occurs near sea level along the shores, has removed the supporting material, and blocks fall. It is also possible that in some areas, sea stacks represent the more resistant portions of such a sequence, and the shoreline has retreated by a similar process, leaving the sea stacks to be eroded by dissolution of the brackish surface water, as suggested by the writer herein.

CONCLUSIONS

Groundwater is believed to play a ubiquitous, if mostly minor, role in seacliff erosion. In most instances, the rate of seacliff retreat depends chiefly on other processes, such as marine abrasion at the cliff base; however, the fact that seacliffs continue to retreat, even where their bases are shielded from direct wave attack by artificial protective structures, widening beaches, or perhaps a drop in relative sea level, clearly shows that the nonmarine agencies, including groundwater sapping and solution, are significant.

The effects of emerging groundwater in promoting coastal erosion are most widespread and most obvious in permeable calcareous rocks, particularly in tropical regions where such rocks are widespread. It is now evident that brackish waters, derived from the mixing of normal seawater and groundwater, even where both may be saturated or supersaturated with calcium carbonate, can produce an undersaturated mixture that, in favorable settings, can effectively and rapidly dissolve limy materials where the mixing takes place, producing an array of solutional landforms.

Coastal cliffs cut in basaltic lava flows also yield notable examples of groundwater sapping. Owing to the extreme permeability of such lava flows, where conditions provide abundant groundwater, enormous springs may develop, resulting in rapid erosion by sapping, with the development of reentrant, steep-walled, theater-headed valleys.

Non-shaly clastic sedimentary rocks are exposed in many seacliffs and likewise provide examples of spring sapping, but because of their generally lower permeability, copious discharges of groundwater are rare, and examples of sapping are much less conspicuous and are apt to be largely masked by other erosional processes.

REFERENCES CITED

Back, W., and Hanshaw, B. B., 1983, Effect on sea-level fluctuations on porosity and mineralogic changes in coastal aquifers, *in* Cronin, T. M., Cannon, W. F., and Poore, R. Z., eds., Paleoclimate and mineral deposits: U.S. Geological Survey Circular 822, p. 6–7.

Back, W., Hanshaw, B. B., Pyle, T. E., Plummer, L. N., and Weidie, A. E., 1979, Geochemical significance of groundwater discharge and carbonate solution to the formation of Caleta Xel Ha, Quintana Roo, Mexico: Water Resources Research, v. 15, p. 1521–1535.

Back, W., Hanshaw, B. B., and Van Driel, J. N., 1984, Role of groundwater in shaping the eastern coastline of the Yucatan Peninsula, Mexico, *in* Lafleur, R. G., ed., Groundwater as a geomorphic agent: Boston, Massachusetts, Allen and Unwin, p. 281–293.

Back, W., Hanshaw, B. B., Herman, J. S., and Van Driel, J. N., 1986, Differential dissolution of a Pleistocene reef in the ground-water mixing zone of coastal Yucatan, Mexico: Geology, v. 14, p. 137–140.

Cloud, P. E., 1965, Carbonate precipitation and dissolution in the marine environment, *in* Riley, J. P., and Skirrow, G., eds., Chemical oceanography: London, Academic Press, p. 127–158.

Davies, J. L., 1980, Geomorphic variation in coastal development, 2nd ed.: London, Longman, 212 p.

Emery, K. O., 1962, Marine geology of Guam: U.S. Geological Survey Professional Paper 403B, 76 p.

Fairbridge, R. W., 1948, Notes on the geomorphology of the Pelsart Group of the Houtman's Abrolhos Islands: Journal of the Royal Society of Western Australia, v. 34, p. 35–72.

Folk, R. L., Roberts, H. H., and Moore, C. H., 1973, Black phytokarst from Hell, Cayman Islands, British West Indies: Geological Society of America Bulletin, v. 84, p. 2351–2360.

Guilcher, A., 1958, Coastal and submarine morphology: London, Methuen, 274 p.

Hanshaw, B. B., and Back, W., 1980a, Chemical mass wasting of the northern Yucatan Peninsula by groundwater dissolution: Geology, v. 8, p. 222–224.

—— , 1980b, Chemical reactions in the salt water mixing zone of carbonate aquifers: Geological Society of America Abstracts with Programs, v. 12, p. 441–442.

Herman, J. S., Back, W., and Pomar, L., 1989, Spelegenesis in the groundwater mixing zone; The coastal carbonate aquifers of Mallorca and Menorca, Spain: Barcelona, Spain, 9th International Congress of Speleology Proceedings, v. 1, p. 13–15.

Higgins, C. G., 1980, Nips, notches, and the solution of coastal limestone; An overview of the problem with examples from Greece: Estuarine and Coastal Marine Science, v. 10, p. 15–30.

Jordan, D. G., 1973, A summary of actual and potential water resources, Isla de Mona, Puerto Rico, *in* Mona and Monita Islands: Puerto Rico Environmental Quality Board, v. 2, p. 1–8.

Neumann, A. C., 1968, Biological erosion of limestone coasts, *in* Fairbridge, R. W., ed., Encyclopedia of geomorphology: New York, Reinhold, p. 75–80.

Norris, R. M., 1968, Sea cliff retreat near Santa Barbara, California: California Division of Mines and Geology Mineral Information Service, v. 21, p. 87–91.

—— , 1985, Southern Santa Barbara County; Gaviota Beach to Rincon Point, *in* Griggs, G., and Savoy, L., eds., Living with the California coast: Durham, North Carolina, Duke University Press, p. 250–278.

Plummer, L. N., 1975, Mixing of seawater with calcium carbonate groundwater, *in* Whitten, E.H.T., ed., Quantitative studies in the geological sciences: Geological Society of America Memoir 142, p. 219–236.

Revelle, R. R., and Emery, K. O., 1957, Chemical erosion of beach rock and exposed reef rock: U.S. Geological Survey Professional Paper 260-T, p. 699–709.

Stearns, H. T., 1936, Origin of the large springs and their alcoves along the Snake River in southern Idaho: Journal of Geology, v. 44, p. 429–450.

Stearns, H. T., and Macdonald, G. A., 1946, Geology and ground water resources of the Island of Hawaii: Hawaii Division of Hydrography Bulletin 9, 363 p.

Wentworth, C. K., 1938, Marine bench-forming processes; Waterlevel weathering: Journal of Geomorphology, v. 1, p. 6–32.

MANUSCRIPT ACCEPTED BY THE SOCIETY NOVEMBER 14, 1989

Geological Society of America
Special Paper 252
1990

Chapter 14

Seepage-induced cliff recession and regional denudation

Charles G. Higgins
Department of Geology, University of California, Davis, California 95616
  with case studies by
W. R. Osterkamp
U.S. Geological Survey, MS 413, Box 25046, Denver Federal Center, Denver, Colorado 80225
Charles G. Higgins
Department of Geology, University of California, Davis, California 95616

INTRODUCTION

Groundwater seepage consists chiefly of intergranular flow, as opposed to the channelized throughflow involved in piping (Howard, 1988). *Seepage erosion* results from entrainment and transport of loose grains by effluent flow in seeps and springs. In contrast, *seepage weathering,* as used here, refers to the effects of a group of processes that are concentrated at sites of seepage moisture. These processes intensify the breakdown of rock materials in the seepage zones, preparing them for removal by erosional agents. Where the loosened debris is carried away from a springhead or from the base of a slope, the undermined area fails and the slope retreats. This undermining is called *sapping.*

In Chapter 11 of this volume, Victor Baker and his colleagues describe how a combination of localized seepage weathering and erosion and the resultant spring sapping can produce steep-headed alcoves and box canyons, and how these processes can affect the development of larger valleys and drainage networks. In Chapter 12, James Robb applies these concepts to the development of some submarine features on the continental shelves. In Chapter 13, Robert Norris sketches the contributions of diffuse rather than concentrated groundwater seepage in the development and retreat of some seacliffs. In this chapter, I discuss how diffuse seepage combined with other agencies can result in large-scale recession of terrestrial cliffs, with resultant regional denudation.

SEEPAGE EROSION, SEEPAGE WEATHERING, AND CLIFF SAPPING

Some examples; Small features

The development of valleys and cliffs solely by seepage erosion and sapping can be illustrated during a falling tide on some beaches, where the outflow of beach groundwater may etch miniature drainage networks in the foreshore. These little features are small-scale replicas of larger landforms (Higgins, 1982, 1984). Figures 1 and 2 show two examples, formed entirely by groundwater outflow within a few hours on a beach near San Diego, California.

Where for any reason groundwater outflow is concentrated, spring sapping may help develop a headward-growing gully or valley. This effect is shown in the upper left of the miniature drainage system of Figure 1. There, subsurface flow was diverted around a large boulder, visible at left center. In contrast, where groundwater emerges more uniformly along a slope, the result is a line of water-table seeps. The resulting diffuse outflow and erosion may sap the valley walls, causing them to recede in more or less parallel lines as backwearing cliffs. This can be seen at the top and right of Figure 1, where relatively nonconcentrated outflow had formed a scalloped, embayed escarpment. Below the little cliff, the braided outflow channels, little terraces, and abandoned watercourses all are consequences of late-stage erosional development with a falling water table (Higgins, 1984).

Fallen blocks of sand along the cliff are also a consequence of cliff sapping. The cliff is weakened by saturation of the sand at its base, and its strength is further sapped as groundwater outflow at the base entrains the grains and carries them away, down the braided channelways. This undermines the cliff, so that the cohesive damp sand above fails by slumping or toppling. As the fallen blocks then become saturated, they themselves disintegrate and disappear. All stages of wasting of such blocks can be seen along the little cliff in Figure 1.

Such failures are even better shown in Figure 2, a close view of another miniature cliff developed entirely by outflow of groundwater on the same beach. Near the head of the alcove at left, two arcuate slumps are seen to be wasting away as outflow entrains grains from their saturated lower edges. Just left of cen-

Higgins, C. G., 1990, Seepage-induced cliff recession and regional denudation, with case studies by Osterkamp, W. R., and Higgins, C. G., *in* Higgins, C. G., and Coates, D. R., eds., Groundwater geomorphology; The role of subsurface water in Earth-surface processes and landforms: Boulder, Colorado, Geological Society of America Special Paper 252.

ter, two other disintegrating masses appear to have been slab failures. Undermining and collapse cause the cliff to recede, maintaining a near-vertical, backwearing slope.

The chief cause of the undermining in both of these examples is *seepage erosion* as defined by Hutchinson:

... when groundwater discharges at a free face and the seepage drag may be large enough to dislodge individual particles of the soil, thus permitting their removal. This phenomenon is largely confined to soils in the coarse silt to fine sand range. . . .The resultant back-sapping tends to undermine the superincumbent strata and produce their eventual collapse (Hutchinson, 1968, p. 691).

A different example of small-scale retreating cliffs was described by Reeves and Reeves (1971) in the channel seepage flats of the Osage Plains of north-central Texas. There, where stream incision is impeded by a local perched water table, groundwater outflow at the base of the valley walls causes undercutting and slumping of overlying soft materials. These are later removed by surface sheet wash. The walls then retreat as low cliffs to produce wide valley flats. The effects of this seepage erosion may be aided by salt crystal wedging, as the flats are uniformly associated with gypsum crusts.

Some examples; Larger features

The significance of the little cliffs, alcoves, and fallen blocks described above is that they resemble, and may serve as analogs for, larger landforms developed in more resistant materials. For example, Figure 3 is an aerial view of an embayed escarpment at the edge of the Danakil Basin, part of the Afar Depression of Ethiopia. The general form of the cliffs, with braided outflow channels at their base, closely resembles the scene in Figure 1. However, these cliffs are considerably larger; the photographer recollects that the flight height was 500 m or more and that the 35-mm camera had a 50-mm lens (Georg Gerster, personal communication, 1984), yielding a view that is about 350 m wide. Despite the difference in scale, the presence of active springs along the base of the cliffs suggests that the cliffs and alcoves are formed in large part by seepage effects akin to those of the little beach features.

Related are some "seepage steps" on valley sides in parts of the New Forest, Hampshire, England, that have been described by Tuckfield (1973). These are steep little ledges with scarps 1 to 6 m high in Paleogene clayey sandstones, bordered below by low convex debris slopes. Tuckfield attributes these features to freeze-thaw action and progressive landslips that result from concentrated seepage of groundwater. When observed in the field, many of the debris slopes were noticeably moist, and boreholes showed the water table to be just below the base of the steps. In some cases the water table appeared to be controlled by clay/sand lithologic contacts, but this was not true everywhere, and Tuckfield concluded that "the position of the step is determined directly by the level of the water table and indirectly by the clay-sandstone junction" (Tuckfield, 1973, p. 367). He observed

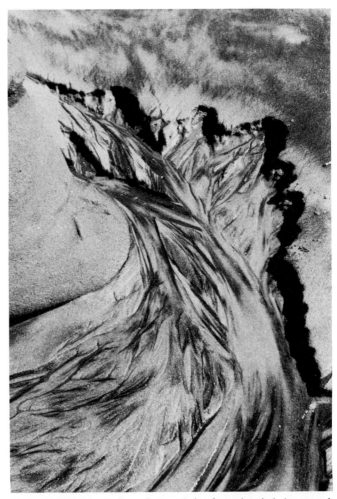

Figure 1. Miniature drainage features being formed entirely by groundwater outflow during a falling tide on the foreshore of Bermuda Avenue Beach, San Diego, California, 2 February 1971. Width of view is about 1 m. At upper left an extended valley has formed where subsurface flow is concentrated around a boulder. To the right, relatively diffuse flow has resulted in an "escarpment" with alcoves. This photograph was taken 30 minutes after the one reproduced in Higgins (1984, Fig. 2.7; also Higgins and others, 1988, Fig. 11), and shows some late-stage modifications.

active recession of some scarps—one eroded back 12 cm in four years—but he thought that others are now relict.

In a related occurrence on hillsides near Marietta, Ohio, Jacobson and Pomeroy (1987) noted that sandstone ledges are maintained by seepage-induced slope failures in the underlying red mudstone. Similar hillside steps were observed by Townshend (1970) in the northeastern Mato Grosso, Brazil. Erosion-pin studies confirmed the steps' active recession, which Townshend attributed to seepage-induced sheetwash and gullying at the base of the steps.

Still larger examples formed in consolidated sandstone are provided by the cliffs of Chaco Canyon National Park, New Mexico. Figure 4 is an air view of Mockingbird Canyon, a tributary to Chaco Canyon. Its scalloped walls resemble, in a general way, the miniature cliffs developed on the beach and in the

Figure 2. Active seepage and sapping at the base of a miniature cliff and its alcoves, Bermuda Avenue Beach, April 1971. Width of view is about 0.5 m. Note slumps and fallen masses of sand along the "cliffs."

Danakil Basin, and it is tempting to conclude that they were formed in a similar manner. This notion is strengthened by the fact that Mockingbird Canyon exhibits other characteristics of valleys formed chiefly by seepage erosion and sapping, with theater-headed alcoves and little evidence of overflow at the valley heads (see Chapter 11, this volume).

Kirk Bryan (1928) was the first to recognize that groundwater seepage and sapping have played a major part in the development of undercuts, or "niches," as he called them, at the base of the cliffs of Cliff House Sandstone in Chaco Canyon, part of which is visible at right in the distance in Figure 4. The development of these niches, with the resulting loss of support, has led to the opening of stress-release joints, the collapse of large blocks, and the retreat of the cliff face.

The well-recorded fall of Threatening Rock at Pueblo Bonito in Chaco Canyon in 1941 (Schumm and Chorley, 1964) was one consequence of this cliff-retreat process, where stress-release jointing helped set the stage for infiltration and cliff failure. An earlier blockfall just behind Pueblo Bonito is clearly shown in a low-oblique aerial photograph taken in 1929 by Charles A. Lindberg and reproduced in Wells and others (1983, p. 218). Other blockfalls have been documented there by archaeological studies, which show that "the fall of the sandstone blocks is a geologically frequent event in Chaco Canyon" (Schumm and Chorley, 1964, p. 1052). A high-oblique photograph of some cliffs about 2 km downstream, reproduced in Wells and others (1983, p. 210), shows several large niches or caves at the base of the cliff with a jumble of fallen blocks in front, clearly illustrating the progression of cliff recession.

Another low-oblique aerial photograph by Lindberg, reproduced in Wells and others (1983, p. 240), shows a similar jumble of rockfall blocks and debris at the base of a cliff near the Wijiji ruins in Chaco Canyon, 7 km upstream from the mouth of Mockingbird Canyon. This site is seen in the far left distance in Figure 4. At Fajada Butte, visible in the far right distance in Figure 4, a prehistoric blockfall from a shallow alcove or niche at the base of the upper unit of the Cliff House Sandstone provided the setting for the noted Anasazi solar markers (Newman and others, 1982). A more recent rockfall there "indicates that the process of disaggregation is continuing" (Newman and others, 1982, p. 1036).

Another example of seepage-influenced cliff development is provided by the Goshen Hole Rim escarpment, part of which is illustrated in Figure 5. This cliff borders the Goshen Hole depression in eastern Wyoming. It is formed of resistant Miocene Arikaree Formation sandstone overlying soft Oligocene White River Formation siltstone. Osterkamp (1987, p. 189) lists five indications that escarpment retreat there has occurred "principally by spring sapping, seepage erosion, and related processes": internal drainage of the upland surface (preventing overwash of the cliffs), the impossibility of fluvial undercutting of the cliffs by local perennial streams, linear trends of topographic features (indicating joint control of seepage-induced sapping), high groundwater transmissivities in the siltstone, and the presence of springs and seeps at the base of the scarp. He attributes the erosion in part to

Figure 3. Outflow from springs at the base of an embayed escarpment of the Danakil Basin, Afar Depression, Ethiopia. Width of view is about 350 m. From Gerster (1976, Fig. 30); aerial photograph copyright by Georg Gerster, reproduced with permission.

Figure 4. Box canyons and cliffs attributable to seepage and sapping in Chaco Culture National Historical Park, 150 km northwest of Albuquerque, New Mexico. Aerial view toward SSE; width of view in foreground is about 2.5 km. Mockingbird Canyon is in foreground; Chaco Wash and Fajada Butte are in distance. Main Chaco Canyon is to the right of this scene. Photograph by Evelyn Newman, U.S. Geological Survey.

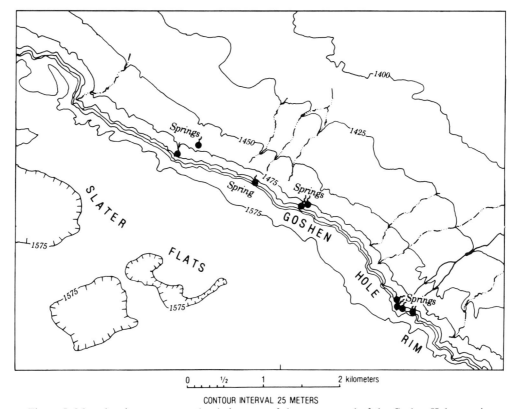

Figure 5. Map showing escarpment bordering part of the western end of the Goshen Hole area in Wyoming. Note springs clustered along the base of the escarpment. Osterkamp (1987) attributes the scarp to sapping where springs and seepage weathering are concentrated in Oligocene siltstone underlying Miocene sandstone caprock. After Osterkamp (1987); modified from U.S. Geological Survey 7½-minute Dickinson Hill Quadrangle map, 1963.

joints and stress-release fractures that provide avenues for groundwater flow. Seepage piping and dissolution of calcium carbonate cement in the White River Formation also help to undermine the cliff and lead to collapse.

Additional examples of seepage-induced cliffs are cited in the discussion and case studies below.

The processes of seepage weathering and cliff sapping

Even though the canyons and cliffs of the American Southwest and elsewhere do resemble the miniature features formed by seepage erosion on some beaches, the sapping process in consolidated rock cannot be the same as that which is effective in damp sand. Simple entrainment and transport of loose grains by effluent water—the process here called *seepage erosion*—is not sufficient. In rock the process must be more complex. There, the seepage serves not as the primary sapping agent, but as a facilitator of a group of preparatory processes that must first disaggregate the material so that some other agent or group of agents can remove the weathering products at the base of the slope. Thus, undermining at the base of the cliffs and alcoves requires first a concentration of chemical and mechanical weathering that is a consequence of the localized seepage.

H. E. Gregory was one of the first to recognize this relation. In northern Arizona he noted that "seeps . . . serve to disintegrate the rock and to increase the amount of fine waste available for transportation" (Gregory, 1917, p. 130). In niches at the heads of box canyons, he noted that waterfall overflow does not even wet the walls, but that after a storm,

water may be seen to ooze from the porous rock or to trickle from bedding planes. . . . Efflorescence of lime or alkalai or salt coating the recessed wall indicates that the supply of water is intermittently renewed. By . . . weathering processes the wall of a canyon is horizontally notched, cavities are formed, the rim is undermined, and the cliff recedes (Gregory, 1917, p. 133).

Gregory observed similar effects of groundwater seepage in forming the large cliff-base niches used as rock shelters by early inhabitants of the region. There, he attributed much of the weathering of the rock to seepage solution of the cement.

Kirk Bryan proposed a similar origin by "differential sapping" for niches associated with rock shelters in Tertiary arkosic conglomerate near Phoenix, in southern Arizona, stressing solution of the cement with resultant "crumbling of the back wall and . . . scaling of the roof" (Bryan, 1925, p. 93). In the Cretaceous sandstone of Chaco Canyon he noted that dissolution is accompanied by other seepage-induced weathering processes: "The crystallization of salts on the surface portion of a rock due to the evaporation of water that has seeped from the interior undoubtedly tends to disrupt the rock" (Bryan, 1928, p. 131). He also credited "the disruptive action of frost" as one of the seepage effects that produce "concentrated weathering or differential sapping."

In a passage that presages the discussion of subsurface flow nets by Terzaghi and Peck (1948, Fig. 207; see Dunne, Chapter 1, this volume), Bryan explained how seepage might be concentrated at niche sites:

However feeble the cause which locates the original slight cavity, once initiated the existence of this cavity provides a shorter outlet for the water from an increased volume of the rock. The niche, as it grows, becomes the place of minimal distance of flow for the water that has soaked into a continuously larger mass of rock (Bryan, 1928, p. 132).

In a posthumous paper, Bryan discussed both the stratigraphic control of the niches and the removal of the weathered detritus:

Erosion of the lower part of the Cliff House sandstone is largely due to differential sapping at its base. Rainwater entering the sandstone above emerges below. If the base is above the level of the floodplain, the water emerges at the top of the friable sandstones, coal, or shale of the somewhat variable underlying Menefee formation. This material is decomposed and carried away partly by this seepage water and partly by direct rainwash. As a result the cliff is undermined and blocks break off along characteristic joint planes.... [Where] the base of the sandstone is below the level of the alluvial plain, as in the north wall of the canyon near Pueblo Bonito... water absorbed by the overlying sandstone emerges at or near the level of the alluvial plain. It dissolves the cement of the rock and appears as an efflorescence of a white salt. The sandstone becomes friable and grains are loosened from the surface. These grains fall by gravity, especially during windstorms, or are loosened and carried off by the sheet of water that covers the face of the rock in rains. Thus cavities or niches... are formed.

With the formation of these cavities the rock splits on its characteristic vertical joints; as loosened blocks fall, the vertical face of the cliff is renewed. Narrow slabs several hundred feet long, partly loosened, are fairly common features, and one directly back of Pueblo Bonito has excited much interest because massive masonry below it shows that the prehistoric peoples attempted to brace the slab against falling (Bryan, 1954, p. 19).

This slab did fall, on January 22, 1941, as documented by Schumm and Chorley (1964), whose analysis of records of its earlier slow movement suggests that it had been sliding away from the cliff on a moistened shale surface:

These data indicate that the movement of the rock was an exponential function of total precipitation and that the rate of movement was seasonal, being relatively... slow during the summer and relatively fast during the winter. Frost action and wetting of the shale by snow melt are considered to be important factors causing movement of the blocks (Schumm and Chorley, 1964, p. 1041 and 1053).

Possibly the first to call for seepage frost action in the origin of niches and arches in sandstone was F. S. Dellenbaugh (1898), who observed the process in the Canyon of Desolation, along the Green River. Schumm and Chorley (1966) conducted weathering experiments on six different cliff-forming sandstones of the Colorado Plateau region, and found intensified disaggregation during freeze-thaw periods.

The importance of frost action in basal sapping has also been cited by Robinson (1970), who wrote that it and "dissolution along seeps" promote the mechanical disintegration of the Navajo sandstone and the undercutting of arches and columns in Zion Canyon, Utah. A similar role of needle ice was observed by Sharp (1976) in Oklahoma on the west flank of a ridge of Ordovician chert outcrops. He found the effects "confined to groundwater discharge areas" and concluded that they depend "upon high precipitation followed by freezing conditions," a sporadic occurrence, but frequent enough that "this mechanism may be the predominant form of mass wasting in local areas" (Sharp, 1976, p. 487). In Wales, Lawler (1984) found that bank recession along the River Ilston was greatest between December and April, "the period of most intense freeze-thaw action." In-situ freezing and needle-ice development there were related to "vertical distribution of temperature... moisture availability and bank material grain size, bulk density and porosity" (Lawler, 1984). The chief role of this frost action, as it is elsewhere, is to disrupt the bank material and prepare it for removal by other agencies.

Laity's (1983) petrographic study of the Navajo Sandstone at a seepage face in Zion National Park, Utah, showed that behind an efflorescent crust of gypsum and sandy calcareous material, the sandstone is so impregnated and expanded by fine-grained calcite that all grain-to-grain contact has been lost as the sand grains have been wedged apart. This decreases the strength of the rock and promotes spalling and scaling of thin sheets. She suggested that winter freeze-thaw cycles enhance removal of weakened sheets by icicle plucking. Earlier, Strahler (1963) had also noted the role of granular disintegration by salt-crystal growth in developing the niches of the southwestern United States. Nocita (1987) has pointed out that such salt-crystal weathering operates in three different ways; not just by the wedging action of the crystal growth, but also by forces generated by hydration and thermal expansion of the salts. He suggested that the latter is the most important in hot, arid environments.

Howard and Kochel (1988) have suggested that such "salt fretting" is responsible for alveolar weathering, or tafoni, and that these features are commonly associated with and best developed in niches or alcoves in Utah because "surface runoff [inhibits] salt fretting simply by solution and removal of salts brought to the surface by evaporating groundwater" (Howard and Kochel, 1988, p. 28). They identify several factors that control the intergranular deposition of minerals by evaporating seepage: type and concentration of salts, rate and variance of water discharge to the surface, distribution of pore sizes and their interconnectivity, presence and size of fractures, temperature regime at the rock face, and frequency of wetting of the rock face by rain or submergence. They further suggest that tafoni are simply minor "dry" sapping features, whereas "large alcoves are generally associated with a more regional groundwater flow and exhibit faces that are at least seasonally wet" (p. 29). Thus the occurrence of tafoni on the walls of large niches indicate that the "wet" sapping responsible for the development of the niches has largely ceased, most likely because of climate change.

Except for the studies cited above, there has been little detailed research on the processes involved in cliff-base weathering and sapping. Where the rock or cement is calcareous, as in the Chalk downs of England, the chief process may be dissolution, or simply "the continual action of springs" (Small, 1961, p. 80), or "spring sapping" (Sparks, 1960, p. 139). Elsewhere, authors have called for a variety of chemical and mechanical processes (e.g., Smith, 1978) or have stressed one particular process such as solution and leaching of calcareous cement (e.g., Campbell, 1973). Most of these studies have been largely descriptive and general. Until we have improved knowledge of the mechanics involved, our understanding of this important geomorphic phenomenon will remain sketchy and inadequate.

In summary, seepage erosion by itself can fashion miniature landforms in unconsolidated sediments, but this is not sufficient in lithified materials. There, the seepage serves chiefly to promote disaggregation by concentrated weathering followed by selective erosion at the base of the cliff or alcove. This combination of processes, including both the preparatory weathering and subsequent erosion is here called *seepage weathering*. At most places where it occurs, the weathering probably results from a number of interactive processes promoted by the localized moisture. These may include intensified chemical weathering, solution, and leaching within the seepage zone, coupled with enhanced mechanical weathering of the periodically damp or saturated rock face. The physical effects may include granular disintegration, splitting, spalling, or flaking owing to wetting/drying, heating/cooling, chemical alteration, salt-crystal wedging, root wedging and animal activity, rainbeat, and in temperate to cold regions, various forms of congelifraction, including needle-ice wedging.

Where the hydraulic gradient is insufficient and the seepage outflow is too feeble to entrain and carry away the weathered particles, erosional removal must owe to other agents such as wind, slope wash, and channeled rill and stream wash. The combined effect of all of these is to sap the base of a cliff so that it retreats by backwearing.

The relative roles of seepage and surface wash

Some landscapes, such as the miniature beach examples cited above, are formed entirely by seepage-induced sapping; others are formed entirely by surface-runoff processes. Between these extremes, many landscapes are the result of a combination of seepage and surface wash. In such cases the effects of runoff may overshadow those of seepage, and it may be impossible to separate and evaluate the role played by subsurface water in cliff evolution and retreat.

H. C. Stetson seems to have been the first to propose extensive retreat of scarps "through the agency of groundwater sapping (Stetson, 1936, p. 353), and he was certainly the first to suggest that similar processes might be responsible for topography on the continental shelf (see Chapter 12, this volume). To illustrate this, he reproduced a high-oblique aerial photograph by Barnum Brown of an escarpment "about 15 miles southeast of Cameron, Arizona, showing the processes at work which may have produced similar-appearing submarine valleys on Georges Bank" (Stetson, 1936, Plate 1, Fig. 2). That photograph is reproduced here as Figure 6.

The escarpment pictured is part of the Tloi Eechii Cliffs, formed in east-dipping sandstones of the Triassic Chinle Formation. It is 30 to 120 m high, with a sinuous, embayed face. This sinuosity may be a consequence of the gentle inclination of the nearly flat-lying strata and of the large groundwater-recharge and surface-runoff area provided by Ward Terrace to the east. The role of sapping in the development of these cliff has been questioned because the high drainage density of the lower slopes makes it appear that they are eroded chiefly by surface wash. Stetson offered no field observations on active processes there.

In their sinuosity and fine erosional texture these cliffs resemble the embayed eastern caprock escarpment of northern Texas, discussed in Osterkamp's case study below, and illustrated there in Figures 8, 9, and 10. The high degree of dissection of the caprock escarpment reflects in part the regional eastward dip of the capping Tertiary strata and the extensive groundwater-recharge area provided by the Southern High Plains. However, as noted by Gustavson and Simpkins (1989), maintenance and recession of this escarpment owes not just to surface runoff or to groundwater sapping, but to a combination of processes, including sheetwash, rillwash, stream erosion, piping, slope failures, subsurface salt dissolution, and sapping induced by springs and seeps. The authors found active seepage and sapping at the base of sandstone units along the cliffs, and reiterated an earlier statement that "geomorphic processes related to ground-water discharge, such as springs and seeps, dominate landform development on the southern Great Plains" (Gustavson and Simpkins, 1989, p. 2), but they did not attempt to weigh the relative importance of the various processes responsible for escarpment development. Doubtless all play a part, and all affect the cliffs in different ways.

At both the Tloi Eechii Cliffs and the caprock escarpment, the fine-textured erosional topography of the lower slopes suggests the effects of surface wash. Some of the sheetwash and rill erosion may result from groundwater outflow from permeable units in the cliffs, as observed in Brazil by Townshend (1970), but much or most may be from overland runoff. Nevertheless, the angular appearance of the upper parts of the cliffs and their abrupt contact with the edge of the plateau suggest that they are periodically freshened by slope failures induced by cliff-base seepage weathering and sapping. Without sapping to produce cliff retreat, the cliffs would simply wear down in place. Without surface wash to carry away the fallen material from the base of the cliffs, the cliffs would be buried by their own debris and

Figure 6. Aerial view of embayed Tloi Eechii escarpment southeast of Cameron, Arizona (at top center of photo), attributed to "groundwater sapping" by Stetson (1936, p. 353), but more likely a result of both seepage weathering and surface wash. Painted Desert to left and Ward Terrace to right. Width of view in foreground is about 9 km. Photograph by Barnum Brown, 1934, negative no. 126931, courtesy Department of Library Services, American Museum of Natural History, New York.

retreat would cease. Both processes play essential roles in the recession of the escarpments.

RATE AND EXTENT OF SCARP RETREAT

The studies cited above show that cliff undermining as a consequence of seepage effects can promote cliff retreat, at least locally. Howard (1988) reminds us that in consolidated rocks, scarp recession is limited by the rate of weathering at the cliff face, particularly at the seepage face. The rate of such weathering, the appearance of the scarp, and the distance that it may recede depend on several factors, including climate, lithology, stratigraphy, structure, and the erosional contributions by surface wash. Some of these variables are considered in a simulation model by Howard and Hornberger (1987). Howard has also suggested that "a reasonable process assumption might be that sapping rates are proportional to a power of seepage rates, with, perhaps, a critical, threshold seepage rate below which no sapping occurs" (Howard, 1988, p. 4). Recession rates may also be expected to vary through time with changes in climate and other variables such as scarp height (Smith, 1983), groundwater depletion, and dissolution-induced subsidence (Gustavson and Simpkins, 1989).

Effects of climate

Bryan (1954) acknowledged that cliff retreat at Chaco Canyon has been slow, and that little seems to have happened there since the canyon was inhabited by the Anasazi Indians. That occupation lasted from about A.D. 400 to 1300 (Sofaer and others, 1979), so sapping activity there may have been only minimal for the last thousand years or so. However, Bryan suggested that "when past climates were wetter than the present but still relatively arid, cliff recession was doubtless accelerated" (Bryan, 1954, p. 18). Such conditions may have occurred in late Pleistocene or earliest Holocene time, before about 8,000 years ago, when the climate of the Southwest is thought to have been less arid (Van Devender and Spauling, 1979).

Jennings' (1979) observations of the Arnhem Land plateau in northern Australia similarly suggest that seepage weathering and cliff retreat there is no longer as vigorous as in the past, and that the landscape is now mainly relict. The same conclusions can be drawn from observations in many other regions, such as the Gilf Kebir plateau in southwestern Egypt (case study below), and the Colorado Plateau, where Ahnert's (1960) studies of cliff sites suggest that recession was chiefly active during Pleistocene "pluvial climates." Howard and Kochel (1988) suggest that development of alveolar tafoni on niche walls indicates that backwasting there has largely ceased in the present climate, and that seepage-aided sapping was more active during the late Pleistocene. Similarly, Smith (1978) found that limestone cliffs sapped by basal seepage in the northwestern Sahara "are not weathering under present climatic conditions" (p. 41), and so "must have originated during a former, presumably moister, period" (p. 40).

Even where one can observe that cliff recession was more pronounced in the past, one cannot be sure about the conditions that fostered it. The efficacy of seepage-induced recession in humid, or "pluvial" climates has been questioned. Under humid conditions, increased cliff-base vegetation could both dewater the seepage zone and protect it from the destructive effects of salt-crystal and needle-ice growth, thereby effectively stabilizing the cliffs. On the contrary, more humid conditions could simply produce differences in the cliff-forming processes, with greater emphasis on chemical decay by organic acids, vegetative root-wedging, seepage-induced slope failure and slumping, and enhanced removal of fallen debris by surface runoff.

Bryan suggested that recession may have been optimal under conditions moister (and perhaps somewhat cooler) than now, but still semiarid. Then, both salt- and ice-crystal growth could disrupt the rock in the seepage zone more effectively. Colder temperatures alone, or a change not in total precipitation but in its intensity or seasonality might also serve to increase recession rates. In short, although seepage and cliff recession appear to have been more pronounced at many sites in the past, we can only surmise that increased moisture was a major factor.

The apparent diminution, or even lack, of modern seepage activity thus has several consequences for geomorphic studies. For one, we cannot now observe and evaluate the cliff-forming processes at their full strength and effectiveness. For another, we do not know the optimal conditions for seepage-induced recession. Our understanding of the processes involved and their results must be based on studies of the remaining sites where some seepage and sapping still occur, although with diminished vigor. Some of the latter are preserved along the Southern High Plains escarpments of Texas and New Mexico, as discussed by Gustavson and Simpkins (1989) and in Osterkamp's case study below.

Effects of lithology, stratigraphy, and structure

The alcoves of the Snake River in southern Idaho are developed by seepage weathering in basalt flows (Stearns, 1936), and the Saharan cliffs discussed by Smith (1978) are of limestone, but most of the examples of cliff sapping cited above occur in porous sedimentary rocks. The great scarps of the American Southwest are mainly of sandstone (Bryan, 1925; Schumm and Chorley, 1966), and most are underlain by shale (Howard and Kochel, 1988).

Ahnert (1960) found considerable variation in cliff morphology at several sites on the Colorado Plateau. He attributed this to differences in seepage action and sapping that resulted from such factors as groundwater catchment area and storage volume, thickness of the permeable sandstone unit, presence or absence of an impeding stratum at the base of the sandstone, and whether or not the base is exposed or buried. Bryan (1954) had commented on the latter factor, as noted above, and both writers observed that scarp development also depends on the dip of the strata with respect to the cliff face. Ahnert cited similar observations in western Texas by Mortensen (1953), who found that a deeply embayed and dissected "Achterstufe," or "back scarp," is formed where the dip is toward the cliff face. Back scarps also tend to be more pronounced than the straighter "Frontstufen," or "front scarps" that form where the strata dip into the cliff face. The strongly embayed caprock escarpment of the Southern High Plains, discussed in Osterkamp's case study below, is an example of a back scarp. The Tloi Eechii Cliffs resemble a back scarp although they are technically a front scarp. Their strong dissection and embayment probably owe to the very gentle inclination of the strata and the large recharge/drainage area above them.

Ahnert noted that sandstone back scarps in the Zuñi Mountains are "deeply indented by valleys whose heads . . . have a broad semicircular shape with steep, often vertical rear walls . . . [that] greatly resemble those created by contact springs and seepage emerging along the bedding plane between pervious rock and underlying weak impervious rock in humid central Europe" (Ahnert, 1960, p. 144). Similar observations on the effect of dip direction on valley development have been made by Laity and Malin (1983) in the Glen Canyon region (summarized in Laity's case study, Chapter 11, this volume) and by Howard and Kochel (1988, p. 35–41).

In a study of sandstone cliffs at Canyonlands National Park, Nicholas and Dixon (1986) found that differences in scarp retreat rates at promontories and re-entrants are mainly controlled by joint spacing and orientation. The strata's compressive strength, slake-durability, porosity, and cementation seem to have only minor influence.

Oberlander (1977) has suggested that cliff morphology in the Colorado Plateau region is more complex than previous authors have considered. He identified at least six different kinds of cliffs, or "walls," developed in the Jurassic Entrada Sandstone of the Arches region in southeastern Utah. There he found that development of alcoves and major cliff faces is commonly related to thin shale partings in the Dewey Bridge Member at the base. Although it would seem that these shales should form aquicludes and promote seepage weathering, Oberlander observed that "there is no spring action associated with these alcoves, now or in the past" (1977, p. 100). Instead, he attributed much of the undercutting to surface erosion of some unspecified kind. Oberlander's observations suggest that at some sites seepage may play only a minor role in cliff development and/or that the traces of seepage weathering may be effaced after the climate changes and seepage is no longer effective.

Scarp retreat rates

In some of the studies cited above, the cliffs form the walls of canyons, where they cannot have retreated very far. In other cases, as in the great sandstone escarpments of the American Southwest discussed by Ahnert and Oberlander, scarp retreat may have covered many kilometers.

The Tloi Eechii escarpment lies 3½ to 8 km from the Little Colorado River. The cliffs must have retreated this distance if their development was initiated by incision of the river. The Adeii Eechii Cliffs, which lie to the east of the view shown in Figure 6 and are formed in sandstones of the Glen Canyon Group, may have developed in the same manner. If so, they have retreated as much as 15 km from the river. According to Lucchitta (1979), the most recent entrenchment of the Colorado River and its major tributaries began less than 5.5 m.y. ago. If cliff recession began at that time, the recession rate of the Tloi Eechii cliffs has averaged 0.64 to 1.5 m/1,000 yr (meters per thousand years) or more. If recession of the Adeii Eechii cliffs began at the same time, their rate has averaged at least 2.7 m/1,000 yr. These figures would be larger if, as would have been likely, cliff recession did not begin until river incision reached the water table or the base of the sandstone units. However, the figures are consistent with other estimates of drylands cliff-recession rates of 0.2 to 2 m/1,000 yr reported by Cole and Mayer (1982, Table 1).

Considerably greater distances have been proposed for retreat of the eastern caprock escarpment of Texas. Gustavson and his co-workers have proposed that these cliffs "may have retreated as much as 322 km (200 mi) since the end of deposition [of the Ogallala Formation] late in the Pliocene Epoch" (Gustavson and others, 1981, p. 418). This yields an average rate of retreat of 110 m/1,000 yr. Using other data, Gustavson and Simpkins (1989) calculate rates of 180 m/1,000 yr since Kansan time, or 120 to 140 m/1,000 yr since the deposition of a terrace at an archaeological site dated at 8,000 to 9,500 B.P. Their erosion-pin studies indicate modern rates of 10 to 20 m/1,000 yr, which are compatible with the long-term rates if one considers that spring discharge there "has decreased as much as an order of magnitude over the last 50 yr" (Gustavson and Simpkins, 1989, p. 1).

In his case study below, Osterkamp reviews these and other estimates of retreat of the caprock escarpment, and concludes that the recession must have been somewhat slower: 12 to 45 m/1,000 yr for the eastern escarpment and about 42 m/1,000 yr for the northern one. These figures are consistent with the erosion-pin studies and seem reasonable for long-term rates, in part because most other calculations are based on the assumption that scarp recession began at the limit of deposition and covered the entire distance to the present cliffs. This may have been the case at the northern escarpment, which seems to have been formed by incision of the Canadian River and to have retreated only 18 to 50 km to its present position. However, development of the eastern escarpment may have been more complex.

The Rolling Plains to the east are creased with stream valleys (see Fig. 7 in Osterkamp's case study), and it seems likely that scarp retreat there was actually begun by the incision of a number of these east-flowing streams, including the Red, White, and Pease Rivers, and their north- and south-flowing tributaries. The resulting valley walls then receded not only to the west, but also to the north and south, consuming the uplands between the streams but leaving outlying remnants far to the east of the present escarpment.

In their general model for relative rates of scarp recession, Howard and Kochel (1988) note that a predicted very rapid rate of horizontal retreat for low dips "does not occur because the escarpment becomes segmented by erosion along drainage lines into isolated mesas and buttes whose local relief, distance from the main escarpment, and rate of backwasting increase through time." Gustavson and Simpkins (1989) discuss and illustrate stages in such a progressive development of the eastern caprock escarpment, where segments have been eroded unevenly to produce a highly crenulated rim and detached outliers (see Fig. 10 in Osterkamp's case study). Such scarp initiation by stream incision implies that the present cliffs have not retreated all the way from the limit of Ogallala deposition, but only a fraction of that distance. If so, the rate of scarp retreat must have been relatively modest—perhaps even less than Osterkamp's figures indicate. In any case, this process of piecemeal cliff sapping and scarp retreat seems to have removed the Ogallala Formation from a large area.

Elsewhere, as in the Western Desert of Egypt and Libya (see my case study below), it is similarly difficult to determine rates of retreat of cliffs that were initiated by drainage-network incision. However, the results are the same—namely, regional denudation on a large scale by lateral removal of the cliff-forming materials.

CASE STUDY: SEEPAGE WEATHERING AND SAPPING OF THE SOUTHERN HIGH PLAINS ESCARPMENTS, TEXAS AND NEW MEXICO

W. R. Osterkamp

The Southern High Plains of Texas and New Mexico is a regional landform whose geomorphic development has possibly been controlled as strongly by groundwater processes as has that of any large-scale feature of North America. Although eolian processes continue to modify the plateau surface, the general form, erosion rates, and processes of canyon encroachment appear to be determined largely by the movement of groundwater. Both deep and relatively shallow systems of groundwater circulation have contributed to the geomorphic evolution of the Southern High Plains (Gustavson and Simpkins, 1989).

Setting and characteristics

The Southern High Plains of Texas and New Mexico is a remnant plateau, about 80,000 km^2 in area (Fig. 7), the surface of which is partly determined by the southernmost remaining deposits of the Ogallala Formation. The plateau grades gently southward into the Edwards Plateau, and on the north and west is bounded by escarpments of up to 200 m relief that rim the

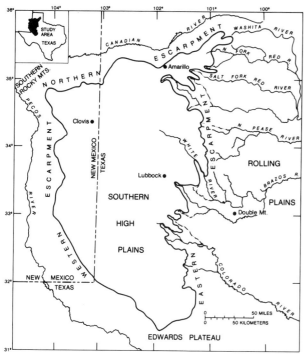

Figure 7. Map of Southern High Plains, escarpments, and surrounding features, with Rolling Plains to the east.

Canadian and Pecos River valleys. The eastern escarpment, which in places approaches a height of 300 m above the Rolling Plains to the east, is indented by tributaries of the east-flowing Red, Pease, Brazos, and Colorado Rivers.

The upper drainage networks of these rivers are poorly developed on the Southern High Plains, where all streamflow is ephemeral and channels occur mostly as narrow, elongate draws trending southeast. Most of the plateau lacks integrated drainage, and most adjacent through-flowing channels do not share common drainage divides, but may be separated by 50 km or more of undissected plains (Gustavson and Finley, 1985). Runoff from areas without fluvial drainage, over 90 percent of the plateau surface (Osterkamp and Wood, 1984, p. 181), collects in the more than 30,000 ephemeral-lake basins, or playas, that dot the Southern High Plains surface. Development of these basins is discussed by Wood (Chapter 4, this volume).

Most groundwater recharge to the Miocene Ogallala Formation, the principal geologic unit of the uppermost major aquifer in this area (Gutentag and others, 1984), is by infiltration of runoff ponded in the playas (Wood and Osterkamp, 1984). The Ogallala Formation is as much as 150 m thick and is formed of fluvial sandstone and gravel beds that grade upward to eolian sandstone and siltstone. Pedogenic caliche layers are numerous in the Ogallala Formation and in the overlying Pleistocene eolian deposits. The most conspicuous of these calcrete horizons is the "caprock" caliche, which forms the upper surface of the Ogallala Formation and largely preserves the plateau topography of the area.

Analyses based on extrapolated gradients and petrographic examinations (Byrd, 1971, p. 22; Menzer and Slaughter, 1971, p. 221) suggest that Ogallala sediment originally may have spread eastward as far as 400 km beyond the present eastern escarpment of the Southern High Plains during late Miocene time. Double Mountain is an erosional feature about 90 km east of the escarpment rim. Walker (1978, p. 24) considered the gravel that caps this butte to be an outlier of the lowermost Ogallala Formation. Whatever may have been the original areal extent of Ogallala deposition, by late Miocene to early Pliocene time, Ogallala beds of the Texas–New Mexico area formed an unbroken alluvial-eolian apron from the Rocky Mountains to points well east of the present eastern escarpment. This preceded the development of the upper Pecos River drainage network and its pirating of sediment moving from the west.

A second aquifer system of the Southern High Plains area, the deep-brine aquifer (Orr and Kreitler, 1985), circulates water through Paleozoic carbonate rocks. Beginning probably in early Pliocene time, seepage from the Paleozoic rocks into overlying Permian-age evaporite beds caused local dissolution followed by structural collapse along parts of the present Pecos and Canadian River valleys and in parts of the Rolling Plains (Fig. 7; Gustavson and Simpkins, 1989; Gustavson and Budnik, 1985; Gustavson and Finley, 1985), as mentioned in Chapter 10 of this volume. These collapse basins led to: (1) headward extension of the Pecos and Canadian Rivers, (2) development of the western and northern escarpments of the Southern High Plains, and (3) piracy and diversion of water and sediment supplies flowing eastward from the mountains on the west. The cutoff of water and sediment terminated Ogallala deposition and may have caused stable but water-deficient conditions that led to the development of the massive caprock caliche (Osterkamp and Wood, 1984). Continuing solution of Permian salts by water moving through the deep-brine aquifer caused subsidence and collapse of basins east of the present Southern High Plains (Gustavson and Simpkins, 1989; Gustavson and Budnik, 1985; Gustavson and Finley, 1985; Orr and Kreitler, 1985). The effect of this solution and collapse was local destruction of the Ogallala Formation, thereby possibly making remaining outcrops susceptible to relatively intense fluvial erosion and to seepage erosion as groundwater drained from the freshly incised beds (Osterkamp and Wood, 1984).

Groundwater discharge at the escarpments

Ever since the initial development of the escarpments bounding the Southern High Plains, streamflow and fluvial erosion on the plateau have been very limited, especially at the western escarpment (Osterkamp and Wood, 1984). However, Quaternary recharge to the High Plains aquifer from the playa basins has generally maintained a dynamic groundwater system with relatively high water levels in the Ogallala Formation, except near escarpments where drainage occurs. Recent numerical-modeling studies by Knowles and others (1984) and Luckey and others (1986), suggest an average recharge rate to the High Plains

Figure 8. Oblique aerial photograph of a particularly embayed segment of the eastern caprock escarpment and adjacent part of the Southern High Plains northeast of Lubbock, Texas. Note the theater shapes of the stream heads and the lack of fluvial channels leading to them from the High Plains surface. These features indicate development by seepage-induced sapping, although the fine-textured drainage on the lower slopes suggests that surface runoff may also help to undermine the cliffs. Several playas are visible on the plateau surface west of the escarpment.

aquifer of about 0.3 cm/yr. It is assumed that prior to recent groundwater development, almost all of this water discharged as springs and seeps along the lower portions of the eastern and northern escarpments.

White and others (1946, p. 390) estimated average annual predevelopment spring and seepage losses from a 120-km length of the eastern escarpment to be 24×10^6 m^3. Extrapolating to all escarpments of the Southern High Plains, this estimate accounts for roughly half of the calculated recharge. The remainder may be lost as undetected seepage. Brine springs and seeps also discharge from Permian and possibly Triassic rocks beneath the escarpment rims (Gustavson and Simpkins, 1989), indicating continuing circulation of water through the deep-brine aquifer. It is not known whether the data of White and others (1946) included water moving upward from Paleozoic beds.

Support for the estimates of White and others (1946) is provided by Brune (1981), who compiled 53 spring discharges, mostly for the predevelopment period, from the eastern and northern escarpments. The 53 discharges total a flow rate of nearly 15×10^6 m^3/yr, again suggesting substantial infiltration from the playas through a generally thick unsaturated zone to the saturated zones of the High Plains aquifer, and implying generally eastward movement of groundwater with eventual discharge from sandstone and gravel beds of the lower Ogallala Formation and underlying Triassic rocks along the escarpments.

Most springs discharging along the eastern escarpment occur in steep box canyons, where evidence of spring sapping is extensive (Gustavson and others, 1981; Gustavson, 1983). In most cases the contributing surficial drainage area above a spring is negligible, indicating that channels below the springs must be maintained largely by spring effluent. The typical occurrence of playa basins within short distances of the escarpment edges (Fig. 8) demonstrates that fluvial runoff over the plateau rim is inconsequential (Osterkamp and Wood, 1984). Some runoff does occur, however, especially on the escarpment face, so that the badlands-like dissection below the rim (Fig. 8) owes at least in part to erosion by surface wash.

Field observations along the eastern escarpment between Amarillo and Lubbock (Fig. 7) indicate that processes of groundwater erosion have been active recently, although the rates of erosion appear to have declined owing to reductions in groundwater level by extensive pumping. Among the indications of recent sapping and seepage erosion are: (1) continuing springflow and seepage at some sites, albeit at reduced rates; (2) precipitation of salts, mostly as calcium bicarbonate and sulfate, at sites of dispersed groundwater discharge and evaporation; (3) arches, rincons, and other escarpment faces whose slopes approach and locally exceed the vertical, and which exhibit evidence of under-

Figure 9. Vertical aerial photograph of a portion of the Southern High Plains and the eastern escarpment about 60 km east of Lubbock, Texas. Width of view 4.7 km. Playa basins, indicated by arrows, show that the theater-headed channel network below the escarpment receives little direct runoff from the plateau. Fine-textured drainage below the rim may be formed partly by spring and seepage outflow and partly by surface runoff. Photograph by U.S. Department of Agriculture, 1970. The southern third of the area shown in this picture is depicted on the map in Figure 10.

cutting, leading to collapse by mass movement; (4) solution openings, in places associated with nearly vertical fractures in well-developed caliches of the basal Ogallala Formation; and (5) natural pipes, as much as ½ m in diameter, in poorly indurated, fine-grained beds of the Ogallala Formation; some pipes also exhibit an association with near-vertical fractures.

On a larger scale, springflow and geomorphic data provide striking evidence that groundwater erosion has been active in modern times around the Southern High Plains, and thereby suggest that it may have been more active during cooler and wetter periods (Pleistocene) of possibly higher groundwater levels. Along a 12-km north-south length of the eastern escarpment, part of which is shown in Figure 8, White and others (1946), as part of their study, measured discharges at 14 springs in the North Pease River drainage northeast of Lubbock, Texas (Fig. 7). All of the springs were within 2 km of the escarpment rim. Owing to extensive groundwater withdrawals, most of the springs no longer flow and are not shown on recent topographic maps. The 12-km length of escarpment is scalloped by 35 headwater channels of the stream net into which all of the springs discharged, and by possibly 10 times as many tributary incisions that also start abruptly below the escarpment edge. On the High Plains surface along the 12-km length, however, no streams flow toward the eastern escarpment, and most surface runoff within 3 km of the rim is ponded in about 25 playa depressions. Throughout the area, pronounced northeast and southeast trends of stream channels, aligned playas, and swales feeding the playa lakes are suggestive of structural control.

The abrupt edge of the pockmarked High Plains surface, strongly indented by gullies and ephemeral channels but un-

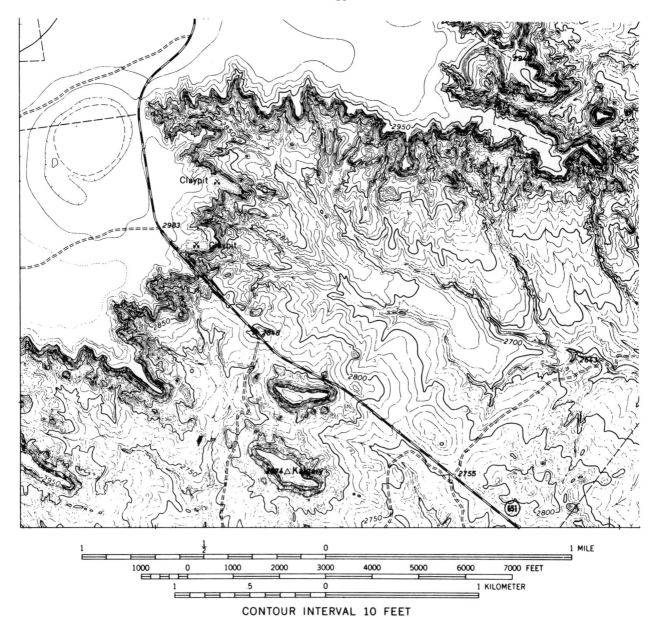

Figure 10. A portion of the eastern escarpment about 60 km east of Lubbock, Texas. The upper third of this map area is shown on the aerial photograph in Figure 9. Note the playa basin adjacent to the escarpment rim. Localized sapping-induced recession of valley heads along WNW-trending structures has produced pronounced irregularities in the rim. The map illustrates all stages in the development of flat-topped outliers, from long promontories and nearly detached projections of the plateau (upper right) to isolated residuals such as the two hills in the lower center. Further reduction of such outliers by seepage and wash yields conical and irregular hills such as those in the lower right. Part of the U.S. Geological Survey 7½-minute Collett Springs Quadrangle map, 1966.

notched by channels heading on the plateau surface, is illustrated in the oblique aerial photograph of Figure 8. The lack of a drainage network on the High Plains surface, but the occurrence of playa basins extending nearly to the escarpment edges, is apparent in a vertical aerial view of an area east of Lubbock and south of the North Pease River basin (Fig. 9). In this representative segment of the eastern escarpment, theater-headed channels impinge on the High Plains surface, and here too there is no evidence of surficial water movement over the escarpment edge. Headward extension of the fluvial system appears to have developed mainly by groundwater and related mass-movement processes, with removal of erosion products by local runoff during high-intensity storms.

Escarpment retreat

The escarpment along the west side of the Southern High Plains is generally smooth, in places nearly straight. There, the theater-headed draws encroaching on a jagged rim, which are products of and indicative of back scarp sapping and seepage erosion, are largely lacking due to the relatively minor westward component of groundwater movement and discharge. In comparison, great irregularity is apparent along the northern and, especially, the eastern escarpments (Figs. 8 and 9). There, preferred paths of groundwater movement concentrate sapping and result in the theater shapes and channel spacings that Laity and Malin (1985) identify as suggestive of sapping (also see Chapter 11, this volume). Along some parts of the eastern escarpment the retreating rim has become so embayed that detached outliers have been left behind (Fig. 10).

Long-term retreat rates for the northern and eastern escarpments are difficult to evaluate because the positions and timing of initial evaporite dissolution and subsequent collapse in the Ogallala Formation for these areas are poorly dated. Gustavson and Budnik (1985), however, report a late Pliocene age for lacustrine deposits associated with at least one subsidence basin near the eastern escarpment. Based on that age and on the distance of assumed retreat from outliers of Ogallala Formation east of the Southern High Plains, regarded here as relic landforms, rates of retreat ranging from 12 to 45 m/1,000 yr are indicated. For comparison, Simpkins and Baumgardner (1982) calculate a maximum retreat rate of 190 m/1,000 yr for the eastern escarpment following incision of the 620,000-yr-old Lava Creek B ash bed of the Pleistocene Seymour Formation. Based on terrace incision, Gustavson and others (1980) estimate retreat rates for the northern escarpment of 31 to 42 m/1,000 yr during the last 620,000 yr, and rates of as much as 180 m/1,000 yr for the same period east of the Southern High Plains (Gustavson and others, 1981). A recent paper (Gustavson and Simpkins, 1989) summarizes several studies of modern retreat rates conducted in the Southern High Plains that range from 10 to 20 m/1,000 yr.

Assuming initial incision of the Canadian River valley in early to mid-Pliocene time, the estimates for the northern escarpment may be high. They suggest as much as 26 km of retreat in the last 620,000 yr, but essentially none prior to that time because the Canadian River incisement is typically about 50 km wide. Likewise, the higher retreat rates cited above for the eastern escarpment are based on specific sites, and may be substantially greater than mean retreat rates. The estimate of 12 to 45 m/1,000 yr considers distances between the escarpment and several outliers of Ogallala Formation, and therefore may reflect a more typical range of erosion rates. These estimates suggest a mean retreat of the eastern escarpment of as much as 28 km since deposition of the Lava Creek B ash bed, and no more than 110 km since mid-Pliocene time. Considering the likelihood that fluvial and subsidence processes helped degrade the High Plains surface, this estimate seems compatible with a mid- to late Pliocene age for the isolation of Double Mountain, 90 km east of the present escarpment edge. These figures are for overall recession of the caprock escarpment front, not for retreat of the individual cliffs that make it up. If cliff retreat was initiated by incision of the major streams of the Rolling Plains, the distances that individual cliffs have retreated and their rates must have been considerably less. Regardless of the accuracy of these estimates, it appears likely that much of the retreat of the escarpments bordering the Southern High Plains during Quaternary time has been the result of groundwater erosion.

CASE STUDY: THE GILF KEBIR AND THE WESTERN DESERT OF EGYPT

Charles G. Higgins

Even more extensive scarp retreat and regional denudation have occurred in southwestern Egypt and adjoining parts of Libya, where a great blanket of Jurassic-Cretaceous sediments that include the Gilf and Nubia Sandstones (Issawi, 1982) has been stripped from a region at least 450 km wide. The geologic map of Egypt (Anonymous, 1981) shows that remnants of these strata still remain in places such as the Kharga uplift on the east and the Gilf Kebir on the west.

One of the earliest interpretations of large-scale cliff retreat through seepage-induced sapping concerned the Gilf Kebir. Even though active seepage and its effects seem largely to have ceased there, and even though the evidence of past activity is faint or circumstantial, it is worth recounting R. F. Peel's attempt to decipher the origin of the cliffs and valleys in terms of seepage-induced sapping at a time when that process was essentially unknown.

The Gilf Kebir is a great plateau, about 250 km long north to south, with a narrow waist that divides it into a northern and a southern part. Its flat upland surface slopes gently northward, reflecting the dip of the strata, so that on the south and west the steep cliffs rise more than 330 m above the desert floor, but become progressively lower to the north. At the northern end the upland merges with the plains underlying the Great Sand Sea, which extends northward nearly unbroken for 450 km. The southern half of the Gilf Kebir is shown in Figure 11; part of its eastern escarpment is shown in Figure 12.

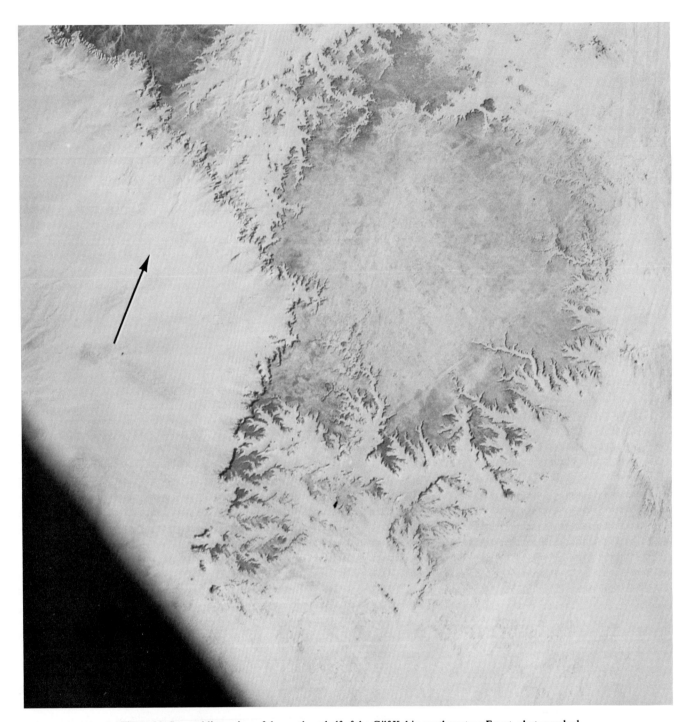

Figure 11. Steep oblique view of the southern half of the Gilf Kebir, southwestern Egypt, photographed from Apollo 7 (picture number AS7-5-1622). Arrow indicates approximate north, as determined from Peel's map (in Bagnold and others, 1939). Width of view at top edge is about 140 km.

Figure 12. Eastern escarpment of the Gilf Kebir. Climbing dunes testify that scarp recession has greatly decreased or ceased, and that wind action is now a dominant process there. Photograph by T. A. Maxwell, National Air and Space Museum, October, 1978.

Deep, gorge-like *wadis,* or box canyons, penetrate the margins of the plateau. Those in the southern Gilf are as much as 60 km long, and one in the north, shown on Bagnold's map (Bagnold and others, 1939), is 120 km long. According to Peel (1941), the slopes of the steep wadi walls and the surrounding cliffs average 35°. As seen in Figure 11, the wadis widen dramatically near their mouths with no decline in steepness of their walls, suggesting that the same process that forms the valley walls has also formed the marginal cliffs. Bagnold estimated "a total cliff frontage of more than 2,000 miles" including the walls of all the wadis (Bagnold and others, 1939, p. 283). The Gilf Kebir is waterless and remote, and few scientists have visited the area. The chief reports on its geomorphology are results of expeditions there in 1938 and 1978.

The 1938 expedition

The first to write at length on the geomorphology of the Gilf Kebir was the late R. F. Peel, who spent two months there as a member of the 1938 expedition led by R. A. Bagnold. Impressed by "the abrupt distinction between desert floor and plateau surface," and seeking an explanation for the development of the great wadis and cliffs that separate them, Peel reported that "the general appearance suggests very strongly that the plateau surface is not being, and has not been for a long period, attacked directly from above, but that all attack has come from below, the desert plains extending themselves into the wadis, which enlarge themselves and destroy the plateau by lateral encroachment and undercutting of the cliffs" (Peel, 1941, p. 8, 13).

Development of the wadis. Peel was unable to explain the development of the wadis by conventional surface wash:

there is a surprising lack of evidence of water flowing over the plateau surface to drain into the wadi head. Often the 'lip' is completely unworn, and drops sheer from the plateau surface to the cliffs above the wadi floor. . . .indeed, apart from a few shallow surface runnels, there is little evidence of water action on the plateau surface at all (Peel, 1941, p. 12, 13).

Other features of the wadis, such as their flat floors with abrupt cusped walls, irregular long profiles, and angulate drainage patterns, are all characteristic of valleys elsewhere that have been attributed to seepage erosion and sapping (see Chapter 11, this volume). From these indications, Peel concluded, "The present wadis show features which do not appear to be explicable on a theory of stream erosion alone. . . .They appear to have grown essentially by lateral cutting, and by the retreat of the cliffs by undercutting and basal sapping" (Peel, 1941, p. 16).

In an earlier report Peel wrote that he had considered all the mechanisms for basal lateral erosion and scarp retreat that had been summarized in an earlier article by J. L. Rich—mechanisms such as "weathering of scarps and removal of the debris by occasional rain wash and wind deflation . . . sheet-floods, and . . . heavily loaded streams, swinging widely in the lateral plain"— and that "all three may reasonably be invoked to explain the wadis," except that several features of the canyons "cannot be satisfactorily explained even by admitting all these processes" (Peel, 1939, p. 305). Later he added, "It is difficult to envisage this process being carried out by streams or sheetfloods in the wadis alone, for the cliffs show the same features along the open edge of the Gilf and in the broad 'bays,' which could never have harbored effective streams" (Peel, 1941, p. 16–17). He summarized the dilemma this way: "The essence of the problem seems to be to discover some agency which might gnaw away the cliffs from below in rather an irregular manner" (Peel, 1939, p. 305).

Just as Stetson credited Kirk Bryan for the suggestion that "groundwater sapping" might explain the submarine topography of Georges Bank, Peel credited Bagnold for suggesting that spring sapping could have formed the cliffs and wadis of the Gilf. The inspiration for this idea seems to have been the cliff-base "cavities" or niches found "at many points round the base of the cliffs," some with early folk-art drawings on their walls, and some "showing traces of former spring action. Peel summarized the postulated sapping process as follows:

If rainfall on the plateau surface were to sink straight into the porous sandstones, and reappear near the base of the cliffs as copious springs, irregular undercutting and retreat of the cliffs would undoubtedly follow. This would help explain the lack of drainage channels scoring the lips of the wadi cliffs and the plateau surface, the irregularity in the outline of the wadis, and the fact that they begin from abrupt cliffs. If this mechanism were allowed, recession of the cliffs and extension of the wadis would take place essentially by sapping and undercutting of the cliffs, the scree material produced being removed *en masse* by sheet-flooding after rains, and possibly by slow wind deflation in dry seasons (Peel, 1939, p. 305).

He was unable to test this hypothesis because it did not rain while he was there, nor had it rained "for a very long period," but some of the cliff-base caves showed "dry springs," and Count L. E. de Almásy, who had been there in 1932 and 1933, identified some as containing intermittent springs. Peel also noticed and photographed a plexus of dry streamlets heading at the cliff bases, and reasoned that "water had issued thence in considerable quantities during the last rains" (Peel, 1941, p. 18).

Peel evidently believed that some degree of wadi development, or at least channeling of the alluvial floors, is still continuing, but he commented that "it would appear that the bulk of the wadi-cutting took place at a remote period, when rainfall was more abundant and more frequent, for Neolithic implements can occasionally be picked up in the wadi beds apparently *in situ*" (Peel, 1941, p. 16). Moreover, he reported that in some places the wadi floors are now blocked by sand dunes and veneered with sediment that is locally "cemented into a compact reddish mass by iron compounds," that is, it is apparently a paleosol old enough to have developed an oxic horizon. Concentration of iron compounds in the alluvium of the wadis suggests soil formation during a time of more humid climate.

Scarp retreat and regional denudation. The plateau does not everywhere end abruptly at the marginal cliffs; in places, especially along the northwest and northeast sides, it dribbles away in "a maze of detached fragments and dissected hills" (Peel, 1941, p. 8). Peel identified this terrain as "plains and inselberge" and described its relation to the plateau:

This level [plateau] surface is apparently continued in the smaller plateaux and flat-topped hills which partially surround the Gilf, those examined revealing an identical structure and rising, so far as could be judged by eye, to just about the same elevation. These features would suggest that the Gilf and its satellites represent surviving fragments of a former land-surface now dissected and largely destroyed by the subaerial denudation which has produced the present desert plains (Peel, 1939, p. 301).

One of these desert plains is the Selima Sand Sheet, which adjoins the plateau and its outliers on the east, and which Sandford, quoted by Peel, termed a "desert peneplain":

Covering an area of at least 3,000 square miles [7,770 km^2], this region appears to the eye absolutely flat and featureless save for an occasional line of sand dunes....The solid rock is everywhere covered with a uniform sheet of wind-blown sand, which is probably nowhere more than a few feet thick: the sandstone beneath it would appear to have been worn down almost to a true plane (Peel, 1941, p. 6–7).

Peel clearly saw the large-scale implications of the cliff-recession mechanism that he had proposed, and he interpreted the regional topographic relations as the consequences of this mechanism. His hypothesis for the evolution of this desert landscape is perhaps the strongest statement on regional denudation by lateral scarp retreat in all geomorphic literature:

The writer conceives highlevel [plateau] erosion surface, wadis, inselbergs, and desert peneplain as genetically connected features in a general cycle of topographical evolution.
 . . . during some epochs of the Quaternary, at least, pluvial conditions seem to have prevailed. Rain water may even then have largely sunk away and evaporated, but some at least would run over the surface and would score drainage channels which would enlarge and deepen. Eventually however drier conditions with occasional rainstorms seem to have ensued. It is conceived that as a result of this change, direct surface run-off would more and more be confined to previously existing channels, the intervening sections of the original surface being left high and dry, and also that the postulated process of basal sapping would become more and more important in the evolution of the valleys....Vertical cutting . . . would to a large extent be replaced by lateral undercutting.
 The cliffs would everywhere retreat; the wadis expand into broad flat-floored basins and eventually plains scattered with small residual masses of plateau. The latter would progressively shrink with the undercutting of the cliffs all round them, and eventually only a central core would be left, and this would have a conical form because of the uniform slope of the cliffs (Peel, 1941, p. 18–19).

Peel then elaborated on the progressive development of the distinctive conical hills, some of which are shown in Figure 13:

... every shade of transition can be noticed between these and the small isolated cones of the open desert. Whatever the size, the shape remains remarkably uniform, and the sides retain the normal angle of desert cliffs, i.e., about 30–35°. The retention of these steep slopes in spite of general wasting and shrinkage in size argues that the same dominant process controls the evolution of plateau cliffs, wadis, and inselbergs, and this can only be concentration of attack at the cliff base (Peel, 1941, p. 20).

He surmised that the final reduction of the hills would be aided

Figure 13. "Inselberg" outliers of the eastern escarpment of the Gilf Kebir. In the foreground, a lorry of the Sudan Defense Force abandoned in 1943. Photograph by T. A. Maxwell, October, 1978.

by wind-driven sand blast, evidenced by steeper slopes on the windward northern sides, and resulting in "the eventual reduction of the hill to a shattered pile of stones which would gradually waste away *in situ,* leaving the typical open desert plain" (Peel, 1941, p. 21). In this view the flat floor of the Selima Sand Sea is the final product of the destruction of the upland surface of the plateau. If true, then seepage weathering has been a major factor in the stripping of a great blanket of Mesozoic sediments, not only from the immediate area of the Selima Sand Sheet, but from much of the Western and Libyan Deserts, a region about the size of Germany, covering 500,000 km^2 or more.

The 1978 expedition and recent studies in the Western Desert

A much larger party organized by Farouk El-Baz reconnoitered the Gilf Kebir during five days in October 1978 (El-Baz and others, 1980; El-Baz and Maxwell, 1982). Despite the brevity of the study, their findings tended to confirm the observations and many of the conclusions of Bagnold's group 40 years earlier. In addition, new information on past climates in the Western Desert has been obtained through other field studies in the Western Desert, including those of Vance Haynes and the Combined Prehistoric Expeditions of Wendorf and Schild in the 1970s.

Age of the wadis and cliffs. Peel recognized that the climate is now more arid than it was when the rock paintings were made on the walls of the niches and when Neolithic artifacts were left in the wadis, but he believed that spring sapping still continues there on a reduced scale. More recent studies of archaeological sites in the Western Desert suggest that most of the sapping and landscape development predates even the oldest human occupation. Haynes (1982a, b) presents this chronology:

The present hyperarid climate began about 5,000 B.P. Before that there appear to have been "three major periods of prehistoric human occupation associated with pluvial periods and separated by hyperarid periods at least as dry (less than 1 mm of annual rainfall) and windy as today" (Haynes, 1982a, p. 630). The earliest recognizable occupation in the region was Acheulean, dated about 200,000 ± 100,000 B.P. The Acheulean people lived around oases that appeared when the water table rose above the floors of deflation basins that had been scooped out of the surface of the desert peneplain that extends over most of the Western Desert. This indicates that general aridity must have existed for some period even before the Acheuleans. Acheulean artifacts at the base of the eastern cliffs of the Gilf Kebir show that scarp retreat there mainly predates that occupation. The much later Neolithic occupation sites in some of the wadis reflect a relatively recent but brief interval of humid climate (McHugh, 1982). Haynes concluded, "For the past 200,000 years, running water has played a minor role in shaping the surface of the land" (Haynes, 1982b, p. 96).

Grolier and Schultejann (1982) dated the weathered older alluvium on the wadi floors as "late Tertiary(?) to late Pleistocene." These sediments postdate recession of the wadis and cliffs, the major development of which must have been much older, probably in Pliocene time or earlier, and long before human occupation of the region.

Development of the wadis and cliffs. McCauley and others (1982a) identified traces of the interfluves of an ancient drainage network on a Landsat image of the southern Gilf Kebir, and concluded that these were late Tertiary features formed by running water as parts of "integrated, mature stream networks . . . that extended far beyond the present plateau escarpment on the surrounding desert plains" (p. 207). These "fossil' Gilf networks" were "formed under climatic conditions sufficiently humid to provide rainfall, which collected in runoff that carved integrated valleys in the bedrock of the Gilf region" (p. 214–215). When the regional climate became more arid in late Tertiary or early Quaternary time, the mode of erosion changed:

Cliffs undercut by spring sapping, sheetwash, and episodic flash flooding retreated by mass wasting ... [and] divides between adjacent basins were breached [so that].... Scattered inselbergs dot the pediplain formed by retreat of the Gilf scarps (McCauley and others, 1982a, p. 215).

This evolutionary picture resembles that reconstructed for the eastern caprock escarpment of the Southern High Plains, described above, where incision of a surface drainage system initiated seepage-induced sapping and recession of cliffs and valley heads. Regarding the observations of the 1978 Gilf Kebir expedition, T. A. Maxwell (1982) generally agreed with Peel that the steep cliffs and theater heads of the wadis, the angularity of the drainage pattern with its stubby tributaries, and the lack of feeder drainages on the plateau above suggest that much of the development and extension of the valleys may best be explained by sapping from below, not by wearing down from above.

Case-study summary

Wind is now the dominant geomorphic agent in southwestern Egypt, piling huge climbing dunes against the Gilf Kebir cliffs (Fig. 12), but cliff-base niches and "dry springs" testify to the former activity of seepage and sapping. Because these processes are now largely dormant in that region, their actions and effects must be studied elsewhere—in less arid regions such as the Colorado Plateau or Southern High Plains, for example—where seepage-induced sapping is still effective. However, even though groundwater seepage is now rare or absent at the Gilf Kebir, its cliffs remain as residuals standing high above the denuded expanse of the Western Desert to show the grand scale on which these processes may operate when conditions are favorable.

GEOMORPHIC IMPLICATIONS OF SEEPAGE WEATHERING

From the beginning, modern geomorphic theory has stressed the role of running water, and especially of streams, in the development of erosional landscapes. In this view, streams create relief by incising their channels, and the resulting slopes are then graded by a combination of processes, including slopewash and mass wasting. Such slope erosion is generally most intense where the inclination is steepest, so that the steeper parts tend to wear down faster than the other parts. The result is to reduce the overall inclination so that the slope declines by downwearing.

On the other hand, where valleys and slopes are "cut out from below, rather than let down from above," to borrow Ronald Peel's phrase, the progression of landscape development appears to be quite different. There are several ways in which erosion can be concentrated at the base of a slope: where seacliffs are attacked by waves or eroded at the waterline by intertidal organisms or carbonate dissolution (see Chapter 13, this volume), or where valley walls are undercut by glaciers or streams. The complex of processes here called seepage weathering can also focus erosion at the base of slopes, and under favorable conditions their effects can promote extensive backwearing. Some of the consequences of these processes and their results are examined below. These include seepage control of knicklines and regional denudation by scarp recession.

Seepage control of knicklines

Knicklines are abrupt breaks in slope at the back of an erosional surface such as a pediment. It has long been recognized that in layered strata such breaks can be controlled by the relative resistance of the beds. Other factors can also be responsible. Bryan stated that "the abrupt change in gradient at the base of hills and mountains—the Knickpuncke of German writers—is the result of the change from unconcentrated rainwash to the more efficient flow of ephemeral streams in channels" (Bryan, 1940, p. 260). Such an effect may explain some knicklines, but not others. Moss (1977) has shown that some knicklines at the rear of pediments eroded in granite reflect differential erosion across a weathering front. Seepage effects have also been credited with creating and maintaining knicklines by undercutting the base of the steeper upper slope.

This was observed 30 years ago by Schumm (1956) on slopes developed on the Chadron Formation in Badlands National Monument, South Dakota. There, a general change from steeper slopes on the overlying Brule Formation to gentler ones on the Chadron owes in the conventional way to lithologic differences. On a smaller scale, though, knicklines are formed on the two formations where "miniature pediments lie at the bases of retreating badland slopes" (Schumm, 1956, p. 700). On the relatively impermeable Brule, the pediment angle results from changes in the overland flow, as demonstrated by Smith (1958). However, Schumm reported that "on pediments formed at the base of Chadron residuals the slope retreat is aided by basal sapping of the slope by the appearance of subsurface flow at the junction of slope and pediment surface" (Schumm, 1956, p. 700). He tested this mechanism experimentally in the field, and found that "water poured onto the surface of the Chadron formation follows subsurface channels to the base of the slope where it reappears on the pediment surface, sapping back the base of the slope" (caption for Plate 1, Fig. D in Schumm, 1956). He felt that this process is a significant factor in the retreat of the badlands slopes.

On a larger scale, scarps in the American Southwest also owe to several different processes. At some sites, erosion is controlled by lithologic differences, but in many of these cases the actual erosional mechanism owes to seepage that is concentrated along a contact between a permeable unit and an underlying less permeable one. Baker has summarized:

Terrestrial scarp recession by sapping occurs most readily where a massive, resistant lithology overlies a relatively weak incompetent one. Undermining of the resistant layer occurs by weathering and groundwater flow along the contact between the two lithologies. Where scarp steepness is maintained, scarp recession will leave a smooth, nearly flat surface, often termed a 'pediment' (Baker, 1980, p. 286).

At other sites, as Bryan (1954) observed, erosion may be concentrated at the cliff base even though the contact lies below the level of the valley bottom or pediment. In these cases, seepage is, or used to be, determined by the local topography rather than by rock units. B. J. Smith (1978, p. 40) reported such a situation on limestone hamadas of the northwestern Sahara, where "basal sapping by seepage erosion and spring-sapping are the major processes responsible for the sharp junction between cliff and debris slope," the maintenance of which "is essential for the continued backwearing of the cliffs and their retention as landscape elements." Similarly at the Gilf Kebir, as discussed above, seepage lines rather than stratigraphy seem to have controlled the junction of the cliffs with the wadi floors and the surrounding desert plain. On the desert plain to the east, "the flat surface truncates the general bedding as well as shallow structures" (Haynes, 1982b, p. 108). These examples suggest that seepage levels may substitute for structural or stratigraphic control of some knicklines and erosion surfaces.

Scarp retreat and regional denudation

As noted above, the Tloi Eechii Cliffs (Fig. 6) appear to have retreated some 3.5 to 8 km from the Little Colorado River in Arizona, and the Adeii Eechii Cliffs may have retreated as much as 15 km. The cliffs' irregular outline and fine erosional texture suggest that their recession owes both to surface wash and seepage-induced sapping. The Southern High Plains escarpments seem to have retreated much longer distances, possibly more than 100 km. However, as discussed above, this does not mean that the existing cliffs have moved so far. Instead, it is likely that the region was first incised by an integrated drainage network, and that the valley walls began to recede by a combination of surface runoff and seepage-induced sapping when valley incision tapped the water table. In this scenario, individual scarp retreat may have been relatively minor, but the overall effect has been regional denudation, stripping the Ogallala Formation from a broad region and leaving only scattered outliers.

A similar landscape has developed through similar processes in northern Australia, where Baker reported scarp recession by sapping:

The Arnhem Land Plateau of northern Australia is an extensive surface of resistant sandstone surrounded by a steep escarpment from 30 to 330 meters in height. The Middle Proterozoic Kombolgie Formation provides the resistant caprock for the escarpment. Since the Cretaceous, erosion has been attacking a variety of weaker rocks at the base of the escarpment, undermining the caprock and inducing retreat. Spring sapping, cavernous weathering, and rock spalling produce debris from the sandstone cliffs. This debris mantles slopes, inhibiting further erosion until it is reduced in size and removed by wind or sheetwash. The escarpment assumes a variety of active or stable configurations depending on the jointing of the sandstone, exposure of weak rocks at the scarp base, and faulting of the scarp boundary. Local enhancement of erosion along joints, faults, and anticlines has produced large embayments into the escarpment and many sandstone outliers on the broad, flat pediplain that was left by scarp retreat (Baker, 1980, p. 286–287).

Jennings' (1979) observations in this region suggest that seepage weathering and cliff retreat may no longer be as vigorous as in the past, and that the landscape, like many others that owe to these processes, is now mainly relict.

On an even larger scale, much the same seems to have occurred in the Western Desert of Egypt. There Jurassic and Cretaceous strata have been stripped from a region at least 400 km wide, leaving only a few large residuals, such as the Gilf Kebir, and scattered small outliers. The Gilf cliffs may not have retreated all the way from Kharga or even halfway across the desert. It is more likely that erosion of the Gilf Sandstone began with the incision of a regional drainage system, much as proposed by McCauley and others (1982a). Seepage-aided lateral erosion then could have begun when the valleys reached the water table, and the valley walls began to retreat to the positions of the present scarps. Relics of the incised drainage network seem to have been detected on the desert floor beneath the windblown sands of the Selima Sand Sheet on images obtained by the Shuttle Imaging Radar Experiment in 1981 (McCauley and others, 1982b).

However, even if the scarps have receded only tens rather than hundreds of kilometers, the result has been the removal of the Gilf and Nubia Sandstones from a large part of southwestern Egypt and adjacent parts of Libya. Significantly, this denudation seems to have been accomplished in large part not by conventional downwearing but by the lateral recession of cliffs as a consequence, at least in part, of seepage weathering and cliff-base sapping. These processes can be especially effective where lateral retreat begins at the walls of valleys in a drainage network. Then sapping can gnaw away the upland from both sides of the divides, eventually leaving behind a wide expanse of stripped terrain with perhaps only a few remnant outliers.

Development and enlargement of deflation basins. The scarp-retreat mechanism outlined above may also play a vital part in the development of steep-walled closed depressions in arid and semiarid regions. Excavation of such depressions has generally been attributed partly or entirely to deflation by wind, removing loose material down to a wetted capillary fringe above the water table. This periodic moisture is also responsible for disaggregating the substrate through salt-crystal wedging and other related weathering processes. Haynes describes these effects in some shallow depressions in the Arba'in Desert, that part of the Western Desert east of the Gilf Kebir:

As the land surface is lowered by wind erosion, evaporation from the water table occurs. Salts form, expand, and further weaken the rock at the surface, allowing more of it to blow away which enhances further evaporation. Thus, the ultimate limit on wind scour and deflation is the water table. . . .This is very likely the explanation for the extreme flatness of the sand sheet areas of the Arba'in Desert because the water table is relatively flat where it occurs in porous sandstone such as the Nubia Formation (Haynes, 1982b, p. 96, 109).

This role of groundwater in desert morphology has long been known, but the role of lateral erosion by seepage-aided scarp retreat in widening these basins and forming their steep slopes has rarely been recognized.

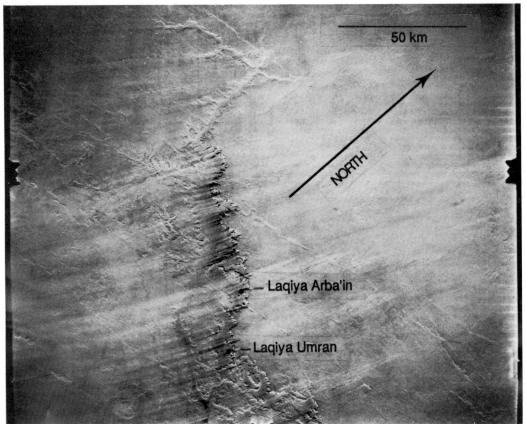

Figure 14. High-altitude vertical photograph of the Laqiya region in northern Sudan. The deflation basin of Arba'in Valley, center, extends northward as a narrow wadi that appears to be an enlarged joint. Other joint-controlled linear wadis and lines of disconnected hollows trend ENE. Piping and seepage may have played a part in this erosion, but the light-toned streaks of wind-blown sand over all testify to the region's present aridity. Width of view about 190 km. Photograph by the STS 9/Spacelab 1 metric camera, 5 December 1983; processed by DFVLR for ESA/EARTHNET; print courtesy of T. A. Maxwell.

In the Western Desert "deflational hollows range in size from meters to many tens of kilometers" (Haynes, 1982b, p. 94). The largest, at Kharga, is 220 km long and as much as 40 km wide. Haynes (1982b) noted that it, like most of the smaller depressions in the region, held lakes in earlier times, suggesting that its base corresponds to former water tables.

Six hundred kilometers to the southwest, at the southern edge of the Selima Sand Sheet in northern Sudan, the Laqiya area illustrates early stages in the development of such depressions. There, as shown in Figure 14, several linear wadis are incised into a flat, stony tableland overlooking the Arba'in Valley, a narrow depression to the south. Two of the wadis and several lines of elongate hollows trending about N70°E show the effects of joint control. Haynes (1982c) reported that the walls of these wadis are relatively steep and as much as 30 m high. The photograph shows that the edge of the upland drops sharply in a line of crenulated cliffs that continue back into the wadis, suggesting a common origin. Their resemblance to the lines of cliffs and wadis of the Gilf Kebir further suggests that they have developed in a similar manner through seepage weathering, sapping, and retreat, with the wind carrying away the weathered and detached debris. The lines of hollows may represent early forerunners of the wadis.

Haynes visited the Laqiya region in 1981 and found a late Acheulean site in one of the wadis, showing that this landscape had developed essentially its present form before late Pleistocene time. He commented:

The unintegrated branches indicate that it is not derived from a stream course, but may have formed by deflation as the water table fluctuated in a sandstone fracture weakened at the surface by leaching of cement or salt weathering (Haynes, 1982b, p. 106).

. . . a sapping mechanism like that proposed by Peel (1941) for the wadis of the Gilf Kebir is also a possibility (Haynes, 1982c, p. 300).

The photograph shows the landscape to be thinly overlain with streaks of windblow sand, suggesting that its major features were developed before aridity enveloped the region. This appears to be true also of other sites in northern Africa and in lands elsewhere. B. J. Smith's comments on the age of the scarp-foot recesses in the northwestern Sahara are quoted above. Similar relict conditions are shown in aerial photographs published by von Bandat (1962, Fig. 11-6) and H.T.U. Smith (1972). Von Bandat's stereopair shows an irregular scarp associated with deflation basins in the northern Sahara. Some of the basins and parts of the scarp are overlain by blown sand, which could not have been there when scarp recession was active. Smith's photos show a variety of scarps and basins in the Saharan region, also overprinted with blown sand, showing that topographic development there also predated the modern aridity.

The same appears to be true of the large "P'ang Kiang" hollows, or deflation basins of the Mongolian desert, described by Berkey and Morris (1927). The largest are as much as 8 km in diameter and 120 m deep. Berkey and Morris wrote that the deepening of the hollows is controlled or retarded in part by the water table, but they thought that the bordering steep slopes:

are dissected by rainwash and by rills from the upland. . . .Gullies are rarely seen on the floor of the hollow, and still more rarely on the upland. The gullies along the bluffs are very short, heading a little way behind the edge of the scarp and dying out a short distance beyond its foot (Berkey and Morris, 1927, p. 338–339).

However, if these slopes were affected solely by surface wash, they would wear down as well as back. Scarp retreat calls for concentration of erosional sapping at the base, which in this setting may be aided by seepage erosion. That this is likely is indicated by the short, steep-headed "gullies" (Berkey and Morris, 1927, Fig. 142) that seem to resemble those formed by seepage (see Chapter 6, this volume) and the evidence of occasional shallow lakes in the basins. As elsewhere, the action of the formative processes is doubtless much diminished since the region became dry.

Analogous landscapes of Mars. Broad plains bordered by embayed steep cliffs that rise to an older upland surface also occur on Mars. This "fretted terrain" with cliffs 1 to 2 km high, is best developed between 30 to 45°N and 280 to 350°W (Carr, 1981), and was first shown on imagery returned by Mariner 9 and later by the Viking orbiters. Many features of the Martian low plains and their bordering escarpments, isolated residuals, and "broad, flat-floored, sinuous channels . . . reaching deep into the uplands" (Carr, 1981, p. 71) resemble parts of the Gilf Kebir and Western Desert. McCauley and others (1982a) have made this resemblance especially clear:

The fretted terrain along the northern margin of the martian highlands is a belt of dissected plateaus, protected by caprocks, from which relict fluvial channels debouch onto the smooth, flat northern plains. The plains are partly veneered with eolian deposits and dotted with clusters of inselbergs, including mesas and conical hills that resemble in general those of the Gilf region (McCauley and others, 1982a, p. 208).

This resemblance includes the now-relict character of the Martian landforms, for it appears that their formative processes have not been active for several hundred million years. One of the first, and best, reports on these features was by R. P. Sharp, who postulated that the plains had evolved:

by recession of a steep bounding escarpment, leaving a smooth lowland floor at a remarkably uniform level. Escarpment recession is speculatively attributed to undermining by evaporation of ground ice exposed within an escarpment face, or, under a different environment, by ground water emerging at its foot. The uniform floor level may reflect the original depth of frozen ground. Removal of debris shed by the receding escarpments could be by eolian deflation . . . or by fluvial transport (Sharp, 1973, p. 4073).

Most later studies have not changed this assessment of scarp retreat by some kind of basal sapping, except to stress the apparent role of a thick zone of ground ice. How much, if any, of the cliff recession may have owed to seepage outflow of liquid water is not known, although backwearing of the equally high walls of portions of the Kasei Vallis outflow channel (see Chapter 11, this volume) has been attributed to mass wasting and sapping by water on the basis of the low gradients of the tributary trenches.

Some kind of piping or vertical and lateral migration of saturated substrate seems to have helped form some joint-controlled linear depressions on the upland adjoining Tithonium Chasma of the Valles Marineris. Such processes may also have been responsible for Hebes Chasma, a huge depression north of Tithonium Chasma (Higgins, 1984, Fig. 2.18). Similar lines of discontinuous hollows in the upland of the Laqiya region of northern Sudan suggest that similar processes may once have helped shape that landscape.

However, tempting as it is to equate Martian landscapes with terrestrial ones, their origins may or may not have been similar. Haynes warns:

the physical-chemical conditions on Mars are so different from those on Earth that exact analogs seem unlikely. Considerable caution, therefore, should be exercised in proposing terrestrial analogs of Martian processes, landforms, and mineralogy (Haynes, 1982a, p. 117).

Nevertheless, the search for explanations for the origin of Martian landscapes has already made a lasting contribution to terrestrial geomorphology in inspiring a number of investigators to examine regions on Earth where spring sapping and seepage weathering have played major roles in shaping the landscape. It is safe to say that much of the present interest in groundwater geomorphology has stemmed from these efforts to understand the valleys and plains of Mars.

The role of backwearing in geomorphic theory

A long-standing controversy in geomorphology concerns the behavior of slopes during the erosional development of landscape. In the evolutionary model developed by William Morris Davis (1899), rapid uplift is accompanied by stream incision and valley deepening. This creates slopes that are initially steep, but which are then worn down under the combined influence of surface wash and mass wasting. The slopes decline because their bases remain more or less fixed while the upper parts are eroded back (Fig. 15B). Such development is called downwearing, and

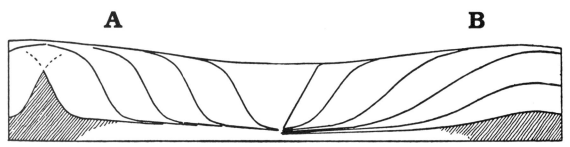

Figure 15. "Profiles of a widening Valley." **A.** Backwearing and parallel retreat of a steep valley side slope, which "may have been Penck's view of the case." **B.** Downwearing of a valley side slope, as favored by W. M. Davis. (From Davis, 1932, Fig. 4.)

its end form is a "peneplain," perhaps with occasional convex residual hills, or "monadnocks."

This view was challenged in 1924 by Walther Penck, who proposed a scheme of landscape development that depended largely on the type and rate of uplift. Penck postulated that under certain conditions steep valley slopes ("Steilwande") would not wear down and become rounded, but would retreat with essentially fixed inclination until the initial upland would be consumed and all the land reduced (Fig. 15A). Such development is called backwearing, and its end form is an "Endrumpf" with isolated steep-sided residual hills, or "Inselberge." Later writers have tended to call such an end form a "pediplain."

A chief contribution of Penck's work was to direct attention to slope development, particularly to the development and recession of rectilinear slopes—those with straight profiles. Some early studies (Lehmann, 1933; Bakker and Le Heux, 1947, 1950) focused on the geometry of scarp development and retreat. Scheidegger's (1961) broader analysis yielded some cases of "parallel slope recession" with others in which the slope angles decreased. These studies prompted other writers to apply similar geometrical analyses to the cliffs and benchlands of the Colorado Plateau (Cunningham and Griba, 1973; Aronsson, with Linde, 1982). These studies largely ignore the processes responsible for the slopes; the treatments are almost entirely graphical and theoretical. Penck had supposed that slope retreat was effected largely by mass wasting of thin slices, a view adopted by Koons (1955) in the Colorado Plateau.

The contrast between the developmental concepts of Davis and Penck was underscored by Davis in his own synopsis of Penck's views (Davis, 1932), but his review did not resolve the differences. A controversy over slope development continued through the 1930s and 1940s. Then, in 1952, a key element in the debate was provided when R.A.G. Savigear demonstrated that backwearing and parallel cliff retreat is promoted by unimpeded removal of colluvium from the base of a slope, whereas downwearing in place occurs where slope-foot debris accumulates. That is, slopes wear back where there is an agent to remove the detritus from the base, and they wear down where there isn't.

This analysis goes far toward explaining why some slopes wear down while others wear back, but it is incomplete. What is missing is a mechanism for concentrating erosion at the slope base. Without that, erosion will tend to be more effective on the steeper, more exposed hill shoulders and upper parts, so that even with removal of colluvium from the base, the overall slope will decline even while its base may be retreating.

Savigear's (1952) own field study of slope retreat and slope decline was conducted along a reach of coast in South Wales. There, wave action had maintained a steep seacliff until the progressive growth of a sand spit diminished the effects of wave attack and halted the undercutting. As a result the cliff began to wear down in place. This alteration of slope behavior is significant, but it seems even more significant that the accumulation of scarp-foot colluvium that caused the alteration was itself a result of the cessation of active undercutting by the sea.

Wave action can concentrate erosion at a cliff base, as can lateral cutting by streams and glaciers. However, most of the debate about scarp retreat has concerned landforms at sites where scarp steepening cannot have been caused by any of these factors. Under some conditions, surface wash can selectively erode weak strata so as to undermine overlying stronger units. However, in cliffs formed of permeable strata, basal erosion may owe largely to the effects of seepage weathering. The resulting concentration of erosion causes the cliff to be sapped at the base so that the slope retreats at a constant steep angle, provided that the fallen scarp-front debris is periodically removed, largely by surface processes. This combination of effects can provide the means of maintaining the scarp geometry required by the developmental models of Walther Penck and his followers.

It may now be seen that slopes steepen and wear back where erosion is concentrated at the base and the erosion products are removed, whereas slopes wear down where concentrated scarp-foot erosion is lacking and colluvium accumulates. At many sites, seepage weathering can provide a means of concentrating erosion at the base of a slope, and thus may be an important factor in the long-standing controversy over backwearing versus downwearing.

CONCLUSIONS

Seepage weathering is a varied group of interacting processes; it occurs where diffuse groundwater seepage emerges at the base of a slope. The result is to undermine the slope, promoting its retreat with constant angle and leaving behind a broad pediplain or inselberg landscape. Such retreating cliffs can be seen at all scales, from tiny features on beaches to great escarpments in the Colorado Plateau, the Western Desert of Egypt, and possibly even on Mars. Indeed, in some regions, seepage-induced scarp recession appears to have covered long distances and accomplished the denudation of large areas.

Conjecture has played a large part in the recreation of the origin of some of the landforms and landscapes discussed here. This is inescapable where the forms are relict and the processes responsible for them are no longer operating or are now working much below their former strength. However, if these interpretations are correct, then extensive regions that were once covered by near-horizontal resistant strata and that are now bordered by distant lines of cliffs may owe their denudation more to backwearing by seepage weathering and its attendant processes than to the conventionally accepted processes of surface runoff and downwearing.

For two chief reasons these processes have not been generally recognized in the origin of such landscapes. First, many or most of the landscapes formed by them are now relict. Active seepage weathering can be observed at some sites, but at most, especially in now-arid regions, it is either no longer active or is much diminished. This has greatly hampered our understanding of the processes involved and has hindered our appreciation of their importance, perhaps even dominance, in landscapes developed primarily by scarp retreat. Second, these processes have been generally overlooked, partly because the "fluvial doctrine" has conditioned geomorphologists to believe that surface runoff and fluvial processes are the primary sculptors of landforms, and partly because the effects of seepage weathering, even where still active, tend to be obscured by those of other, more obvious processes. This simply means that one must look more carefully to discover the important role that subsurface water has played in aiding cliff recession and shaping some parts of the Earth's great erosional surfaces.

REFERENCES CITED

Ahnert, F., 1960, The influence of Pleistocene climates upon the morphology of cuesta scarps on the Colorado Plateau: Association of American Geographers Annals, v. 50, p. 139–156.

Anonymous, 1981, Geologic map of Egypt: Cairo, Egyptian Geological Survey and Mining Authority, scale 1:2,000,000.

Aronsson, G., with Linde, K., 1982, Grand Canyon; A quantitative approach to the erosion and weathering of a stratified bedrock: Earth Surface Processes and Landforms, v. 7, p. 589–599.

Bagnold, R. A., Myers, O. H., Peel, R. F., and Winkler, H. A., 1939, An expedition to the Gilf Kebir and 'Uweinat, 1938, Geographical Journal, v. 93, p. 281–312.

Baker, V. R., 1980, Some terrestrial analogs to dry valley systems on Mars: National Aeronautics and Space Administration Technical Memorandum 81776, p. 286–288.

Bakker, J. P., and Le Heux, J.W.N., 1947, 1950, Theory on central rectilinear recession of slopes, 1-4: Koninklijke Nederlandsche Akademie van Wetenschappen, Afdeeling Natuurkunde, Proceedings, series B, v. 50, p. 959–966 and 1154–1162; v. 53, p. 1073–1084 and 1364–1374.

Berkey, C. P., and Morris, F. K., 1927, Geology of Mongolia, in Central Asiatic expeditions, Natural history of central Asia, v. 2: New York, American Museum of Natural History, 475 p.

Brune, G., 1981, Springs of Texas, v. 1: Fort Worth, Texas, Branch-Smith, Inc., 566 p.

Bryan, K., 1925, The Papago Country, Arizona; A geographic, geologic, and hydrologic reconnaissance with a guide to desert watering places: U.S. Geological Survey Water-Supply Paper 499, 436 p.

—— , 1928, Niches and other cavities in sandstone at Chaco Canyon, New Mexico: Zeitschrift für Geomorphologie, v. 3, no. 3, p. 125–140.

—— , 1940, The retreat of slopes: Association of American Geographers Annals, v. 30, p. 254–268.

—— , 1954, The geology of Chaco Canyon, New Mexico, in relation to the life and remains of the prehistoric peoples of Pueblo Bonito: Smithsonian Miscellaneous Collections, v. 122, no. 7, 65 p.

Byrd, C. L., 1971, Origin and history of the Uvalde gravel of central Texas: Waco, Texas, Baylor University Geological Studies Bulletin 20, 43 p.

Campbell, I. A., 1973, Control of canyon and meander forms by jointing: Area, v. 5, p. 291–296.

Carr, M. H., 1981, The surface of Mars: New Haven, Connecticut, Yale University Press, 232 p.

Cole, K. L., and Mayer, L., 1982, Use of packrat middens to determine rates of cliff retreat in the eastern Grand Canyon, Arizona: Geology, v. 10, p. 597–599.

Cunningham, F. F., and Griba, W., 1973, A model of slope development and its application to the Grand Canyon, Arizona, U.S.A.: Zeitschrift für Geomorphologie, v. 17, p. 43–77.

Davis, W. M., 1899, The geographical cycle: Geographical Journal, v. 14, p. 481–504.

—— , 1932, Piedmont benchlands and primärrümpfe: Geological Society of America Bulletin, v. 43, p. 399–440.

Dellenbaugh, F. S., 1898, The causes of natural arches: Science, n.s., v. 7, no. 177, p. 714.

El-Ba, F., and Maxwell, T. A., eds., 1982, Desert landforms of southwest Egypt; A basis for comparison with Mars:National Aeronautics and Space Administration CR-3611, 372 p.

El-Baz, F., and 15 others, 1980, Journey to the Gilf Kebir and Uweinat, southwest Egypt, 1978: The Geographical Journal, v. 146, p. 51–93.

Gerster, G., 1976, Grand design: New York, Paddington Press, 312 p.

Gregory, H. E., 1917, Geology of the Navajo Country; A reconnaissance of parts of Arizona, New Mexico, and Utah: U.S. Geological Survey Professional Paper 93, 161 p.

Grolier, M. J., and Schultejann, P. A., 1982, Geology of the southern Gilf Kebir Plateau and vicinity, Western Desert, Egypt, in El-Baz, F., and Maxwell, T. A., eds., Desert landforms of southwest Egypt; A basis for comparison with Mars: National Aeronautics and Space Administration CR-3611, p. 189–206.

Gustavson, T. C., 1983, Diminished spring discharge; Its effect on erosion rates in the Texas panhandle, in Geology and geohydrology of the Palo Duro Basin, Texas Panhandle: Austin, University of Texas Bureau of Economic Geology Geological Circular 83-4, p. 128–132.

Gustavson, T. C., and Budnik, R. T., 1985, Structural influences on geomorphic processes and physiographic features, Texas Panhandle; Technical issues in siting a nuclear-waste repository: Geology, v. 13, p. 173–176.

Gustavson, T. C., and Finley, R. J., 1985, Late Cenozoic geomorphic evolution of the Texas panhandle and northeastern New Mexico; Case studies of structural controls of regional drainage development: Austin, University of Texas Bureau of Economic Geology Report of Investigations 148, 42 p.

Gustavson, T. C., and Simpkins, 1989, Geomorphic processes and rates of retreat affecting the Caprock Escarpment, Texas Panhandle: Austin, University of Texas Bureau of Economic Geology Report of Investigations 180, 49 p.

Gustafson, T. C., Finley, R. J., and Baumgardner, R. W., Jr., 1980, Preliminary rates of slope retreat and salt dissolution along the eastern Caprock Escarpment of the Southern High Plains and in the Canadian River valley, *in* Geology and geohydrology of the Palo Duro Basin, Texas panhandle: Austin, University of Texas Bureau of Economic Geology Geological Circular 80-7, p. 76–82.

—— , 1981, Retreat of the Caprock Escarpment and denudation of the Rolling Plains in the Texas panhandle: Association of Engineering Geologists Bulletin, v. 18, no. 4, p. 413–422.

Gutentag, E. D., Heimes, F. J., Krothe, N. C., Luckey, R. R., and Weeks, J. B., 1984, Geohydrology of the High Plains Aquifer in parts of Colorado, Kansas, Nebraska, New Mexico, Oklahoma, South Dakota, Texas, and Wyoming: U.S. Geological Survey Professional Paper 1400-B, 63 p.

Haynes, C. V., Jr., 1982a, Great Sand Sea and Selima Sand Sheet, eastern Sahara; Geochronology of desertification: Science, v. 217, p. 629–633.

—— , 1982b, The darb El-Arba'in Desert; A product of Quaternary climate, *in* El-Bas, F., and Maxwell, T. A., eds., Desert landforms of southwest egypt; A basis for comparison with Mars: National Aeronautic and Space Administration CR-3611, p. 91–117.

—— , 1982c, Quaternary geochronology of the Western Desert, *in* Proceedings of the International Symposium on Remote Sensing of Environment, Cairo, Egypt, 1982: Ann Arbor, Environmental Research Institute of Michigan, p. 297–311.

Higgins, C. G., 1982, Drainage systems developed by sapping on Earth and Mars: Geology, v. 10, p. 147–152.

—— , 1984, Piping and sapping; Development of landforms by groundwater outflow, *in* LaFleur, R. G., ed., Groundwater as a geomorphic agent: Boston, Massachusetts, Allen and Unwin, p. 18–58.

Higgins, C. G., and 12 others, 1988, Landform development, *in* Back, W., Rosenshein, J., and Seaber, P., eds., Hydrogeology: Boulder, Colorado, Geological Society of America, The Geology of North America, v. O-2, p. 383–400.

Howard, A. D., 1988, Introduction; Groundwater sapping on Earth and Mars, *in* Howard, A. D., Kochel, R. C., and Holt, H. E., eds., Sapping features of the Colorado Plateau: National Aeronautics and Space Administration Special Publication SP-491, p. 1–5.

Howard, A. D., and Hornberger, G. M., 1987, Simulation model of scarp erosion by groundwater sapping: EOS Transactions of the American Geophysical Union, v. 68, p. 1273.

Howard, A. D., and Kochel, R. C., 1988, Introduction to cuesta landforms and sapping processes on the Colorado Plateau, *in* Howard, A. D., Kochel, R. C., and Holt, H. E., eds., Sapping features of the Colorado Plateau: National Aeronautics and Space Administration Special Publication SP-491, p. 6–56.

Hutchinson, J. N., 1968, Mass movement, *in* Fairbridge, R. W., ed., Encyclopedia of geomorphology: New York, Reinhold, p. 688–695.

Issawi, B., 1982, Geology of the southwestern desert of Egypt, *in* El-Baz, F., and Maxwell, T. A., eds., Desert landforms of southwest Egypt; A basis for comparison with Mars: National Aeronautics and Space Administration CR-3611, p. 57–66.

Jacobson, R. B., and Pomeroy, J. S., 1987, Slope failures in the Appalachian Plateau, *in* Graf, W. L., ed., Geomorphic systems of North America: Boulder, Colorado, Geological Society of America, Centennial Special Volume 2, p. 21–29.

Jennings, J. N., 1979, Arnhem Land; City that never was: Geographical Magazine, v. 51, p. 822–827.

Knowles, T., Nordstrom, P., and Klemt, W. B., 1984, Evaluating the groundwater resources of the High Plains of Texas, v. 1: Texas Department of Water Resources Report 288, 113 p.

Koons, D., 1955, Cliff retreat in the southwestern United States: American Journal of Science, v. 253, p. 44–52.

Laity, J. E., 1983, Diagenetic controls on groundwater sapping and valley formation, Colorado Plateau, revealed by optical and electron microscopy: Physical Geography, v. 4, no. 2, p. 103–125.

Laity, J. E., and Malin, M. C., 1985, Sapping processes and the development of theater-headed valley networks on the Colorado Plateau: Geological Society of America Bulletin, v. 96, p. 203–217.

Lawler, D. M., 1984, Processes of river bank erosion; The River Ilston, South Wales [Ph.D. thesis]: Swansea, University of Wales, 518 p.

Lehmann, O., 1933, Morphologische Theorie der Verwitterung von Steinschlagwänden: Vierteljahrsschrift der Naturforschende Gesellschaft in Zürich, v. 78, p. 83–126.

Lucchitta, I., 1979, Late Cenozoic uplift of the southwest Colorado Plateau and adjacent lower Colorado River region: Tectonophysics, v. 61, p. 63–95.

Luckey, R. R., Gutentag, E. D., Heimes, F. J., and Weeks, J. B., 1986, Digital simulation of ground-water flow in the High Plains Aquifer in parts of Colorado, Kansas, Nebraska, New Mexico, Oklahoma, South Dakota, Texas, and Wyoming: U.S. Geological Survey Professional Paper 1400-D, 57 p.

Maxwell, T. A., 1982, Erosional patterns of the Gilf Kebir Plateau and implications of the origin of martian canyonlands, *in* El-Baz, F., and Maxwell, T. A., eds., Desert landforms of southwest Egypt; A basis for comparison with Mars: National Aeronautics and Space Administration CR-3611, p. 281–300.

McCauley, J. F., Breed, C. S., and Grolier, M. J., 1982a, The interplay of fluvial, mass-wasting, and eolian processes in the eastern Gilf Kebir region, *in* El-Baz, F., and Maxwell, T. A., eds., Desert landforms of southwest Egypt; A basis for comparison with Mars: National Aeronautics and Space Administration CR-3611, p. 207–239.

McCauley, J. F., and 7 others, 1982b, Subsurface valleys and geoarchaeology of the eastern Sahara revealed by shuttle radar: Science, v. 218, p. 1004–1020.

McHugh, W. P., 1982, Archaeological investigations in the Gilf Kebir and the Abu Hussein dunefield, *in* El-Baz, F., and Maxwell, T. A., eds., Desert landforms of southwest Egypt; A basis for comparison with Mars: National Aeronautics and Space Administration CR-3611, p. 301–334.

Menzer, F. J., and Slaughter, B. H., 1971, Upland gravels in Dallas County, Texas, and their bearing on the former extent of the High Plains Physiographic Provinece: Texas Journal of Science, v. 22, p. 217–222.

Mortensen, H., 1953, Neues zum Problem der Schichtstufenlandschaft: Nachrichten der Akademie der Wissenschaften in Göttingen, Mathematisch-Physikalische Klasse, Mathematisch-Physikalisch-Chemische Abteilung, Jahr 1953, p. 3–22.

Moss, J. H., 1977, The formation of pediments; Scarp backwearing or surface downwasting?, *in* Doehring, D. O., ed., Geomorphology in arid regions; Proceedings of the 8th Annual Geomorphology Symposium: Binghamton, State University of New York Publications in Geomorphology, p. 51–78.

Newman, E. B., Mark, R. K., and Vivian, R. G., 1982, Anasazi solar marker; The use of a natural rockfall: Science, v. 217, p. 1036–1038.

Nocita, B. W., 1987, Thermal expansion of salts as an effective physical weathering process in hot desert environments: Geological Society of America Abstracts with Programs, v. 19, p. 790.

Nicholas, R. M., and Dixon, J. C., 1986, Sandstone scarp form and retreat in the Land of Standing Rocks, Canyonlands National Park, Utah: Zeitschrift für Geomorphologie, v. 30, p. 167–187.

Oberlander, T. M., 1977, Origin of segmented cliffs in massive sandstone of southeastern Utah, *in* Doehring, D. O., ed., Geomorphology in arid regions; Proceedings of the 8th Annual Geomorphology Symposium: Binghamton, State University of New York Publications in Geomorphology, p. 79–114.

Orr, E. D., and Kreitler, C. W., 1985, Interpretations of pressure-depth data from confined underpressured aquifers exemplified by the deep-brine aquifer, Palo Duro Basin, Texas: Water Resources Research, v. 21, no. 4, p. 533–544.

Osterkamp, W. R., 1987, Groundwater; An agent of geomorphic change, *in* Graf, W. L., ed., Geomorphic systems of North America: Boulder, Colorado, Geological Society of America, Centennial Special Volume 2, p. 188–195.

Osterkamp, W. R., and Wood, W. W., 1984, Development and escarpment retreat of the Southern High Plains, *in* Proceedings Ogallala Aquifer Symposium 2: Lubbock, Texas Tech University Water Resources Center, p. 177–193.

Peel, R. F., 1939, The Gilf Kebir, *in* Bagnold, R. A., and others, An expedition to the Gilf Kebir and 'Uweinat, 1938: The Geographical Journal, v. 93, p. 281–312.

——, 1941, Denudational landforms of the central Libyan Desert: Journal of Geomorphology, v. 4, p. 3–23.

Penck, W., 1924, Die Morphologische Analyse: Stuttgart, J. Engelhorn, 283 p.

Reeves, C. C., Jr., and Reeves, D. W., 1971, Pleistocene and Recent channel seepage flats in north-central Texas: Geological Society of America Bulletin, v. 82, p. 3511–3518.

Robinson, E. S., 1970, Mechanical disintegration of the Navajo Sandstone in Zion Canyon, Utah: Geological Society of America Bulletin, v. 81, p. 2799–2806.

Savigear, R.A.G., 1952, Some observations on slope development in South Wales: Transactions of the Institute of British Geographers Publication 18, p. 31–52.

Scheidegger, A. E., 1961, Mathematical models of slope development: Geological Society of America Bulletin, v. 72, p. 37–50.

Schumm, S. A., 1956, The role of creep and rainwash on the retreat of badland slopes: American Journal of Science, v. 254, p. 693–706.

Schumm, S. A., and Chorley, R. J., 1964, The fall of Threatening Rock: American Journal of Science, v. 262, p. 1041–1054.

——, 1966, Talus weathering and scarp recession in the Colorado Plateau: Zeitschrift für Geomorphologie, v. 10, p. 11–36.

Sharp, J. M., Jr., 1976, Ground-water sapping by freeze-and-thaw: Zeitschrift für Geomorphologie, v. 20, p. 484–487.

Sharp, R. P., 1973, Mars; Fretted and chaotic terrains: Journal of Geophysical Research, v. 78, p. 4073–4083.

Simpkins, W. W., and Baumgardner, R. W., Jr., 1982, Stream incision and scarp retreat rates based on volcanic ash date from the Seymour Formation, in Geology and geohydrology of the Palo Duro Basin, Texas panhandle: Austin, University Texas Bureau of Economic Geology Geological Circular 82-7, p. 160–163.

Small, R. J., 1961, The morphology of Chalk escarpments; A critical discussion: Transactions and Papers of the Institue of British Geographers Publication 29, p. 71–90.

Smith, B. J., 1978, The origin and geomorphic implications of cliff foot recesses and tafoni on limestone hamadas in the northwest Sahara: Zeitschrift für Geomorphologie, v. 22, p. 21–43.

——, 1983, Comment on 'Use of packrat middens to determine rates of cliff retreat in the eastern Grand Canyon, Arizona': Geology, v. 11, p. 494.

Smith, H.T.U., 1972, Playas and related phenomena in the Saharan region, in Reeves, C. C., Jr., ed., Playa Lake Symposium, 1970: Lubbock, Texas, International Center for Arid and Semi-arid Land Studies Publication 4, p. 63–87.

Smith, K. G., 1958, Erosional processes and landforms in Badlands National Monument, South Dakota: Geological Society of America Bulletin, v. 69, p. 975–1008.

Sofaer, A., Zinser, V., and Sinclair, R. M., 1979, A unique solar marking construct: Science, v. 206, p. 283–291.

Sparks, B. W., 1960, Geomorphology: New York, John Wiley and Sons, 371 p.

Stearns, H. T., 1936, Origin of the large springs and their alcoves along the Snake River in southern Idaho: Journal of Geology, v. 44, p. 429–450.

Stetson, H. C., 1936, Geology and paleontology of the Georges Bank canyons; Part 1, Geology: Geological Society of America Bulletin, v. 47, p. 339–366.

Strahler, A. N., 1963, Physical geography, 3rd ed.: New York, John Wiley and Sons, 733 p.

Terzaghi, K., and Peck, R. B., 1948, Soil mechanics in engineering practice: New York, John Wiley and Sons, 566 p.

Townshend, J.R.G., 1970, Geology, slope form, and slope process, and their relation to the occurrence of laterite: The Geographical Journal, v. 136, p. 392–399.

Tuckfield, C. G., 1973, Seepage steps in the New Forest, Hampshire, England: Water Resources Research, v. 9, p. 367–377.

Van Devender, T. R., and Spaulding, W. G., 1979, Development of vegetation and climate in the southwestern United States: Science, v. 204, p. 701–710.

von Bandat, H. F., 1962, Aerogeology: Houston, Texas, Gulf Publishing Co., 350 p.

Walker, J. R., 1978, Geomorphic evolution of the Southern High Plains: Waco, Texas, Baylor University Geological Studies Bulletin 35, 32 p.

Wells, S. G., Love, D. W., and Gardner, T. W., eds., 1983, Chaco Canyon Country; A field guide to the geomorphology, Quaternary geology, paleoecology, and environmental geology of northwestern New Mexico: American Geomorphological Field Group 1983 Field Trip Guidebook, 253 p.

White, W. N., Broadhurst, W. L., and Lang, J. W., 1946, Ground water in the High Plains of Texas: U.S. Geological Survey Water-Supply Paper 889-F, p. 381–421.

Wood, W. W., and Osterkamp, W. R., 1984, Recharge to the Ogallala Aquifer from playa lake basins on the Llano Estacado; An outrageous proposal?, in Proceedings Ogallala Aquifer Symposium 2: Lubbock, Texas Tech University Water Resources Center, p. 337–349.

MANUSCRIPT ACCEPTED BY THE SOCIETY NOVEMBER 14, 1989

Printed in U.S.A.

Chapter 15

Groundwater and fluvial processes; Selected observations

Edward A. Keller
Environmental Studies Program and Department of Geological Sciences, University of California, Santa Barbara, California 93106
G. Mathias Kondolf
Department of Landscape Architecture, University of California, Berkeley, California 94720
with case studies by
D. J. Hagerty
Department of Civil Engineering, Speed Scientific School, University of Louisville, Louisville, Kentucky 40292
G. Mathias Kondolf

INTRODUCTION

Geomorphic models have traditionally emphasized the role of surface runoff in drainage network development and in relations between stream channel form and process (e.g., Horton, 1945). According to these models, drainage develops principally by surface erosion, but subsurface erosion is the dominant mechanism for drainage network extension in some localities (Small, 1965; Dunne, 1980; Higgins, 1984; and Swanson and others, 1989). Although surface channels are the most visible portion of a drainage network, groundwater may exert a significant influence on stream channel form and process, especially as it relates to riparian and aquatic biota.

Groundwater, here defined to include intragravel water below the streambed, plays an important role in all three phases of the fluvial system: (1) the fluid water, (2) the channel and riparian corridor, and (3) the drainage network. When discussing the flow of subsurface water in a channel, positive seepage refers to subsurface water that enters the stream; negative seepage refers to surface flow that infiltrates the streambed.

The purpose of this chapter is to review and discuss how groundwater influences the first two phases of the fluvial system; groundwater influences on development of the drainage network are considered elsewhere in this volume. It is emphasized that the nature and extent of possible interactions between surface-subsurface flow and channel form and process remain a part of fluvial studies that have been little researched, and as a result constitute a potentially fruitful area for future work.

GROUNDWATER INFLUENCES ON CHANNEL FLOW

At a basic level, groundwater largely determines whether a particular stream reach is classified as perennial, intermittent, or ephemeral. Perennial streams are fed more or less continuously either by a shallow water table or by discrete point sources of groundwater. Intermittent stream reaches receive groundwater flow only part of the year, and flow ceases when the water table drops below the channel (e.g., Hill and Lehre, this volume). Ephemeral channels are, by definition, everywhere above the water table and so do not have a component of flow derived from that source. Instead these streams typically experience flow loss in the downstream direction as water seeps into the channel bed.

Many different and often complex interactions between stream, soil, and groundwater are possible. A recent study of the Mattole River basin in northern California used water chemistry (chloride and silica content) and environmental isotopes (oxygen-18, tritium, and deuterium) to evaluate relative abundance of major components (rainfall, surface runoff, soil water, and groundwater) of stormflow (Kennedy and others, 1986). Analysis of rainfall and river discharge from a January 1972 storm that produced 250 mm of precipitation in 4 days (80 percent in 42 hr) and a resulting flow event that increased river discharge from 22 m^3/s to a peak of 1,300 m^3/s provides interesting insight into runoff processes. The storm produced a large volume of runoff in a short period of time for the 620-km^2 basin, which is characterized by relatively steep slopes, rocks with low permeability, and a relatively deep (15 to 30 m) water table. Surprisingly the chemistry and isotopic composition of the stormflow from the January event remained remarkably stable, suggesting that the major component of stormflow was displaced prestorm soil water rather than surface runoff or groundwater discharge. Because the runoff occurred within hours to a few tens of hours following precipitation, slope processes that convey large amounts of water laterally through soil are required (Kennedy and others, 1986). These authors concluded that more research is necessary to better

---

Keller, E. A., and Kondolf, G. M., 1990, Groundwater and fluvial processes; Selected observations, with case studies by Hagerty, D. J., and Kondolf, G. M., *in* Higgins, C. G., and Coates, D. R., eds., Groundwater geomorphology; The role of subsurface water in Earth-surface processes and landforms: Boulder, Colorado, Geological Society of America Special Paper 252.

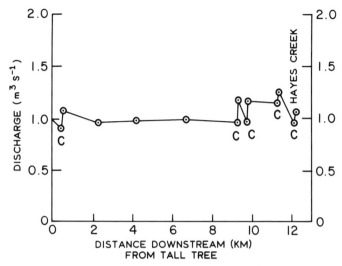

Figure 1. Downstream changes in baseflow, Redwood Creek, California, on August 19, 1983. C denotes location where a point source of groundwater discharges into the streambed. The location of the reach is in Redwood National Park, near the Tall Trees Grove downstream to the confluence of Hayes Creek near Orick. (Modified after Moses, 1984.)

understand relations between overland flow and subsurface flow in generating stormflow. Furthermore, a good deal of variability in stormflow can be expected due to large variability in basin characteristics such as topography, geology, soils, and hydrology.

Point sources of groundwater are often lithologically controlled, but need not be. Many groundwater paths lie entirely in the alluvium. Streamflow may infiltrate into the bed and banks, flow through alluvial deposits, and reemerge downstream. In many cases, however, water is permanently lost to or gained from the regional groundwater flow system. Huff and others (1982) reported streamflow increases of 43 percent from point sources in a small bedrock-controlled channel in eastern Tennessee. In a different, lithologically controlled setting, Kondolf and others (1989) document baseflow increases of up to 277 percent over a 5-km reach of Pine Creek in the eastern Sierra Nevada, California. The importance of groundwater in locally augmenting surface flow was illustrated by Moses (1984), who measured downstream flow increases of up to 20 percent from point sources of groundwater in Redwood Creek, northern California (Fig. 1).

In valleys with highly permeable alluvial banks, bank storage (water stored in alluvial banks at high flow) can be important in attenuating flood peaks (Todd, 1955; Pinder and Sauer, 1971), and as a source of baseflow after the flood peak. In the Carmel River, near Monterey, California, baseflow was found to increase by up to 15 percent over a 10-km reach from bank storage for 2 months after the last flood peak (Kondolf and others, 1989). On the Terror River on Kodiak Island, Alaska, Trihey (1981) documented changes in flow over an alluvial reach, upper Bear Meadow. During the first flows of the spring (after the river ice has broken up), the Terror River loses flow to recharge (negative seepage) the alluvium underlying upper Bear Meadow. By late May, recharge is complete and streamflow is either constant or increasing across upper Bear Meadow (Trihey, 1981). The influence of river stage on alluvial water tables is well illustrated by well-level records along a losing reach of the Mokelumne River, California. As illustrated in Figure 2, wells close to the river respond rapidly and directly to stage changes; more distant wells show more delayed and indirect responses (Piper and others, 1939).

The Arkansas River in southwestern Kansas illustrates the interaction of streamflow with a regional groundwater system, the High Plains aquifer. Prior to the late 1960s, the Arkansas River was a gaining stream (discharge increased due to positive seepage) in the reach from Lakin to Garden City, Kansas (L. E. Stullken, personal communication, September 1985). However, the underlying, partially confined High Plains aquifer has been intensively pumped for irrigation in recent years, resulting in widespread drawdown of 6 to 24 m in the aquifer from 1974 to 1980. The overlying shallow alluvial aquifer dropped more than 4 m from 1974 to 1980 (Dunlap and others, 1985). As a result, the Arkansas River now loses water (negative seepage) to the groundwater system.

INFLUENCES OF GROUNDWATER ON CHANNEL FORM AND PROCESS

The influences of groundwater on a stream channel may be direct and/or indirect. Direct influences include the effect of seepage flow on bank erosion, bedform development, and sediment transport. Indirect influences are varied and include the role of groundwater in maintaining riparian vegetation, which in turn,

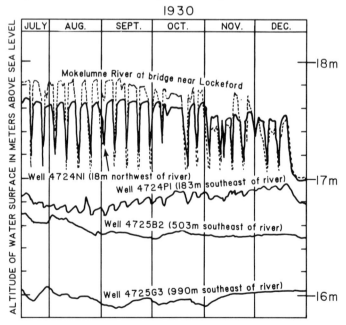

Figure 2. Hydrographs for the Mokelumne River at the bridge near Lockeford, California, and for four wells at increasing distance from the riverbank, 1930 (from Piper and others, 1939).

enhances bank stability. Direct influences on bedforms and sediment transport are small. They can be demonstrated in the laboratory but are hard to detect in the field, because the forces involved are small relative to the tractive forces involved in sediment transport. In contrast, bank collapse following high flow related to seepage can exceed erosion effected by the flows themselves. Indirect influences can be substantial, but their identification may be confounded by concurrent changes of other variables in the fluvial system.

Direct influences: Channel roughness, bedform, and sediment transport

Channel roughness: Groundwater flow could be expected to directly affect channel roughness, bedform development, and sediment transport. Nakagawa and Tsujimoto's (1984) theoretical analysis indicated that flow resistance is increased by infiltration of streamflow into the bed and decreased by emergence of groundwater from the bed.

Bedform and sediment transport: A bedform in a natural stream or river may be defined as an irregularity in the bed formed by the interaction between flowing water and moving sediment (Simons and Richardson, 1966). A classification or characterization of bedforms in natural streams and rivers must be "born from the river," derived from the natural fluvial processes that produce these forms. This would exclude classification based on flume studies, an idea apparently consistent with Neill's (1969) conclusion that excessive study and analysis of bedforms from flume studies results in oversimplification of these forms. Bedforms have been defined or classified on the basis of flow regime (Simons and Richardson, 1966), relative size (Neill, 1969), and morphology (Leopold and others, 1964; Keller, 1971). Terminology for description of bedforms in stream channels is not yet completely standardized. An attempt to do so by the ASCE Task Force (1966) resulted in a fairly good descriptive classification of many common bedforms, but neglected some major forms in sinuous channels, such as pools and riffles. Figure 3 shows a new classification of bedforms suggested for sinuous alluvial channels. The classification is not intended for channels with significant bedrock control that cause expansion and contraction of the channel, or large-scale obstruction of flow and stalling of the thalweg (see Lisle, 1986). The proposed classification has two major subdivisions: channel-forming bedforms and channel-altering bedforms. Channel-forming bedforms are those that control the development of the channel pattern, whereas channel-altering bedforms generally do not control the channel pattern and are often superimposed on the larger channel-forming bedforms. Channel-forming bedforms are further subdivided into first- and second-order forms and the channel-altering bedforms are subdivided into third- and fourth-order forms.

First-order bedforms are asymmetric shoals or alternate bars, which slope first toward one bank then toward the other. These shoals appear to be the primary bedform in the evolution of alluvial channels (Keller, 1972). In many situations asymmetric

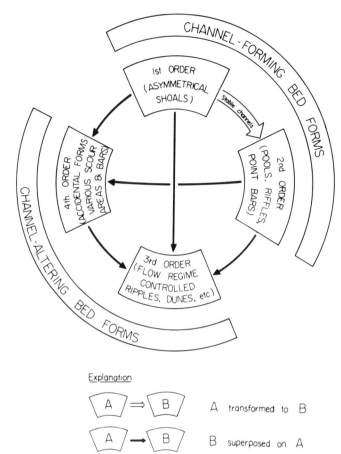

Figure 3. Classification of bedforms for alluvial stream channels.

shoals will be transformed to point bars as pools and riffles develop. If the stream channel is relatively unstable and contains fine-grained bed material or steep slope, pools and riffles may not develop. In this case, asymmetric shoals may be masked by channel-altering bedforms. However, these conditions are not always clearly defined and an intermediate zone exists for channels that have poorly developed pools and riffles.

Second-order bedforms consist of bar-pool topography (pools, riffles, and point bars). This topography may be altered or partially masked by third- or fourth-order bedforms.

Third-order bedforms are those forms shown to be controlled by flow regime. Simons and Richardson (1966) defined flow regime as a range of flows characterized by similar bedforms, resistance to flow, and mode of sediment transport. Two flow regimes, upper and lower, separated by a transition zone are recognized in alluvial channels. Bedforms of the lower-flow regime are ripples, ripples on dunes, and dunes, whereas the bed forms of the upper-flow regime are plane bed, antidunes, and chutes and pools. These forms are not generally significant in controlling the development of the channel, but they may significantly alter the channel morphology. In channels with bar-pool topography, ripples and ripples on dunes can sometimes be seen superimposed on the bars or pools, slowly moving through these larger channel-forming forms in response to relatively low flows.

Fourth-order bedforms are accidental forms produced by local obstruction to flow. They consist of variously shaped scour areas or bars. An example of an accidental form is a scour area and bar area produced when a large tree falls into a stream and obstructs flow. Scour often takes place upstream of the tree, and a depositional bar may form downstream. Fourth-order forms are usually observed at low-flow conditions and high flow tends to remove obstructions and destroy the bedforms unless the roughness elements are large and stable. In alluvial channels (with the exception of small streams in forested areas) these forms do not control (except locally) the development of the channel, but they can significantly alter it. Third-order forms may be superimposed on fourth-order forms (e.g., ripples on a bar produced by a tree trunk obstructing flow), but the converse is not generally true.

Effects of inflow of subsurface water into gravel-bed streams with well-developed pools and riffles probably has little effect on the morphology of these bedforms. This results because seepage forces are small relative to other fluvial forces associated with flow of water in the main channel. However, in sand-bed channels with a well-developed set of channel-altering bedforms such as riffles or dunes, the inflow of subsurface water may change or modify bedforms and, thus, resistance to flow (Richardson and Richardson, 1985). This may result in part due to interactions that take place between the flow of water seeping out of the bed and flow in the main channel. Richardson and Richardson (1985) point out that an exaggerated boundary layer or wedge forms between the channel bed and where water emerges from the bed. This layer is evidently larger than the boundary layer that would exist without emergence of subsurface flow, a phenomenon reported earlier by Watters and Rao (1971).

Experiments in a natural channel have not been designed to test the significance of groundwater entering the stream or channel water infiltrating into the local groundwater system. However, a recent flume study (Richardson and Richardson, 1985) for a sand-bed channel provides some interesting information. In that study, it was determined that inflow of subsurface water into the channel did significantly facilitate transformation of bedforms and, thus, bed roughness. The seepage caused forms such as dunes to become longer and flatter in the channel reach where subsurface water emerged. One of the more interesting results that Richardson and Richardson (1985) found was that unit stream power (defined as the product of bed shear stress and mean velocity) consistently increased where subsurface water was introduced into the sand-bed channel. In their most dramatic case, stream power increased approximately 12 times, causing a bedform change from plane bed to ripples to dunes. Change in unit stream power was found to be most pronounced where initial unit stream power prior to introduction of subsurface water into the channel was relatively low. The significant variables that effect unit stream power are depth and width of flow, mean velocity of flow, and energy slope. Unit stream power is particularly sensitive to changes in the energy slope (approximately the slope of the water surface), and Richardson and Richardson (1985) reported that the water-surface slope locally increased by 10 to 110 percent where subsurface water was injected into the stream from below. The flume experiments also suggested an increase in velocity along the reach where subsurface water was emerging. Effects were most pronounced for relatively low flows in the lower-flow regime where velocity was observed to increase by as much as 23 percent. Experiments in the upper-flow regime showed that inflow of subsurface water had only a small effect on velocity.

Simons and Richardson (1966) suggested that emerging groundwater (positive seepage) should produce a seepage force that, by reducing the effective weight of bed material, should lead to easier entrainment. Field observations on a high-sediment-load Alaskan stream by Harrison and Clayton (1970) suggested that gaining reaches (those with positive seepage) were more competent than losing reaches (those with negative seepage). Laboratory experiments designed to test this observation demonstrated no direct effects of seepage. Harrison and Clayton (1970) credited the differences in competency to the formation of a mud seal that formed when water and fine sediment infiltrated into the losing reaches. A mud seal develops just below the streambed where fine suspended sediment is trapped when streamwater infiltrates the bed, rendering it resistant to erosion (Harrison and Clayton, 1970). In their recent flume experiments, Richardson and Richardson (1985) did detect higher sediment-transport rates associated with emerging groundwater (positive seepage) in the lower-flow regime, but a possible inhibition of sediment transport in the upper-flow regime.

Because seepage forces tend to be very small relative to the tractive forces involved in sediment transport, it appears that emerging groundwater (positive seepage) has little effect on sediment transport (K. S. Richard, personal communication, 1983). However, the effects of downwelling (negative seepage) may be more pronounced if the mud seals observed by Harrison and Clayton (1970) develop. Despite these findings, the role of the subsurface flow of water on sediment transport remains largely unknown and untested in natural streams.

Observation of rivers in western North Dakota by Hamilton (1970) suggested that sandbars may have a subsurface flow system characterized by negative seepage on the upstream portion of the bar and positive seepage at the downstream end. As a result, the upstream end of the bar should be relatively resistant to erosion compared to the downstream end where the sediment is looser and presumably less resistant to erosion. Additional research is necessary to confirm this hypothesis—the concept has interesting implications for bar formation and downstream migration of bars.

Direct influence: Bank erosion

Erosion of river banks is an important geomorphic process in the fluvial system. This is because bank erosion is one of the fundamental processes involved in channel migration and the formation of flood plains by lateral migration (Leopold and others, 1964; Hooke, 1979). A number of physical, biological, and climatic factors may influence bank erosion. These include, among others (Knighton, 1984): the nature and extent of riparian vegetation, and the number and size of animal burrows; channel geometry, including width and depth of the channel and height and angle of the banks; climate as characterized by the duration, intensity, and amount of rainfall, along with the number of freeze-thaw cycles per year; composition of the bank material, particularly grain-size distribution, shear strength, and stratification of sediments in the bank; characteristics of flow, including the magnitude, frequency, and duration of discharge along with the distribution of stream power, shear stress, and intensity of turbulence; and, finally, the presence of water in the banks, particularly as it relates to seepage forces, piping, and moisture content of the bank materials.

One of the more important factors in bank erosion is the characteristics of flow, as it is the water that transports eroded bank materials. Of the remaining factors, the water content of bank materials and condition of riparian vegetation are related to groundwater processes that may affect rates of bank erosion. The role of spring sapping and tunneling in facilitating local bank erosion has been alluded to by several studies, including that of Twidale (1964). Twidale's study of a river in South Australia illustrates that sapping and tunneling can produce local areas of acclerated bank erosion, producing small embayments along the bank. This process occurs where point sources of groundwater emerge from the streambanks. In another study, Lawler (1987) studied channel-bank erosion in Wales and concluded that migration of subsurface moisture to freezing points and frost action were very significant in preparing riverbank materials for erosion. Lacking such preparation, little bank erosion occurred.

Seepage effects on channel-bank stability were investigated in a large flume by Burgi and Karaki (1971). Their experiments for a poorly sorted sand-bed channel suggest that both positive groundwater seepage and velocity of channel flow are important in bank erosion. At relatively low-flow velocity of channel flow (<0.3 m/s), bank erosion was found to be sensitive to positive seepage and to vary directly with local hydraulic gradient. For the flume experiments, a threshold velocity of flow (approximately 0.3 m/s) apparently exists, which if exceeded, causes change in the processes of bank erosion. Above the threshold the dominant cause of bank erosion is the velocity of flow producing shear stress that exceeds the strength of the bank material. Finally, experiments by Burgi and Karaki (1971) suggest that when negative seepage is present, the channel banks are more stable than if no seepage or positive seepage is present. This is attributed to the formation of a silt seal similar to that discussed by Harrison and Clayton (1970).

Dewatering of saturated banks in the aftermath of high flows can produce substantial bank collapse. Leopold and others (1964, p. 446) observed numerous fresh blocks of bank material in gully channels in alluvial valleys following high flows. They reasoned that the blocks fell from the bank under the influence of seepage pressure from drainage of flood-stored water. During the rising limb the bank materials are relatively dry and accepting water, but during the falling limb they are saturated and losing water. This effect has been documented along the Ohio River and is described by Hagerty and others (1983) and in the case study at the end of this chapter. Similarly, Twidale (1964) found that sapping and tunneling led to locally accelerated bank erosion, producing small embayments along the bank; he further speculated that diffused subsurface flow was important in initiating more general bank collapse. Finally, Schumm and Phillips (1986) speculate that some channels on the Canterbury Plain, New Zealand, have a composite origin characterized by initial overland flow, which establishes the channel, followed by enlargement produced by positive seepage and surface flow that removes eroded sediment. Modification of a channel by seepage produces a wide, flat valley with irregular margins.

Indirect influences: Riparian vegetation

One of the most important mechanisms by which groundwater influences channel form is through its influence on riparian vegetation. Riparian vegetation contributes substantially to the ability of streambanks to resist erosion by increasing the shear strength of the bank materials. The effect is most pronounced on smaller systems, where the erosive forces of the flow are less and, thus, the resistance provided by vegetation is relatively more important.

Ziemer (1981) measured in situ soil strength on slopes, and concluded that soil strength increases as total biomass of roots increases. Shear strength of soil results from a combination of cohesive and frictional forces, and the role of the roots may increase the former by providing an apparent cohesion. Ziemer speculates that the increase in soil strength may result from several factors, including: the anchoring of the soil mass by roots that penetrate through the soil into fractures in bedrock; the binding of zones of weak soil with stronger soil through criss-crossing root systems that go through both layers; and interlocking fibrous binding due to the roots within a weak soil mass. These three mechanisms of increasing soil strength through root systems are as applicable to streambanks as other slopes, and root systems where exposed by streambank erosion are more resistant to erosion than are banks free of vegetation.

Maddock (1972) noted that vegetation reduces bank erodibility and that tree-lined channels tend to be narrower than unvegetated channels transporting the same water and sediment load. Charlton and others (1978) reported that channels were 30 percent narrower when lined with riparian trees, and Graf (1978) states that in the Colorado Plateau region, channels have shown an average reduction in width of about 27 percent due to the spread of Tamarisk, which is a shrub or low tree introduced into the American Southwest in the late 1800s.

Undercut banks are less likely to fail when stabilized by root systems. Erosion by direct flow of water is also inhibited by the protective armor provided by roots, stems, and limbs. Vegetation is hydraulically "rough" and so tends to reduce flow velocities that impinge upon the protected bank. The presence of root mats can impart a shear strength to banks composed on unconsolidated, noncohesive sediments—banks that would otherwise have essentially no shear strength. Smith (1976) found that samples of bank material with root mats at least 5 cm thick were 10,000 times more resistant to erosion than samples with no roots.

Riparian vegetation can be affected by changes in groundwater levels in the banks. Water-table declines have resulted in extensive mortality of riparian trees along the Arkansas River (L. E. Stullken, personal communication, Sept., 1985) and the Carmel River (Kondolf and Curry, 1986). In the latter example, extensive bank erosion occurred along the reach affected by drawdown (by pumping) and vegetation die-off, as described in the case history at the end of this chapter.

Loss of riparian vegetation can increase potential damage caused by floods. For example, Taylor (1984) found that damage from the September 1982 floods was more severe on stream reaches in the eastern Sierra Nevada that had previously suffered declines in riparian vegetation from dewatering of channels than on streams with unaffected riparian vegetation.

The potential effect of lowering groundwater level on streambank erosion was dramatically shown by massive bank erosion during the floods of October 1983 in Tucson, Arizona. Investigations following the floods to determine why bank erosion was so extensive concluded that recent (pre-flood) lowering of the groundwater, due to overpumping of aquifers, had caused a reduction in the abundance of riparian vegetation, which had provided a measure of bank stabilization (Saarinen and others, 1984).

The significance of riparian vegetation in stabilizing streambanks in the southwestern United States is emphasized by Baker (1985) in the review of a book on the Colorado River basin by W. L. Graf. In the review, Baker argues that prior to settlement by Anglos in the 1800s, large floods probably had only a small effect on streambanks, which were stabilized by riparian vegetation. Harvesting of riparian trees for lumber and fenceposts by the early settlers destabilized the banks and made them vulnerable to floods of a magnitude that earlier would not have caused significant bank erosion. Then, in the 1900s, exotic species, such as Tamarisk, invaded the region. These exotics displaced native riparian vegetation, but did provide bank stability. Today, in many areas, the water table has dropped well below the channel, and both native and exotic riparian vegetation has disappeared. For example, in 1890 the Santa Cruz River that flows through Tucson, Arizona, had several perennial reaches characterized by riparian vegetation, and surface water with fish and mammal life supported by a near-surface water table. Today the river's course is characterized by a continuous ephemeral channel, and the water table is approximately 100 m below the surface of the land (Saarinen and others, 1984). Although it might be argued that the 1983 floods of the Santa Cruz River in the Tucson area would have caused extensive damage whether riparian vegetation was present or not, it is likely that bank erosion there would have been less severe. Certainly, the existence of riparian vegetation would help to armor streambanks and reduce the rate of bank erosion by floods of smaller magnitudes than the 1983 event.

A rise in the water table may also have adverse effects on riparian vegetation. This is illustrated on Redwood Creek, which drains 720 km^2 of mixed forest land, varying from Douglas Fir in the upper part of the watershed to Redwood trees in the lower. Timber harvesting and high-magnitude storms resulted in accelerated erosion in the watershed in recent years. Delivery of this sediment to the channel of upper Redwood Creek resulted in as much as several meters of aggradation. The aggradation caused a localized rise in the water table along the banks of Redwood Creek, killing Douglas Fir and Redwood trees, which require well-drained soils. These trees had been stabilizing large alluvial terraces; their loss contributed to the onset of extensive bank erosion (Fig. 4; Nolan and Janda, 1979).

The case histories discussed above suggest that either a rise in, or lowering of groundwater level may initiate significant bank erosion, channel widening, and a tendency to braid. The response of the bank vegetation to change in the water table was the

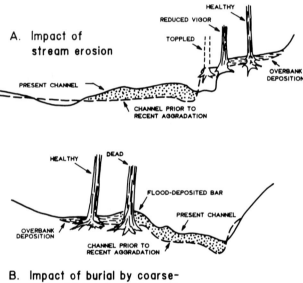

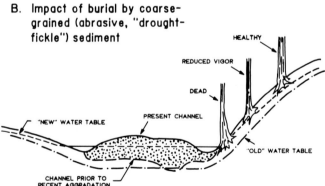

Figure 4. Idealized diagrams illustrating effects of channel aggradation and local rise in groundwater table on trees (Douglas Fir) in the riparian zone (after Nolan and Janda, 1979).

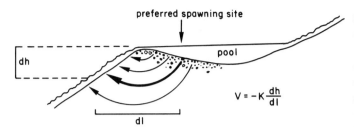

Figure 5. Idealized diagram showing tail-of-pool spawning site with intragravel flow induced by hydraulic gradient from pool surface to point of emergence in riffle. Vertical scale greatly exaggerated.

determining factor. These cases also illustrate that the stability of a river system depends not only on the driving forces, such as high-magnitude storms, but also the resisting forces that defend the channel. Roots of trees in the riparian zone can increase the resisting forces, but are vulnerable to changes in water level. Bank erosion can be particularly severe when there is a combination of high driving forces and reduced resisting forces.

GROUNDWATER AND FISH HABITAT

Surface-groundwater interactions assume a particular importance in determining the quality of spawning and juvenile-rearing habitat for fish of the family Salmonidae (salmon and trout). Salmonids spawn by excavating a depression in the streambed (called a redd), depositing their eggs there, and covering the eggs with gravel. Successful incubation of the buried eggs depends on the throughflow of oxygenated water to bring dissolved oxygen to the eggs and carry off the metabolic wastes. In an apparent measure to maximize intragravel water past incubating eggs, many fish locate their redds where groundwater is upwelling from or downwelling into the streambed.

The upwelling or downwelling may be the result of groundwater flow patterns in the surrounding terrain, but more often is related to position within the longitudinal profile of the stream. Spawning is most often observed in gravels in the tails of pools (Vaux, 1962), where a positive hydraulic head drives oxygenated streamwater into the bed, to emerge downstream within the riffle (Fig. 5). This flow can be considered with reference to Darcy's Law:

$$V = -K \frac{dh}{dl}$$

where V is the Darcy velocity (specific discharge), K is the hydraulic conductivity, and dh/dl is the hydraulic gradient, as illustrated in Figure 5. Local position within the longitudinal profile is important in providing the necessary hydraulic gradient. So long as the gravels remain relatively free of fine sediment, they will be permeable and have high values of hydraulic conductivity, thereby insuring intragravel flow past the incubating eggs. Water that infiltrates the tail of a pool reemerges in the riffle, where spawning is also frequently observed. However, riffle sites are considered less desirable for spawning because the dissolved oxygen of intragravel waters may be somewhat depleted by biological and chemical oxygen demands as it moves through the gravel. This discussion refers to an idealized case; physical conditions at field sites are exceedingly variable, and fish behavior is notoriously unpredictable. Spawning has been observed in a wide range of conditions, and the fish are not always right: many redds have poor survival because of inadequate intragravel flow. Nonetheless, it has often been observed that spawning fish frequently select sites of downwelling or upwelling.

Sites of downwelling are commonly selected for spawning by many salmonids (e.g., Hazzard, 1932; Stuart, 1954; Vronskiy, 1972). In high-latitude channels, upwelling groundwater may be the principal desideratum for spawning salmonids because of the protection it affords against freezing of eggs (e.g., Benson, 1953; Vining and others, 1985). The ecological importance of upwelling is most dramatically illustrated on the Chilkat River in southeastern Alaska, where a reach of the river is kept ice-free throughout the winter by actively upwelling groundwater. Although most species of salmon in Alaska spawn in the summer or early fall (an evident adaptation to the harsh winters), a large run (100,000 to 500,000) of chum salmon (*Oncorhynchus keta*) takes advantage of the ice-free reach of the Chilkat River to spawn in the fall and early winter. Weak spawned-out salmon and their carcasses provide a uniquely reliable winter food source for American bald eagles (*Haliaeetus leucocephalus*). Because of the extraordinary importance of this food source in winter, the most critical period for their survival, eagles come to the Chilkat River from breeding areas hundreds of kilometers away; the population of eagles peaks at about 3,200 to 3,600 in November (Fig. 6; Hansen and others, 1984). Thus, the largest population of eagles remaining in the world is dependent on upwelling groundwater.

Groundwater often plays an important role in the quality of juvenile-rearing habitat for fish. Juvenile salmonids require pool environments with adequately low temperatures to survive the summer months. Denton (1974) identified summer pool environment as a principal limiting factor to steelhead trout and salmon productivity in many streams of northern California. In Redwood Creek, recent aggradation has fully or partially filled many pools, reducing the available habitat for juvenile rearing and resulting in higher water temperatures. In the lower 20 km of the stream, a number of the larger pools are associated with redwood trunks that accumulate along the bank in log jams and induce local scour. Some of these pools are sites of upwelling groundwater that is colder than the rest of the stream (Keller and others, 1990). These cold pools are crucial refugia for juvenile salmonids during years of low baseflow.

Figure 7 shows a cold pool as it existed in the summer of 1981 at the junction of Hayes Creek and Redwood Creek, located approximately 4 km upstream from the town of Orick. Although the lower portion of Hayes Creek is dry in the summer, it contributes subsurface water to the main channel of Redwood

Figure 6. Upwelling groundwater along part of the Chilkat River, Alaska, prevents freeze up of the river and provides suitable habitat for as many as 500,000 spawning chum salmon. More than 3,000 bald eagles gather here each fall and winter to feed on dead and dying salmon. (Photograph courtesy of the National Audubon Society.)

Creek. The temperature of the groundwater entering the channel was approximately 11° to 12°C, in marked contrast to mainstream temperatures of Redwood Creek, which on warm afternoons may exceed 20°C. In 1981, the cold water apparently was prevented from mixing with the warmer water in the mainstream of Redwood Creek by both the organic debris and midchannel bar (Fig. 7). Thus, a plume of cold water lingered in the bottom of the pool. As cold groundwater (with its relatively low dissolved-oxygen content) slowly mixes with the warmer, more oxygenated water of Redwood Creek, the oxygen concentration of cold pools increases as shown in Figure 7. The low dissolved-oxygen levels, 1.6 to 4.8 ppm (Fig. 7), are consistent with a groundwater source for the colder water. Sources other than groundwater (e.g., cold water due to density stratification or shading) should have dissolved-oxygen levels closer to main stream values of about 8 ppm.

During seasons of low flow in Redwood Creek, the cold pools apparently persist. Figure 8 shows thermograph data for two locations in the Hayes Creek cold pools and maximum-minimum temperature data from a location in the main stream for the period September 22–25, 1981. At no time during the observation period was the water in the cold pool as warm as that in the main stream, and differences in temperature varied systematically with distance from the source of cold water entering the pool. During warm afternoons, the temperature differences between the main stream and cold pools were most pronounced; during early morning hours, differences were less (Fig. 8). The horizontal portions of the graph (station #1) are presumably related to a problem associated with the recorder.

During the winter of 1981–82, the cold pool at Hayes Creek increased in area, depth, and volume. The mid-channel bar near the mouth of Hayes Creek and some of the large organic debris were removed, producing significant changes in the summer low-flow morphology. The gravel bar and organic debris no longer isolated the effluent groundwater, and as a result, the temperature differences in the pool were not as great as those observed during the previous summer (Keller and others, 1990). Therefore, some cold pools, because they are dependent on subtle interactions between erosion and deposition that control pool morphology and groundwater flow, are likely to be ephemeral features. This was confirmed by a more recent study of cold pools in lower Redwood Creek, which concluded that cold pools remain relatively stable in response to winter flows with recurrence intervals ≤3 yr but may be destroyed by channel changes associated with flows with recurrence intervals of ~5 yr (Ozaki, 1987). In other cases, large cold pools may be characterized by stable-channel morphology (e.g., a pool defended by a large outcrop of resistant bedrock) and a stable source of effluent groundwater. Such pools are likely to be a much more permanent part of the fluvial system.

Flow measurements on Redwood Creek above and below the Hayes Creek cold pool (Keller and others, 1990; Moses, 1984) are presented in Table 1. The percent increase in flow at the Hayes Creek cold pool is well in excess of possible measurement error of 5 to 10 percent. The data show that as discharge in the main stream decreases, the amount of groundwater entering the main stream also decreases, but the percentage of flow contributed by groundwater sources increases significantly. Therefore, at particularly low flow (0.4 $m^3 s^{-1}$), groundwater through the Hayes Creek reach increases main stream discharge by approximately 22 percent. Some of the increase in discharge was due to seepage out of a large gravel bar in Redwood Creek. The Hayes Creek pool is located on the outside of a bend in Redwood Creek associated with a large upstream and adjacent gravel bar. It was observed that cool water seeped from the bar into the pool. The source of the water is assumed to be Redwood Creek. As the water moves through the bar in the down-valley direction, it is cooled. It is speculated that this source of water did not produce the cold pool, because similar geometry occurs at many pools in Redwood Creek but there are few cold pools.

If more cold pools existed in Redwood Creek, fish produc-

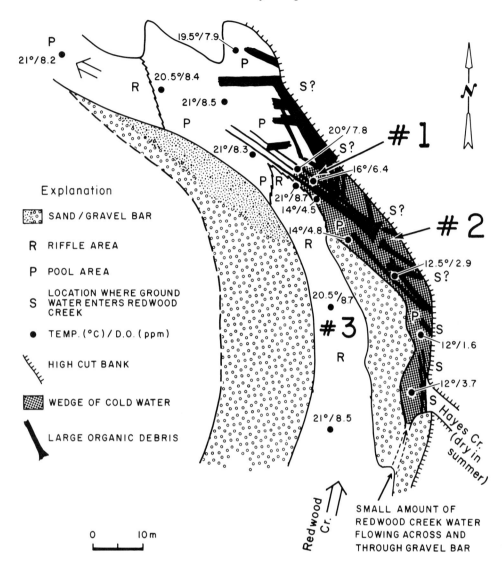

Figure 7. Morphological map and water-temperature measurements, Hayes Creek cold pool, August, 1981 (after Keller and others, 1990).

tivity might be greater. In particular, the distribution of young Coho salmon is restricted in some years when none have been found rearing in the Redwood Creek embayment or any of the tributaries (excluding Prairie Creek). Thus, in some years, production of Coho salmon in Redwood Creek may be closely tied to the existence of cold pools (Keller and others, 1990). Because some anadromous fish enter river systems during the summer when mainstream water temperatures are quite high, cold pools might be critical refugia for migrating fish and holding sites for those waiting to spawn. In fact, fish migrating in the summer may move upstream from one cold pool to the next, thus avoiding high temperatures that exist during the day over most of the stream (Keller and others, 1990).

CASE HISTORY: BANK FAILURES CAUSED BY GROUNDWATER FLOW IN ALLUVIUM

D. J. Hagerty

INTRODUCTION

Considerable argument has occurred in recent years about erosion/deposition on the banks of the Ohio River in the midwestern United States. Riparian landowners, research scientists, and engineers have presented differing opinions concerning the causes of erosion, the intensity (rate) of erosion, and the balance (or lack thereof) between erosion and deposition. River-level

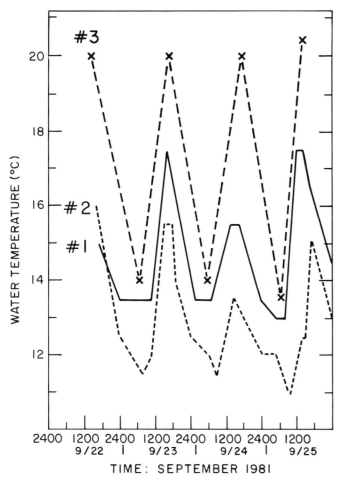

Figure 8. Water temperatures in Hayes Creek cold pool, September 22–25, 1981. Temperatures at stations #1 and #2 were taken with continuously recording thermographs placed in the cold pool. Temperatures at station #3 were measured with a maximum-minimum thermometer in the main stream. Locations of the stations are shown in Figure 7. (After Keller and others, 1990.)

TABLE 1. INCREASE IN DISCHARGE OF REDWOOD CREEK, DUE TO EFFLUENT SUBSURFACE WATER AT THE HAYES CREEK COLD POOL*

	Date Measured†	Upstream Discharge (m^3s^{-1})	Downstream Discharge (m^3s^{-1})	Increase (m^3s^{-1})	Increase§ (%)
Hayes Creek	Sept. 4, 1981	0.40	0.49	0.09	22
	August 10, 1982	0.69	0.82	0.13	19
	July 9, 1982	1.77	1.97	0.20	11

*Data from Keller and others, 1990.
†On August 23, 1982, the discharge upstream and downstream of the Hayes Creek cold pool remained nearly constant at 0.45 m^3s^{-1}.
§Discharge can be measured at ± 5 to 10 percent. Some of the increase in discharge was probably due to seepage through a large upstream gravel bar, rather than groundwater from Hayes Creek. See text for further explanation.

changes caused by construction of navigation dams, waves created by vessels and/or wind, natural tractive forces, and slope-failure mechanisms (creep, drawdown slumping, and other destabilizing processes) all have been cited as causes of erosion. Recent investigations provide an explanation for this puzzle; namely, groundwater flow. The influence of groundwater flow on the banks of the Ohio River was not recognized by land owners or the fluvial geomorphologists and engineers involved in the controversy. Erosion is a complex phenomenon, involving the interaction of many factors. The detailed mechanisms by which bank failures occur still are understood only sketchily at best, as shown by recent analyses of river-meander formation (Ikeda and others, 1981; Kitanidis and Kennedy, 1984; Blondeaux and Seminara, 1985). The formulations given in these and many other descriptions of bank erosion imply that the erosive process is continuous, whereas in reality the erosion occurs as distinct episodes resulting from loss of structural integrity in the bank material.

The author became familiar with many of the details of the mechanisms causing these episodic failures as a result of studies on Ohio River banks. The initial emphasis in these studies was to evaluate the effects of construction and operation of navigation dams on the Ohio River on streambank erosion (Hagerty and others, 1981). During these studies, reconnaissance surveys of more than 7,000 km of streambanks were conducted in the Ohio River basin; a number of reaches were surveyed numerous times. Later field studies were conducted at more than 120 sites, at which 50 had actively eroding alluvial banks. In addition to the general investigations, detailed studies were made at a small number of sites where weekly or monthly visits were made to detect episodic changes (Weigel and Hagerty, 1983; Hagerty and others, 1983). In all of these studies of alluvial banks in the Ohio River system, a number of common features were found:

1. The alluvium was very distinctly stratified, with cohesive layers between relatively thin sand lenses, as is considered typical for much of the Ohio River alluvium (Walker, 1957).

2. The bank topography consisted of nearly vertical faces in the cohesive layers, separated by very gently sloping berms at the face of the sandy lenses.

3. At the face of many of the sand layers, distinct holes or long slit-like cavities were found where sand had been removed from under the overlying cohesive layer, and detached slabs of the cohesive sediment had fallen away from the face of the bank.

4. Abundant evidence was found of groundwater outflow from the sand-layer faces exposed in the eroded banks, with rills and gullies created downslope by this flowing water.

In addition to the characteristics described above, at many

of the sites some rather puzzling features were found. For example, during inspections soon after floods at many sites, fresh scarps and other features indicated that bank failures had occurred during or after those floods. Engineers usually ascribe such failures to "drawdown" effects of falling stream levels. However, failure debris from the collapses was not plentiful on the lower portions of the banks as would be expected from a slumping failure triggered by a rapid fall in river level. When many of the same sites were visited later, long after the flood event, additional failures had obviously occurred since the last inspection. Debris from these failures was present in abundance below the failure locations. At many of these same sites, stream velocities had been measured during the flood events. Near the locations of bank failure the velocities were generally less than 1 m/s, indicating very limited capacity of the stream to remove in situ soils by tractive forces. Evidence of the failure of bank zones after flood recession and ample evidence of copious flow of groundwater out of the banks toward the stream were found on practically all types of bank alignments, including sites on the insides of bends and along straight reaches where meander-growth theory would indicate that minimal erosion or even deposition should have occurred. Finally, on several occasions during inspections by the author, slabs and chunks of bank material separated and fell from the banks, even though the inspections were made at time periods long after flood events had occurred.

During other inspections, following periods of precipitation when the stream did not rise significantly, failures occurred at elevations far above the maximum stage reached during the preceding period of precipitation.

For example, at one site studied in 1981–1982 (Hagerty and others, 1983), the changes in river elevation between the beginning of the study in mid-September and a visit on January 20 were not sufficient to bring floodwaters in contact with the upper bank zones. During this time period, most of the changes that occurred in the upper bank coincided with periods of precipitation and were caused by flow of water out of the face of the bank in sandy zones, on the upper surfaces of more silty layers in the bank alluvium. This flow of water out of the bank face carried the sandy materials out of the face in a gradual grain-by-grain removal. Also, freezing and thawing action at this site, beginning in December 1981, caused some slight change in the appearance of the upper bank zones. Conditions at the beginning of this study are shown in Figure 9A, taken on September 25, 1981. Comparison with Figure 9B, taken on January 20, 1982, indicates very little change in bank conditions during the first 17 weeks of study at this site. When floodwaters occasionally had covered the lower-bank zones, the major changes observed were a reworking and shifting of the sandy materials on the gently sloping lower-bank zones, as can be seen in the lower portions of Figures 9A and 9B. Then a significant flood event occurred between January 20 and January 27, 1982, with a crest approximately 2 m below the top of the bank. This flood event was preceded by a significant episode of precipitation. The late stages of this precipitation event appeared as snowfall. When this site was visited on January 27, significant changes in the bank face were noted. Slabs had toppled out of the bank face, as shown in Figure 9C, and were covered with a dusting of snow on their upper elevations, an indication that they had fallen before the floodwaters crested. Changes continued to occur during February and March 1982 during high-water events. These changes are quite obvious in Figure 9D, taken on April 7. Bank retreat of as much as 2 m occurred on this site.

EXPLANATION OF FAILURES

The explanation for the occurrence of failures when and where they were noted along the Ohio River banks was found in mechanisms of bank failure associated with piping and seepage erosion. In this context, piping has been used to describe the removal of material by percolating waters to produce tubular conduits, in agreement with the definition given by Mears (1968). Seepage erosion denotes removal of soil grains over a broad areal extent by general groundwater outflow (see Dunne, this volume). Both forms of bank-material removal have been noted by the author along the Ohio River.

Examination of the Ohio River banks that failed due to piping or seepage erosion reveals that the failure surfaces typically consist of nearly vertical scarps. The layered nature of the bank alluvium produces at many sites a terraced appearance resulting from successive failures of the different strata. The sand seams, sloping gently toward the stream, provide the basal failure planes. Most often, surficial failures of thin slabs of material occur at the bank faces. Less frequently, large blocks of alluvium are found displaced horizontally in what have been called "wedge" failures (Springer and others, 1985). These various modes of failures are idealized in Figure 10. The localized failure modes shown in Figure 10 have been adopted from those given by Thorne and Tovey (1981) in their study of failures along the Severn River in Wales. Along the Ohio River, tension failures are extremely rare except for thin zones very near the tops of the banks where grass roots provide tensile reinforcement. The cantilever and shear modes of local failure were found very frequently on Ohio River banks; the wedge failure mode shown in Figure 10 was found occasionally along the Ohio River but not nearly as frequently as the local cantilever and shear failures.

THE EVANSVILLE SITE

To illustrate the types of failures shown in Figure 10, reference can be made to a site located on the right (north) bank of the Ohio River a short distance downstream from Evansville, Indiana (Fig. 11). The materials in the bank at this location consist of a sequence of alluvial soils generally more cohesive near the top of the bank than near the water's edge. Figure 11 shows the exit location of an overflow channel created by the Ohio River when, during periods of flooding, relatively silty and sandy materials were deposited in a slough. Stiffer, more cohesive soils occur both

Figure 9. Changes in Ohio River bank: (1) September 25, 1981; (B) January 20, 1982; (3) January 27, 1982; (D) April 7, 1982 (from Hagerty and others, 1983).

upstream and downstream from the area shown in Figure 11. For example, Figure 12 shows the riverbank upstream from the area shown in Figure 11. Here, stiff cohesive layers provide a greater measure of resistance to erosion by the river and groundwater flow. Nevertheless, linear features can be seen in Figure 12, where groundwater flowing out of thin sand seams has caused undermining and collapse of the cohesive layers. Figure 13 shows a detail of the site shown in Figure 11, where the bank is of more easily eroded sandy and silty materials. At this location the removal of the sandy seams by flowing groundwater has been even more significant. The materials in the upper portions of the bank are more fine-grained and possess more cohesion. They are more resistant to the effects of tractive forces in the stream as well as to the seeping groundwater. Nevertheless the removal of sand along thin layers has caused undermining of the cohesive zones and their subsequent collapse. The man standing near the top of the bank in Figure 13 is on a bench created at the location of a sand seam in the bank. Figure 14 shows the removal of more cohesive alluvium in the area immediately downstream from the reach shown in Figure 11. Even in these stiffer, more cohesive units, groundwater emerging from the riverbanks has removed thin sand layers and produced collapse, as shown in the photographs. A more graphic example of this type of failure, which corresponds to the failure mode shown in Figure 10C, is shown in Figure 15. This site is immediately beneath the feet of the man shown near the top of the bank in Figure 13. A detached slab of cohesive material is clearly visible.

The conditions shown in Figure 9 and 11 through 15 are typical of many of the sites investigated on the Ohio River. In many instances, sandy layers were found to be removed to distances of 1.5 to 2.0 m back into the bank face, and detached slabs of cohesive horizons as much as 0.5 m in width were found below their former locations. These failures occurred as late as 7 weeks after flood recession occurred.

Figure 12. Area upstream from the site shown in Figure 11. Stiff cohesive bank materials here strongly resist erosion.

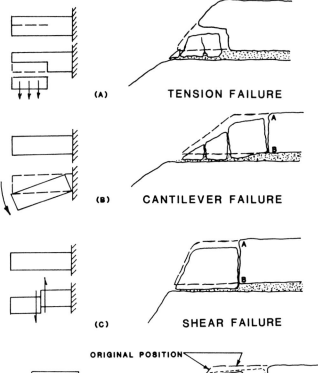

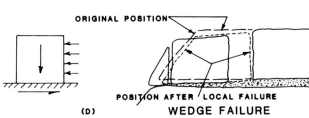

Figure 10. Alluvial-bank failure modes caused by groundwater flow out of bank.

Figure 11. Alluvial streambank on the Ohio River near Evansville, Indiana, where groundwater outflow is an important erosional factor.

Figure 13. Cohesive material (steep scarps) and silty sand seams (gentle slopes) show the influence of groundwater outflow at the site shown in Figure 11.

Figure 14. Area downstream from the site shown in Figures 11 and 12. Cohesive materials are undermined and failing as a result of removal of sandy seams caused by groundwater outflow.

STABILITY ANALYSES

Subsequent to the field investigations mentioned previously, the author and colleagues at the University of Louisville developed numerical, computer-based models of the failure modes shown in Figure 10 (see Springer and others, 1985; Ullrich and others, 1986). The mathematical stability analyses produced the results that had been found during the field studies. Both the field and analytical studies indicate that the formation of nearly vertical cracks is very significant to bank stability during subsequent floods or rainfall events. During field inspections, it was noted that such cracks were widespread behind failing banks, and that such cracks tended to fill rapidly from overland flow during precipitation events. Precipitation or surface runoff infiltrating such cracks can flow downward into the sand seams in the banks and travel through those seams to exit from the bank face, producing removal of the sand from the face. Additionally, water filling the cracks can produce significant hydrostatic pressure and cause the bank material in front of the crack to be displaced toward the stream. This type of action was found to be significant in failure of alluvial banks along the Ohio River. To further illustrate this type of behavior, a hypothetical sequence of events can be described with reference to Figure 14.

HYPOTHETICAL SCENARIO

The hypothetical bank shown in Figure 16 consists of layers of cohesive clay and silt (A, B, and C) interspersed with thin seams of sand. This hypothetical bank has been greatly simplified compared to most actual situations in order to highlight the important features of the failure mechanism. The groundwater table is assumed to be initially below layer C, and the streambank is treated as having been graded to the plane (as is often done in attempts to stabilize the bank). In the sequence of events shown in Figure 16, soon after grading is completed, a flood occurs and inundates the site, with water partially recharging the sand layers (Fig. 16B). After the floodwaters begin to recede and drop below the level of the sand layers in the bank (Fig. 16C), tensile failure occurs in the top bank stratum, and groundwater flowing out of the bank removes sand from all three seams. In the hypothetical bank, layer B has sufficient shearing strength to delay failure until the sand has been removed for a considerable distance from the middle sand seam, at which point layer B eventually fails in cantilever action as shown in Figure 16C. Minor cantilever failures had already occurred in this layer to remove the thin triangular edge of the layer, which had been produced during grading. Similar minor failures occurred in layer C during the flood and soon afterward, during the early stages of seepage erosion in the lowest sand seam. For illustrative purposes, the thickness of layer C is shown greater than the other cohesive layers, and the cohesion is assumed to be lower; these conditions would lead to shear failures and the detachment of slabs, i.e., wedges of cohesive silt and clay having a low width to height ratio. These failures occur when the lowest sand layer is removed below cohesive layer C, as shown in Figure 16C. Seepage erosion continues as groundwater continues to flow out of the bank, but removal continues at a

Figure 15. Failure zone located just below the feet of the man seen standing near the top of the bank in Figure 13. Note detached slab of cohesive material.

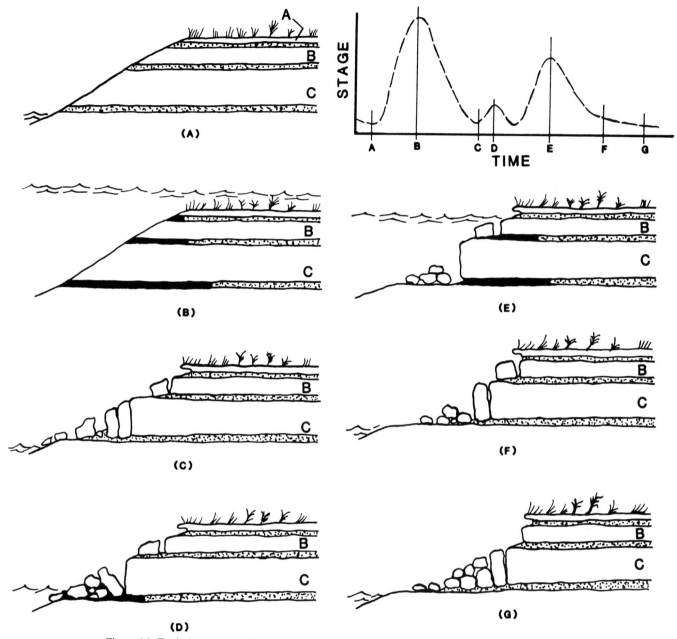

Figure 16. Typical sequence of bank failure and erosion caused by interaction of groundwater and stream.

slower rate because the hydraulic gradients in the seams are steadily decreasing.

A minor flood then occurs as a result of localized rainfall, and the river stage rises slightly as shown in the stage diagram in Figure 16. Some of the fallen blocks are removed by flood flow and are degraded by precipitation. Water seeps into the lowest sand seam during the flood, as shown in Figure 16D. Rain water and runoff flowing into cracks in the bank topple the slabs, which previously rested in front of layer C (compare Figs. 16C and 16D). Precipitation infiltrates upper-bank zones and causes minor sand flows from beneath layer B, and a minor displacement of the fallen block previously located adjacent to the face of layer B.

After a drop in river stage, major rainfall in upstream tributary areas causes the stream to rise to the top of layer B as shown in Figure 16E. During this high-water event, water seeps into the two lower sand layers. Also, blocks of cohesive soil, which had fallen onto the lower bank, are softened and removed by the flow, as is much of the sand that had been sapped out onto the benches in front of the seams. Near-bank flow is likely to be very turbulent because of the blocks and slabs that have fallen previously from the faces of the cohesive layers, and thus is quite erosive. After the flood recedes, sand removal begins again and produces more failures when water flowing out of the sand seams undermines layers B and C, as shown in Figure 16F.

In the months that follow, as water continues to flow out of the sand seams at a steadily decreasing rate, additional failures occur, as shown in Figure 16G. This sequence of events is entirely hypothetical and highly simplified, but the resultant sequence of conditions shown in Figure 16G closely resembles those found by the author at many sites along the Ohio River, and by other investigators on other streams (Kesel and Baumann, 1981).

CASE STUDY SUMMARY

The hypothetical scenario shown in Figure 16 furnishes explanations for many of the puzzling features found at field sites along the Ohio River. These modes of failure present implications for the management of bank erosion along many river reaches in alluvial materials. The sapping mechanisms explain the occurrence of bank failures long after floods have receded. Additionally, the outflow of water and sand from the sand seams during the fall in flood waters, together with the erosive capacity of the flood flow, removes from the bank zone debris that resulted from failures occurring during earlier portions of the flood event. This explains the relatively clean appearance of many sites shortly after flood retreat, as well as the presence of failure debris on those same sites at later times. Because the failure mechanism is initiated in large part by water seeping into sand seams during floods, erosion is as likely to occur on the insides of bends as on the outside of the bends. However, on the outside of bends, tractive forces are more intense and are more likely to remove the debris from failures. Additionally, although sediment deposited along the insides of bends by receding floodwaters can be reworked and carried toward the stream by groundwater flowing out of the sand seams, the increment of sediments tends to counteract the removal process produced by groundwater flow toward the stream. Deposition may be dominant at such a site, and the material lost by seepage may simply reduce the net gain in sediments along the insides of stream bends. On the outside of bends, the action of groundwater flow out of sandy layers is likely to augment significantly the erosive tendencies of the tractive forces of the streamflow. All of the field observations made by the author along the Ohio River, as well as the quantitative stability analyses, indicate that erosion of alluvial streambanks is a complex process involving an interaction of hydraulic and geotechnical factors. This interaction is poorly understood; much work remains to be done in fully documenting, describing, and analyzing the constituent phenomena. However, it is abundantly clear that groundwater flow out of streambanks can be significant in causing erosion and collapse of those banks.

CASE HISTORY: BANK EROSION FROM WATER-TABLE DRAWDOWN, COASTAL CALIFORNIA

G. Mathias Kondolf

INTRODUCTION

The Carmel is a lovely little river. It isn't very long but in its course it has everything a river should have. It rises in the mountains, and tumbles down a while, runs through shallows, is dammed to make a lake, spills over the dam, crackles among round boulders, wanders lazily under sycamores, spills into pools where trout live, drops in against banks where crayfish live. In the winter it becomes a torrent, a mean little fierce river, and in the summer it is a place for children to wade in and for fishermen to wander in. Frogs blink from its bank and the deep ferns grow beside it. Deer and foxes come to drink from it, secretly in the morning and evening, and now and then a mountain lion crouched flat laps its water. The farms of the rich little valley back up to the river and take its water for the orchards and the vegetables. The quail call beside it and the wild doves come whistling in at dusk. Raccoons pace its edges looking for frogs. It's everything a river should be.

A few miles up the valley the river cuts in under a high cliff from which vines and ferns hang down. At the base of this cliff there is a pool, green and deep, and on the other side of the pool there is a little sandy place where it is good to sit and to cook your dinner (Steinbeck, 1945, p. 77-78).

The Carmel River and its valley have undergone profound changes in the decades since John Steinbeck wrote *Cannery Row*. The Carmel Basin is the primary water source for the Monterey Peninsula, and demand for water has grown with the population. Locally extensive channel widening began in 1978 and 1980 along a reach affected by water-table drawdown and bank devegetation near municipal supply wells. Where Steinbeck described the pool, green and deep, there is now a wide channel floored by sand and gravel. The available historical evidence indicates that the bank erosion at this site was caused by the decreased resistance to erosion of banks devegetated as a consequence of water-table drawdown.

SITE DESCRIPTION AND RECENT GEOMORPHIC HISTORY

The Carmel River drains 660 km^2 and has an average discharge of 2.9 m^{3}s^{-1}. It rises in the rugged Santa Lucia Range but traverses an alluvial valley for its lower 24 km before debouching into the Pacific at Carmel. A major flood in 1911 (peak discharge over 570 m^3/s at an upstream gage) caused extensive bank erosion and channel shifting of as much as 0.5 km along the entire length of the alluvial reach. No major floods have occurred since 1914. In 1921, construction of a dam cut off the sediment supply from half of the drainage basin. Recovery from an aggradational flood and adjustment to dam closure tend to produce similar channel changes, so it is difficult to separate their effects in this case. In any case, recovery from the flood and adjustment to dam construction proceeded differently in the upper and lower parts of the 24-km alluvial valley. Upstream of a bedrock narrows, 15 km from the mouth, where the river is steeper (average slope = 0.005), channel change has been continuous up to the present. In contrast, by the date of the first aerial photography in 1939, the lowermost 15 km (average slope = 0.003) had reached a configuration that was to remain essentially unchanged for 4 decades. In this reach (the Lower Carmel River), the channel became incised into the 1911 flood deposits, leaving a 4-m terrace that can be traced unambiguously for 9 km. Incision was accompanied by channel narrowing and development of a dense riparian corridor.

BANK DEVEGETATION AND EROSION

Steinbeck's description remained accurate for the Lower Carmel River until growing demand for water and shrinking reservoir capacity led to extensive pumping of alluvial wells (Fig. 17). Residents' complaints of dying trees near the Berwick wells (among the first pumped) appeared in local papers in the 1960s. As pumping increased and more wells were drilled, the die-off of vegetation became more widespread. Time-series maps prepared from aerial photography by Groeneveld and Griepentrog (1985) show the decline of the riparian forest in the region of high-capacity export wells (Fig. 18).

Pumping reached a peak in the drought year of 1976; by 1977, production was sharply lower because the aquifer (unconfined sand and gravel about 30 m deep and 0.5 km wide) was depleted near the well field; drawdown exceeded 10 m along 4 km of the river (Fig. 19). It is notable that downstream of the well field, the water table was close to normal dry-season level, and the bank vegetation remained healthy.

Low-magnitude floods in 1978 and 1980 (8- and 6-yr events, respectively) produced massive bank erosion in the pumped reach. Resurvey of a cross section near Steinbeck's supper spot documented channel widening from 25 to 65 m from 1965 to 1982 (Fig. 20). The degradation apparent in this cross section occurred after 1980 and was subsequent to—not the cause of—the erosion. Severe bank erosion was limited to the 0.6-km reach above Schulte Road bridge (Fig. 19). Downstream of the pumped reach, the channel remained stable.

CASE STUDY CONCLUSION

The Lower Carmel River is a potentially unstable system on which the presence of healthy bank vegetation can make the

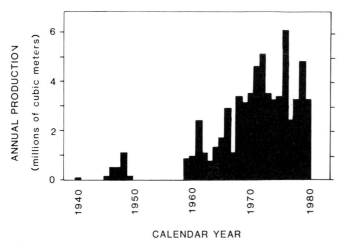

Figure 17. Annual production, Lower Carmel River municipal supply wells through 1980. Source: California-American Water Company production records. (From Kondolf and Curry, 1986; used by permission of John Wiley and Sons.)

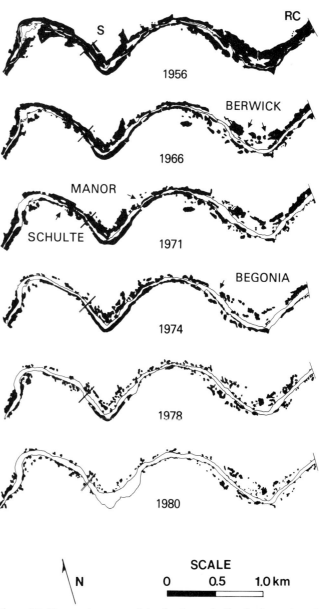

Figure 18. Time-series maps of riparian forest decline in the region of water supply wells along the Lower Carmel River. Lines perpendicular to flow denote divisions into subsections according to locations of wells. (From Groeneveld and Griepentrog, 1985, p. 46.).

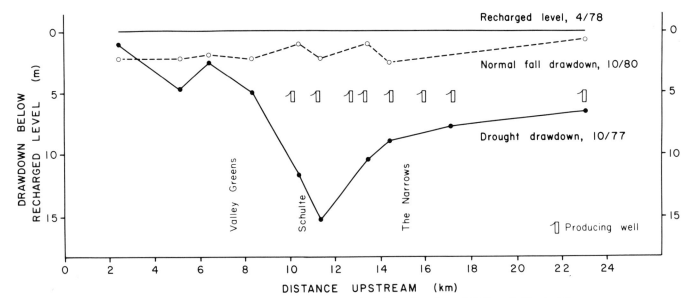

Figure 19. Water-table drawdown and occurrence of bank erosion, Carmel Valley, California. The extreme drawdown of the second drought year (1977) contrasts sharply with the autumn drawdown in a normal flow year (1980), and the normal recharged level following a normal winter flow season (1978). The reaches of the Carmel suffering severe drawdown and die-off of phreatophytes over the 1976–1977 drought were sites of bank erosion in 1978 and 1980. In this figure, the association of bank erosion (measured from cross sections and aerial photographs) and drawdown is shown with bank erosion as measured on an arbitrary scale of: severe (typically 30 m); moderate (locally variable; many reaches unaffected, many reaches with erosion, typically 10 to 20 m, but locally more or less); or none (typically no erosion, but local erosion of generally 10 m or less). Sources of data: Monterey County Flood Control Records, California-American Water Company, and field observations. (From Kondolf and Curry, 1986; used by permission of John Wiley and Sons.)

difference between erosion and stability. Bank materials (deposits from the 1911 flood) consist of noncohesive interbedded sands, gravels, and silts that offer little resistance to erosion when exposed. As shown in Figure 21, the Lower Carmel plots in the braided field of the slope versus bankfull discharge plot of Leopold and others (1964), suggesting its vulnerability to disruption. In this century, massive erosion has occurred twice: (1) in 1911, along the entire alluvial Carmel River, caused by application of the large erosive force of a 100-yr flood; and (2) beginning in 1978, but restricted to the reach affected by water-table drawdown and caused by lowering the bank's resistance by devegetation.

Other hypotheses that could be advanced to explain the recent erosion fail to account for its coincidence in space and time with pumping. For example, Carlson and Rozelle (1978, p. 7.1) concluded that incision, caused primarily by construction of the San Clemente Dam, was the "underlying cause of enhanced bank erosion." It is clear that the Lower Carmel River became incised after closure of the dam in 1921, but the incision was essentially complete by about 1940. This behavior is consistent with that observed in other rivers where incision below dams is rapid at first, then usually insignificant after about 20 yr (Williams and Wolman, 1984; Leopold and others, 1964). It is difficult to accept that the incision could be the cause of erosion that did not begin until 4 decades later.

Ironically, the bucolic riparian environment that Steinbeck described, with its narrow channel, deep pools, and lush vegetation, existed in part because of the artificial reduction in sediment load effected by the dam. If the channel had been carrying predam sediment loads from the steep watershed above the dam, it probably would have been considerably wider and somewhat less scenic.

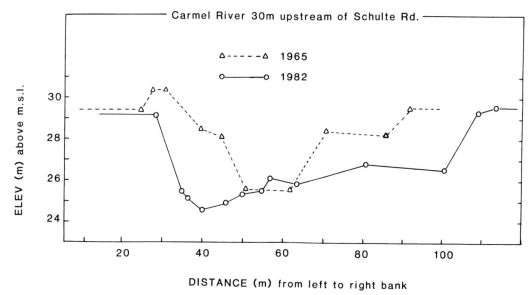

Figure 20. Sequential changes in channel cross secton of the Carmel River 30 m upstream of Schulte Road bridge. 1965 data from U.S. Army Corps of Engineers, San Francisco Office. 1965 section was not monumented, so location may not be precise. (From Kondolf and Curry, 1986; used by permission of John Wiley and Sons.)

CHAPTER SUMMARY AND CONCLUSIONS

The nature and extent of possible interactions between surface-subsurface flow and channel form and process remains a part of fluvial geomorphology that has been little researched, and as a result this area is potentially fruitful for future work. At a very basic level, groundwater determines whether a stream reach is classified as perennial, intermittent, or ephemeral. Perennial reaches are often thought to be fed more or less continuously by groundwater that intersects the channel, producing flow the entire year. This concept needs further evaluation because some channels have been shown to receive groundwater inputs at discrete point sources rather than from a more diffuse source. Once groundwater enters the channel, it experiences complex interactions between surface and subsurface flow. That is, water in its journey through the stream system may submerge (negative seepage) and reemerge (positive seepage) through alluvial deposits and the more general groundwater flow system. In intermittent channel reaches, the stream receives groundwater flow only part of the year, and flow may cease when the groundwater table drops below the level of the channel. Change from perennial to intermittent flow should be reflected by change in channel geometry related to bank processes as controlled by riparian vegetation and moisture content of the bank materials. This aspect of fluvial geomorphology has not been studied adequately. Ephemeral channels are, by definition, everywhere above the groundwater table and do not have a component of flow derived from that source. However, channel form and processes in these streams may also be affected by groundwater flow because the discharge tends to decrease in the downstream direction as water seeps (negative seepage) into the channel bed and banks to eventually reach the groundwater table. Fine material in the flow may be carried through the gravels, possibly affecting competence of the stream to transport bedload.

The influence of groundwater on stream channel form and process may be direct or indirect. Direct influences include the effects of seepage of flow on bank erosion, bedform development, and sediment transport. Indirect influences are varied and include the role of groundwater in maintaining riparian vegetation, which enhances bank stability. Indirect influences are also important in formation and maintenance of the biological environment in the riparian zone, particularly fish habitat.

Experiments in natural channels have not been designed to test the significance of groundwater entering the stream or chan-

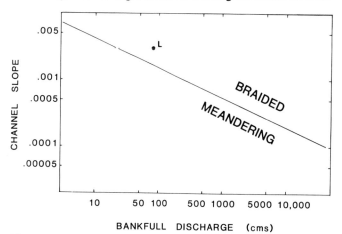

Figure 21. Channel patterns plotted for slope versus bankfull discharge, with Lower Carmel River (L) plotted on graph. Adapted from Leopold and others, 1964, p. 293. (From Kondolf and Curry, 1986; used by permission of John Wiley and Sons.)

nel water infiltrating into the local groundwater system. However, recent flume experiments by Richardson and Richardson (1985) for a sand-bed channel provide interesting information for speculation. That study indicated that introduction of emerging subsurface water (positive seepage) resulted in an increase in unit stream power and a tendency for bedforms to elongate and flatten. In one case, stream power increased twelve fold and bedforms changed from plane bed to ripples to dunes. In addition, water-surface slope and velocity were observed to be higher under conditions of emerging subsurface water. Effects were most pronounced at relatively low flows in the lower flow regime and negligible at flows in the upper flow regime.

Positive seepage forces tend to be very small relative to the tractive forces involved in sediment transport. Therefore, it appears that emerging groundwater has little effect on sediment transport. However, the effects of downwelling (negative seepage) may be more pronounced, particularly if a mud seal forms that renders the streambed more resistant to erosion.

Erosion of riverbanks is an important geomorphic process in the fluvial system and a number of physical, biological, and climatic factors may influence bank erosion. Of particular importance here are effects of water content of bank materials, frost action, and nature and extent of riparian vegetation. For examples, dewatering of saturated banks following high flows can produce substantial bank collapse; sagging and tunneling along streambanks may locally accelerate bank erosion, producing embayments along the bank. These processes may lead to considerable modification of the channel boundaries.

One of the most important mechanisms by which groundwater influences channel form is through its influence on riparian vegetation. This results because riparian vegetation contributes substantially to the ability of the streambank to resist erosion by increasing shear strength of the bank materials. Therefore, streambanks that contain abundant roots are more resistant to erosion than are banks free of vegetation. Bank vegetation may also affect channel width and, in general, tree-lined streambanks are narrower than those without trees. Erosion by direct flow of water is inhibited by protective armor provided by roots, stems, and limbs. Vegetation is hydrologically rough and so tends to reduce flow velocities that impinge on banks.

Riparian vegetation can be significantly affected by changes in groundwater levels in the banks. Depending on local conditions, either a rise in or a lowering of groundwater table may result in damage to riparian vegetation and loss of trees. This may render the banks more susceptible to erosion and damage from flooding. This was dramatically shown during the 1983 floods in Tucson, Arizona. Overpumping of aquifers there had caused reduction in the abundance of riparian vegetation that provided a measure of bank stabilization. Although it can be argued that the 1983 floods of the Santa Cruz River in Tucson would have caused extensive damage with or without riparian vegetation, it is likely that bank erosion there would have been less severe had there been more riparian vegetation to provide shear strength to the bank materials and roughness elements to decrease the velocity of floodwaters.

Consideration of relations between riparian vegetation and bank stability suggests that the stability of a river system depends not only on the driving forces, such as high-magnitude storms, but also on the resisting forces that defend the channel. Roots of trees in the riparian zone can increase resistant forces but are vulnerable to changes in water level. Bank erosion may be particularly severe where there is a combination of high driving forces and reduced resisting forces.

Surface-groundwater interactions are particularly important in determining the quality of spawning and juvenile-rearing habitat for Salmonid fish. Areas of upwelling or downwelling water in the streambed are of particular importance for spawning in that these interactions carry rich oxygenated water to developing fish eggs. Areas of upwelling of cool groundwater may also produce cold pools that serve as refugia for juvenile fish, who must reside in the channel for a year or two before returning to the ocean. Mainstream temperatures in some streams approach the lethal level for Salmonid fish during the summer, and at these times the cold pools are crucial habitats. These pools also serve as refugia for migrating fish that enter the stream channels during the warm summer months. Migrating fish in the summer may move upstream from one cold pool to the next, thus avoiding high temperatures that exist during the day over most of the stream.

ACKNOWLEDGMENTS

L. E. Stullken of the USGS Water Resources Division field office in Garden City, Kansas, provided information on hydrology and riparian vegetation conditions along the Arkansas River of southwestern Kansas. Critical review and suggestions for improvement of the manuscript by B. Hill, K. J. Gregory and T. E. Lisle are acknowledged and appreciated.

REFERENCES CITED

ASCE Task Force, 1966, Nomenclature for bedforms in alluvial channels: Proceedings of the American Society of Civil Engineers, v. 92, no. HY3, p. 51–64.

Baker, V. R., 1985, Fluvial geomorphology, review by W. L. Graf of book, The Colorado River, published by the Association of American Geographers: Science, v. 229, p. 376–377.

Benson, N. G., 1953, The importance of groundwater to trout populations in the Pigeon River, Michigan, in Transactions of the 18th North American Wildlife Conference: p. 260–281.

Blondeaux, P., and Seminara, G., 1985, A unified bar-bend theory of river meanders: Journal of Fluid Mechanics, v. 157, p. 449–470.

Burgi, P. H., and Karaki, S., 1971, Seepage effects on channel bank stability: Proceedings of the American Society of Civil Engineers 7968, IR1, Journal of the Irrigation and Drainage Division, p. 59–72.

Carlson, F. R., and Rozelle, K. D., 1978, Carmel Valley vegetation study: M. Hill, Inc., report to Monterey County Flood Control and Water Conservation District, Salinas, California, chapter 2.

Charlton, F. G., Brown, P. M., and Benson, R. W., 1978, The hydraulic geometry of gravel bed rivers in Britain: Wallingford, Hydraulics Research Station Internal Report 180, 48 p.

Denton, D. N., 1974, Water management for fishery enhancement on north coastal streams: California Department of Water Resources, 107 p.

Dunlap, L. E., Lindgren, R. J., and Sauer, C. G., 1985, Geohydrology and model analysis of stream-aquifer system along the Arkansas River in Kearny and Finney Counties, southwestern Kansas: U.S. Geological Survey Water-Supply Paper 2253, 52 p.

Dunne, T., 1980, Formation and controls of channel networks: Progress in Physical Geography, v. 4, p. 211–239.

Graf, W. L., 1978, Fluvial adjustments to the spread of Tamarisk in the Colorado Plateau region: Geological Society of America Bulletin, v. 89, p. 1491–1501.

Groenveld, D., and Griepentrog, T., 1985, Interdependence of groundwater, riparian vegetation, and streambank stability; A case study, in Proceedings of the Symposium on Riparian Ecosystems and their Management, Tucson, Arizona, April 16-18, 1985: Ft. Collins, Colorado, U.S. Forest Service General Technical Report RM-120, p. 44–48.

Hagerty, D. J., Ullrich, C. R., and Spoor, M. F., 1981, Bank failure and erosion on the Ohio River: Engineering Geology, v. 19, p. 119–132.

Hagerty, D. J., Sharifounnasab, M., and Spoor, M. F., 1983, Riverbank erosion; A case study: Association of Engineering Geologists Bulletin, v. 20, no. 4, p. 411–437.

Hamilton, T. M., 1970, Channel-scarp formation in western North Dakota: U.S. geological Survey Professional Paper 700-C, p. C229–C232.

Hansen, A. J., Boeker, E. L., Hodges, J. I., and Cline, D. R., 1984, Bald eagles of the Chilkat Valley, Alaska: Ecology, behavior, and management; Final report of the Chikat River Cooperative Bald Eagle Study: National Audubon Society and U.S. Fish and Wildlife Service.

Harrison, S. S., and Clayton, L., 1970, Effects of ground-water seepage on fluvial processes: Geological Society of America Bulletin, v. 81, p. 1217–1226.

Hazzard, A. S., 1932, Some phases of the life history of the eastern brook trout, *Salvelinus frontinalis*: Transactions of the American Fish Society, v. 62, p. 344–350.

Higgins, C. G., 1984, Piping and sapping; Development of landforms by groundwater outflow, in LaFleur, R. G., ed., Groundwater as a geomorphic agent: Boston, Massachusetts, Allen and Unwin, p. 18–58.

Hooke, J. M., 1979, An analysis of the processes of riverbank erosion: Journal of Hydrology, v. 42, p. 39–62.

Horton, A., 1945, Erosional development of streams and their drainage basins; Hydrophysical approach to quantitative morphology: Geological Society of America Bulletin, v. 56, p. 275–370.

Huff, D. D., O'Neill, R. V., Emanuel, W. R., Elwood, J. W., and Newbold, J. D., 1982, Flow variability and hillslope hydrology: Earth Surface Processes and Landforms, v. 7, p. 91–94.

Ikeda, S., Parker, G., and Swai, K., 1981, Bend theory of river meanders; Part 1, Linear development: Journal of FluidMechanics, v. 112, p. 363–377.

Keller, E. A., 1971, Pools, riffles, and meanders; Discussion: Geological Society of America Bulletin, v. 82, p. 279–280.

—— , 1972, Development of alluvial stream channels; A five-stage model: Geological Society of America Bulletin, v. 83, p. 1531–1536.

Keller, E. A., Hofstra, T. D., and Moses, C., 1990, Summer "cold pools" in Redwood Creek near Orick, California, in Nolan, K. M., Kelsey, H. M., and Marron, D. C., eds., Process and aquatic habitat in the Redwood Creek Basin, northwestern California: U.S. Geological Survey Professional paper (in press).

Kennedy, V. C., Kendall, C., Zellweger, G. W., Wyerman, T. A., and Avanzino, R. J., 1986, Determination of the components of stormflow using water chemistry and environmental isotopes, Mattole River basin, California: Journal of Hydrology, v. 84, p. 107–140.

Kesel, R. H., and Baumann, R. H., 1981, Bluff erosion of a Mississippi River meander at Port Hudson, Louisiana: Physical Geography, v. 2, no. 1, p. 62–82.

Kitanidis, P., and Kennedy, J. F., 1984, Secondary current and river-meander formation: Journal of Fluid Mechanics, v. 144, p. 217–229.

Knighton, D., 1984, Fluvial forms and processes: London, Edward Arnold Publishers, 218 p.

Kondolf, G. M., and Curry, R. R., 1986, Channel erosion along the Carmel River, Monterey County, California: Earth Surface Processes and Landforms, v. 11, no. 3, p. 307–319.

Lawler, D. M., 1987, Bank erosion and frost action; An example from South Wales, in Gardiner, V., ed., International geomorphology 1968, Part 1: London, John Wiley and Sons.

Leopold, L. B., Wolman, M. G., and Miller, J. P., 1964, Fluvial processes in geomorphology: San Francisco, California, W. H. Freeman and Co., 522 p.

Lisle, T. E., 1986, Stabilization of a gravel channel by large streamside obstructions and bedrock bends, Jacoby Creek, northwestern California: Geological Society of America Bulletin, v. 97, p. 999–1011.

Maddock, T., Jr., 1972, Hydraulic behavior of stream channels, in Transactions of the 37th North American Wildlife and Natural Resources Conference: Wildlife Management Institute, Washington, D.C., p. 366–374.

Mears, J. B., 1968, Piping, in Fairbridge, R. W., Encyclopedia of geomorphology: Stroudsberg, Pennsylvania, Dowden, Hutchison and Ross, p. 849.

Moses, C. G., 1984, Pool morphology of Redwood Creek, California [M.S. thesis]: Santa Barbara, University of California, 117 p.

Nakagawa, H., and Tsujmoto, T., 1984, Interaction between flow over a granular permeable bed and seepage flow; A theoretical analysis: Journal of Hydroscience and Hydraulic Engineering, v. 2, p. 1–10.

Neill, C. K., 1969, Bed forms in the Lower Red Deer River, Alberta: Journal of Hydrology, v. 7, p. 58–85.

Nolan, K. M., and Janda, R. J., 1979, Recent history of the main channel of Redwood Creek, California, in Guidebook for a field trip to observe natural and management-related erosion in Franciscan terrain of northern California; Cordilleran section of the Geological Society of America: San Jose, California, p. x1–x16.

Ozaki, V. L., 1987, Geomorphic and hydrologic condition for cold pool formation on Redwood Creek, California, in Beschta, R. L., Blinn, T., Grant, G. E., Swanson, F. J., and Ice, G. G., eds., Erosion and sedimentation in the Pacific Rim: International Association of Scientific Hydrological Sciences Publication 165, p. 415–416.

Pinder, G. F., and Sauer, S. P., 1971, Numerical simulation of flood wave-modification due to bank storage effects: Water Resources Research, v. 7, p. 63–70.

Piper, A. M., Gale, H. S., Thomas, H. E., and Robinson, T. W., 1939, Geology and groundwater hydrology of the Mokelumne area, California: U.S. Geological Survey Water-Supply Paper 780, 230 p.

Richardson, J. R., and Richardson, E. V., 1985, Inflow seepage influence on straight alluvial channels: Journal of Hydraulic Engineering, v. 111, no. 8, p. 1133–1147.

Saarinen, T. F., Baker, R. V., Durrenberger, R., and Maddock, T., Jr., 1984, The Tuscon, Arizona, flood of October 1983: National Academy Press, 112 p.

Schumm, S. A., and Phillips, L., 1986, Composite channels of the Canterbury Plain, New Zealand; A Martian analog?: Geology, v. 14, p. 326–329.

Simons, D. B., and Richardson, E. V., 1966, Resistance to flow in alluvial channels: U.S. Geological Survey Professional Paper 422-J, 61 p.

Small, R. J., 1965, The role of spring sapping in the formation of chalk escarpment valleys: Southampton Research Series in Geography, v. 1, p. 3–29.

Smith, D. G., 1976, Effect of vegetation on lateral migration of a glacial meltwater river: Geological Society of America Bulletin, v. 87, p. 857–860.

Springer, F. M., Jr., Ullrich, C. R., and Hagerty, D. J., 1985, Streambank stability: American Society of Civil Engineers, Geotechnical Engineering Division Journal, v. 111, no. G5, p. 624–640.

Steinbeck, J., 1945, Cannery Row: New York, Viking Press, p. 77–78.

Stuart, T. A., 1954, Spawning sites for trout: Nature, v. 173, p. 354.

Swanson, M. L., Kondolf, G. M., and Boison, P. J., 1989, An example of rapid gully initiation and extension by subsurface erosion, coastal San Mateo County, California: Geomorphology, v. 2, p. 393–403.

Taylor, D. W., 1984, Ecological effects of flooding on riparian vegetation, Inyo National Forest, California: Report to the U.S. Forest Service, Contract 40-91W2-3-0301, 56 p.

Thorne, C. R.,a nd Tovey, N. K., 1981, Stability of composite riverbanks: Earth Surface Processes and Landforms, v. 6, p. 469–484.

Todd, D. K., 1955, Groundwater flow in relation to a flooding stream: Proceedings of the American Society of Civil Engineers, v. 81, p. 628.

Trihey, E. W., 1981, Reach-specific streamflow analysis for the Terror and Kizhuyak Rivers, *in* An assessment of environmental effects of construction and operation of the proposed Terror Lake hydroelectric facility, Kodiak, Alaska; Instream flow studies: Anchorage, University of Alaska, final report by Arctic Environmental Information and Data Center, Appendix 2.

Twidale, C. R., 1964, Erosion of an alluvial bank at Berwood, South Australia: Zeitschrift fur Geomorphologie, v. 8, p. 189–211.

Ullrich, C. R., Hagerty, D. J., and Holmberg, R. W., 1986, Surfical failures of alluvial streambanks: Canadian Geotechnical Journal, v. 23, p. 304–316.

Vaux, W. G., 1962, Interchange of stream and intragravel water in a salmon spawning riffle: U.S. Fish and Wildlife Service Special Scientific Report Fisheries 405, 11 p.

Vining, L. J., Blakely, J. S., and Freeman, B. M., 1985, An evaluation of the incubation life-phase of chum salmon in the middle Susitna River, Alaska: Alaska Department of Fish and Game Winter Aquatic Investigations, September 1983–May 1984, Report 5, v. 1.

Vronskiy, B. B., 1972, Reproductive biology of the Kamchatka River chinook salmon (*Oncorphynchus tschawytscha* [Walbaum]): Journal of Ichthyology, v. 12, p. 259–273.

Walker, E. H., 1957, The deep channel and alluvial deposits of the Ohio Valley in Kentucky: U.S. Geological Survey Water-Supply Paper 1411, 25 p.

Watters, C. G., and Rao, M., 1971, Hydrodynamic effects of seepage on bed particles: American Society of Civil Engineers Proceedings Paper 7873, Hydraulics Division Journal, v. 97, no. HY3, p. 421–439.

Weigel, T. A., and Hagerty, D. J., 1983, Riverbank change; Sixmile Island, Ohio River, U.S.A.: Engineering Geology, v. 19, no. 2, p. 119–132.

Williams, G. O., and Wolman, M. G., 1984, Downstream effects of dams on alluvial rivers: U.S. Geological Survey Professional Paper 1286, 83 p.

Zeimer, R. R., 1981, Roots and the stability of forested slopes, *in* Davies, T.R.H., and Pearce, A. J., eds., Erosion and sediment transport in Pacific Rim steeplands: International Association of Hydrological Sciences Publication, p. 343–361.

MANUSCRIPT ACCEPTED BY THE SOCIETY NOVEMBER 14, 1989

Chapter 16

Geomorphic controls of groundwater hydrology

Donald R. Coates
Department of Geological Sciences, State University of New York at Binghamton, Binghamton, New York 13901

INTRODUCTION

In its broadest context, geomorphology is the study of landforms and those processes and surface materials involved in topographic transformation. The landscape is a collage of landforms whose characteristics have evolved through the interaction of the Earth's crust and those surface processes that act over time to create changes. Groundwater is an interactive component of surface and near-surface terrain; geomorphology controls groundwater behavior, and at the same time, groundwater fashions terrain. There are many variables that exert control of groundwater hydrology. These include relief and hillslope differences, the properties of surface material, weathering phenomena, climate, and time.

GENERAL TERRAIN FEATURES

Hills and valleys are a feature of nearly all landscapes. Regardless of how they originated, their present form exerts the principal control on the direction of groundwater flow, influences the distribution of groundwater quality, and in some regions provides a convenient guide for the location of large-yield wells. Hillslopes are the dominant components of a landform, and the landscape is a series of interconnected slopes. Slopes have three principal attributes: length, steepness, and shape. So a landscape may be composed of short or long slopes, gentle or steep slopes, or have contours that are convex, concave, planar, or any combination of these. When underlying rocks are similar, each different type of slope translates into different groundwater hydrological conditions. For example, the steeper the slope, the steeper the water table, because the character of the water table generally reflects topographic slope, which it mimics, but with a somewhat flatter gradient. Slope steepness affects other aspects of groundwater conditions, such as its velocity. This can be shown by one form of Darcy's Law wherein V = K I, in which V is velocity, K is hydraulic conductivity, and I is slope of the water table. Therefore, topographic slope is a vital influence in determining rate of groundwater movement (Fig. 1).

An important reason why the character of hills, valleys, and intervening slopes can be determinants in well production is that topographic conditions can be a reflection of such underlying causes as rock resistance to weathering and erosion, and structural features of the rocks. Rock fractures, because of their secondary porosity, are favored positions for groundwater movement. When aligned along preferred lineament directions, such features can indicate sites for groundwater development. The discovery of such localities is done by fracture-trace analysis,

DARCY'S LAW

V = K I

V = Velocity

K = Hydraulic conductivity

I = Water table slope

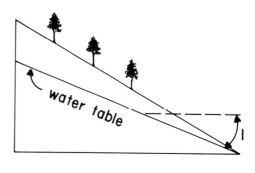

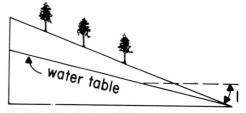

Figure 1. Sketches showing how differences in topographic slope affect water table slope and Darcy's Law.

Coates, D. R., 1990, Geomorphic controls of groundwater hydrology, *in* Higgins, C. G., and Coates, D. R., eds., Groundwater geomorphology; The role of subsurface water in Earth-surface processes and landforms: Boulder, Colorado, Geological Society of America Special Paper 252.

which maps the structures because they occur in zones of weakness where subsequent weathering and erosion have topographically etched them into such diagnostic landforms as gaps in ridges, sags and aligned depressions, and straight valley segments. Statistical studies in the carbonate terrane near University Park, Pennsylvania, have shown that high-yielding wells are more likely to be located along lineaments than at other places in the topography (Siddiqui and Parizek, 1971). Similar studies in New England also reveal that the best bedrock wells are positioned along fracture traces, or at their intersections (Gary Smith, personal communication, 1979). Thus, there can be a correlation of topography with fracturing and well yields.

In the crystalline terrane of Statesville, North Carolina, Le Grand (1954) found that topographic slope was strongly correlated with well yield (Fig. 2). The greatest well production occurs in valleys and broad ravines, with lowest yields from wells at or near the crests of hills. He also found that wells drilled in flat upland areas produced more water than wells farther down on the hillslope. In a larger-based study, using well data from the Piedmont and Blue Ridge physiographic provinces, Le Grand (1967) developed a rating system whereby some prediction could be made of potential well yields when topographic slope was coupled with soil depths and rock outcrops. He assigned point values on the basis of position of slope in the following manner (see also Fig. 3):

Points	Topography
0	steep ridge top
2	upland steep slope
4	pronounced rounded upland
5	midpoint ridge slope
7	gentle upland slope
8	broad flat upland
9	lower part of upland slope
12	valley bottom or flood plain
15	draw in narrow catchment area
18	draw in large catchment area

In similar fashion, Le Grand (1967) also gave point values to soil and bedrock conditions (see also Fig. 4):

Points	Character of Soil and Rock
0 to 2	bare rock, almost no soil
2 to 6	very thin soil, some rock outcrops
6 to 9	soil thin, a few rock outcrops
9 to 12	moderately thick soil, no fresh outcrops
12 to 15	thick soil, no rock outcrops

By totaling these two sets of point values, Le Grand (1967) was able to construct a probability set of curves that represented the potential for obtaining well yields of a certain caliber (Fig. 5). Once again, wells on flat uplands and in valleys tend to yield larger amounts of water than wells on valley sides or sharp hilltops. The relatively lower yields of water on or near the steeper

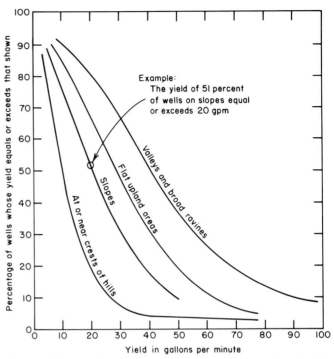

Figure 2. Topographic influence on well yields in North Carolina (Le Grand, 1954).

slopes is explained by the fact that erosion removed much of the weathered and more permeable rock. Water levels are also farther below the surface because groundwater flows to discharge points in the adjacent lowlands. Wells located on concave slopes are commonly more productive than wells on convex or straight slopes. The broad, but slightly concave slopes near saddles in a gently rolling upland area proved to be especially good sites for potentially high-producing wells.

Hillslope properties can be an important factor that also influences groundwater quality, as shown in case studies near Syracuse and Binghamton, New York. Bedrock wells in the lowlands and the lower part of adjacent hills contain more dissolved solids than their counterpart wells drilled farther upslope. The cause for the lower quality water is a function of the topography. The groundwater flowlines arc from the upper slopes and exit at lower elevations. Thus, groundwater near hilltops and upper slopes is younger and has had a shorter residence time to dissolve mineral matter from the host rocks than those waters near the base of hills.

The depth of the water table is primarily a function of topographic and climatic factors. In all terrains, depth is greatest at higher elevations and least in valleys and basins. Humid areas also have shallower water tables than dry-land environments. The specific conditions that determine the position of the water table in lowlands include topographic relief, drainage basin size, amount and composition of valley fill, and climate. Although an effluent system (with groundwater recharging streamflow) is more apt to occur in humid regions, many other geomorphic factors are also important. For example, in temperate areas it is unusual for second- or third-order basins to have perennial streams, and with the exception of unusually fine-grained bedrock, it generally requires at least fourth-order basins to have strong baseflow conditions (recharge of streams by groundwater). In arid areas, the water table is often so deep that only the major through-flowing rivers have permanent flow with drainage basin order of at least seven. Most streams in arid areas exhibit influent flow, wherein streamflow infiltrates and recharges the groundwater zone because the water table is topographically lower than the stream channel. The permeability of rocks is higher near land surface than at greater depth, so depth of the water table can influence the productivity of wells. Frommurze (1937) reported larger well yields in moist regions than from similar rock types in dry lands, because the most permeable rocks in drylands are above the water table.

Topographic slope and groundwater hydrology are closely related in coastal and small island environments. In these settings there is an interface within the groundwater system between salt water and fresh water. The Ghyben-Herzberg relation approximates the position of the salt water–fresh water boundary for hydrostatic equilibrium at the contact zone. The equation that defines this principle is:

$$Z_s = \frac{P_f}{P_s - P_f} Z_w \qquad (1)$$

where Z_s = depth to salt water below sea level, P_f = fresh water density, P_s = salt water density, and Z_w = height of water table above sea level. This means that for every meter the water table is

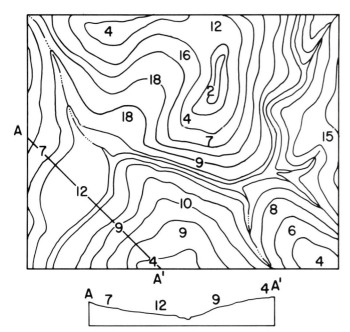

Figure 3. Contour map showing rating system (see text) for influence of topographic conditions on well yields. Higher well yields can be expected at sites with higher numbers (after Le Grand, 1967).

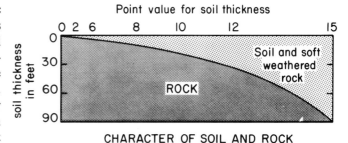

Figure 4. Rating system for influence of soil thickness on well yield (after Le Grand, 1967).

above sea level, the depth to the salt water–fresh water interface will be about 40 m. However, this relation does not allow for vertical components of flow, discharge of fresh water into the sea floor, or diffusion at the interface. More precise equations can be derived using such steady-state models as those of Hubbert (1940), which account for more of the complex variables.

GROUNDWATER PROVINCES

To assess the groundwater conditions throughout the United States, many workers have found it convenient to classify them into regional systems that possess somewhat similar environments. These schemes usually bear a strong relation to the physiographic provinces of the country as delineated by Fenneman (1923). The uniting feature of such provinces is similarity in topography and climate, although rock type can also be important as in the volcanic terranes. Meinzer (1923), following Fenneman's (1923) lead, divided the conterminous United States into 21 groundwater provinces, a slight contraction from the 24 physiographic provinces of Fenneman. Thomas (1951, 1952) has shown the importance in using groundwater regions for strategies in groundwater management and conservation. Other authors, such as Ries and Watson (1914), McGuinness (1963), and Fetter (1980) have used the following 10 regions to define groundwater provinces of the 48 states: (1) western mountain ranges, (2) alluvial basins, (3) Columbia Lava Plateau, (4) Colorado Plateau and Wyoming Basin, (5) High Plains, (6) unglaciated Appalachian region, (7) unglaciated central region, (8) glaciated Appalachian region, (9) glaciated central region, and (10) Atlantic and Gulf Coast Plain. In addition, the Hawaiian Islands should be designated as the "volcanic islands," and Alaska the "permafrost region."

Heath (1984) has shown the groundwater importance of geomorphology and the unconsolidated deposits formed by geomorphic processes. He has identified the most important mapping

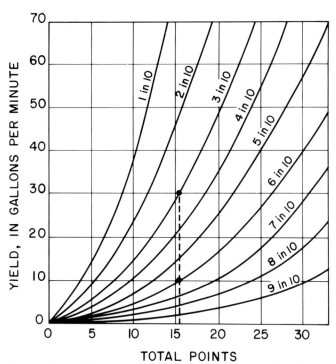

Figure 5. Probability of obtaining a certain yield from a well at different sites having a total-point rating. For example, a site with 16 points has a 3 in 10 chance of yielding at least 30 gal/min and a 6 in 10 chance of yielding 10 gal/min (after Le Grand, 1967).

units in the following way: fine-grained floodplain deposits; coarse-grained floodplain deposits; lake deposits; wind deposits (loess); coarse-grained alluvial fan and plain deposits; fine-grained ice-laid deposits; coarse-grained ice-laid deposits; fine-grained residuum; coarse-grained residuum; bedrock; fine-grained marine deposits; coarse-grained marine deposits; and marine limestone.

When geomorphology is linked with climate, particular sets of groundwater conditions can occur. Thus, effluent streams, sustained by groundwater discharge, are the rule in humid regions, whereas influent streams, with infiltration into the groundwater zone, are most common in dry-land environments. In the semiarid intermontane basins of the western U.S., groundwater flow systems and the availability of water for wells is dependent on geomorphic and lithologic conditions. Here, sand and gravel aquifers produce the only reliable supply of water for large-scale development, and their location is dependent on geomorphic settings.

Although alluvial fans and pediments are not totally restricted to dry climate regions, they are most plentiful in semiarid regions such as those in the Basin and Range physiographic province of the West. Here, in the piedmont areas, these two contrasting landforms give rise to entirely different types of aquifers. Wells drilled to similar depths on alluvial fans and pediments encounter very different yields, because alluvial fans are composed of unconsolidated sediments, whereas the substrate of pediments is bedrock with much lower yields. Proper identification of these different geomorphic landforms is important for well siting and development. In addition, the best wells on alluvial fans are those at positions where minimum thicknesses of mudflows and debris flows are encountered. The best wells in regions of pedimentation are located immediately basinward of the farthest extension of the pediment subsurface bench, because the basin-fill sediments are coarsest at this break in profile (Coates and Cushman, 1955).

The intermontane basins of such states as Utah, Nevada, Arizona, and New Mexico evolved through a complex set of geomorphologic processes. Although the initial differential relief of mountains and basins was tectonically produced, the sculpture and sedimentologic history was the result of geomorphic forces. During tectonism, disparities between base levels of adjoining basins was common. Many basins had interior drainage until such a time as the sediment in-filling could produce an integrated drainage system of the region. The axes of such basins consist of sediments that were produced under playa lake conditions with evaporites and massive clay-silt sequences (Coates, 1952). Therefore, the highest production wells, and those with the minimum amount of dissolved solids in the water, are those drilled at peripheral positions in the basins rather than in the center. Also, in these intermontane basin regions, there can be two different groundwater flow systems (Eakin and others, 1976). One is flow between the basins on a regional basis; the other is subsurface flow within a single basin. Therefore, the combination of geomorphology, structure, and climate have controlled groundwater conditions in such terrains.

When evaluating groundwater conditions in a single groundwater province, or physiographic region, it is important to realize there are still many heterogeneities. So, although there may be general principles that pertain to the groundwater systems throughout the region, there may also be many diversities. For example, Coates (1982) has separated the glaciated Appalachian plateau of New York and Pennsylvania into 11 subregions or sections (Fig. 6). Each section differs geomorphically from the others and, in turn, exhibits differences in groundwater hydrology, as will be shown by some examples below.

BEDROCK CONDITIONS

Bedrock is a fundamental consideration in geomorphic and hydrologic studies because it not only can influence topographic characteristics but also is a major control of drainage density. Furthermore, rock type can determine how the geomorphic processes of weathering and erosion provide feedback mechanisms in the control of groundwater hydrology. These conditions are illustrated by conditions at the Georgia Nuclear Laboratory locality. Metamorphic crystalline rocks underlie this site in the Piedmont physiographic province. Here a zone of deeply weathered rock, or saprolite, contains alteration products that mimic the underlying solid rock. The direction and rate of groundwater movement is a function of the mineral composition of the parent rock, the structural alignment of minerals, shear zones, and joints, which have

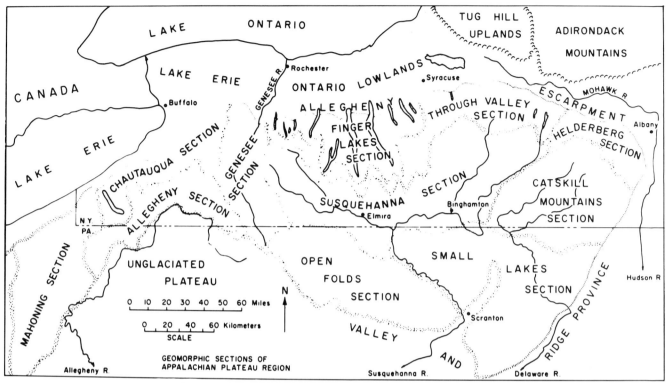

Figure 6. Map of geomorphic sections of the glaciated Appalachian Province with adjacent provinces (Coates, 1982).

all been differentially affected by weathering processes. Thus, the permeability of samples from the upper part of the saprolite is 25 to 100 times greater when oriented parallel to the schistosity than when normal to the fabric. The porosity of the saprolite is greatest at depths of 9 to 12 m (30 to 40 ft) (Stewart, 1962) where it is as much as 54 percent. Porosity of unweathered rock is about 5 percent. Specific yield to wells increases to depths of 12 m (40 ft) then decreases in the more intact bedrock.

Geomorphology, lithology, and hydrology have a symbiotic relation when water is introduced into a carbonate terrane. Limestone affects the behavior of water, and in turn, water changes the configuration of the host materials through which it moves. Unlike most other rock types, underground water alters the very environment that contains it. The water transport system influences not only the subsurface skeletal constitution of the area, but in time the feedback mechanisms reach the land surface and form a unique set of terrain conditions known as "karst topography." A somewhat analogous situation can occur in massive stagnant glacial ice where a "pseudokarst" landscape may develop.

Sweeting (1973) divides karst into three types: (1) holokarst is the traditional true karst where solutional landforms dominate and the drainage is internal: (2) fluviokarst is terrain more influenced by surface rivers; and (3) glaciokarst has been modified by glacial erosion processes. There are several different classification schemes for describing hydrologic characteristics and the cave passages through which they move. Sweeting (1973) groups caves into three different sets: phreatic, vadose, and vertical. The first two types are based on the apparent genesis of the passage, and the last involves a bias toward the degree of its exploration. However, these terms do not describe the primary function of a cave system, which is to transport water from insurgences to resurgences. Jennings (1971) divided different cave types into four sets: inflow, outflow, through flow, and between flow. Mylroie (1984) provided a rather comprehensive classification of karst forms on the basis of their position and function in the hydrologic regime:

I. Surficial karst forms. A. Exposed. B. Mantled.

II. Interface forms. A. Insurgences: (1) gravity spring, (2) artesian spring, (3) overflow spring. B. Intersection forms: (1) vertical, (2) lateral.

III. Subsurface forms. A. Active cave passages: (1) tributary passage, (2) master cave passage, (3) diversion passage, (4) tapoff passage, (5) abduction passage. B. Abandoned cave passage.

The utility of such a classification is that it demonstrates the interaction of geomorphic controls with the hydrologic system.

In a discussion of the relation of carbonate rocks, geomorphology, and hydrology, it is important to note that the flow rate of water is one of the most significant variables because it determines the supply of the dissolved ions to the mineral surface and regulates the removal of products. With rapid movement of water there is not much time for dissolution to occur; therefore, slower rates can enhance the possibility of greater reaction of the wall rocks. However, there is an optimal flow rate whereby there is sufficient time for water to produce dissolution but also sufficient motion for the dissolved products to be removed. There are form differences when vadose passages are compared with phreatic ones. The shapes are a function of the initial fracture and porous media properties, the amount of water within the openings, and the rate of dissolution or downcutting. Thus the passages may vary from vertical slits in the vadose zone to circular and keyhole-type geometry in the phreatic zone. During their formative stages, each possesses its own hydrologic regime.

Drainage density

Drainage density is the ratio of the sum of all channel lengths to the area in which the lengths have been measured. It is usually expressed in terms of miles or kilometers of channel lengths per square mile or square kilometer. Variations in drainage density may be caused by such factors as climate, vegetation, relief, and lithology. Thus, drainage density in badland topography of drylands is much greater than for similar lithologies in a forested or well-watered terrain. Chorley and Morgan (1962) compared drainage density of two areas of similar crystalline and metamorphic rocks; the Unaka Mountains of Tennessee and North Carolina, and Dartmoor, England. The large difference in drainage density (11.2 for the Unakas, 3.4 for Dartmoor) was attributed to differences in rainfall intensity and hillslope steepness.

Carlston (1963) was the first person to show a relation of drainage density to lithology and baseflow conditions, as well as to surface water. He studied 13 small, almost monolithic basins in the nonglaciated parts of the eastern U.S. At his study sites the basins had a precipitation range of 102 to 127 cm/yr (40 to 50 in/yr) and a size difference from 13 to 144 km^2 (5 to 55.7 mi^2). He calculated their drainage density using U.S. Geological Survey 1:24,000-scale topographic maps; these ratios varied from 3.0 to 9.5. Carlston (1963) demonstrated a close correlation of baseflow per unit area to drainage density (Fig. 7). He developed the mathematical basis for this phenomenmon by using a series of equations and derivations from the Jacob water table model and Horton's overland flow model (Fig. 8).

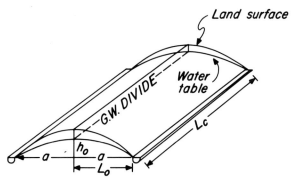

Figure 8. Groundwater table baseflow model after Jacob (1943).

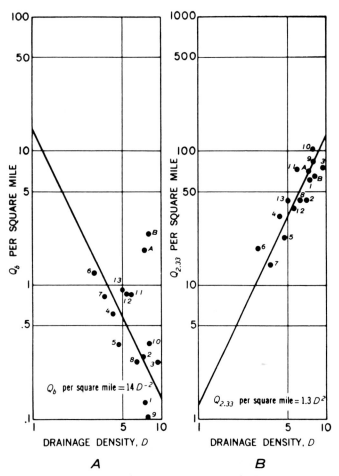

Figure 7. The relation of drainage density to baseflow and floods. A. Baseflow (Q_b). B. Mean annual flood ($Q_{2.33}$) (Carlston, 1963).

$$T = \frac{L_o^2 W}{2 h_o} \quad (2)$$

where T = transmissibility (or the volume rate of flow through a vertical strip of the aquifer of unit width under a hydraulic gradient of unit slope), h_o = height of water table above the draining stream, L_o = length of overland flow, and W = rate of recharge to the water table.

Horton (1945) showed that the average length of overland flow is approximately equal to half the average distance between stream channels and, thereby, is also nearly equal to half the reciprocal of drainage density (D). Thus:

$$L_o = \frac{1}{2 D} \quad (3)$$

and substituting in Equation 2:

$$T = \frac{W D^{-2}}{8 h_o} \quad (4)$$

Thus, baseflow (Q_b) can be shown to be directly proportional to the rate of recharge to the water table (W) as long as the water table height is above the draining system (h_o) and below the land surface. If W increases, then h_o will increase proportionally, causing lateral groundwater flow to increase so that $Q_b = W \cdot L_o$. Therefore, because these are all related, groundwater discharge as measured by baseflow (Q_b) will also vary according to Equation 4, and can be written in the following form:

$$Q_b \propto D^{-2} \qquad (5)$$

These relations show that as transmissibility decreases, the amount or rate of groundwater movement through the system will decrease, and a proportionately greater percentage of precipitation will flow directly into the streams over the land surface.

It is shown that Equation 4 can be derived from Jacob's basic Equation 2, so that transmissibility varies inversely with drainage density squared. Therefore, when transmissibility increases, drainage density decreases, and when transmissibility decreases, drainage density increases. The other aspect of this relation further shows, as in Figure 7, that as transmissibility decreases there is an increase in surface flow. This aspect of the surficial processes reacting to precipitation and the infiltration capacity of earth materials has been termed "terrane transmissibility" by Carlston (1963, p. 5). Thus, there is an integrated hydrogeologic system that relates surface water flow, groundwater flow, transmissibility of the environment, and drainage density.

Coates (1971) compared the hydrogeomorphologic properties of drainage basins in the Catskill Mountain section with those in the Susquehanna section (Fig. 6). The 13 study basins in the Catskills ranged in size from 36.5 to 624 km^2 (14.1 to 241 mi^2), and the Susquehanna basins were 71.5 to 730 km^2 (27.6 to 282 mi^2). The average drainage density for the Catskill basins is 5.4 (on the basis of stream mile lengths per square mile area), whereas it is 7.5 for the Susquehanna basins. This means there are 39 percent more streams per unit area in the Susquehanna section. This large difference is attributable to the contrasts in bedrock. The Catskill region is largely composed of sandstone and coarse-grained clastic sedimentary rocks of Devonian age, but the Susquehanna basins consist mostly of fine-grained Devonian shales and siltstones. When baseflow of these two regions is compared, $Q_b 90$ for the Catskills averages 0.22 cfs/mi^2, but it is only 0.12 cfs/mi^2 for Susquehanna basins. These differences are even more important when the area of valley-fill unconsolidated materials is considered. For example, Flint (1968) showed that the amount of valley fill in glaciated terrane controlled streamflow conditions, i.e., larger amounts of baseflow are correlated with larger amounts of valley fill. However, just the reverse is true for the Catskill-Susquehanna comparison, 5.5 and 13.1 percent, respectively. Thus, the same inverse relation exists for these New York regions as Carlston (1963) reported, namely, high baseflow conditions are associated with low drainage-density figures. Table 1 shows a comparison of similar-size drainage basins. Mill Brook is in the Catskills and Karr Valley is in the Susquehanna section. In spite of the much greater amount of valley-fill sediments in Karr Valley, its $Q_b 90$ is only about one-tenth as great as the low flow discharge is for Mill Brook. Again, such a large difference is a reflection of the drainage density, which is much lower for Mill Brook, 5.4 versus 8.1.

TABLE 1. HYDROGEOLOGICAL COMPARISON OF TWO BASINS IN THE GLACIATED APPLACHIAN PLATEAU*

Basin and geomorphic section	Basin area (mi^2)	Drainage density	Q90	Valley-fill sediments as percent of basin
Mill Creek Catskill Mountain Section	25.0	5.4	0.270	2.9
Karr Valley Creek Susquehanna Section	27.6	8.1	0.028	9.4

*After Coates (1971).

GEOMORPHIC PROCESSES

The function of the surface geomorphic processes is to erode, transport, and deposit earth materials. In accomplishing this work, sedimentary deposits are created and new landforms are developed. Each of the processes (fluvial, gravity, wind, coastal, and glacial) provides its own distinctive array of sediments and features. Furthermore, each geomorphic process has the capacity to cause either fine-grained materials or coarser-grained sediments to form. When placed into an environment that ultimately becomes part of the groundwater zone, the fine-grained materials may constitute an aquiclude or aquitard, whereas the coarser-grained sediments become part of the aquifer.

The alluvium of river valleys and tectonic basins is the principal aquifer of many groundwater systems throughout the world. In the American Southwest these unconsolidated materials provide many of the groundwater resources for such states as New Mexico, Arizona, Utah, Nevada, and southern California. In addition, in other parts of the nation, the valley-fill materials constitute important aquifers. The regime of the rivers and the depositional history of the sediments determine whether the materials will be fine or coarse grained. Thus, the sand and gravel of point bars and traction materials in the channel serve as environments that can become good aquifers. The overbank silts and backswamp silts and clays in the floodplain contain material that is too fine grained to become important groundwater sources.

In similar fashion, each of the other geomorphic processes can produce either sediments that may inhibit groundwater flow, or that become important aquifers. For example, the loess deposits throughout the central Mississippi Basin region are devoid of good aquifers, whereas sand dunes, as in the 51,000 km^2 Sand

Hill Region of Nebraska, have a well-tuned groundwater regime that provides water throughout the area.

In the following section, many of the case histories are drawn from examples in glaciated terrane, so a brief introduction about the glacial geomorphic process is in order. Geomorphic controls have played a large role in forming and localizing good aquifers throughout much of the 3.7 million km^2 of the U.S. that have been glaciated. New York is somewhat representative of the relation of glacial processes, glacial sedimentation, and landforms, with groundwater conditions. The preglacial bedrock surface was modified by glaciation. Those valleys that were aligned parallel to glacial motion were greatly enlarged and deepened, whereas valleys with other orientations were not so greatly eroded. Most valleys were partially filled with glaciofluvial, glaciolacustrine, and till deposits. Even after complete deglaciation, some troughs still retain lakes, such as in the Finger Lake section (Fig. 6). The sedimentary heritage of the Pleistocene Ice Age is marked by till in the uplands, along many valley sides, and certain parts of the valleys. In addition there is a very rich series of meltwater-stratified deposits, and some of these provide the best aquifers in the state of New York.

The meltwater streams left well-sorted sand and gravel along the ice margin and downgradient from the margin. These kame, kame terrace, and outwash deposits constitute the source of water from some of the most productive wells in New York. For example, the Ranney well in Endicott is capable of producing 15.14 m^3/min (4,000 gal/min).

The outwash and kames occur in the majority of all valleys in New York. The outwash covers many valley floors and in many places overlies glacial-lake deposits. Where there are multiple sequences, the outwash from an earlier ice advance may be covered by later glacial lacustrine events, and even capped by another series of outwash deposits. Kames commonly extend from valley sides to the base of the valley fill and are hydraulically connected with the buried outwash deposits. The glacial lake beds are usually nonproductive, but may contribute water slowly to the adjacent aquifers.

LANDFORMS AND LANDSCAPES

Specific landforms can exert a dominant control on groundwater hydrologic conditions. Alluvial fans and pediments form the principal terrain along mountain fronts in the southwestern U.S. It is vital to differentiate such features if a successful water resource program is to be undertaken. The chance of developing important water supplies from pediment areas is slight because bedrock occurs at shallow depths. However, the deep sands and gravels of alluvial fans can yield significant amounts of groundwater to wells.

Kaye (1976) provided a revealing case study of what happens when there is a failure to properly identify geomorphic landforms and recognize the character of materials that compose them. Engineers who designed several new buildings in the Beacon Hill area of Boston assumed the hills were drumlins and that the sediment composing them was clay-rich glacial till. Foundations were designed accordingly. It had been interpreted that the till would be largely impermeable and so would not present any serious water-flow problems. The till was believed to be very compact and capable of withstanding great loading stresses. Unfortunately, these were not the conditions that were encountered during the digging and construction of the foundations for several buildings. Because of these erroneous calculations, some of the buildings developed severe construction problems, delays, and cost overruns. For example, at the Boston Common Garage, engineers had predicted the 13.5-m-deep excavation would contain thick clay till and would be dry. Instead the site encountered large groundwater flows that required costly construction methods and delayed the project for many months. At the Leverett Saltonstall State Office Building it had been anticipated that the pile foundation would penetrate a tough and durable till. Unfortunately the piles never reached the predicted resistance and failed because the granular materials that were actually encountered were not supportive.

The principal point to be learned from this case study is the verity of the law of equifinality, i.e., what may appear to be similar landforms can be created by entirely different processes. Thus, instead of drumlins, the Beacon Hill area is part of an end moraine, and instead of the more traditional glacial till substrate, the materials consisted of beds of clay, sand, and gravel with an imbricated series of thrust planes wherein the groundwater hydrology was a vital factor in its behavior under stress. Furthermore, the granular materials provided an entirely different water-flow regime than what had been predicted for the till.

The effect of basin shape and geometry on surface-water hydrology was investigated by Black (1972) in a study of experimental watersheds at Coshocton, Ohio. He used as a determinant a parameter he defined as "eccentricity," the measurement of basin length to the center of mass and at right angles to the center line. He found a correlation wherein basins with higher eccentricity values also had higher peaks of stream discharge. Kowall (1974) studied baseflow conditions for several drainages in Pennsylvania where the principal geomorphic difference was the eccentricity of the basins. He compared drainages in the Appalachian Plateau Province with dendritic stream networks with the trellis-pattern watersheds of the Folded Appalachian Province. These two regions have essentially the same climate, and both contain primarily Paleozoic clastic sedimentary rocks. The size of basins studied ranged from 12.7 to 704 km^2 (4.93 to 272 mi^2). When all basins were averaged for each region the trellis terrane of the folded area had longer sustained baseflow conditions than the dentritic terrane, as measured by both baseflow duration equations and baseflow recession curves. Table 2 shows a comparison of two basins, each of which is typical for their particular physiographic province. Kinzua Creek in the Appalachian Plateau has a much lower Q_b90 than that of Stony Creek, in the Folded Appalachians. This is despite the fact that Kinzua Creek has flatter hillslopes and the drainage basin contains only about one-half the relief of Stony Creek. When Kowall (1974) applied

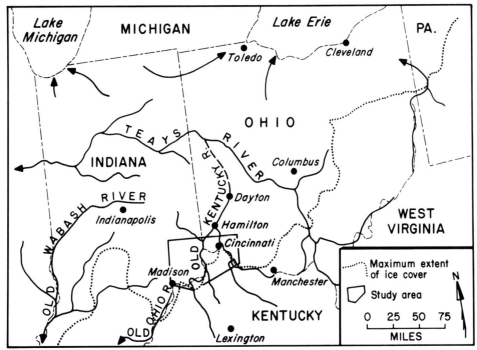

Figure 9. Preglacial drainage of the Teays River system in Ohio, Indiana, and northern Kentucky (after Wayne, 1956).

an eccentricity (E factor) to the basins in his study, basins with greater eccentricity values had larger baseflow discharge. He attributed this to the differences in structure and fractures that overprinted any difference that would have been caused by basin shape.

Another geomorphic parameter that Kowall (1977) evaluated was the hypsometric integral, a calculated measure of the percent of mass of a drainage basin that still remains in an undissected condition. When watersheds within the same physiographic province were compared, such as those in the Appalachian Plateau, there was a positive and direct correlation of low baseflow with the size of the hypsometric integral. Thus, stronger baseflow conditions were related to those basins with higher integrals, or amount of undissected uplands.

Buried valleys

Many geologic forces, including volcanic activity, can bury former stream valleys, but here we will consider only buried valleys filled by river or glacial deposits. Buried valleys in glacial terranes can be important aquifers (Norris and White, 1961; White, 1982). The largest buried valley system in the U.S. is that of the ancestral Teays River. Prior to Pleistocene glaciation it was the major stream system that drained Ohio, Indiana, Kentucky, and West Virginia. It originated in the Appalachian highlands and flowed west across most of Ohio, Indiana, Illinois, and into the Mississippi Valley embayment (Fig. 9). The valleys that were formed across most of Ohio and Indiana are now largely buried by a thick mantle of glacial sediment. Modification of the original valleys even occurred at the outer margin of the glacial drift and to the south of the glacial limits. With the onslaught of the glaciers, many of the valleys became clogged with varying thicknesses of till, and during deglaciation, meltwater deposits continued to form in the valleys. In addition, a new fluvial array of drainages was developed that became integrated into the present Ohio River system (Teller, 1973).

The character of the glacial drift that filled the Teays valleys is highly variable, ranging from till to stratified clay, silt, sand, and

TABLE 2. HYDROGEOMORPHOLOGICAL COMPARISON OF TWO BASINS IN PENNSYLVANIA IN DIFFERENT PHYSIOGRAPHIC PROVINCES*

Basin and physiographic province	Basin area (mi^2)	E factor†	Topographic slope (%)	Q90 (cfs/mi^2)	Basin relief (feet)
Kinzua Creek Unglaciated Appalachian Plateau	46.4	0.71	7.5	0.19	775
Stony Creek Folded Appalachian	35.0	5.47	17.0	0.36	1340

*After Kowall, 1974.
†Black, 1972.

gravel. Impondment of many river reaches resulted in formation of proglacial lakes with deposits of glaciolacustrine silts and clays. During the Yarmouth interglacial stage, part of the Teays system underwent a drastic change in southwestern Ohio, and is called the "deep stage." During this time, the bedrock valleys were deeply entrenched, and part of this valley system passed through the present site of Dayton, Ohio. Subsequently these valleys were buried with a wide range of heterogeneous materials that include till units alternating with outwash sequences. This sediment package contains some of the best aquifers for the city of Dayton (Norris and Spieker, 1966). Those aquifer layers below the till horizons are recharged where the till is missing, either due to nondeposition or to river-channel erosion. Although the potentiometric surface in the lowermost aquifer lies below the bed of the present Miami River, its level rises when the Miami River rises, thereby showing a good hydraulic connection between the river and aquifer. However, at other sites within these valley train aquifers the impermeable till units may be so extensive that they retard recharge to the deeper wells, thereby setting practical limits to the quantity of water available from induced infiltration. Norris and Spieker (1962) proposed a method to determine these relations. Groundwater derived from stream infiltration sources will vary seasonally in temperature, but where there are semiconfining beds, the seasonal temperature fluctuations will be damped and penetrate to lesser depths. The method consists of plotting the vertical temperature gradients measured in wells that penetrate the till units. Winslow and others (1965) showed the importance of temperature in determining permeability of riverbed sediments and groundwater conditions in the Mohawk Valley, New York. When the temperature was above freezing, water viscosity was sufficiently low to allow for river infiltration through the river sediments, thereby recharging the aquifer. However, when temperatures were below 0°C, recharge was reduced and the Schenectady and Rotterdam well fields pumped water from deeper zones of groundwater storage.

Other studies have also demonstrated the importance of buried valleys as potential sites for groundwater development (Frye and Walters, 1950; Cagle, 1969). Dreeszen and Burchett (1971) described and mapped the buried valleys in parts of Iowa, Kansas, Missouri, and Nebraska. They discuss the evolution of the Grand River system, which has little relation to present drainage. Here the continental ice sheets, streams, and wind combined forces during the Pleistocene to modify the topography in the region. Thus, glacial, alluvial, and eolian deposits now mantle much of the bedrock topography and fill many of the old valleys to depths as much as 100 m.

There are several buried valleys throughout the Binghamton metropolitan region. They were caused by course changes of the Chenango and Susquehanna Rivers. The best-documented case study was done on the most important aquifer of this region, which is called the "Clinton Street–Ballpark aquifer" (Randall, 1977). Figure 10 shows its location. It occupies the ancestral valley of the Susquehanna-Chenango River system prior to its glacial diversion to a more southerly course on the east side of the umlaufberg (a bedrock inlier within the system of valleys). The Clinton Street–Ballpark aquifer underlies 8 km^2 of urban land in the heart of the Binghamton region and consists chiefly of permeable sand and gravel. Transmissivity generally exceeds 900 m^2/day. In places, water movement is restricted by local silt lenses. The east and west ends of the aquifer are in contact with the major rivers, but the central part of the aquifer is separated from the rivers by either bedrock or till subcrops. More than 29 billion liters (7.6 billion gallons) is pumped from the aquifer annually.

New studies continue to show the importance of buried valleys to groundwater hydrology in the Northeast. Veeger (1986) studied the buried valley deposits in the western part of Oneida County, New York. Unlike many other New York buried valleys, those in Oneida County contain as much as 61+ m (200+ ft) of unconsolidated material in which the upper 46 m (150 ft) is composed of kame terrace and alluvial fan sediment, ablation till and glaciolacustrine deposits. The sand and gravel units provide the best aquifers, whereas the lacustrine beds are aquitards. Buried valleys in southern and central Maine were mapped by Tolman and others (1986). They showed how the location of such valleys was imported in environmental planning, i.e., providing a buffer type of action for groundwater flow and a recharge source for downstream hydrologic systems. However, it must be pointed out that not all buried valleys are potential aquifer sites. For example, there are numerous buried valleys throughout the Finger Lake region of New York whose valley fill consists primarily of till and other low-permeability sediments. At many of these sites, postglacial erosion has fashioned new bedrock gorges, such as Watkins Glen in Seneca Lake, and Enfield Glen in Cayuga Lake.

TOPOGRAPHIC ASPECT AND ORIENTATION

The Appalachian Plateau has often been described as a maturely dissected upland terrane with well-rounded hills interspersed among rivers with floodplains. In the Susquehanna section (Fig. 6), Coates (1966) called attention to the large-scale asymmetry of the upland hills. Throughout this region the bedrock hills, with relief that ranges from 100 to more than 150 m, have north-facing slopes of 24 percent compared with south-facing slopes of 12 percent. A study of 400 bedrock wells on these hills showed that the asymmetry and slope orientation is a reflection of the thickness of till. Whereas till on south-facing slopes ranges to depths of 80 m and averages of about 30 m; the depth to bedrock on north-facing slopes averages only about 3 m. Thus, the hills have been called till shadow hills (Fig. 11). The glacial processes that produced this asymmetry have affected not only new alignments of surface waters, but also the characteristics of drilled wells. The former thalwegs of the east-west–oriented streams have been buried, and the postglacial streams occupy channels that are higher and displaced to the south. In these upland hills, water supply to wells must come from bedrock aquifers because the till does not yield sufficient quantity for

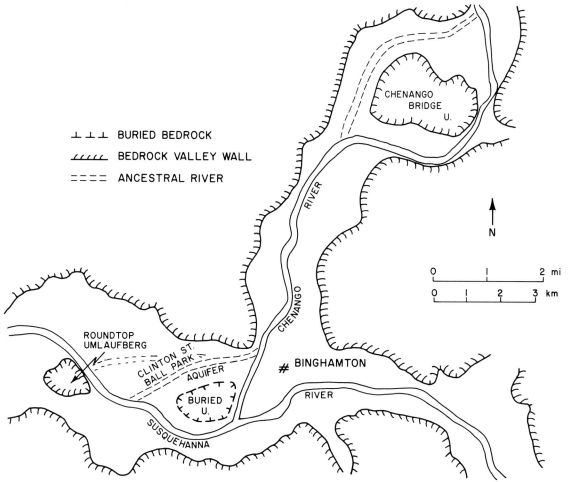

Figure 10. Map showing location of Clinton Street–Ballpark aquifer, Binghamton, New York. Double dashed lines show ancestral position of the Chenango River. U = umlaufberg, a bedrock inlier.

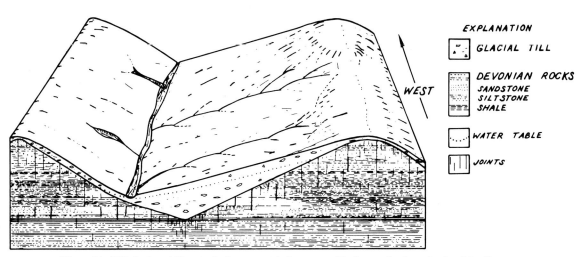

Figure 11. Till shadow hill, a typical asymmetrical topographic feature in the uplands of the Susquehanna section, New York (Coates, 1982).

household use. In the shale and siltstone of this region it usually requires a minimum well penetration into rock of at least 50 m to produce 17.5 to 35 l/min (5 to 10 gal/min). This groundwater condition saves homeowners on north slopes much money because their wells do not have to be drilled so deep or contain as much casing as their counterparts on the south slopes.

Surface water and groundwater relations

The azimuthal relations and alignments of the landscape can influence hydrologic features and conditions. For example, in the northern hemisphere, south and southwestern slopes receive greater insolation than north and northeastern slopes. This produces differences in soils, vegetation, and water retention and infiltration rates. Such factors create feedback systems that affect the amount of water entering the groundwater zone. There are additional direction alignments that are produced in glaciated terranes.

Within a 32-km radius of the Binghamton metropolitan area, the 25 largest lakes all occur as source areas for south-flowing streams. In a study of 350 lakes and upland wetlands in the Small Lakes section (Fig. 6), 86 percent of these features were situated near the headwaters of south-flowing streams. These surface-groundwater relations are so consistent that they can be predicted and incorporated into a useful model that explains the style of glaciation throughout the section (Conners, 1969; King and Coates, 1973; Coates and Kirkland, 1974). The glacial processes have been responsible for determining characteristics of the groundwater hydrology. Knowledge of these factors has been important in the design of the 21 dams emplaced north of the Binghamton urban area under the Watershed Management and Protection Act of 1954. Thus, geomorphology has had economic, engineering, and land-use-planning significance.

The lowest and successful bid for the construction of one of the dams was submitted by Old Forge Construction Company, a firm based in the Adirondack Mountains. Construction began in April 1974 with the clearing and grubbing of forested slopes and vegetation on the valley bottom, in the area that was destined to become the reservoir pool. Unfortunately the work did not progress on the time schedule set by Old Forge. Instead there were numerous delays and equipment breakdowns because the character of the soils and terrain was more formidable than anticipated. By late May, the company was so aggrieved by the conditions and time lost that additional funds in excess of the contract were requested from Broome County and the Soil Conservation Service (U.S. Department of Agriculture). In July 1974, an on-site inspection was arranged for all interested parties and their consultants to evaluate the nature of the problems. The troubles stemmed from discharge of springs into the wetland environment, and the character of the substrate that consisted of till and muck soils. This combination of features continually mired the construction equipment and made tree removal arduous. When Broome County officials were unwilling to allocate an additional $125,000 to the project, Old Forge brought a lawsuit against the county for recovery of cost overruns (Old Forge versus Broome County, New York). In the court record it was shown that Old Forge had not made a geomorphic or hydrologic study of the site. The company had anticipated conditions similar to those encountered in the Adirondacks, believing the locale would be dry and underlain by sandy soils. In my court testimony, I demonstrated that the Broome County Soils Survey prepared by the Soil Conservation Service showed a wetland with muck and till soils. Thus, an earth scientist could have predicted the conditions that were actually encountered. The jury dismissed the case and assigned no monetary damages to the company. Those doing engineering work should have a basic understanding of how geomorphology affects hydrologic conditions, and be aware of the literature that has been written about a project site, as illustrated by this case.

Hydrogeomorphology of Susquehanna section basins

In a 1971 study, Coates analyzed the geomorphic character and streamflow conditions for 97 drainage basins in the Susquehanna River basin of southern New York and northern Pennsylvania. The watershed areas range in size from 1.8 to 992 km^2 (0.7 to 383 mi^2) with a mean size of 123.2 km^2 (76.6 mi^2). This region is characterized by clastic, Paleozoic-age, sedimentary rock units that are essentially flat lying with no structural complications. The rocks are mostly fine-grained shales and siltstones. The basins occur in a climate that is mostly uniform in temperature and precipitation. The area is all part of the glaciated Appalachian plateau. The principal focus of the investigation was to determine the geomorphic and geologic controls of low baseflow, because an earlier investigation had indicated that traditional comparison of variables showed unsystematic correlations. Instead it appeared that the azimuthal orientation of the principal valley might influence hydrologic behavior.

To test this hypothesis, the 97 basins were grouped into seven different classes on the basis of streamflow direction and valley morphology, and an eighth class contained the remaining basins (Table 3). Figure 12 diagrammatically depicts some of the characteristics for these diverse orientation groups.

Class 1. Basins with headwaters that occur north of a major through valley (a valley that crosscuts the Appalachian Escarpment). The principal rivers flow in a southerly direction in valleys that are usually wide and that contain extensive valley-fill sediments.

Class 2. The master rivers have southerly flow in wide valleys, but the headwaters occur in a small col instead of existing in a through valley system as in Class 1.

Class 3. Basins that have been less glacially modified because all drainage is within the plateau. The flow direction is southerly but the valleys have a smaller percentage of area covered by valley-fill sediments.

Class 4. Basins with east-flowing streams. Such a direction is transverse to the regional fluvial fabric. The valleys were not modified by glacial erosion because they were athwart the major direction of glacial motion.

TABLE 3. HYDROGEOMORPHOLOGICAL COMPARISON OF DRAINAGE BASIN ORIENTATION CLASSES*

Class Number	1	2	3	4	5	6	7	8	Mean of all basins
Number of basins	15	10	24	11	11	8	14	4	97 (total)
Basin area (mi^2)	101	140	34	66	89	23	119	26	76
Q_{cfs}	1.57	1.41	1.48	1.31	1.29	1.46	1.50	1.67	1.46
Q_{95}/Q_{cfsm}	0.155	0.093	0.041	0.027	0.021	0.051	0.100	0.040	0.069
Q_{98}/Q_{cfsm}	0.107	0.069	0.015	0.016	0.009	0.022	0.071	0.018	0.043

*After Coates, 1971.
Q = stream discharge; cfs = cubic feet per second; Q_{95}/Q_{cfsm} and Q_{98}/Q_{cfsm} = percent of time flow equaled or exceeded on the basis of discharge per square mile.

Class 5. Basins with north-flowing streams. This flow direction is also transverse to regional landscape trend, but mostly parallel with glacial ice-flow direction.

Class 6. These are essentially upland basins with streams that drain areas between major glacial sluiceways. Thus, the basin size is smaller than average, is situated where the headwater locality is not well developed, and possesses smaller amounts of valley-fill sediments.

Class 7. Basins with west- and southwest-flowing streams that exist in wide valleys. These trends are conformable with the regional fabric of the landscape, and the valleys were sites for major outwash deposits. Thus, the valley-fill areas are larger than the regional average.

Class 8. The orientation and valley properties of these basins are diverse and do not fall into any of the above categories. Thus, these streams have dissimilar geometries.

Inspection of Table 3 shows there are vast differences in baseflow discharge between the eight azimuthal classes. Q_{95} and Q_{98} (that discharge equaled or exceeded 95 percent or 98 percent of the time) were selected as baseflow parameters to assure that the discharge was almost totally from groundwater flow into the streams. The classes with the highest baseflow are those directly affected by glacial action and in a compatible mode with ice flow, such as classes 1, 2, and 7. Lowest baseflow conditions exist in classes 4 and 5, those basins with flow transverse to the fabric of the landscape.

The most dramatic differences occur between classes 1, a major south-flowing stream, and 5, major north-flowing streams. Although the size of these two basin groups is somewhat similar, 261 and 230 km^2 (101 and 89 mi^2), respectively, their baseflow quantities are vastly different. Thus, Q_{98} (when measured on a per square mile basis) is 0.107 for south-flowing streams, but only 0.009 for north-flowing counterparts. This means that south-flowing drainages deliver more than ten times as much groundwater discharge during low flow as do north-flowing drainages. The reason for this vast discrepancy is the composition of the valley-fill materials, which in turn can be attributed to the deglaciation history of the valleys. Figures 13 and 14 are representative of conditions for these two azimuthal class groups. Figure 13 shows the occurrence of widespread sand and gravel deposits in the valley fill of south-flowing streams. In contrast, Figure 14 shows that north-flowing streams are underlain by valley fill that

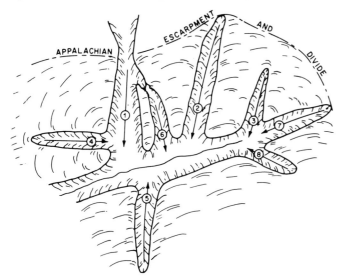

Figure 12. Diagrammatic sketch showing morphology/orientation classes (see text) for drainages in the Susquehanna section (Coates, 1971).

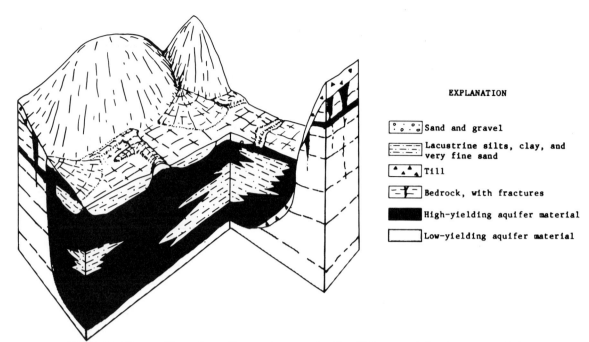

Figure 13. Diagram illustrating sediment character of valley fill in a south-flowing stream of the Susquehanna section (MacNish and Randall, 1982).

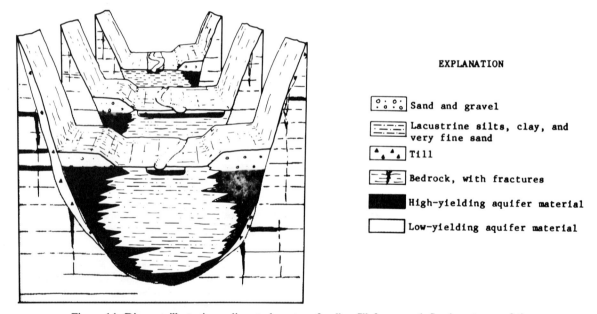

Figure 14. Diagram illustrating sediment character of valley fill for a north-flowing stream of the Susquehanna section (MacNish and Randall, 1982).

contains a large percentage of fine-grained deposits. During recession of the ice margin in the north-flowing streams, mostly fine-grained sediments and lacustrine materials were emplaced into proglacial lakes that had been impounded by the ice margin. However, in the south-flowing streams, the glacial meltwaters flushed away the fine sediments, removing them from the valley, and leaving behind the lag deposits of sand and gravel. These coarser-grained materials are much more permeable, provide improved aquifer conditions, and produce larger groundwater flow rates than the finer materials in the north-draining valleys. Thus, the major geomorphic control of groundwater hydrology is associated with the amount and type of valley-fill sediments, and these characteristics, in turn, are dictated by the azimuthal orientation of the drainage basin.

CONCLUSIONS

This chapter emphasizes how the geomorphology of an area can determine its groundwater hydrology. Landforms, surface processes, and materials can all play a role in controlling groundwater characteristics. Topographic slope can influence the depth and position of the water table, and the velocity of groundwater flow. Drainage density can be an important indicator of variables that affect both baseflow and flood behavior of streams. Examples are cited that illustrate how glacial processes control groundwater conditions. Buried valleys can provide significant aquifer systems, and such groundwaters are a major source of supply for such cities as Dayton, Ohio, and Binghamton, New York. The azimuthal orientation of the major valleys in the Susquehanna River basin significantly control the strength of baseflow discharge. A study of 97 drainage basins shows that south-flowing streams have ten times the amount of baseflow as do their north-flowing counterparts.

Several case histories are cited that show the practical necessity for understanding the geomorphic flavor of the landscape when doing engineering geology projects. Construction delays and cost overruns occurred in Boston, Massachusetts, and Binghamton, New York, when the engineering design and predictions did not adequately account for the geomorphic conditions at the project site.

REFERENCES CITED

Black, P. E., 1972, Hydrograph response to geomorphic model watershed characteristics and precipitation variables: Journal of Hydrology, v. 17, p. 309–329.

Cagle, J. W., 1969, Availability of groundwater in Wayne County, Iowa: Iowa Geological Survey Water Atlas, v. 3, 33 p.

Carlston, C. W., 1963, Drainage density and streamflow: U.S. Geological Survey Professional Paper 422-C, 8 p.

Chorley, R. J., and Morgan, M. A., 1962, Comparison of morphometric features, Unaka Mountains, Tennessee and North Carolina, and Dartmoor, England: Geological Society of America Bulletin, v. 73, p. 17–34.

Coates, D. R., 1952, Groundwater in Willcox Basin, Cochise and Graham Counties, in Halpenny, L. C., and others, Groundwater in the Gila River Basin and adjacent areas, Arizona: U.S. Geological Survey Open-File Report, p. 177–186.

—— , 1966, Glaciated Appalachian Plateau; Till shadow on hills: Science, v. 152, p. 1617–1619.

—— , 1971, Hydrogeomorphology of Susquehanna and Delaware Basins, in Morisawa, M., ed., Quantitative geomorphology: Binghamton, State University of New York Publications in Geomorphology, p. 272–306.

—— , 1982, Reappraisal of the glaciated Appalachian Plateau, in Coates, D. R., ed., Glacial geomorphology: London: George Allen and Unwin, p. 205–243.

Coates, D. R., and Cushman, R. L., 1955, Geology and groundwater resources of the Douglas Basin, Arizona: U.S. Geological Survey Water-Supply Paper 1354, 56 p.

Coates, D. R., and Kirkland, J. T., 1974, Application of glacial models for large-scale terrain derangements, in Mahaney, W. S., ed., Quaternary environments: York University Geographical Monograph 5, p. 99–136.

Conners, J. A., 1969, Geomorphology of the Genegantslet basin of New York [M.A. thesis]: Binghamton, State University of New York, 154 p.

Dreeszen, V. H., and Burchett, R. R., 1971, Buried valleys in the lower part of the Missouri River Basin, in Pleistocene stratigraphy of Missouri River Valley along the Kansas–Missouri border: Geological Survey of Kansas Special Distribution Publication 53, p. 21–25.

Eakin, T. R., Price, D., and Harrill, J. R., 1976, Summary appraisal of the nation's ground-water resources; Great Basin region: U.S. Geological Survey Professional Paper 813-G, 37 p.

Fenneman, N. M., chairman, 1923, Map of the United States showing physiographic divisions and physiographic provinces: Association of American Geographers.

Fetter, C. W., 1980, Applied hydrogeology: Columbus, Ohio, Charles E. Merrill Publishing Co., 488 p.

Flint, J. J., 1968, Hydrogeology and geomorphic properties of small basins between Endicott and Elmira, New York [M.A. thesis]: Binghamton, State University of New York, 74 p.

Frommurze, H. F., 1937, The water-bearing properties of the more important geological formations in the Union of South Africa: Union of South Africa Geological Survey Memoir 34, 186 p.

Frye, J. C., and Walters, K. L., 1950, Subsurface reconnaissance of glacial deposits in northeastern Kansas: Kansas Geological Survey Bulletin 86, part 6, p. 143–158.

Heath, R. C., 1984, Ground-water regions of the United States: U.S. Geological Survey Water-Supply Paper 2242, 78 p.

Horton, R. E., 1945, Erosional development of streams and their drainage basins; Hydrophysical approach to quantitative morphology: Geological Society of America Bulletin, v. 82, p. 1355–1376.

Hubbert, M. K., 1940, The theory of groundwater motion: Journal of Geology, v. 48, p. 785–944.

Jacob, C. E., 1943, Correlation of ground-water levels and precipitation on Long Island, New York; Part 1, Theory: American Geophysical Union Transactions, p. 564–573.

Jennings, J. N., 1971, Karst: Cambridge, Massachusetts Institute of Technology Press, 252 p.

Kaye, C. A., 1976, Beacon Hill end moraine, Boston; New explanation of an important urban feature, in Coates, D. R., ed., Urban geomorphology: Geological Society of America Special Paper 174, p. 7–20.

King, C.A.M., and Coates, D. R., 1973, Glacio-periglacial landforms within the Susquehanna–Great Bend area of New York and Pennsylvania: Quaternary Research, v. 3, p. 600–620.

Kowall, S. J., 1974, Hydrogeology and geomorphology of two structurally dissimilar terranes in Pennsylvania [Ph.D. thesis]: Binghamton, State University of New York, 143 p.

—— , 1977, The hypsometric integral and low streamflow in two Pennsylvania provinces: Water Resources Research, v. 12, no. 3, p. 497–502.

Le Grand, H. E., 1954, Geology and groundwater in the Statesville area, North Carolina: North Carolina Department of Conservation and Development, Division of Mineral Resources Bulletin 68, 68 p.

—— , 1967, Ground water of the Piedmont and Blue Ridge Provinces in southeastern States: U.S. Geological Survey Circular 538, 11 p.

MacNish, R. D., and Randall, A. D., 1982, Stratified-drift aquifers in the Susquehanna River Basin, New York: New York State Department of Environmental Conservation Bulletin 75, 68 p.

McGuinness, C. L., 1963, The role of ground water in the national water situation: U.S. Geological Survey Water-Supply Paper 1800, 1121 p.

Meinzer, O. E., 1923, The occurrence of ground water in the United States: U.S. Geological Survey Water-Supply Paper 489, 321 p.

Mylroie, J. E., 1984, Hydrologic classification of caves and karst, in LaFleur, R. G., ed., Groundwater as a geomorphic agent: London, George Allen and Unwin, p. 157–172.

Norris, S. E., and Spieker, A. M., 1962, Temperature-depth relations in wells as indicators of semiconfining beds in valley-train aquifers: U.S. Geological Survey Professional Paper 450-B, p. 103–105.

—— , 1966, Ground-water resources of the Dayton area, Ohio: U.S. Geological Survey Water-Supply Paper 1808, 167 p.

Norris, S. E., and White, G. W., 1961, Hydrologic significance of buried valleys in glacial drift: U.S. Geological Survey Professional Paper 424-B, p. 34–35.

Randall, A. D., 1977, The Clinton Street–Ball Park aquifer in Binghamton and Johnson City, New York: New York State Department of Environmental Conservation Bulletin 73, 87 p.

Ries, H., and Watson, T. L., 1914, Engineering geology: New York, John Wiley and Sons, p. 330–337.

Siddiqui, S. H., and Parizek, R. R., 1971, Hydrogeologic factors influencing well yields in folded and faulted carbonate rocks in central Pennsylvania: Water Resources Research, v. 7, p. 1295–1312.

Stewart, J. W., 1962, Water-yielding potential of weathered crystalline rocks at the Georgia Nuclear Laboratory: U.S. Geological Survey Professional Paper 450-B, p. 106–107.

Sweeting, M. M., 1973, Karst landforms; New York, Columbia University Press, 362 p.

Teller, J. T., 1973, Preglacial (Teays) and early glacial drainage in the Cincinnati area, Ohio, Kentucky, and Indiana: Geological Society of America Bulletin, v. 84, p. 3677–3688.

Thomas, H. E., 1951, The conservation of groundwater: New York, McGraw-Hill, 327 p.

—— , 1952, Ground-water regions of the United States; Their storage facilities, *in* The physical and economic foundation of natural resources, v. 3: U.S. Congress House Interior and Insular Affairs Committee, p. 3–78.

Tolman, A. L., Kelly, J. T., and Lepage, C. A., 1986, Buried valleys of southern and central Maine: Geological Society of America Abstracts with Programs, v. 18, p. 72.

Veeger, A. I., 1986, Hydrogeology of Wisconsin and deposits in a valley fill; Southern portion of the Tug Hill aquifer, Oneida County, New York: Geological Society of America Abstracts with Programs, v. 18, p. 74.

Wayne, W. J., 1956, Thickness of drift and bedrock physiography of Indiana north of the Wisconsin glacial boundary: Indiana Geological Survey Report of Program 7, 70 p.

White, G. W., 1982, Buried glacial geomorphology, *in* Coates, D. R., ed., Glacial geomorphology: London, George Allen and Unwin, p. 331–349.

Winslow, J. D., Stewart, H. G., Johnston, R. H., and Crain, L. J., 1965, Ground-water resources of eastern Schenectady County, New York: State of New York Conservation Department Water Resources Commission Bulletin 75, 148 p.

MANUSCRIPT ACCEPTED BY THE SOCIETY NOVEMBER 14, 1989

Index

[Italic page numbers indicate major references]

abrasion, marine, 286, 290
Acheuleans, 309
Achterstufe, 299
acids
  carbonic, 106, 183, 188
  organic, 59, 183
  sulfuric, 183, 189
accretionary wedge, 274, 275
Adeii Eechii Cliffs, 299, 311
Adirondack Mountains, New York, 59
Afar Depression, 292
Afghanistan, 81
Africa, 41, 114
agglomerates, 90
aggregates, 24, 84, 100, 101, 102, 112
Agula-Nanka region, Nigeria, *148*
alases, 212, *217*
Alaska, 88, 211, *325*
Alaska Arctic Coastal Plain, 217
Alberta, southern, 112
alcoves, 291, 292, 296, 299
  formation, 242, 284
  multiple, 71
  theater-headed, 292
algae, 242
Alkali Creek, Colorado, 85
allasy, 212, 217
alluviation, 181
alluvium, 81, 85, 89, 90, 106, 144, 146, 221, 222, 227, 286, 308, 320, *327*, 329, *330*, 347
Alps, 161
Anasazi Indians, 298
anastomoses, *195*
anhydrite, 187
anisotropy, 184
anticline, 8, 14
Antilles, 30
Antrim coast, Ireland, *69*
Apache Junction area, 221, 225
Appalachian channel networks, 13
Appalachian Escarpment, 352
Appalachian Highlands, 167
Appalachian Piedmont, 241
Appalachian Plateau, 62, 159, 348, 350
Appalachian Valleys, 159
aquicludes 3, 7, 15, 16, 19, 299, 347
aquifers, 13, 15, 19, 104, *105*, 241, *242*, 250, 289, *301*, 335, 347, *350*
  alluvial, 320
  artesian, 268
  carbonate, 162, 189
  Chalk, 8
  Clinton Street–Ballpark, *350*
  colluvial, 5
  confined, 7, 24
  deep-brine, 301, 302
  dolomite, 199
  freshwater, 267
  gravel, 344

High Plains, 103, 105, 301, 302, 320
homogeneous isotropic, 3, 14, 19
karst, *178*, *183*, *189*, 193
permeable, 2
sand, 344
sandstone, 244
unconfined, 7
aquitards, 1, 2, 8, 22, 247
Arabia, eastern, 35
Arba-in Desert, 311
Arba-in Valley, 311
Arches region, Utah, 299
Arikaree Formation, 293
Arizona, 80, 92, *219*, 324
  south-central, *221*, 231
Arkansas River, 320, 324
Arnhem Land Plateau, 298, 311
Arroyo de los Frijoles, 146
Arva Valley, 223
ash, volcanic, 99, 100, 106
Aumont district, France, 37
Australia, 35, 92, 114
  central, 33
  northern, 298, 311
  southeast, 115
avalanches, 9
  debris, *58*, *62*, *68*
  rock, 57
  slush, 66
  soil, 238
Aztec Wash, Colorado, 95

backwearing, *313*
bacteria, 47
  aerobic, 102
  anaerobic, 189
Badger Mountain, 149
Badlands National Monument, 310
badlands, 81, *87*, 91, 93, 94, *112*, 133
Bahamas, 161, 198, 272
Balcones Escarpment, 169
Baldwin Hills Reservoir, 83
Balearic Islands, Spain, 288
Bananal, Brazil, 8
banks
  alluvial, *328*
  collapse, 323
  composition, 323
  devegetation, *335*
  erosion, *322*, *324*, *334*
  failures, *327*, *329*
  retreat, 329
  storage, 320
Banks Peninsula, New Zealand, 133
Barbados, *68*
basalts, 15, 35, 37, 69, 238, 247, 249, 250
baseflow, *320*, *352*
Basin and Range fault, 229
Basin and Range Province, 81, 221, 344

basins, 41
  alluvial, 231
  collapse, 301
  deflation, *311*
  drainage, 342, 347, 349
  geometry, *348*
  intermontane, *344*
  morphology, *252*
  morphometry, 256
  playa, *102*, 301, 302, 303
  playa lake, 95, *101*
  rock, 46
  semiarid, 344
  shape, *348*
Bayfield, Colorado, 94
beach groundwater, 291
Beacon Hill, *348*
Bearn Meadow, 320
bedding 193, 194
bedform, *321*
  classifications, *321*
bedrock, 30, 37, 44, *56*, 59, 65, *73*, 81, 88, 99, 146, 149, 157, *158*, 164, 167, 179, 182, 195, 199, *344*, 348, 350
  buried, 227, 229
  channels, 260
  collapse, *197*
  depressions, *171*
  hollows, 8
  meanders, 238
  slope failure, *56*
  spurs, 88
  wells, 342, 350
Benson, Arizona, *89*
bentonite beds, 91
Bermuda, 188, 198
Berwick wells, 335
Bieszcady Mountains, 81
Big Basin, Oregon, 96
Big Level, West Virginia, 167
Big Muddy Valley, Saskatchewan, 81
Bighorn Basin, 189
Bighorn Reservoir, 72
Bighorn River, 189
Binghamton, New York, 342, 350, 352
biosparite, 193
biota, 44
bioturbation, 20, 24
Blackwater Draw Formation, 102
Blanco fracture zone, 275
Blanco Ridge, 275
blockfalls, 293
blow holes, 128
Blue Ridge province, 342
Blue Spring Cave, *205*
bog bursts, 132
Bolivia, 81
bornhardts, 46
Boston, *348*
boulders, 37, 44, 46
  disintegration, 51

358 Index

Bradshaw Surface, 35
Brazil, 30, 38, 144
Brazons River, 300
breccias, 90, 106
Brecon Beacons, South Wales, 119, 121
brines, 276
    discharge, 267, 270
    springs, 302
    sulfide-rich, 277
British Isles, 130
Brule Formation, 310
Burnsville, Cover, 168
burrow, animal, 6, 20, 22, 82, 84, 323

calcarenite, 42
calcite, 10, 91, 161, 163, 168, 186, 187, 188, 237, 242, 287, 296
calcium, 68, 187
calcium carbonates, 83, 102, 283, 284
calcrete, 33, *34*
caliche, 34, 102, 106, 146, 301
California
    central, 42, 64
    coastal, 286, *334*
    northern, 319
    northwestern, *65*
    southern, 59
Cameron, Arizona, 297
Canada, 21, 67, 215
Canadian River valley, 300, 305
Canterbury Plain, New Zealand, 147, 323
Canyon of Desolation, 296
Canyonlands National Park, 299
canyons
    box, 302, *305*
    submarine, 270, 275
    theater-headed, *273*
    vadose, 203
Cañada de la Cueva, 146
caprock, 204
carbon dioxide, 102, 106, 162, *163*, 169, 183, *186*, 189, 199, 284
carbonate cement, 83, 102, 242
carbonate escarpments, *272*
carbonate minerals, 161
carbonates, 29, 102, 106, 164, 179, *183*, *272*, 345
carbonation, 29
Carlsbad Caverns, 189
Carmel Basin, 334
Carmel River, 320, 324, *334*
Carnatic plain, 42
Carpathians, 133
Casa Grande area, 227
Casa Grande Mountains, 225
Cascade Mountains, Oregon, *68*
Cascadia Channel, 275
case studies, *64*, *66*, *68*, *73*, *96*, *101*, *149*, *200*, *241*, *245*, *257*, *287*, *300*, *305*, *327*, *334*
Castleguard Cave, 188, 199
catchments, *120*, *171*
Catskill basins, 347
cave development, 79, 177, 188, 194, 195, 203
caverns, 89
    subterranean, 23

caves, 78, *96*, 81, 177, *179*, 185, 190, 193, *194*, 202, *203*, 288
    anastromotic, 195
    branchwork, *190*
    halocline, 161
    maze, *190*, 194
    origin, 177, 188
    passages, *195*
    patterns, *190*
    phreatic, 345
    pseudokarst, *96*
    radose, 345
    sediments, 184
    springs, 194
    streams, 191, 194, 196
    vertical, 345
    water-table, 161
cavities, solutional, 3, 6
cells, catchment, *171*
cement, 187, 260, 295, 297
Centerville, Texas, 69
Central Arizona Project aqueduct, 222, 227
Centre County, Pennsylvania, 166
Chaco Canyon National Park, New Mexico, 292
Chaco Canyon, *292*, 298
Chadron Formation, 310
chalk, 69
Chalk aquifer, 8
Chalk downs, England, 297
chamber, sapping, 12
Chandler area, Arizona, 230
channel networks, 8, 13, 15, *19*, 143
    Appalachian, 13
    evolution, 238
channel process, groundwater influence, *320*
channels, 18, 25, *259*, 300, 302, 350
    alluvial, 321, 322
    beach-face, 241
    bedform classification, *321*
    bedrock, 260
    braided outflow, 292
    classes, *245*
    development, 235
    extension, *16*
    flow, 11, 112, 323
    form, groundwater influence, *320*
    formation, *19*
    geometry, 323
    gully, 149
    initiation, *13*
    migration, 322
    outflow, 291
    precipitation, 120, 122
    roughness, *321*
    sand-bed, 322
    scarps, 140, 144, 146
    stream, 17, 25, 120
    submarine, 268
    subsurface, 112
    subterranean, 80
    surface, 319
    surface-water, 84
    theater-headed, 304
Chester Escarpment, 202, 203
Chester Upland, 202, *203*
Chesterian Series, 202

Chézy formula, 22
Chilkat River, *325*
chimneys, solution, 168
China, 78, *94*, 159, 160, 172
Chinle Formation, 88, *297*
Chinle Valley, 89
chlorite, 68
Chusha Mountains, New Mexico, 95
chute, *58*
circulation, seawater, 272
Claremont, New Hampshire, 148
clay coatings, 91
clay loam, 149
clay matrix, 65
clay soil, 103
clays, 23, 34, 69, 77, 85, 89, 92, 99, 114, 115, 242, 270, 271, 276, 332, 347, 350
    dispersive, 23, *24*, *111*
    expansive, 111
    hydrophillic, 29
    kaolinite, 68
    liquefied, 68
    minerals, 24, 54, 114, 135
    montmorillonitic, 23, 66, 97, 100
    natural-levee, 69
    pure, 92
    quick, *67*
    sensitive, *67*
    smectite, 91, 101, 106
    smectoid, 111, 120
clayslides, sensitive, *67*
claystone, 89, 106, 271
Cliff House Sandstone, 293
cliffheads, recession, 309
cliffs
    age, *309*
    development, *291*, *293*, *309*
    erosion, 283
    morphology, 299
    recession, seepage-induced, *291*
    retreat, 19, *298*, *305*
    sapping, *291*, *295*
    seepage-influenced, 293
climate, *94*, *115*, 199, 236, 298, 309, 323, 344
    hyperarid, 309
    monsoon, 111
Clinton Street–Ballpark aquifer, 350
Clovis, New Mexico, 103
clustering, *134*
    peak, 160
Coastal Plain, 271
Coeymans Formation, 193
cohesive alluvium, *330*
collapse
    bedrock, *197*
    roof, 126
    surface, 29, 51
colluvium, 8, 11, 12, 16, 18, 24, 59, 62, 73, 90, 106, 144, 249, 314
Colombia, 81
Colorado, 7, 81
Colorado Front Range, 147
Colorado Plateau, 15, 22, *235*, *241*, 245, 296, 298, *299*, 314, 323
    case study, *241*
Colorado Plateau sandstones, 237
Colorado River, 300

Colorado River basin, 324
Colorado River entrenchment, 300
Columbia Ice Field, 188, 199
Columbia Plateau, 15
Columbia River Valley, 71
compaction, 227
 differential, 229
Comprehensive Classification System, 30
compression, 268
 tectonic, 268
 hydraulic, 2, 3, 7, 8, 10, 13, 14, 19, 22, 95, 119, 124, 132, 190, 194, 262
conduits, 22
 development, 198
 formation, 9
 phreatic, *182*, 195
 relict, 181, 182
 solution, 177, *178*, *184*, *189*, *196*
 subterranean, 15, 55
 tubular, 329
 vadose, *182*
continental margin, 267
 erosion, *270*
continental shelves, sea levels, *270*
continental slope, 267, *271*
Contra Costa County, *64*
convection, 268
convergence, 14, 15, 17, 199
 flow, 239, 260
coral cap, 68
corestones, *37*, 41
Cornfield Wash site, 92
corrosion, 112
Coshocton, Ohio, *348*
Coulomb failure, 9, 11, *12*, 18
cracking, mass-movement, 135
cracks, 9, 22, 25, 116
 desiccation, *85*, 106, 133, 135
 development, *84*, 112
 dislocation, 100
 shrinkage, 6, 23
 soil, *84*
 tension, 141
 vertical, 332
Crawford Upland, 202, *203*
creep, 24, 51, 52, *56*, *68*, 72, 82, 144
 frost, 212
 soil, 21, *56*, 80
crystal growth, 237, 296
Cuba, 160
Cunyarie Rocks, 36
currents, turbidity, 268, 275
cutoffs, subterranean meander, *198*

Daly Basin, Northern Territory, 35
Danakil Basin, 292
Darcy's Law, 2, 164, *184*, *325*, 341
Darling Range, 41
Dartmoor, England, *346*
Davilla Hill landslide, *64*
Davis, California, 93, 141
Dayton, Ohio, 350
deaeration, 23
Death Valley, 89
debris
 avalanches, *58*, *62*, *68*
 failure, 145, 329, 334
 flows, 9, 11, 21, 59, *63*, 132, 268, 275
 rock, 158
 slides, *54*
 slopes, 292
 soil, 145
 torrent, *68*
deflocculation, 91, *92*, 95, 114
degassing, *187*
Denmark coast, *66*
denudation, 185
 equilibrium, 166
 karst, *164*
 rates, *164*, *167*
 regional, *291*, *305*, *308*, *311*
 vegetation, 94
deposition, 327
deposits
 alluvial, 221, 337
 glacial, 349
 hydrothermal, 276
 lacustrine, 305
 loess, 78, 347
 river, 349
 unconsolidated, 343
depressions, 198
 bedrock, *171*
 closed, *159*, *170*, 311
 microtopographic, *131*
 sinkhole-like, 273
 surface, 196
 thermokarst, 217
desalination, 115
desilication, 33
detritus, *36*, 44
 weathered, 296
devegetation, 22, 79
 bank, *335*
development
 gully, *139*
 soil, 29
 valley network, *235*
Devil's Marbles complex, Australia, 47
dewatering, 148, *274*, 277, *323*
 fluids, 274, 275
 routes, 275
 sediments, 229
Dewey Bridge Member, 299
dégagés, 37
diapirs
 formation, 275
 surficial, 275
diffusion, 187
Dinaric Karst, 158
disaggregation, 29
discharge, 15, 16, 17, 24, 184, 277
 brines, 267, 270
 groundwater, 7, *152*, 243, 267, 270, 272, 283, 285, *301*, 344
 hydrothermal, 277
 peak, 122
 rates, 7
 river, 319
 seepage, *243*
 spring, 255, 302, 352
 surface gas, 274
 throughflow, 111
 tunnel, 24
discontinuities, *134*

dispersion, *24*, 92, 111
dispersion zone, 289
displacement, 229
 piping, 77, 81
 pseudokarst, *7*
dissection, 247
dissolution, 29, *102*, 106, 157, 162, *164*, *182*, 186, 268, *272*, 288, 297
 calcite, 163
 carbonate escarpments, *272*
 carbonates, 269
 differential, 289
 dolomite, 162
 evaporites, 219
 groundwater, 179
 limestones, 162, *185*, 219, 287
divergence, 15
dolerite, 37
doline, *159*, *170*
dolomite, 34, 85, 157, 158, 161, 187, 199, 202
Dominica, 111, 117, *119*
doppelten Einebnungsflächen, 37
Double Mountain, 301, 305
downslope movement, *51*, *69*
downslope, stress, 12
downwearing, 313, 314
drainage, 69, 95
 hillslope, *126*
 pipeflow, 120
drainage density, *18*, *346*
drains
 horizontal, *73*
 vertical, 168
Drakensberg Mountains, 128
drift, glacial 349
dripstones, 167
drylands, 77, 94
Ducktown, Tennessee, 148
dunes, 101, 105
Dupuit-Forchheimer model, *5*, *16*
Durham, New Hampshire, 148
duricrusts, *33*, 41, 44, 114

earth fissures, 222, *223*, *231*, 272
 formation, *219*, *221*
earthflow, 63, 64, 67, 68, 69
 slump, 68, *71*
earthquakes, 275
East Brady, Pennsylvania, 62
East Coast margin, United States, 270
East Twins, England, 130
Ecuador, 81
Edwards Plateau, 300
effects, common-ion, *187*
efforescence, crystalline, 283
Egypt, 298, 300, *305*, 310, 311
Ellison's Cave, Georgia, *194*
Eloy, Arizona, 221
Eloy-Picacho area, 222
elutriation, 83
eluviation, 32, *82*, 105, 106, 115, 116, 126
embayment, 15
emergence, groundwater, 13, 242, 291, *322*, 338
Encounter Bay, South Australia, 44
Endrumpf, 313

England, southeastern, 8
English basin, 130
English Chalk, 14
English Peak District, 114
English Pennines, 132
Entrada Sandstone, 299
entrainment, 9, 142
environment
　problems, 206
　submarine, 267
epigene agents, 43, 46
erosion, *1*, 15, 24, 99, 140, 148, *152*, *237*, *242*, *270*, *283*, *310*, 327, *335*, 342
　alluvial stream banks, 334
　bank, *322*, *324*, *334*
　basal lateral, 307, 311, 314
　chemical, 268
　cliff, 283
　concentrated, 126
　current, 272
　fluvial, 301
　glaciers, 112
　groundwater, 242, 283, 286, *302*
　hillside, 62
　hillslope, 131
　karstic, 273
　mechanical, 127, 129, 268
　model, *262*
　peat, 28
　piping, 106, 114, *125*, *130*
　pothole, 80
　rates, *128*
　rill, 297
　sapping, 238
　seepage, 9, 12, *13*, 19, 77, 140, *142*, 146, 147, 235, 268, *291*, 295, 297, 307, 310, 329
　seepage-face, *83*, *84*, 87
　sheet, 33
　sheetrock, 131
　slope, 310
　soil, 196
　solutional, *127*, 130
　spring, 82
　submarine, 267
　subsurface, 77, 81, 82, 107, 319
　subsurface flow, *8*
　subterranean, 100
　surface, 1, 310
　tunnel, 9, 80, 81, 82, 268
　tunnel-gully, 79
Escalante River basin, 238
escarpment, 19, 35
　carbonate, *272*
　Goshen Hole Rim, 293
　Gulf of Mexico, 272
　retreat, *304*
　Southern High Plains, *300*, 311
　Texas Caprock, 297, 299, 300
　Tloi Eechii, 299
　West Florida, 270, 272, *273*, 277
Ethiopia, 292
Evansville site, Indiana, *329*
evaporates, *183*, 269, 272
evaporation, 10
　rates, 59
evaporite beds, 301
evapotranspiration, 52, 199

evolution
　landform, 13
　landscape, *44*
　solution conduits, *189*
　valley, *255*
Evrard Ranges, Australia, 38
exfiltration, 6, 8, 10, 19, 21
Exmoor, England, 33
experiments
　flowtank, 261
　flume, 322, 323, 338
　sapping erosion, *257*
extension, stream channel, 132
extrusion, 67
Eyre Peninsula, South Australia, 36, 41, 42, 47

face
　free, 22
　hillside, *51*, *69*
　mechanism, 334
　riverbank, 132
　shear, 132
　slab, 291
　slope, 11, *59*, 69, 132, 269
Fajada Butte, 293
Falkland Islands, 134
fans, alluvial, 344, 348
fault scarps, 229
faults, 276
　Picacho, 230
　surface, *230*
fauna, soil, 6
fengcong, 160
fenglin, 160
ferricrete, *33*
Finger Lake section, 348
fish habitat, *325*
fissures, earth, 222, *223*, *231*, 272
fissuring, multiple, 229
Flinders Ranges, 34
flocculation, 91
flood peak, 320
flood regime, *321*
floodplains, 116, 323, 247, 350
floods, 190, 197, 324, 329, 334
　flash, 87
Flores thrust zone, 275
Florida, 159, 198, 271, 272, 273
Florida-Bahamas platform, 270, 272
Florida boulder zone, 272
Florida Canyon, 273
Florida peninsula, 270
flows
　basalt, 299
　channel, 11, 112, 323
　concentrations, 19
　conduit, 179
　convergence, 239, 260
　debris, 9, 11, 21, 59, *63*, 132, 268, 275
　diffuse, 193
　ephemerel, 142
　fluidized sediment, 268
　grain, 268
　groundwater, 181, 184, 236, *319*, *327*, 341, 348
　high, 323
　intergranular, 291

　laminar, 178, *184*
　lava, 246, 286
　lines, 4, 5
　low, 338
　macropore, 6
　model, *262*
　nets, 8
　network, 262
　overland, 130, 139, 148, *152*, 197, 310, 320, 323
　paths, 143, *168*, 178, 180, 184, 189, 193, 195
　plug, 63
　quasi-plastic, 72
　rates, 259, 260
　regional graoundwater, 2
　saturation overland, 6, 11, 112, 121
　sediment gravity, *268*
　shallow subsurface, 3
　submarine, *268*
　subsurface, *1*, *8*, 82, 84, 127, 140, 146, 153, 193, 235, 310, 320
　subsystems, 3
　surface, 142, 146, 291, 323
　subterranean path, 8
　systems, *19*
　tunnel, 242
　turbulent, 178, *185*, 189
　velocities, 184, 324
flowslides, retrogressive, 67
flowstone, 167
fluvial processes, *319*
fluviokarst, *158*, 345
flux
　carbon dioxide gas, *105*
　groundwater, 8, 106, 270
　particulate, *105*
　rate, 2
　recharge, 104, 106
　sediment, 106
　water, 1, 14, 106
Folded Appalachians, 348
forest, peak, 160
fractures, 36, 37, *170*, 179, 193, 199, 242, 341
　discordant, 193
　rock, 16
fracturing, 23, 24
Franklin D. Roosevelt Lake, 71
Fraser Canyon, 69
freezing, 212
Frontstufen, 299
frost, 295, 323
　action, 10, 338
　creep, 212
　heaving, 51, *212*
　seepage, *296*
　shattering, 30
　soil, 151
frost table, 211
Furnace Creek Formation, 89

Galveston, Texas, 219
gases, 29, 47
　natural, 230
　subsurface, 274
Gatesburg dolomite, 166
Gefügerelief, 37
geliflution, 56

geochemistry, *161*
geomorphology, *1, 199, 211, 341, 343*
Georges Bank, 270
Georgia, 194
geothite, 276
Gepatsch Dam, Austria, 72
Ghats, 42
Ghyben-Herzberg lens, 284, 285
Gilf Kebir, case study, *305*
Gilf Kebir plateau, 298, *305*, 311
Gilf Sandstone, 305, 311
Girkin Formation, 202
glaciation, 205, 270, 348, 349, 352
glaciers, 349
   erosion, 112
   semi-permanent, 199
glaciokarst, 345
Glen Canyon Group, 299
Glen Canyon region, *241*, 299
Golden Valley Formation, 94
Goshen Hole Rim escarpment, 293
gradients
   capillary, 23
   differential pressure, 277
   geothermal, 188, 272
   hydraulic, 8, 11, 14, 15, 16, 66, 112, 115, 127, 143, 184, 186, 191, 243, 244, 260, 261
   hydraulic head, 5, 11
   osmotic potential, 23
   pore-pressure, 24
   surface-flow, 145
grain release, 242
Grand Coulee Dam, 71
Grand River system, 350
granite, 36, 37, 115, 237
Grant County, Oregon, *96*
gravel, 184, 206, 301, 347, 350
gravity, 51, 85, 182, 268, 347
Grayson County, Kentucky, 203
Great Bend region, New York, 56
Great Sand Sea, 305
Great Smoky Mountains, Tennessee, 59
Great Valley, 167
Greece, 285
Green River area, 296
Greenland, 56, 67
ground ice, 216, 313
ground subsidence, *215*
groundwater, 8, *44*, 52, *143*, 190, *221*, 241, *261, 270, 283, 320, 325, 341, 352*
   beach, 291
   continental-margin, 267
   defined, 319
   discharge, 7, *152*, 243, 267, 270, 272, 283, 285, *301*, 344
   dissolution, 179
   emergence, 13, 242, 291, *322*, 338
   erosion, 242, 283, 286, *302*
   flow, *2*, 184, *327*, 341, 348
   flux, 8, 270
   hydraulics, *184*
   hydrology, *1, 341*
   inflow, 8
   mixing zones, 287
   mobility, 270
   movement, *1*
   outflow, 8, 235, 245, 291, 328, *329*
   perched, 8, 88, 241, 247, 255
   processes, *177, 182, 196, 267, 319*
   provinces, *343*
   recharge, *1*, 16, 19, 102, *103*, 106, 177, *190*, 310
   sapping, *134*, 238, *242*, 244, *245*, 257, 271, 286, 290
   seepage, *291*, 295
   shallow, 29
   storage, *1*
   submarine, 267, *268*
   table, 71
   transport, 270
   weathering, 46
   withdrawal, *219*
grus, 42
Guadalupe Mountains, 189
Gulf of Maine, 270
Gulf of Mexico, 270, 277
Gulf of Mexico escarpment, 272
gullies, *139*, 313
   banks, erosion, *152*
   box, 79
   channel, 149
   continuous, 140
   control, *148*
   development, *139, 148*, 261
   discontinuous, 140, *144, 147*
   extension, 133
   formation, *86, 141*
   heads, 88
   headward-growing, 291
   initiation, *149*
   rill-like, *141*
   scarps, 140, 223
   sizes, *140*
   steep-walled, 77, 78, 80
   subsidence, 77
   types, *140*
   upland, *147*
   valley-bottom, 140, *144, 147*
   valley-head, 140, *143*
   valley-side, 140, *143*
   wall, 84, 153
gullies by sinking, 77
gullyhead retreat, 147
gullying, 24, 115, 131, 139, 141
gutters, 36, 37, 41
gyperete, 33, 34, 35
gypsum, 187, 198, 202, 203, 283, 296
talline, 34, 35

habitat, fish, *325*
Hachioji basin, Japan, 119
Halawa Valley, 255
Haliacetus leucacephalus, 325
hardpans, 33, 116
Harquahala Plains, 222
Harris Wash, 238
Harrisburg Surface, 167
Harrodsburg Formation, 202
Hawaii, 22, 286
Hawaiian Island valleys, 235
   case study, *245*
   geomorphic development, *245*
Hayes Creek, 325, *326*
coldpool, *326*
headcuts, 140, 141, 146
heads, amphitheater, 21, 246
heat, geothermal, 268, 272
heaving, *83*, 112
frost, *212*
Hebes Chasma, 313
Hells' Gate Bluffs, 69
Hereford, Texas, 102
Hidden River Cave System, 205
High Plains, 80, 102
High Plains aquifer, 103, 105, 301, 302, 320
Highland Rim, 159, 167
highways, *73*
   construction, 69
hills, residual, 313
hillsides
   erosion, 62
   evolution, model, 127
   failure, *51, 52, 69*
   hydrology, 127
   instability, *52*
   stability, 52, 57, 72
hillslopes, 21, 58, 341
   drainage, *126*
   erosion, 131
   humid, 127
   nonplanar, 4
   planar, 4, 11
   retreat, 21
hilltops, 342
hollows
   bedrock, 8
   colluvial, *58*
   development, 126
holokarst, *158*, 345
Hong Kong landslide, 54
Honokane Valley, 255, 256
Honshu, Japan, 117
horizons, impermeable, 44
Horton overland flow, 1
humid lands, *111, 116*
hummocks, early, *215*
Hungary, 188
Huntly-Kawhai Subdivision, 79
Hurricane Camille, 62
Hwang-Ho River depists, China, 95
Hydratationssprengung, 30
hydration, 23, 29, 47, 162
Hydrauger method, 73
hydraulics, groundwater, *184*
hydrogen sulfide, 183, 189, 203, 272
hydrogeomorphology, *352*
hydrolysis, 29, 47

ice
   crystals, 10, 313
   lenses, 212
   needle, 296
   pingo, 212
   subsurface, 212
   wedges, 212
Idaho, southern, 299
Illinois Basin, *200*
   case study, *200*
Illinois Basin karst, case study, *200*
illite, 68, 91, 100
illuviation, 32

Indiana, *200, 329*
  southern, 159
Indian Creek, 204
Indonesia, 30, 275
infiltration, 1, 36, 69, 95, 102, 112, *124*, 144, 146, 148, 164, 167, 196, 199, 244, 247, 301, 344, 350
  diffuse, 167, *191*
inflow, *322*
inselbergs, 36, 37, *38*, 42, 44, 88, 89, 308, 313
insolation, 30
interflow, 4, 52
ionic concentrations, 127
Iowa, 145
iron compounds, 308
iron-montmorillonite, 276
iron oxide, 33, 91
ironstone, 35
Israel, 148

Japan, 119, 219
John Day Country, 95, *96*
John Day Formation, 89, 100
Johnstown, Pennsylvania, 62
jointing
  columnar, 33
  stress-release, 293
joints
  high-angle, 16
  master, 15
  microsheeting, 242
  structural, 100
  tectonic, 6
  vertical, 13, 146, 272
Judito Wash, 87

kadinite, 33
Kadrak Island, Alaska, 320
Kaei Vallis outflow channel, 313
kames, *348*
Kamiesberge, Namagualand, 38
Kamloops, British Columbia, 90
Kane Caves, Wyoming, 189
Kangaroo Island, Australia, 35
kankar, 34
Kansas, 80, 320
Kansas High Plains, 80
Kansu Province, China, 95
kaolinite, 68, 91, 114
Karr Valley, 347
karren, 37, 159
karst, 106, *164, 198, 200*
  alpine, *161*, 199
  aquifer, *178, 183, 189*, 193
  coastal, *161*
  cockpit, *160*
  cone, *160, 172*
  denudation, *164*
  development, *199*, 202
  doline, *159*, 167, 173
  forms, *345*
  groundwater environmental problems, *206*
  groundwater geomorphology, *199*
  groundwater hydraulics, *184*
  groundwater processes, *196*
  groundwater systems, *178*
  landforms, 78, 106, *157*
  landscapes, *157*
  mogote, 160
  pavement, *158*
  processes, *161*, 177, *178, 203*
  rock type classification, *157*
  sinkhole, *159*
  springs, 181, 199
  subsurface, *167*
  subsurface water, *187*
  surface, *166, 203*
  terranes, *177*
  topography, 345
  tower, *160, 172*
  tropical, 172
  types, *345*
  valleys, development, *196*
  water table, *180*
Kayenta Formation, 241, 244
Kentucky, 159, 194, *200*
Kharga, Egypt, 311
Kharga uplift, 305
Kilauea volcano, 246
kinetics, *162*
  carbonate-dissolution, *163*
  limestone dissolution, *185*
  mass transport, *163*
  reaction, 186
Kinzua Creek, 348
knicklines, *310*
Kohala area, Hawaii, 246, *249, 252*
Kohala Mountain, 286
Kohala volcano, 247, 249, 255
Kombolquie formation, 311
krasnozems, 116
kunkar, 34
Kutz Canyon, New Mexico, *89*

Labrador Peninsula, 41
Lac Laflamme basin, *119*, 124
Lake Eyre Basin, 35
Lake Ontario, 131
lake water, 103
lakes, 352
  playa, 105
  proglacial, 350, 354
  saline, 101, 103
  thaw, 212, *217*
land subsidence, *219, 221*, 231
landforms, 13, *29*, 37, 56, 67, 106, *345*
  aggradational, *213*
  cold regions, *211*
  concavo-convex, 57
  degradational, *215*
  evolution, 13
  initiation, shallow subsurface, I36
lands, humid, *111, 116*
landscapes, *348*
  defined, 341
  development, humid, *125*
  evolution, 44
  inselberg, 314
  karst, *157*
  pediplain, 314
  sea-bottom, 267
landslides, 54, *57, 62, 64, 66*, 69, *71*, 131
  blocks, cracks, 6
  Davilla Hill, *64*
  Hong Kong, 54
  Minor Creek, *65*
  San Dimas, 54
  stabilization, 73
  types, *63, 68*
landsliding, 9, 51, 52, 54, *88*
  multiple, 69
landslips, 131
  progessive, 292
Laqiya area, 311, 312, 313
Las Vegas Valley, Nevada, 219
laterite, *33*, 41, 44, 114
Latosols, 114
Laurentian Shield, 285
Lava Creek B ash bed, 305
lavas, 246, 286
  basaltic, 283
Lea County, New Mexico, 103
leaching, 16, 32, 67, 68, 114, 115, 297
Leda clay, 67
level, phreatic, 114
Libya, 300
Libyan Desert, 308
lime, 33, 35
limestone, 8, 36, 37, 85, 157, 158, 160, 167, *172*, 188, 194, *198*, 200, 202, 203, 205, 289, 298, 299, 345
  dissolution, *185*
  hamadas, 310
liquefaction, 9, *11*
  spontaneous, 62
Little Colorado River, 311
Little Level, West Virginia, 167
Llano Estacado, *101*, 105
  Texas segment, 103
Llano Estacado basins, case study, *101*
loam, clay, 149
lobes, soliflution, 57
loess, 77, 81, 85, 90, 106, 133, 140, 144, 145, 146, 148, 347
Long Beach, California, 86
Lookout Creek earthflow, 68
Loop current, 273
Los Angeles area, California, 54
Lost River, 205
Lower Carmel River, *335*
Lubbock area, Texas, 102, 103, 303

macropores, *6*, 9, 24, 112, 115
  development, 111, 134
Madagascar, 30
Madeira abyssal plain, 276
Maesnant Basin, Wales, 117, *120, 128*, 130, 131, 132
Maesnant catchment, *120*
Maesnant stream, *120*
Maesnant subcatchment, 124
magnesium, 68, 100, 187
Mahogany Meadow, Nevada, *149*
  casestudy, *149*
Malaysia, 159
Mali, central, 94
Mammoth Cave, Kentucky, 166, 168, 191, 200, 203
Mancos Shale, 89
manganese oxide, 91

mantle, weathered, *30*
Maranboy Surface, 35
Marengo Cave, Indiana, 203
margin, continental, 267
Maricopa-Stanfield area, 222
Marietta, Ohio, 292
Mars
   channel classes, *245*
   fretted terrain, *313*
   valleys, 21, *238*, 245
Maryland, 167
Masenant Basin, Wales, 134
mass movement, *51*
mass wasting, 10, 21, *51*, 90, 157, 196, 249, 268, 270, 272, 275, 277, 310, 313
Massif Central, France, 37
Mato Grosso, Brazil, 292
Mattole River basin, 319
Mauna Kea volcano, 246, 249, 255
Mauna Loa volcano, 246, 255
maze
   anastomotic, 195
   flood-water, *191*
meanders, bedrock, 238
Mediterranean Sea, 269
megafauna, 277
melting, 2, 6, 215, 216
meltwaters, 199, 354
Mendip Hills, 130
methods, water-control, 72
Mexico City, 219
Miami River, 350
microclimate, 237
microfracturing, 102
micropiping, 105, 106
micropores, 142
microrelief, 46
microtopography, 122, 126, *128*
midslopes, 116
Mill Brook Valley, 247
mineral accumulations, *33*
minerals
   carbonate, 161, 288
   clay, 114, 135
   silicate, 106
   sulfide, 276
Minor Creek landslide, *65*
   case study, *65*
Mississippi Basin, central, 347
Missouri, 144, 159
Mitchell Plain, 202, *205*
mixing zones, *188*, *287*, *289*
mobility, groundwarer, *270*
Mockingbird Canyon, 292
models
   antecedent soil water status, 54
   Dupuit-Forchheimer, *5*, *16*
   equilibrium, *165*
   erosion, *262*
   flow, *262*
   groundwater sapping, *257*
   hillside evolution, 127
   New Jersey subshelf hydrology, 274
   pipe initiation, 115
   valley development sapping, *238*
   valley development simulation, *262*
Moenkopi Formation, 88
Mohawk Valley, New York, 350

moisture, *52*, 62
Mokelumne River, 320
Molodai Island, Hawaii, 80, *252*, *255*
Mona Island, 289
monadnocks, 313
Mongolian desert, 312
monkstones, 37
Monterey, California, 320, 334
montmorillonite, 99, 114
Morocco, 144
morphology, *235*
   basin, *252*
   cliff, 299
   sand surface, 261
morphometry, basin, 256
morros, 38
mounds, *215*
   thermokarst, *217*
Mount Kaijende, New Guinea, 159
mudflows, 24
   alpine, 66
   categories, *66*
mudseal, 322, 338
mudslide, 63, *66*, 69
   defined, 66
mudstone, 90, 241, 292

Nacimiento Formation, 89
Nant Gerig, 122, 124, 134
Nant Llwch, Brecon, Beacons, 124
Nantucket Island, 271
nari, 34
Natal, 81
Navajo Sandstone, 10, 238, *241*, 296
Nebraska, 92, 140, 144, 348
Nebraska loess, 81
Nelson County, Virginia, 62
Netherlands, 134
Nevada, *149*
New England, 342
New England continental shelf, 270
New Forest, Hampshire, England, 292
New Jersey, 270, *271*
New Mexico, 80, 92, *101*, 140, 146, 30
New Plateau, *41*
New South Wales, Australia, 80, 81, 92, *114*, 130, 146
New York, 193, 352
New Zealand, 67, 79, 130, 146, 270, 323
Nickpoint development, *146*
Nigeria, *148*
North Carolina, 346
North Dakota, western, 322
North Dakota Badlands, 81, 87, 91, 93, 94
North Island, New Zealand, 79, 147
North Pease River, 303
notches, 285
Nova Scotian shelf, 273
Nubia Sandstone, 305, 311

Ochil Hills, Scotland, 132
Officer's Cave, Oregon, 95, *96*
   case study, *96*
Officer's Cave Ridge, *96*
Officer's Cave Stream, *96*
Ogallala Formation, 102, *300*, 305, 311
Ohio River, 132, 205, 323
Ohio River banks, *327*
Oklahoma, 142, 296
Oncorphynchus keta, 325
Oneida County, New York, *350*
Oregon Coast Range, 54
Oregon, easter, 89
organisms, benthic, 277
Osage Plains, 292
osmosis, 23
outflow boundary, 8
outflow channels, 291
outlets, 22
outwash, *348*
oxidation, 29, 105, 187, *189*
oxisols, 33
oxygen, 102

Pacific Northwest, 68
Pahala ash, 247
Painted Desert, Arizona, 88
Pakistan, *89*
paleoclimate, *94*
paleokarst, 200
paleosol, 144
Palo Ubin, 37
palsas, 212, *215*
P`ang Kiang hollows, *312*
Paradise Valley, 221, 222, 223, 229
Paradise Valley Basin, 227
parent material, 32
partings, bedding-plane, 179
Pease River, 200
peats, 30
   erosion, 128
Pecos River valley, 300
pedalfers, 115, 116
pedimentation, 344
pediments, 344, 348
   rock, 46
pedocretes, 33
pedogenesis, *30*
peds, 22
peneplain, 167, 212
   desert, 308
Pennsylvania, 134, 167, 352
   northwestern, 54
   western, 62
Pennyroyal Plateau, 167, 202, *205*
perching, 194
percolation, 1, 2, 7, 9, 24, 52, 144, 329
   rate, 71
   steady-state, 4
permafrost, *211*, 212
permeability, 11, 32, *93*, 111, 112, 115, *124*, 132, 134, 162, 168, 343, 345, 350
perturbation, 15
petroleum, 203
Phoenix, Arizona, 222, 295
Phoenix Mountains, 227
phreatic zone, 1, 8, 162, 181, 190
   caves, 345
   conduits, *182*, 195
   solution, *181*
phytokarst, 161
Picacho Basin, 221, *223*, 231

Picacho fault, 230
Picture Gorge Basalt, 96
Piedmont province, 342, 344
piezometric surface, 6
Pilbara, Australia, 35
Pine Creek, 320
pingos, 212, *215*
pipeflow, *117*
   drainage, 120
   sediment, *128*
   solutes, *130*
   velocities, 127
pipes, 11, 15
   development, 114, 116, 141
   discharges, 122
   dry-land, 112
   enlargement, 87
   ephemeral, 120, 122, 124, 128, *130*
   formation, 9, 102
   initiation, 111, *112*, 115, *116*
   model, 115
   natural, 302
   networks, 121, 131, *133*
   outfalls, 116
   perennial, 122, 124, 127, 128, 130, 133
   roof, 84
   soil, 6, 20, 141, *142*
   vertical, *78*
piping, *9*, 52, *55*, *77*, *81*, *88*, *90*, *102*, *106*, *111*, *126*, *132*, 142, 153, 268, *283*, 286, 329
   control, *95*
   defined, 9, 77
   desiccation-crack, *85*
   development, *85*, *101*
   dry-land, *111*
   erosion, 106, 114, *125*, *130*
   gully formation, *86*
   gully-wall, 84
   humid-land, *111*, *116*
   mechanisms, 114
   occurrences, *78*
   prevention, *95*
   processes, *77*, *116*
   relict, *94*
   soil, 134
pits, thermokarst, *216*
plains
   etch, *41*
   sinkhole, 173, 203, 206
planation
   double, 37
   marine, 42
planes, bedding, 13
plant roots, 11, 15
plasticity, 68
plateau, remnant, *300*
platforms, *41*
playa development, recharge, *103*
playa lake basins, 95, *101*
playa lakes, 105
playas, 301
plucking, 242, 296
plug flow, 63
pockmarks, *273*
podzolics, 115
podzols, 117, 130
Point Baptiste, Dominica, 119, 130

Point Brown, 47
Poland, 130
poljes, 170
Polulu Valley, 255
polygons, ice-wedge, 212, *213*
ponding, flood-water, 190, 197
pools
   cold, 325, *326*
   plunge, 22, 256
Popcorn Spring Cave, 203
pore fluid, 24, 224, 268
pore space, 221
pore water, 1, 68, 268, 272, 274
pores, 1
   granular, 179
   intergranular, 3
   pressures, 4, 7, 11, 52, 66, 367, 268, 274, 277
   soil, 127
   vadose, 187
porosity, 5, 7, 10, 14, 15, 16, 54, 167, 188, 189, 289, 299, 341, 345
Pourtales terrane, 273
precipitation, 32, 52, 62, 65, 102
   channel, 120, 122
   input, 2
pressures
   cleft-water, 54, *56*
   hydraulic, 221
   hydrostatic, 64, 71, 182, 188
   joint, 72
   pore, 4, 7, 11, 52, 66, 267, 274, 277
   pore-water, 1, 54, 62, 64, 65, 67, 68, 69, 71, 148
   seepage, 10, *54*, 112, 153
   soil-water, 6
prisms, accretionary, 274, 275
problems, environmental, *206*
proto-depressions, 106
Pseidoclapia arenaria, 103
pseudilarsts, 106, *128*
pseudo anticlines, 35
pseudokarst, 23, 25, *77*, *86*, *88*, *96*, 106, 345
   cave, *96*
   defined, 78
   development, *85*
   landscapes, 95
   occurrences, *78*
   sinkhole, 89
Pueblo Bonito, 293
Puerto Rico, 159, 172
Puerto Rico karst belt, 160
pumping, 302, 335, 336
Pyrenees, 161

quartz, 29, 33, 34
quartzite, 37
Quebec City, Canada, 68, 117, *119*
Queen Creek area, 221, 230
Queen Elizabeth National Park, Uganda, 81
Qumran Caves, Jordan, 95

radiation, solar, 59
radose, caves, 345
rain forest, 111, 119, 130

rainfalls, 1, 2, 6, 22, 52, *54*, *62*, 66, 69, 72, 84, 105, 111, 115, 120, 125, 199, 244, 246, 319, 323, 333
rainstorms, 7, 52, 63, 86, 112, 227
rampas, 144
Ranney well, 348
raveling, 21
ravines, 139
recession
   gully-head, 87
   seepage-induced, 299
recharge, 1, 102, *103*, *190*, 268, 320
   allogenic, *167*, 191
   authigenic, *167*
   diffuse, 194
   groundwater, *1*, 102, *103*, 106, 177, *190*, 310
   rate, 19, 199
Red River, 300
Red Sea, 276
Redwood Creek, 320, *324*, *326*
regions
   arid, 114, 242
   cold, *21*, *211*
   humid, *111*, 132, 133, *148*, 299, 342, 344
   humid temperate, 130
   semiarid, 114, 115, 149, 153, 286
   subhumid, 114
   tropical, 238, 284
regolith, *29*, *36*, *44*, 62, *196*
   differentiation, *30*
   subsidence, 197
relict, conduits, 181, 182
reservoirs, *71*
residuum, weathered, 19
retreat
   headwart, 241
   scarp, *19*, 261, *305*, 307, *308*, *311*, 313, *314*
Rhodesia, 81
Richards Island region, 215
Richland Creek, 204
Rif Mountains, 144
rillen, 37, 46
rills, *139*
   development, 261
   erosion, 297
Rio Puerto, New Mexico, 87
River Ilston, 296
River Irwell, 132
river discharge, 319
riverbank failure, 132
Rock Spring, Pennsylvania, 166
rocks
   avalanches, 57
   basaltic, 286
   carbonate, *158*, 167, 187, 189, 191, 198, *200*, 272, 283, *284*, 301, 345
   clastic, 77, 78, 100, 283, *286*
   consolidated, 298
   crystalline, 344, 346
   decomposition, 19
   falls, 69
   fractures, 16
   igneous, 283
   insoluble, 189, 191

metamorphic, 346
mushroom, 36
parent, 344
penitent, 37
reef, 179
sedimentary, 25, 94, 114, 193, 283, *286*, 299, 348, 352
siliceous, 183
soluble, 186, 189, 190, 191, *199*
volcanic, 30, *286*
weathering, 29
Rocky Mountains, 151
Rolling Plains, 300
roof cave, 85
roof collapse, 126
roofing, 82
root growth, 20
rootholes, 6, 7, 22
roots, plant, 11, 15
Rough Creek fault zone, 202, 203
Round Rock, Arizona, 89, 107
rubble dams 89
runoff, 1, 6, 22, 80, 84, 95, 102, 106, 117, 132, 165, 225, 297, 301, 302, 311
annual, 19
internal, 167
surface, 141, 147, 148, 319
valleys, 245, *250*, 253, 255, 256

Saguenay River, 58
Sahara, northwestern, 298, 310, 312
St. Frances Dam, 83
St. Jean Vianney landslide, Canada, 68
St. John Vinanney landslide, ease study, 68
St. Lawrence region, Canada, 67
St. Louis Limestone, 202, 203, 204, *205*
St. Lucia, 68
Ste. Genevieve Limestone, Kentucky, 194, 202
Salem Limestone, 202, *205*
Salford, England, 132
salinity, 188, 268, 288
Salmonidae, *325*
Salt River Valley, 221
Salt River Valley basin, 222
salt content, 91, *92*
salt fretting, 296
salts, 23, 29, 32, 44, 47, 114, 115, 237, 242, 295, 296, 301, 302
San Diego area, California, 147, 291
San Dimas landslides, 54
San Francisco Bay levees, 219
San Francisco Bay region, 63
San Joaquin Valley, Califorina, *219*
San Mateo County, califorina, 143
San Pedro Valley, Arizona, 133, 134
San Pedro River bed, 89
San Tan Mountain, 225
San Xavier, Arizona, 145
sand, 69, 148, 184, 206, 260, 261, 291, 330, 347, 350
weeping, 69
sand blast, *308*
Sand Creek, Nebraska, 146
Sand Hill Region, 348

sandbars, 322
sanddunes, 347
sandseams, 334
sandstone, 89, 90, 106, 167, 191, 235, 237, 241, 244, 272, 283, *292*, 296, 297, 299
cliff-forming, 296
Santa Barbara coast, 286
Santa Clara Valley, California, 219
Santa Cruz Formation, 42
Santa Cruz Mountains, California, 63
Santa Cruz River, 324
Santa Lucia Range, 334
Santa Rose rockslide, Mexico, 72
saphrolite, *344*
sapping, 9, *12*, 19, 20, 52, *55*, 114, *134*, 139, *142*, 147, 148, 268, 291, 292, 293, *297*, *300*, 302, 307
basal, 296, 310
chamber, 12
cliff, *291*
differential, 295
erosion, 238
experiments, *257*
groundwater, *134*, 238, *242*, 244, *245*, 271, 286, 290
heatcut, 142
mechanisms, 334
models, *238*
plunge-pool, 142
processes, *236*
seepage-induced, 305, 309, 310, 311
spring, 9, *235*, 245, *283*, 286, 290, 291, 297, 302, 307, 310, 323
valley development, 237, *238*
valleys, *235*, 245
zone, 257
saprolith, 38
sarsenstone, 33
Saskatchewan, 112
saturation, 2, 4, 10, 22, 24, 101, 112, 169, 188, 191
scales, process time, *163*
scallops, solutional, 195
scarps, 18, 46, 62, 139, 141, 143, 153, 223, *230*, 259, 261, 292, 293, 329
back, 299
channel, 140, 144, 146
fault, 229
front, 299
gully, 140, 223
head, 145, 146
recession, 236, 300
retreat, *19*, 261, *305*, 307, 308, 311, 313, *314*
retreat extent, *297*
retreat rate, *297*, *299*
seepage-aided, 311
side, 140
vertical, 329
Schrondweiler baach catchment, Luxembourg, 117
scour, 82, 148, *322*
thermokarst, *9*, 11, *24*
tunnel, 15, *22*, *83*, *84*, 91
sea bottom, discharges, 268

sea level
changes, *270*
drop, 268
regressions, *270*
transgressions, 270
seacliffs, *283*, *284*, *286*
erosion, *283*
retreat, *286*
volcanic, 286
sediments, 20, *54*, 89, 103, *271*, 305, 308, 334
ashy, 90
cave, 184
compaction, 268
dewatering, 229
loading, 268
mineral, 128
pipeflow, *128*
saline, 68
slurries, 54
submarine, 273
subsea, 267
transport, 22, 102, 114, *183*, *321*, 338
unconsolidated, 226, 297
yields, *128*
seepage, 7, 9, 12, 68, 69, 77, *82*, 106, 139, *142*, 293, 297, 302, *310*, 322, *323*
cliff recession, *291*
concentrated, 292
diffuse, 152, 291
discharge, 243
erosion, *9*, 12, *13*, 19, 77, 140, *142*, 146, 147, 235, 268, *291*, 295, 297, 307, 310, 329
evaporating, 296
face, 7, 10, 17, 142, *242*, 277, 298
force, 10, 14
frost, *296*
groundwater, *291*, 295
negative, 319, 338
outflow, 142
positive, 319, 323, 338
pressure, 10, *54*, 112, 153
slow, 116
steps, 292
velocity, 95
weathering, 10, 19, *291*, *295*, 297, 299, *300*, *308*, *310*, 314
zone, 7
seismic activity, 219
Selima Sand Sea, 308
Selima Sand Sheet, 308, 311
Sentinel Butte, 81, 94
Severn River, 329
Seymour Formation, 305
shafts, vertical, 24, 168, *170*, 173
shale, 77, 90, 106, 112, 283, 287, 299, 352
shear
failure, 132
fluid, 24
strength, 268
stress, 268, 323
sheet erosion, 33
sheet wash, 292
Shoals, Indiana, 203

shoals, asymmetric, *321*
shore nips, 285
shorelines, stillstand, 270
Siberia, 211, 217
Sierra Nevada, California, 30, 320, 324
Signal Peak, 225
Signal Peak fissure, 225
Sikkim, 112
silcrete, *33*
silica, 91, 172
silts, 332
   backswamp, 347
   glaciolacustrine, 350
   overbank 347
siltstone, 90, 106, 198, 241, 283, 287, 293, 301, 352
Sinkhole Plain, 167
sinkholes, 25, 78, 84, 96, 100, *159, 170*, 179, 190, 196, 197, 200, 203, *205*
   collapse, *197*
   development, 80
   karst, *159*
   plain, 203
   pseudokarst, 89
   silting, 207
   solutional, 197
sinks, 89
slab failures, 291
slides
   quick-clay, 67
   shallow, 64
   translational slab, 68
sliding, 9, 275
Slithero Clough, Yorkshire, 124
slope decline, *314*
slope development, *126*, 314
slope erosion, 310
slope evolution, *126*
slope failure, 11, *59*, 69, 132, 269
   bedrock, *56*
   mudslide activity, *66*
   reservoirs, *71*
   sediments, *54*
   seepage-induced, 292
   submarine, 268
slope movement, *51*
   classification, *56*
   control, *72*
   materials influence, *58*
   prevention, *72*
   terrane influence, *58*
   types, *52*
slopes
   concave, 74
   continental, 267, *271*
   convergent, 131
   debris, 292
   flared, *36*
   rectilinear, 314
   retreats, 291
   stability, 69
   topographic, *58, 341*
   water table, 341
slopewash, 310
sloughing, 242
slump earthflows, 68, *71*
slump/slides, shallow, 64

slumping, 12, 89, 90, 139, 197, 238, 270, 275, 292
   gully wall, 55
slumps, 66, 68, *71*, 260, 291
   deep, 64
   talus, 71
slurries
   sediment, 54
   viscous, 67
slush avalanches, 66
slushflows, 66
Small Lakes section, 352
smectite, 91, 100
Snake River, 299
Snake River Canyon, 284
Snake River Plateau, 15
snow, 1
snow cover, 199
snowmelt, 7, 11, 52, 68, 106, 143, 151, 152, 199
   rate, 142
sodium, 24, 99, 100, 105, 111, 114, 115
Sohm abyssal plain, 276
soil. loam, 119
soils
   A-horizon, 33, 35, 116, 119, 149
   alluvial, 329
   argillic, 33
   avalanche, 238
   B-horizon, 115, 116, 142, 143
   C-horizon, 167
   cambric, 33
   classification, *30*, 115
   clay, 103
   cohesive, 13, 22, 23, *333*
   cracks, *84*
   creep, 21, *56*, 80
   debris, 145
   development, *29*
   dispersion, 115
   dispersive, 140
   duplex, 32, 86, 115
   erodibility, 111, 114
   erosion, 196
   fauna, 6
   formation, 32
   frost, 151
   gilgai, 29
   homogeneous, 2
   horizons, 93
   isotropic, 2
   kaolinitic, 68, 130
   leaching, 68
   loessial, *133*
   matrix, 112
   moisture, 52
   montmorillonitic, 68
   muck, 352
   nonpiping, *92*
   O-horizon, 128
   order, 32
   packing, 22
   peaty, 112, 117, 124, 130
   pipe, 6, 20, 141, *142*
   piping, *92*, 134
   piping-prone, 95
   podzolic, 115, 132
   pores, 127

   profile, 30, *32*, 112
   shallow, 112
   silty, 90
   skeletal, 30
   smectoid, 130
   sodium-rich, 22, 141
   solodic, 115
   solonetz, 115
   strength, *323*
   susceptible, 116
   taxonomy, 30
   till, 352
   transport, 170
   water, 92
soilwater outflow, 142
soliflution, *56*
solodization, 115
solutes, 23, *103*
   concentrations, *120*
   pipeflow, *130*
   transport, 268
solution, 29, 47, 78, 79, 83, 100, *101, 188*, 219, *283*, 285, 290, 295, 297, 301
   chimneys, 168, *170*, 173
   flood-water, 191
   phreatic, *181*
   rates, 189
   vadose, *181*
South Africa, 128
South Carolina, 148
   Piedmont, 142
South Dakota, 310
South Thompson River, British Columbia, 80
South Wales, 131, 314
Southern High Plains, *101*, 297, 299, *300*
Southern High Plains escarpments, *300*, 311
   case study, *300*
spalling, 10, 296
spalls, 10, 242
speleotherms, travertine, 187
spills, toxic, 207
spitzkarren, 159
spreading centers, mid-ocean, 276
spreads, lateral, 67
spring alcove, 198
spring discharge, 255, 302, 352
spring effluent, 302
spring heads, 19, 237, 239
spring sapping, 9, *235*, 245, *283*, 286, 290, 291, 297, 302, 307, 310, 323
springs, 7, 180, *249*, 286, 292, 293, 297, 308, 310
   alluviated, 181
   brine, 302
   cave, 194
   dry, 310
   hot, 276
   karst, 181, 199
   offshoe, 272
   perched, 182
   submarine, 271
spurs, 19
Ssandinavia, 67
stabilization, hillside, *72*

Staked Plains, 84, 95
Stanton, Texas, 102, 104
steephead, 198
Steilwande, 313
stone rivers, 134
Stony Creek, 348
stormflow, 119, 132, 319, *320*
    mean, *122*
    subsurface, 126
stormwater, 164
streambed, 325
streamflow, 6, 17, 122, *124*, 152, 301, 352
    ephemeral, 300
    generation, *152*
streams
    baseflow, 119
    beaded, 212, *217*
    cave, 191, 194, 196
    channel, 17, 25, 120
    east-flowing, 352
    effluent, 344
    ephemeral, 196, 249, 319, 327
    gravel-bed, 322
    influent, 344
    intermittent, 319, 327
    meltwater, 348
    networks, 19
    north-flowing, 352
    perennial, 319, 327
    power, unit, 322
    sinking, 179, 190, 196, 200, 203
    south-flowing, 352
    southwest-flowing, 352
    stormflow, 119
    surface, 196
    vadose, 194
stress cracks, 84
stresses
    downslope, 12
    drag, 54
    driving, 12
    erosion, *283*
    normal, 12
    shear, 9, 22, 24, 54, 57, 111, 114, 323
    throughflow, 111
study, water-budget, 104
sub-Carpathians, Romania, 131
subduction zone, 268, *274*
subsaturation, 289
subsidence bowl, 222
subsidence, 29, 51, *84*, 116, 126, *196*
    annual, 219
    causes, *221*
    differential, 227, 229, 230
    fluid-withdrawal, 219
    ground, *215*
    land, *219*, *221*, 231
    local, 177
    man-induced, 219
    regolith, 197
subsoils, erodible, 134
subsurface flow, shallow, *3*, 29, *36*
subsystems, flow, 3
Sudan, 81, 313
suffosion, *82*, 197
sugarloafs, 38
sulfides, 276

supersaturation, 187, 288
surface wash, *297*, 311, 313, 314
Susquehanna basins, 347, 352
    hydrogeomorphology, *352*
Susquehanna section, 350
swale, 18, 159
swamp slots, 35
swelling, 24
Swago Creek Basin, 168
Syracuse, New York, 342

tafoni, 36, 237, *296*
taiga, 217
talus accumulation, 21
Tama Hills, Japan, 119, 131
Tanzania, 81
Tasmania, 95, 130
taxonomy, *170*
Teays River system, 205, 350
Teays River valleys, *349*
temperature, 199, 350
Tennessee, 159, 167, 346
    eastern, 320
    highways, *73*
terraces, fluvial, 195
terranes
    carbonate, 345
    crystalline, 342
    glacial, 349
    karst, *177*
    Pourtales, 273
    upland, 350
Terror River, 320
Terzaghian mechanics, 114, 117, 135
Teton Range, 161
Texas, *101*, 142, 292, 297, *300*
thawing, 212
thermokarst, *211*, *216*
    depressions, 217
    mounds, *217*
    pits, *216*
    scour, 9, 11, *24*
Thompson Spring, Pennsylvania, 166
Thousand Springs, 284
Threatening Rock, 293
throughflow, *4*, 52, 111, 112, 119, 127, 130, 131, 132, 148, 179, 325
    convergence, 144
    yields, 125
    subsurface, 146
thufur, *215*
till, 348
Tithonium Chasma, 313
Tloi Eechii Cliffs, *297*, 299, 311
Tloi Eechii escarpment, 299
tombstones, 37
Tongue River Formation, 81
topography
    bar-pool, *321*
    karst, 345
Town Dump Wash, 94
Trans-Canada Highway, 69
transport
    aqueous, 184
    fluvial, 10, 12
    groundwater, *270*
    mass, 162, *163*, 167, 173, 200
    particulate, 105

sediment, 22, 102, 114 *183*, *321*, 338
soil, 170
solute, 268
subsidence, 184
travertine, 34, 35
travertine speleotherms, 187
tree throw, 20, 25
tritium value, 103
tropics, wet-and-dry, 114, 115
Trowbridge Creek valley, 57
Tucson, Arizona, 324
Tucson Basin, 223
tuff, 99, 106, 237
    montmorillonitic, 100
    volcanic, 90
Tuktoyaktuk Peninsula, 215
tunneling, 33, *83*, 323
    natural, 77
tunnels, 24, 25, 33, 80
    discharge, 24
    erosion, 9, 268
    flow, 242
    macroscopic, 262
    margins, 7, 24
    network, 23
    scour, 15, *22*, *83*, *84*, 91
Turtle Cove area, Oregon, 96

Uinta Range, 161
Unaka Mountains, *346*
undersaturation, 199
United Kingdom, 116
University Park, Pennsylvania, 342
Upper Wye catchment, *117*
Utah, 296
    south-central, 238
    southern, 10

vadose canyons, 203
vadose conduits, *182*
vadose passages, 190
vadose perching, 193
vadose pores, 187
vadose solution, *181*
vadose streams, 194
vadose zone, 1, 8, 159, 162, *167*, *172*, 181, 182, 190, 195
Vaiont Dam reservoir, Italy, 71
Valles Marineris, 245, 313
valley heads, 19, *20*, 245, 260
valley networks, 16, *18*, 20, *238*
    development, *235*
    evolution, 18, 262
    morphology, 245
valleys
    amphitheater-head, 20
    buried, *349*
    development, 126, 235, 237, *245*, *291*
    evolution, *255*
    heads, recession, 309
    runoff, 245, *250*, 253, 255, 256
    sapping, *235*, 245
    sapping-dominated, *249*, 253, 256
    submarine, 297
    theater-headed, *241*, 286
vegetation, 29, *94*, 103, 148, 287, 299

denudation, 94
riparian, 323, 338
velocities, flow, 184
Venice, 219
vents, hydrothermal, 277
Vermont, 143
vertisols, 29
Victoria, Australia, 80, 81
Virginia, 58
viscosity, 350
voçorocas, 8
volatils, 200
volcanism, submarine, 276
volcanoes
  mud, 267, 275
  shield, 235, 246

wadis, 305, 312
  age, 309
  development, 307, 309
Waikato River basin, 79
Waimanu Valley, 249, 255, 256
Waipio Canyon, 286
Waipio Valley, 249, 255
Wales, 112, 124, 128, 130, 131, 132, 296, 323, 329
wall retreat, 186
Wasatch Range, 161
wastewater, 287
water
  brackish, 290
  budget, 32
  capillary, 52
  carbon dioxide-rich, 162
  connate, 1
  drip, 169
  effluent, 295
  fluid, 319
  flux, 1, 14
  hydration, 52
  infiltration, 157
  interstitial, 268
  juvenile, 1
  lake, 103
  matrix, 130
  meteoric, 1
  movement, 52
  perched, 52, 235
  percolating, 82
  phreatic, 182, 188, 190, 194
  pore, 1, 68, 268, 272, 274
  quality, 342
  shaft, 169
  shallow subterranean, 29
  soil, 92
  subsurface, 8, 51, 157, 196, 211, 274, 319, 322
  subterranean, 29
  surface, 352
  thermal, 188
  vadose zone, 52, 157, 182, 187, 193, 195, 241
  water table, 1, 4, 6, 15, 54, 68, 72, 143, 145, 180, 189, 194, 198, 204, 216, 221, 230, 237, 238, 247, 250, 262, 292, 311, 319, 324
  depth, 342
  drawdown, 334
  perched, 68, 114, 134, 241, 292
  regional, 134
  steady-state, 5
watersheds, 348, 349, 352
wave front, 68
weathering, 9, 10, 14, 16, 19, 29, 114, 115, 237, 247, 283, 286, 342, 345
  alveolar, 237, 296
  chemical, 172, 239, 249, 295, 297
  cliff-base, 297
  concentrated, 237
  enhanced, 238
  front, 30, 36
  groundwater, 46
  mechanical, 199, 295
  rock, 29
  salt-crystal, 296
  seepage, 10, 19, 291, 297, 299, 300, 308, 310, 314
  wedging, needle-ice, 297
Wells, Maine, 270
wells, 342
  alluvial, 335
  bedrock, 342, 350
  Berwick, 335
  contamination, 206
  high-yielding, 342
  Ranney, 348
  yields, 342
Wesley Chapel Gulf, 205
West Florida escarpment, 270, 272, 273, 277
West Florida margin, 273
West Yorkshire, U.K., 131
Western Desert, 300, 305, 308, 311
  case study, 305
  human occupation, 309
wetlands, upland, 352
Whangarei, Auckland, 80
White River, 300
  East Fork, 198
White River Formation, 94, 293
White Silts, British Columbia, 90
wijijiruins, 293
Wilcox-Bowie-San Simon area, 223
wind, 308, 310, 311, 347
Wisconsin, 134
withdrawal, groundwater, 219
Wither Hills, South Island, 79, 81
Wyandotte Cave, Indiana, 203
Wyoming, 189, 293

Xcaret Cave, 288
Xel Ha Caleta, 287

Yarwondutla Rock, 44, 46
Yellowtail Dam, Montana, 72
Yilgarn Block, Western Australia, 41
Yucatan, 198, 269, 272, 287
Yucatan coast (eastern), Mexico, 287
Yucatan Peninsula, 284
Yugoslavia, 158, 199
Yukutia lowland, Siberia, 217

Zimbabwe, 24, 92, 115, 141, 148
Zion Canyon, Utah, 296
Zion National Park, Utah, 296
Zuni Mountains, 299